Ecology of Bats

This limited facsimile edition has been issued for the purpose of keeping this title available to the scientific community.

Ecology of Bats

Edited by
Thomas H. Kunz
Boston University
Boston, Massachusetts

Plenum Press • New York and London

Library of Congress Cataloging in Publication Data

Main entry under title:

Ecology of bats.

Includes bibliographical references and index.
1. Bats—Ecology. 2. Mammals—Ecology. I. Kunz, Thomas H.
QL737.C5E33 1982 599.4'045 82-10157
ISBN 0-306-40950-X AACR2

233 Spring Street, New York, N.Y. 10013

Printed in the United States of America

To Margaret, Pamela, and David

Contributors

Hans G. Erkert Institut für Biologie III der Universität Tübingen, D-7400 Tübingen 1, Federal Republic of Germany

M. Brock Fenton Department of Biology, Carleton University, Ottawa K1S 5B6, Canada

James S. Findley Department of Biology, University of New Mexico, Albuquerque, New Mexico 87131

Theodore H. Fleming Department of Biology, University of Miami, Coral Gables, Florida 33124

E. Raymond Heithaus Department of Biology, Kenyon College, Gambier, Ohio 43022

Thomas H. Kunz Department of Biology, Boston University, Boston, Massachusetts 02215

Adrian G. Marshall Department of Zoology, University of Aberdeen, Aberdeen AB9 2TN, Scotland

Brian K. McNab Department of Zoology, University of Florida, Gainesville, Florida 32611

P. A. Racey Department of Zoology, University of Aberdeen, Aberdeen AB9 2TN, Scotland

Diane Stevenson Vertebrate Division, Milwaukee Public Museum, Milwaukee, Wisconsin 53233

Merlin D. Tuttle Vertebrate Division, Milwaukee Public Museum, Milwaukee, Wisconsin 53233

Don E. Wilson U.S. Fish and Wildlife Service, National Museum of Natural History, Washington, D.C. 20560

Preface

Among living vertebrates bats and birds are unique in their ability to fly, and it is this common feature that sets them apart ecologically from other groups. Bats are in some ways the nocturnal equivalents of birds, having evolved and radiated into a diversity of forms to fill many of the same niches. The evolution of flight and echolocation in bats was undoubtedly a prime mover in the diversification of feeding and roosting habits, reproductive strategies, and social behaviors. Bats have successfully colonized almost every continential region on earth (except Antarctica), as well as many oceanic islands and archipelagos. They comprise the second largest order of mammals (next to rodents) in number of species and probably exceed all other such groups in overall abundance. Bats exhibit a dietary diversity (including insects, fruits, leaves, flowers, nectar and pollen, fish, other vertebrates, and blood) unparalleled among other living mammals. Their reproductive patterns range from seasonal monestry to polyestry, and mating systems include promiscuity, monogamy, and polygyny.

The vast majority of what we know about the ecology of bats is derived from studies of only a few of the approximately 850 species, yet in the past two decades studies on bats have escalated to a level where many important empirical patterns and processes have been identified. This knowledge has strengthened our understanding of ecological relationships and encouraged hypothesis testing rather than perpetuated a catalog of miscellaneous observations. More than ever before, there is an urgent need to intensify efforts to study the ecology of these unique mammals, both in temperate and tropical regions. The tropical regions of the world support the highest diversity of bat faunas known, but unless steps are taken to decelerate or reverse the current rate of loss of tropical habitats to alternate land-use patterns, the ecological diversity of this fauna may never be fully known.

The objective in organizing this volume was to provide a balanced and authoritative account of major topics on the ecology of bats. Each of the ten chapters has been prepared by one or more acknowledged experts. The range of topics treated is by no means exhaustive; rather, the subjects are representative of a sizable literature and have not been treated in recent works. The authors were encouraged to develop chapters in an ecological framework based on subjects of their own interest and expertise, to emphasize empirical studies as well as theory, and, where appropriate, to include discussions of adaptations and evolutionary trade-offs. It is my hope that this volume will provide points of departure for more rigorous study by ecologists and ethologists interested in bats and encourage others to consider ecological adaptations of bats as models for comparative study.

An attempt has been made to organize the sequence of chapters so that most of them are presented in a logical progression; otherwise they are grouped together by related subjects. An inevitable consequence of a multiauthor volume of this scope is that it leads to some redundancy in subject matter. Where appropriate, cross-references are made to relevant chapters. As in any discipline, new viewpoints emerge that will not always be received with universal agreement. This is as it should be, and consequently I have tried not to interfere with presentations of unorthodox ideas and interpretations, for the sake of encouraging further dialogue. The following comments provide an introduction to the topics included in this volume.

Roosts play a dominant role in the life history of bats, but there have been few attempts to integrate roosting habits, foraging behavior, social behavior, morphology, and energetics as factors influencing the roosting ecology of bats. Taking this approach in preparing Chapter 1, I (Kunz) have summarized the diversity of roosting adaptations, emphasized factors affecting roost fidelity and daily time budgets, and provided the first comprehensive treatment of night-roosting ecology. Roost availability, roost dimensions, energetic considerations, and risks of predation appear to be major determinants of roost use. In recognizing the importance of roosts to successful reproduction, Racey (Chapter 2) draws extensively from the literature and upon his own work to identify factors controlling reproduction in bats. Environmental factors affecting reproductive cycles, rates of embryonic development, and the timing and sequence of births are examined in bats from both temperate and tropical regions. He emphasizes that food supply during lactation and weaning is the single most important evolutionary factor in the timing of reproductive cycles.

Plasticity appears to be one of the most prevalent features of growth and survival in bats. Tuttle and Stevenson (Chapter 3) draw from an extensive literature and point out that both the litter size and developmental state at birth vary with biological and environmental factors. They note that successful growth of pre- and postweaned young varies among species and is markedly affected by feeding success and roost temperatures. In their analyses of bat survival they caution the reader on the limitations of sampling procedures and statistical treatments in published studies.

The importance of body size, food habits, and environmental conditions are emphasized by McNab's treatment of bat energetics and water balance in Chapter 4. Drawing extensively from his own work on the physiological ecology of temperate and tropical bats, he discusses factors influencing energy expenditure in bats, the ecological significance of energetics (including endothermy and the relationship between metabolic rate and life span), energy budgets, and the evolution of bat energetics. He argues that the distributional limits to bats may be imposed by food habits and an energy economy imposed by small body size.

In Chapter 5 Erkert summarizes an extensive literature on activity patterns of bats and, drawing from his own experimental work, interprets the periodicity of activity rhythms as adaptations to environments subjected to a 24-hr rhythm. He suggests that the flight and foraging activities of different species are influenced by the plasticity of a circadian system, the inhibitory activity of light, and the sensitivity of the circadian system to meteorological factors. The central thesis of Chapter 6, by Findley and Wilson, is that morphological traits used by systematists are valid ecological indices. Findley and Wilson emphasize the value of multivariate morphological analyses for quantifying the ecological niches of bats and in making comparisons of species within and among different ecological communities.

Fenton (Chapter 7) integrates field studies of bat echolocation with properties of insect prey in the interpretation of the feeding ecology of insectivorous bats. He suggests that the hearing ability of insects may influence the design of echolocation calls used by bats when feeding. He notes that most insectivorous bats are dietary opportunists. Because of extensive overlaps in habitat use and foraging time and morphological similarities among sympatric species, Fenton argues that there is no convincing evidence that insectivorous bats compete for and thus partition available food resources. Foraging strategies of plant-visiting bats are the subject of Chapter 8, where Fleming examines the importance of food availability and roosting behavior to foraging strategies (including foraging group size and foraging distance). He compares the foraging strategies of four well-studied species of Neotropical plant-visiting bats and emphasizes the importance of integrating studies on roosting behavior, social organization, and feeding behavior to the ultimate understanding of the foraging strategies of bats.

Most mutualistic interactions between plants and plant-visiting bats are unique to tropical regions, and in Chapter 9, Heithaus synthesizes a diverse botanical literature and integrates it with current knowledge of the foraging behavior of plant-visiting bats. He develops a model for coupled speciation between bats and plants, summarizes their complex coadaptations, and discusses the ecological and evolutionary consequences of bat–plant interactions. He suggests that one should avoid assuming that mutualism between bats and plants is coevolved, since preadaptations characterize several modern bat–plant systems.

In spite of the potential importance of parasitism in the lives of bats, relatively little research has been published on this topic. Ecologically, the best-known group of parasites associated with bats is the ectoparasitic insects, which Marshall treats in Chapter 10. He draws from the literature and from his own work on life cycles, host associations, and population dynamics to provide insight into some rather complex host–parasite adaptations unique to these flying mammals. This chapter includes an appendix with suggestions for collecting ectoparasitic insects for ecological study.

I am grateful to the authors of this volume for enthusiastically accepting the challenge to contribute their time and ideas and for tolerating my impatience in meeting deadlines. The staff of the Plenum Publishing Corporation has been extremely helpful in all phases of production. I owe special thanks to Kirk Jensen, who initially agreed to embark upon this venture and to Alan Winick and Daniel Jaul for their indulgence in my inexperience as an editor. I want to thank my most recent and present graduate students, Edythe Anthony, Pete August, Christopher Burnett, Marty Fujita, Karen Hoying, Al Kurta, and Holly Stack, for their assistance in proofreading and in the preparation of indices, and Nancy Tinnel and Christine Zotter, who assisted in clerical matters. Karl F. Koopman kindly verified the familial and species nomenclature of the Chiroptera. Finally, I want to acknowledge my graduate mentor, J. Knox Jones, Jr., who encouraged me to pursue my ecological interests in bats.

Thomas H. Kunz

Boston, Massachusetts

Contents

Chapter 1
Roosting Ecology

Thomas H. Kunz

Chapter 2
Ecology of Bat Reproduction

P. A. Racey

Chapter 5
Ecological Aspects of Bat Activity Rhythms

Hans G. Erkert

Chapter 6
Ecological Significance of Chiropteran Morphology

James S. Findley and Don E. Wilson

Chapter 1

Roosting Ecology of Bats

Thomas H. Kunz

Department of Biology
Boston University
Boston, Massachusetts 02215

1. INTRODUCTION

Bats spend over half their lives subjected to the selective pressures of their roost environment; thus it is not surprising that the conditions and events associated with roosting have played a prominent role in their ecology and evolution. Roosts provide sites for mating, hibernation, and rearing young; they promote social interactions and the digestion of food; and they offer protection from adverse weather and predators. Conditions that balance natality and mortality and enhance survivorship are intimately linked to roost characteristics and are paramount to the success of a species. The roosting ecology of bats can be viewed as a complex interaction of physiological, behavioral, and morphological adaptations and demographic response. The roosting habits of bats may be influenced by roost abundance and availability, risks of predation, the distribution and abundance of food resources, social organization, and an energy economy imposed by body size and the physical environment. For many bats the availability and physical capacity of roosts can set limits on the number and dispersion of roosting bats, and this in turn can influence the type of social organization and foraging strategy employed. For example, some bats as refuging animals (see Hamilton and Watt, 1970) may benefit from improved metabolic economy and information transfer but may be subjected to the added costs associated with increased commuting time, competition for food, and risks of predation.

The purpose of this review is to examine the diversity of roosting adaptations of bats and to consider how they are influenced by opposing selective pressures. I have intentionally avoided the use of a rigid classification of roost types (for a review of roost classifications see Gaisler, 1979); rather I have emphasized adaptations to day- and night-roosting behavior and ecology during the nonhibernating period to achieve an integrated view of the relationships between roosting habits, foraging strategies, energy economy, and social organization. A thorough discussion of the physiological aspects of roosting, including hibernation, has been reviewed elsewhere (Chapter 4). The importance of roosts in the evolution of social behavior was emphasized by Bradbury (1977a).

Generalizations about the roosting ecology of bats are difficult to formulate, because the selective pressures on different species are varied and diverse. The factors affecting bats that roost in protected shelters should differ markedly from those affecting bats that roost externally. Sheltered roosts offer the advantage of relative permanency, microclimate stability, reduced risks of predation, and protection from sunlight and adverse weather. External roosts offer the advantages of being ubiquitous and abundant, yet many are temporary (e.g., foliage) and subject to environmental extremes. The relative number of species using external shelters generally decreases with distance away from the equator, and there is a general tendency for bats that roost in caves and man-made structures to be highly gregarious (Gaisler, 1979).

The associations of bats and roosts range from being obligatory to opportunistic. Some species have become highly dependent on certain types of roosts owing to their morphological and physiological specialization. For example, the highly developed adhesive disks on the feet and wrists of *Thyroptera tricolor* restricts this bat to roosting on the smooth inner surfaces of unfurled leaves (Findley and Wilson, 1974). Opportunistic species typically have generalized roosting habits and wider geographic distributions. An example of such a species is the Neotropical bat *Artibeus jamaicensis*. In regions where caves are present, this bat seeks daytime shelter in small recesses and cavities within caves (Goodwin, 1970). Where caves are absent, for example, on Barro Colorado Island in the Panama Canal Zone, this bat typically seeks shelter in tree cavities and foliage. A polygynous mating system is promoted by the defense of these limited roost resources (tree cavities) by males (Morrison, 1979); but to what extent this type of social organization is influenced by the kind and abundance of roosts in other situations remains to be determined for this and other species.

The evolution of wings and flight have permitted bats to exploit roost environments (and food resources) virtually unavailable to most other vertebrates. The evolution of echolocation was undoubtedly a principle determinant leading to the divergence of roosting (and feeding) habits. Largely because of their ability to echolocate, the Microchiroptera have successfully exploited a variety of internal shelters (e.g., caves, rock crevices, tree cavities, and man-made structures). The Megachiroptera have successfully adapted to a variety of external roosts, but they have been virtually excluded from most internal shelters because of their inability to echolocate. Only members of the megachiropteran genus *Rousettus* have independently acquired an ability to echolocate, which they share with the michrochiropterans, echolocating oil birds (*Steatornis*), and swiftlets (*Collocalia*).

What one ultimately observes in the roosting ecology of a species is tempered by constraints imposed by phylogenetic inertia (see Wilson, 1975), and a compromise of opposing selective pressures derived from roost and nonroost origins. Behavioral, physiological, and morphological characteristics of bats commonly regarded as adaptations for roosting, such as dorsoventral flattening, suction pads and disks on feet and wrists, cryptic markings and postures, clustering, torpor, synchronous nightly departures, and so forth should be regarded as compromises imposed by the manner of flight, body size, predator pressure, energy economy, and the physical environment.

2. DAY ROOSTS

2.1. Adaptations for Roosting

2.1.1. Cave-Dwelling Bats

Bats are the only group of vertebrates that have successfully exploited caves for permanent shelter. Most families and genera of bats include species that regularly or occasionally seek refuge in caves. We know little about the conditions that may have led to the use of caves as roosts, although Jepsen (1970) suggested that bat ancestors may have become cave dwellers to escape predators and to conserve moisture and energy between periods of high activity. Caves typically last for long periods of time and, consequently, many have become traditional locations for resident populations; however, they have the disadvantage of being uncommon in many areas and located at considerable distances from suitable foraging areas (Bradbury, 1977a).

Caves serve as roost sites for solitary bats and groups ranging up to the largest known mammalian aggregations. In the southwestern United States single caves in summer may house up to 20 million individual *Tadarida brasiliensis* (Davis *et al.*, 1962). In tropical Africa the numbers of *Hipposideros caffer* may approach 500,000 (Brosset, 1966), and in Australia maternity colonies of *Miniopterus schreibersii* may number up to 200,000 (Dwyer and Hamilton-Smith, 1965). Many caves provide shelter for more than one species; the combined population of four species of mormoopid bats occupying a cave in Mexico may reach 800,000 individuals (Bateman and Vaughan, 1974). Benefits derived from such large numbers in a single cave must be balanced against the increased costs of intraspecific competition for food, misdirected social behavior, and the increased incidence of parasites and disease.

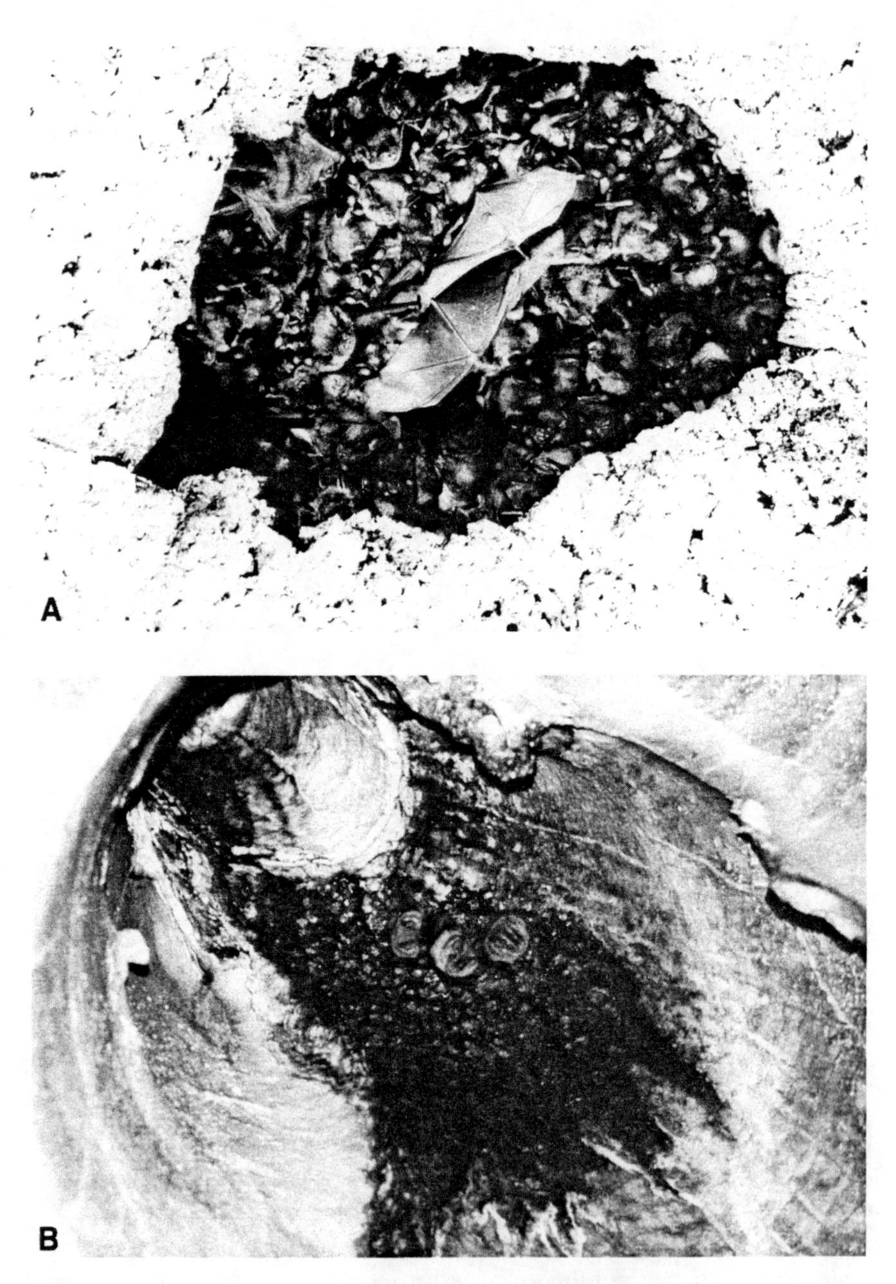

FIGURE 1. (A) Small cluster of *Myotis grisescens* occupying a cave ceiling cavity, where trapped metabolic heat enhances an individual's energy economy in an otherwise cool cave. (Photo by M. D. Tuttle.) (B) Small harem group of *Artibeus jamaicensis* occupying a solution cavity in a cave ceiling. The darkly stained area indicates that this site has had prolonged use. This and other cavities promote population substructuring and facilitate the defense of these roosts by harem males. (Photo by C. D. Burnett.) For a similar example see Fig. 4C in Chapter 8.

The distribution of cave-dwelling bats varies geographically with the distribution of caves and their physical dimensions, topography, and microclimate (Brosset, 1966; Tuttle and Stevenson, 1978). Many areas with caves are unsuitable for bat roosts. Caves with cold, descending chambers are not occupied in summer and seldom in winter, and many cold caves in temperate regions are unsuitable for bats during maternity periods. For example, only 2.4% of the 1635 known caves in Alabama are used by *Myotis grisescens* in summer, and even fewer (0.1%) are used in winter (Tuttle, 1979). In tropical regions cave environments are typically more stable, and they are more uniformally inhabited than in temperate regions (Brosset, 1966).

Caves that offer a wide thermal range combined with structural and elevational complexity provide the greatest diversity of roosting sites (Tuttle and Stevenson, 1978). In tropical regions bats are often distributed internally along gradients of light intensity (Brosset, 1966). Many caves have large, spacious chambers that provide roost sites for large aggregations. The presence of crevices and cavities in cave ceilings and walls can have an important influence on the ecology and social behavior of bats (Fig. 1). They provide roosting sites for a variety of bats (Dalquest and Walton, 1970), they facilitate group substructuring and the defense of roosts or female groups against incursions of conspecifics (Bradbury, 1977a; McCracken and Bradbury, 1981), and they serve as heat traps that enhance metabolic economy (Dwyer and Hamilton-Smith, 1965; Tuttle, 1975; Humphrey, 1975).

2.1.2. Crevice-Dwelling Bats

Rock crevices and narrow spaces beneath exfoliating bark of tree trunks and branches provide a variety of abundant and ubiquitous roost sites for bats. Rock crevices provide relatively permanent roost sites, but spaces beneath loose bark are temporary and thermally more variable and often require bats to make frequent relocations. Crevice dwelling appears to be a prevalent feature of molossids and vespertilionids in arid and semiarid regions (Brosset, 1962d; Brosset, 1966; Barbour and Davis, 1969). Little is known about the roosting ecology of crevice-dwelling bats, because they are difficult to find and often located in inaccessible places.

Crevice-dwelling bats have been most extensively studied in the arid regions of the western United States. Vaughan (1959) found that *Eumops perotis* prefers crevices in vertical or near-vertical cliffs, situated in deep slopes. Roosts characteristically have moderately large openings that can be entered from below. Six is the typical number of bats seeking shelter in these crevices. Occupied crevices are at least twice as wide as a bat's body and the entrances are horizontal and face downward. Vaughan (1959) suggested that the choice of these roost sites may be determined by the bats' need for space and to allow access to roosts from below.

Vertical and horizontal crevices, narrow fissures, and spherical chambers in rock cliffs are common roost sites of *Antrozous pallidus* (Vaughan and O'Shea, 1976). Typically, this bat prefers spacious crevices that permit internal movements and allow departures and returns to occur with outstretched wings. In summer deep, horizontal crevices are preferred, where cliffs serve as massive heat sinks. In spring and autumn bats roosting in vertical crevices are responsive to changes in ambient temperature. Temperatures in these crevices are low and bats remain torpid much of the day. Bats passively arouse as crevices become heated in late afternoon, further minimizing daily energy expenditure.

In the badlands of South Dakota singles and small maternity groups of *Myotis leibii* select small, dry, shallow roosts in horizontal and vertical crevices formed in siltstone boulders and sediment formed from a mixture of clay and volcanic ash (Tuttle and Heaney, 1974). Crevices are seldom more than 2.5 cm wide, ranging in depth from 2.5 to 20.5 cm. Temperatures in these crevices are uniformaly high (26–33°C), averaging 4–5°C lower than outside ambient temperatures. The shallowest roost crevices are located in positions that receive the least amount of direct daily sunlight.

Vertical crevices in steep canyon walls are the most common roost sites of *Pipistrellus hesperus* (Hayward and Cross, 1979). These crevices are typically narrow, approximately 2.5 cm wide, and range in depth from approximately 25 cm to 1 m. Roost temperatures remain considerably below ambient temperatures, as the rock substrate has a buffering effect. This bat is usually solitary, but it occasionally forms small groups whose individuals roost near the openings of crevices, where they can retreat deeper to avoid temperature extremes.

Crevices beneath loose and exfoliating bark of trees provide shelter for small maternity colonies of *Myotis sodalis* (Humphrey *et al.*, 1977). These small groups seek refuge in patches and interconnecting spaces beneath the bark of bitternut and shagbark hickory trees. Spaces beneath loose bark of bitternut hickory trees are apparently preferred in spring and summer, as they are more effective in trapping solar heat and in providing large spaces for rearing young. Areas beneath the exfoliating bark of living shagbark are thermally more stable during cold periods in spring and autumn and thus provide effective alternate roosts at these times.

2.1.3. Use of Tree Cavities

Cavities in the trunks and branches of dead and living trees offer favored shelter for many bats, especially in tropical regions. Such sites are formed from scars on branches and large cavities derived from the rotted interior of old trees. Tree cavities offer protection against fluctuations in ambient temperature and humidity (Verschuren, 1957; Maeda, 1974; Bradbury, 1977a; Morrison, 1979), and, as with hole-nesting birds (see Lack, 1968), they provide protection against predators and adverse weather. A disadvantage of tree cavities is that they offer limited roosting space for colonial species and they eventually rot and fall, requiring the periodic relocation of inhabitants (Bradbury, 1977a).

There are numerous reports of both Old and New World bats occupying tree cavities (Ryberg, 1947; Verschuren, 1957, 1966; Goodwin and Greenhall, 1961; Ognev, 1962; Hall and Dalquest, 1963; Kingdon, 1974; Tuttle, 1976b; Lekagul and McNeely, 1977), but few studies have considered the ecological aspects of these roost sites. In the Old World tropics tree cavities are most commonly used by members of the Nycteridae, Rhinolophidae, Hipposideridae (Brosset, 1966; Rosevear, 1965; Kingdon, 1974). In equatorial Africa small groups of *Nycteris arge, N. grandis, N. major,* and *N. nana* typically prefer trees with entrances located near the base (Rosevear, 1965; Kingdon, 1974). A similar preference is shown by the Neotropical phyllostomids *Trachops cirrhosus* and *Carollia perspicillata* (Fig. 2A). In Kenya the megadermatid *Cardioderma cor* forms groups of up to 80 individuals in hollow baobab trees (Fig. 2B), where roost cavities range from 2.2 to 4 m in height and are 1.8 to 3 m wide, each with a single entrance averaging 25 cm in width and 40 cm in height (Vaughan, 1976).

In the Neotropics tree cavities are used predominantly by phyllostomids, of which 15 genera, representing 28 species, are known to regularly or occasionally use such cavities as day roosts (Tuttle, 1976b). Tree cavities are also used by some molossids, vespertilionids (Walker, 1975), and *Noctilio leporinus* (Goodwin and Greenhall, 1961). The most intensive study of tree-roosting habits of a Neotropical bat was by Morrison (1979), who found that harem groups of *Artibeus jamaicensis* gain access to tree cavities through single openings, approximately 6 cm wide (Fig. 2C). Entry holes ranged in height from 2 to 15 m above the ground, and cavities typically consisted of long cylinders, 12–25 cm in internal diameter, extending 1.5–2 m above the entrance hole.

Tree cavities also provide favored roost sites for several species of Palearctic bats, where individuals typically hang from the upper parts of these cavities (Ryberg, 1947). In Japan Maeda (1974) found that *Nyctalus lasiopterus* occupied tree cavities, the largest of which were used as maternity roosts housing up to 40 individuals; the smallest cavities were used by males. Tree cavities were chosen for the position and shape of the tree and the height of the entrance above the ground. In the Soviet Union at least ten vespertilionid species roost in tree cavities, where two or more species may roost together (Ognev, 1962). *Myotis bechsteini* and *Pipistrellus nathusii* each form pure colonies in tree cavities, but these bats may form mixed colonies in man-made structures. In Finland *M. mystacinus* prefers tree cavities to buildings, and although *M. daubentoni* occasionally roosts in tree cavities, it shows a preference for buildings. In one of the most extensive studies on tree-roosting bats Gaisler *et al.* (1979) examined the

FIGURE 2. (A) A *Trachops cirrhosus* entering (left) and a *Carollia perspicillata* departing (right) from a tree cavity through an opening located at the base of a tree. (Photo by M. D. Tuttle. Courtesy of the National Geographic Society.) (B) Cavities in baobab trees (*Adansonia digitata*) are preferred day-roosting sites of *Cardioderma cor*. (Photo by T. A. Vaughan. With permission from the American Society of Mammalogists.) (C) Tree cavity used by a small harem group of *Artibeus jamaicensis*. (Photo by D. W. Morrison.)

roosting habits of *N. noctula* in central Europe and found that 72.8% of these bats were found in tree cavities (21.8% occurred in buildings and 5.4% in man-made roost boxes and beneath loose bark). From March through October this bat forms maternity groups ranging from 3 to 54 individuals. Roost cavities in trees near the edge of wooded areas, in lowland and sloping regions, are preferred, including 13 different tree species distributed proportionately among the species occurring in the region. The principle criteria for roost selection were that trees had a high trunk below the crown and free (flight) space in front of the entrance. Roost entrances ranged in height from 1 to 16 m above the ground, but most bats occupied cavities at a height of 1.4 m.

That hollow trees are prevalent in nutrient-poor soils in tropical regions prompted Janzen (1976) to postulate that rotted hollow cores may be an adaptive trait selected as a mechanism for nitrogen and mineral trapping resulting from the accumulation of animal feces and subsequent microbial metabolism. If this hypothesis is correct, tree-roosting bats that deposit large quanitites of nitrogen-rich guano (see Hutchinson, 1950) may play an important nutrient role in forest ecosystems.

2.1.4. Foliage- and Other External-Roosting Bats

A fundamental similarity among foliage-roosting bats is that their roost sites are temporary. As leaves die or unfurl they no longer provide suitable roosts. Foliage roosts often promote nomadic populations and low roost fidelity and provide minimal protection from variations in temperature and humidity (Bradbury, 1977a). Most foliage-roosting bats are solitary or form small groups, and, with few exceptions, they are distributed in tropical regions. Large aggregations of foliage-roosting bats are known only among the largest megachiropterans. Several foliage-roosting bats have a highly specialized morphology and behavior that limits their roosting habits to certain types of plants (see Section 2.1.5), some modify leaves into tentlike structures (see Section 2.1.6), and still others rely on some form of crypsis.

Many bats that roost externally on branches, tree trunks, and in foliage have bold, contrasting marks, bright colors, spotted patterns, reticulate wing venation, or assume distinct postures and dispersion patterns that defy visual detection. Such cryptic patterns and behaviors are prevalent among the Pteropodidae, Emballonuridae, Phyllostomidae, and Vespertilionidae and presumably confer some degree of protection from visually oriented predators.

The concealment of many foliage-roosting megachiropterans is enhanced by mottled and broken color patterns and sometimes by motionless postures. In some pteropodids hues of yellow, orange, and red resemble fruits and dry leaves, and contrasting lighter colors around the head and neck suggest a type of countershading (Dobson, 1877; Novick, 1977). The motionless postures often seen in

FIGURE 3. Mother and two young *Lasiurus cinereus* hanging from a spruce tree. (Photo by M. D. Tuttle. Courtesy of the National Geographic Society.)

Pteropus poliocephalus (Nelson, 1965a), *Epomops franqueti* (Jones, 1972), *Lavia frons* (Kingdon, 1974), and *Nyctimene major* (Walker, 1975), engulfed in folded wings, gives the appearance of dead leaves.

Contrasting patches and varying numbers and intensities of facial and dorsal stripes enhance crypsis in several vespertilionids, emballonurids, and phyllostomids. Among Old World vespertilionids cryptic markings are pronounced in *Myotis formosus, Murina aurata,* and in the so-called "painted bats" of the genus *Kerivoula,* many of which have long and woolly pelage, ranging in color from yellow to bright orange and scarlet (Dobson, 1877; Allen, 1939; Wallin, 1969; Walker, 1975). Some of the *Kerivoula* have a grizzled, frosted appearance, with long fingers that contrast against dark wing membranes and which apparently resemble tufts of moss (Walker, 1975). Members of the genus *Glauconycteris* have reticulate markings that resemble leaf venation and bold, contrasting marks on the pelage that suggest disruptive patterns (Rosevear, 1965; Walker, 1975; Novick, 1977). A white dorsal stripe and patches of white on the head and shoulders of the vespertilionid *Scotomanes ornatus* contrasts with an otherwise brownish pelage (Walker, 1975; Lekagul and McNeely, 1977). Cryptic colors and markings are also evident among New World bats of the genus *Lasiurus* which either roost singly or in small groups in foliage, in vines (Constantine, 1966), and in clumps of Spanish moss (Constantine, 1958a). The frosted, grizzled appearance of *L. cinereus* (Fig. 3) may offer protection from visually oriented predators.

Two distinct parallel white dorsal lines and the motionless roosting posture of *Saccopteryx bilineata* may confer protection from predators, as this emballonurid roosts on exposed buttresses and the trunks of trees (Bradbury and Emmons, 1974). Similarly, the brownish, grizzled, yellow-gray color and white, wavy dorsal lines of *Rhynchonycteris naso* resemble lichen when contrasted against the background of tree bark, on which it frequently roosts (Goodwin and Greenhall, 1961; Bradbury and Emmons, 1974). Distinct white facial and dorsal stripes of several phyllostomids, including the tent-making *Artibeus* and *Uroderma,* and members of the genera *Chiroderma, Enchisthenes, Vampyrodes, Vampyrops,* and *Vampyressa* may confer protection from predators. Because many of these bats roost in foliage having either lobed or pinnate leaves (Davis, 1944; Goodwin and Greenhall, 1961; Jimbo and Schwassmann, 1967; Morrison, 1978a, 1980), facial stripes and dorsal markings may complement the disruptive patterns formed by individual leaflets (Foster and Timm, 1976).

Some foliage-roosting bats seek shelter where they are inconspicuous from the ground and nearby branches or roost high in the tree canopy, which reduces their vulnerability to predation or disturbance. Large groups of *Artibeus lituratus* and *Vampyrodes caraccioli* commonly roost high in the tree canopy, where they may be conspicuous from the ground, but those that seek lower sites roost in smaller groups well concealed in understory foliage and vine-entangled subcanopy trees (Morrison, 1980). The North American tree-roosting bats *Lasiurus borealis*, *L. cinereus*, *L. intermedius*, and *L. seminolis*, commonly select roosts that are visible only from below, lack branches from which predators might detect and attack them, and are located over ground cover that minimizes reflected sunlight (Constantine, 1958a, 1966). Constantine (1966) found that the roosts of young and adult female *L. borealis* and *L. cinereus* were located higher in the tree canopy than were solitary males, and he suggested that this behavior may reduce their conspicuousness to predators and provide young bats with greater opportunities to initiate successful flights.

The roosts of most megachiropterans vary from being located in dense foliage in the darkest parts of trees (Ayensu, 1974) to open, conspicuous areas. For example, *Micropteropus pusillus* is often associated with dense foliage, *Epomops franqueti* hangs from small branches near clusters of leaves, and *Eidolon helvum* commonly roosts on sturdy but leafless branches near the trunks of large trees (Jones, 1972). Nelson (1965a) noted that *Pteropus poliocephalus* usually hangs from lower, shaded branches but moves to higher, more open areas of the tree canopy following disturbance. The roosts of *P. vampyrus* may be conspicuous, but they are often located in mangrove swamps, where they are inaccessible to intruders (Goodwin, 1979). Differences in the roosting positions of megachiropterans may be influenced by body size, the degree of crypsis, and flight ability. For example, the large megachiropteran *Hypsignathus monstrosus* roosts on exposed branches, high in forest trees, beneath the dense umbrella canopy, and where there is sufficient flight space between the roost and the understory vegetation (Bradbury, 1977b).

2.1.5. Morphological Specialization

Morphological changes associated with the evolution of flight have had a profound affect on the roosting habits and locomotion of bats (Vaughan, 1970a). For example, the hind limbs of most bats extend dorsolaterally, as if they had been rotated 90° from the position typical of terrestrial mammals. In some families, including the phyllostomids and natalids, this rotation may reach 180°, where it severely restricts quadrupedal locomotion. Vaughan (1970a) recognized three types of morphological modifications of the pelvic and hind limb structures that are related to different roosting habits and terrestrial locomotion in bats. The majority of bats, including most or all of the Emballonuridae, Nycteridae, Natalidae, and Vespertilionidae, have generalized roosting habits. These bats may hang from ceilings in contact clusters, cling to vertical surfaces, or roost in crevices or small holes. Another group, typified by the phyllostomid genera *Choeronycteris*, *Glossophaga*, *Macrotus*, extend their hind limbs to the rear and only their feet are in contact with the roost substrate. These bats usually assume a pendent posture and hang separate from each other, often suspended by one foot.

Bats that hang pendent commonly have large pectoral muscles, short bodies, and deep chests (Vaughan, 1959). Howell and Pylka (1977) suggested that selection for pendent roosting habits may have facilitated flight takeoff, reduced thermal disadvantages that accompanied conduction to a cool substrate (in caves), and lessened accessibility to predators. Roosting pteropodids typically assume a pendent posture and use their feet and wings for crawling. *Pteropus* and *Eidolon* are apparently incapable of crawling on horizontal surfaces (Lawrence and Novick, 1963), but they are agile with both wings and feet when crawling among branches in trees (Nelson, 1965a; Wickler and Seibt, 1976). Other pteropodids such as *Dobsonia peroni* and *Notopteris macdonaldi* are able to exploit shallow, dimly lit caves because of a unique insertion of the wing membrane that allows them to climb vertically and reach cave ceilings (Goodwin, 1979). A third group that includes the molossids often has modified pelvic and pectoral girdles as adaptations for roosting in crevices. Many of these bats have an extraordinary ability to crawl, but some require special roosts that permit vertical drops before taking flight, as in *Otomops wroughtoni* (Brosset, 1962c) and *Eumops perotis* (Vaughan, 1959).

In addition to the selective pressures imposed by echolocation and food habits (Chapter 6), roosting habits have had a marked influence on the evolution of skull shape. Bats that roost by hanging pendent in noncrevice shelters typically have well-rounded skulls (Vaughan, 1970a). Crevice-roosting bats, including molossids and some vespertilionids, show a marked dorsoventral flattening of the skull. Extreme flattening of the skull can be seen in the crevice-roosting molossids *Platymops setiger, P. petrophilus*, and *Neoplatymops mattogrossensis* (Peterson, 1965). Most small bats that seek shelter in crevices show little cranial modification for roosting (Vaughan, 1970a), but there are exceptions. For example, the small vespertilionid *Mimetellus moloneyi* has a strongly flattened skull (Kingdon, 1974; Walker, 1975); similarly, the crania of *Tylonycteris pachypus* and *T. robustula* are strongly flattened, allowing entry through narrow slits to the internode cavities of bamboo culm (Medway and Marshall, 1970, 1972).

Other morphological modifications for roosting include thickened pads or adhesive disks on the thumbs and feet (Table I). The greatest degree of specialization has occurred in the Thyropteridae and Myzopodidae, each having well-developed adhesive disks on the fore and hind limbs. Among the vespertilionids 6 genera and at least 11 species have pads on their thumbs and/or feet. Many if not most of these bats are probably specialized for roosting on the moist surfaces of leaves or within the internode cavities of bamboo.

The highly developed adhesive disks of the Neotropical *Thyroptera discifera* and *T. tricolor* (Fig. 4A) may limit these bats to certain types of roosts. Findley and Wilson (1974) found that *T. tricolor* roosts exclusively in unfurled leaves of *Heliconia* and similar plants in Costa Rica (Fig. 5), and they suggested that its numbers and distribution may be limited by competition for available roost sites. *T. tricolor* apparently prefers the shelter of unfurled leaves in forests, forest clearings, and occasionally along roads and trails, where leaves are shaded for part of the day. This bat roosts singly or in small groups, where individuals arrange themselves in a head-up posture within the leaf. Virtually nothing is known of the roosting habits of *T. discifera*, but, judging from the similarity of

TABLE I
Bats Having Modified Thumb and Foot Pads or Disks

Family	Species	Region	References
Thyropteridae	*Thyroptera discifera*[a]	Neotropical	Walker, 1975
	T. tricolor[a]	Neotropical	Dobson, 1876; Goodwin and Greenhall, 1961; Wimsatt and Villa-R., 1970; Findley and Wilson, 1974; Walker, 1975
Myzopodidae	*Myzopoda aurita*[a]	Ethiopian	Walker, 1975; Schliemann and Mags, 1978
Vespertilionidae	*Eudiscopus* (=*Discopus*) *denticulus*[b]	Indo-Malayan	Osgood, 1932; Hill, 1969; Walker, 1975
	Glischropus javanus	Indo-Malayan	Tate, 1942; Hill, 1969
	G. tylopus	Indo-Malayan	Tate, 1942; Hill, 1969
	Hesperoptenus blanfordi	Indo-Malayan	Hill, 1969; Walker, 1975
	H. tickelli[c]	Indo-Malayan	Hill, 1969; Walker, 1975
	Myotis bocagei	African	Brosset, 1966, 1974, 1976; Walker, 1975
	M. rosseti	Indo-Malayan	Rosevear, 1965; Hill, 1969
	Pipistrellus nanus	African	Dobson, 1876; Verschuren, 1957; Rosevear, 1965; Brosset, 1966; Jones, 1971; Kingdon, 1974; Walker, 1975; LaVal and LaVal, 1977
	P. (=*Glischropus*) *tasmaniensis*	Australian	Walker, 1975
	Tylonycteris pachypus	Indo-Malayan	Dobson, 1876; Medway and Marshall, 1970, 1972; Walker, 1975
	T. robustula	Indo-Malayan	Dobson, 1876; Medway and Marshall, 1970, 1972; Walker, 1975

[a]Specialized disks.
[b]Pads only on hind feet (Walker, 1975).
[c]Friction pads only on thumbs (Walker, 1975), but according to Hill (1969) pads are absent.

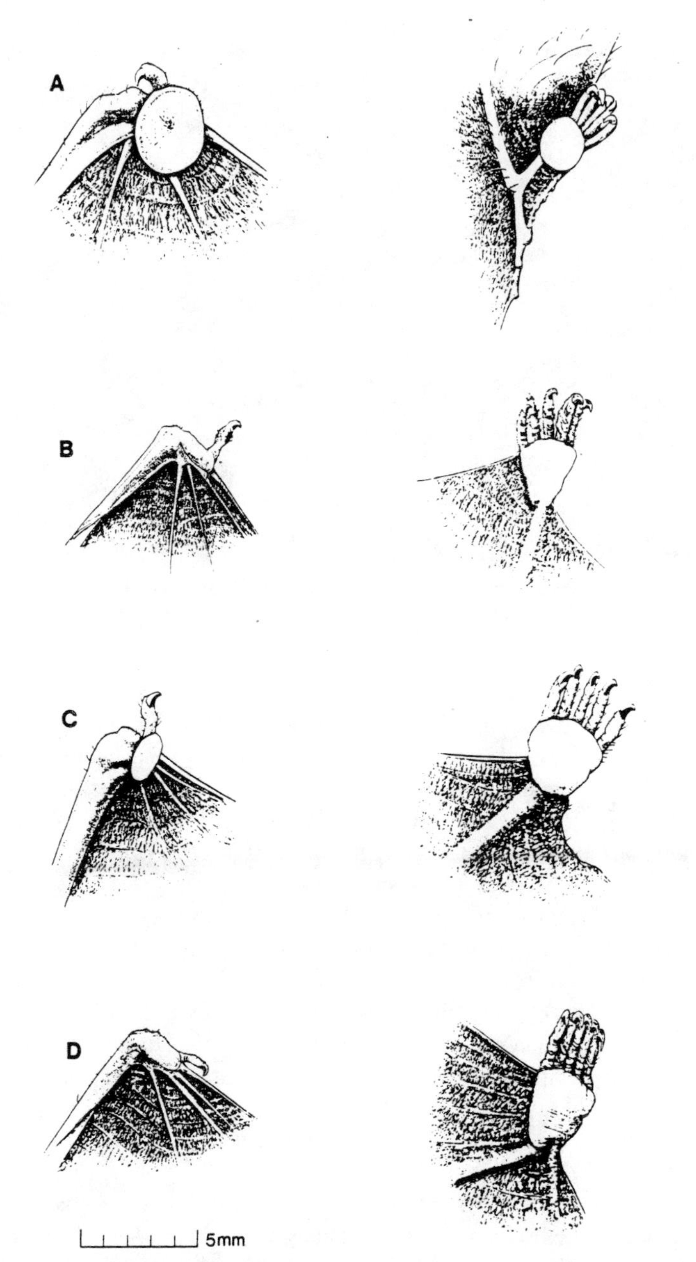

FIGURE 4. Morphological specialization of wrists and feet of (A) *Thyroptera tricolor* (MCZ 28142), (B) *Pipistrellus nanus* (MCZ 14840), (C) *Tylonycteris pachypus* (MCZ 47612), and (D) *Glischropus tylopus* (MCZ 33117). (Drawings prepared by P. Esty from specimens preserved in alcohol.)

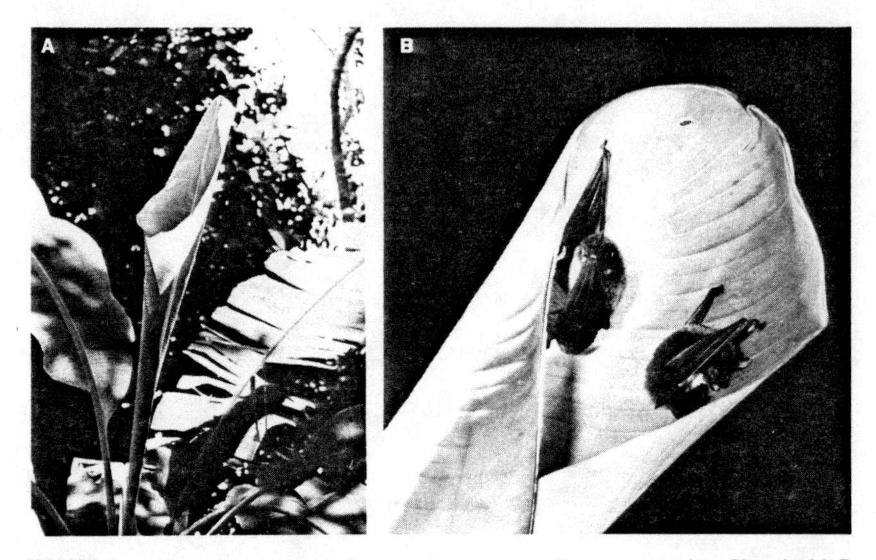

FIGURE 5. (A) Unfurled banana leaf used as a day roost by *Thyroptera tricolor*. (Photo by M. D. Tuttle.) (B) Two *Thyroptera tricolor* clinging to the inner, moist surface near the opening of an unfurled leaf used as a roost. (Photo by M. D. Tuttle. Courtesy of the National Geographic Society.)

its foot and wrist disks, its roosting habits are probably similar to those of *T. tricolor*. The disk-winged bat *Myzopoda aurita,* endemic to Madagascar, sometimes roosts in the leaves of traveler's palm (*Ravenala*) (R. Peterson, in Findley and Wilson, 1974), but not much more is known of its roosting habits.

Among vespertilionids, foot- and thumb pads or wrist callosities are moderately to well developed, but the roosting habits of most species are little known. In northeastern Gabon *Myotis bocagei* regularly roosts singly or in small harem groups in unfurled banana leaves (Brosset, 1966, 1974, 1976). This bat is apparently restricted to riparian habitats and typically roosts within 5–50 m from a river. Sanborn (1949) found this bat roosting in the flower stalks of water arum. In villages where banana plants are cultivated this bat may have a continuous and localized source of unfurled leaves, and small groups may remain resident for several years at these sites (Brosset, 1976). A similar development of fleshy wrist pads and footpads (Fig. 4B) and roosting habits have been reported for *Pipistrellus nanus* (Verschuren, 1957; Rosevear, 1965; Brosset, 1966; Kingdon, 1974). LaVal and LaVal (1977) found that only 20.9% of 565 suitably unfurled leaves were occupied; the highest occupancy rate occurred when females and young roosted together (group size ranged from 2 to 6). LaVal and LaVal (1977) suggested that before the introduction of bananas to Natal, the indigenous *Strelitzia nicolai* may have been the preferred roost of *P. nanus*. In Kenya, where there are no musaceous plants, *P. nanus* commonly roosts between the leaflets of palms growing along rivers and in buildings constructed of palm thatch, where males and females occur in small harem groups (O'Shea, 1980). It seems likely that harem groups of *P. nanus* should also occur in unfurled banana leaves, as in *M. bocagei,* but the census data from LaVal and LaVal (1977) are not compelling.

The fleshy pads on the feet and thumbs of *Tylonycteris pachypus* and *T. robustula* (Fig. 4C) apparently assist the claws in gripping the smooth surface within the internodal cavities of bamboo (Medway and Marshall, 1970, 1972). Access to these cavities is provided by a narrow, slitlike opening (Fig. 6), formed originally by the pupation chamber and emergence hole of a chrysomelid beetle. Internode cavities that are preferred as roosts vary in height from 1 to 10 m above the ground and typically have entrance holes located in the lower half of the cavity. The lower limit of the hole width is determined by body size, with the holes used by *T. robustula* being significantly larger than those used by *T. pachypus.* Both species may use the same roost sites, but usually on separate occasions (Medway and Marshall, 1970). Females of both species outnumber males by approximately 2:1, and females tend to be gregarious at all ages; adult males tend to roost singly or with other females in small harem groups (Medway, 1971; Bradbury, 1977a). Other bats known to roost in the interior of bamboo culm include *Glischropus tylopus* and *Pipistrellus mimus,* but specialized pads are known only for *G. tylopus* (Fig. 4D).

Short, stout legs, hairs on the inner and outer margins (Rosevear, 1965), short interfemoral membranes with hairs on the tail, which may have a tactile function (Lang and Chapin, 1917), appear to be adaptations for crevice roosting in some molossids and a mysticinid. The short, erect, velvetlike pelage, highly modified claws for digging, and the unique wing-folding behavior of *Mysticina tuberculata* appear to be extreme specializations for crevice dwelling (Daniel, 1979).

Bats that use external roosts, including *Murina aurata, Myotis formosus* (Wallin, 1969), *Lavia frons* (Wickler and Uhrig, 1969), *Lasiurus borealis,* and *L. cinereus* (Shump and Shump, 1980), typically have thick, long, woolly pelage and probably benefit from increased insulation. Rosevear (1965) suggested that the pelage of *Lavia frons* may be as important in maintaining body temperature as it is in reducing the effects of direct exposure to sunlight (Wickler and Uhrig, 1969; Jones, 1972).

The importance of well-developed vision for predator surveillance is highlighted by the degree of eye development, especially among externally roosting pteropodids, phyllostomids (Suthers, 1970, 1978), and emballonurids (Suthers, 1970; Bradbury and Emmons, 1974). Some phyllostomids, including *Artibeus hirsutus, A. jamaicensis, A. lituratus, A. phaeotis, A. toltecus, Centurio senex, Chiroderma salvini,* and *Sturnira lilium,* have a transparent dactylopatagium minus, which allows them to observe movements in the vicinity of the roost, even though their wings may be folded over their faces when at rest (Vaughan, 1970b). *Centurio senex* sometimes covers its face with a partially translucent and hairless chinfold, which apparently allows it to see when this structure is stretched over its face (Goodwin and Greenhall, 1961).

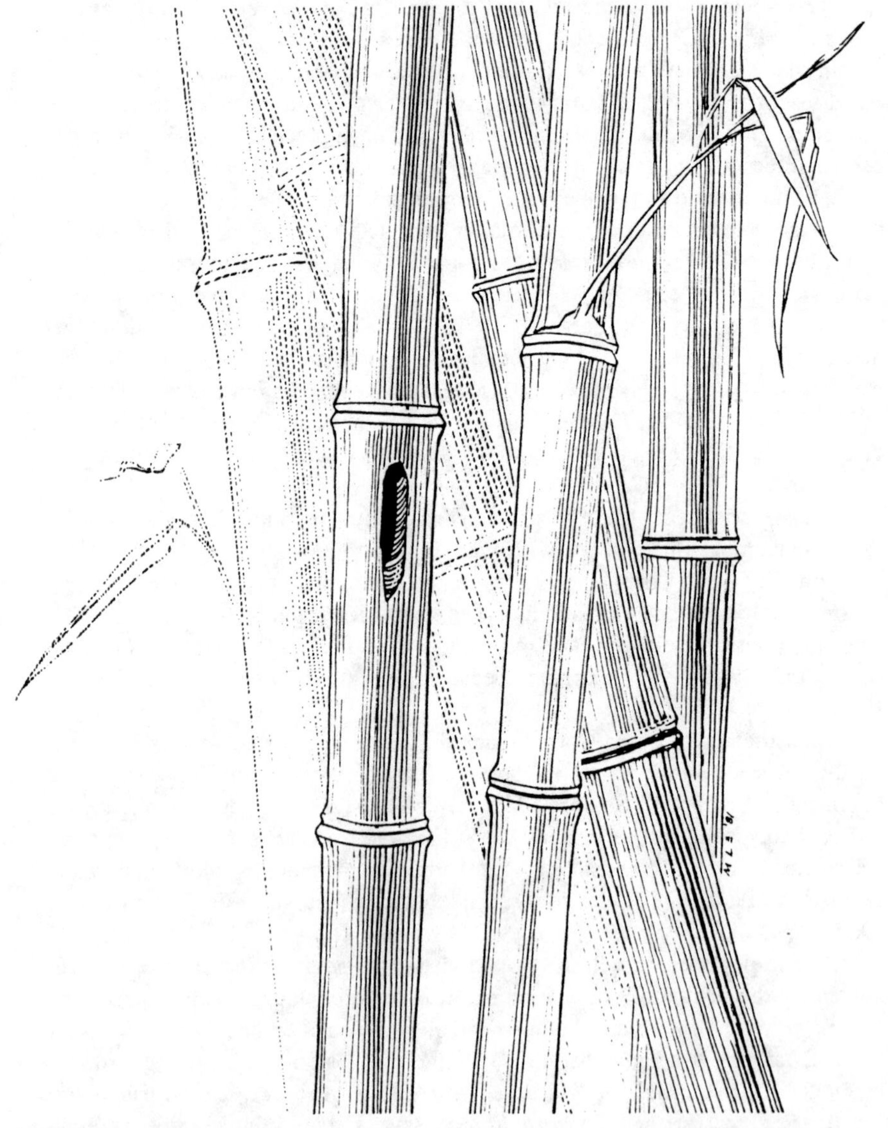

FIGURE 6. Giant bamboo culm used as roosts by *Tylonycteris pachypus* and *T. robustula*. Bats gain entry and exit to an internodal cavity via a slit formed from a pupation chamber of a crysomelid beetle. (Drawn from a photo in Medway and Marshall, 1970.)

TABLE II
Tent-Making Bats and Plants Used in the Construction of Tents

Bats		Plants		
Family	Species	Family	Species	References
Pteropodidae	*Cynopterus sphinx*	Palmaceae	*Corypha* sp.	Goodwin, 1979
Phyllostomidae	*Ectophylla alba*	Musaceae	*Heliconia imbricata*	Timm and Mortimer, 1976
			H. latispatha	Timm and Mortimer, 1976
			H. pogonantha	Timm and Mortimer, 1976
			H. tortuosa	Timm and Mortimer, 1976
			H. sp.	Timm and Mortimer, 1976
	Artibeus cinereus	Palmaceae[a]	Unspecified	Goodwin and Greenhall, 1961
	A. jamaicensis		*Scheelea rostrata*	Foster and Timm, 1976
	A. watsoni		*Geonoma cuneata* (=*ducurrens*)	Chapman, 1932; Ingles, 1953
			G. oxycarpa (=*binervia*)	Chapman, 1932; Ingles, 1953
	Uroderma bilobatum		*Cocos nucifera*	Barbour, 1932; Goodwin and Greenhall, 1961
			Livistona chinensis	Barbour, 1932
			Pritchardia pacifica	Barbour, 1932
			Sabal mauritiiformis (=*glaucescens*)	Barbour, 1932; Goodwin and Greenhall, 1961
	Unspecified		*Asterogyne martiana*	Foster and Timm, 1976
			Bactris wendlandiana	Foster and Timm, 1976
			Geonoma congesta	Foster and Timm, 1976

[a]Taxonomic designation of the American Palmaceae follows Glassman (1972).

2.1.6. Bats That Modify Their Roost Environment

The ultimate adaptation of a species occurs when it manipulates the physical environment to its advantage. At a primitive level colonial bats may inadvertently alter their physical environment by depositing feces and urine or increasing the temperature of the roost as a by-product of metabolism, but the ultimate in roost modification occurs when changes result from self-directed behavior. Such is the case among the "tent-making" pteropodids and phyllostomids (Table II) and a "burrowing" mysticinid.

Five species of phyllostomids and one pteropodid modify the leaves of at least five species of *Heliconia* (Musaceae) and nine species of palm (Palmaceae) for use as roosts. The phyllostomid *Ectophylla alba* typically constructs tents from *Heliconia* (Timm and Mortimer, 1976), but other phyllostomids either use

FIGURE 7. (A) Tent roost of *Ectophylla alba* constructed from a *Heliconia* leaf. (B) Group of *E. alba* hanging from the underside of a *Heliconia* tent. Note the chewed areas on either side of the midrib. (Photo by R. M. Timm. With permission from the Ecological Society of America.)

broadleaf or pinnate palms for tent making (Chapman, 1932; Barbour, 1932; Allen, 1939; Goodwin, 1946; Ingles, 1953; Goodwin and Greenhall, 1961; Timm and Mortimer, 1976; Foster and Timm, 1976). *Ectophylla alba* modifies leaves of *Heliconia* by chewing veins that extend perpendicular to the midrib, causing the sides of the leaf to droop (Fig. 7A). The veins are only partially chewed, leaving interconnected tissue that provide some support for the length of the leaf. Most of the support for the tent comes from the uncut basal and distal parts of the leaf. These bats hang singly or in small groups (2–6) from small claw holes near the center of the leaf (Fig. 7B). *Ectophylla alba* appears to be opportunistic in its selection of leaves, choosing them for making tents from several species of *Heliconia* in proportion to their occurrence in the forest. A preference, however, is shown for leaves that grow horizontally, thus providing maximum protection from sunlight, rain, and predators (Timm and Mortimer, 1976).

Uroderma bilobatum constructs tents from large, stiff palm fronds of *Prichardia pacifica* by biting the ridges of plications on the underside of the leaf until it weakens and droops downward (Fig. 8). This bat roosts singly or in small groups (2–59), often hanging from the cut portion of the leaf. Eventually the distal part of the leaf dries and sloughs off and a new leaf has to be cut (Barbour, 1932). Tents constructed by *Artibeus watsoni* from broadleaf palms are typically formed when this bat makes J-shaped cuts beginning near the distal end of the blade along the rachis and curving back to the base of the blade and out to each side (Fig. 9A). Variations on this pattern exist (Fig. 9B), but whether these reflect differences in plant species or in tent-making behavior is unknown. Pinnate palms are modified into tents when *A. jamaicensis* chews the midrib of leaflets, removing pieces of tissue and leaving small holes (Foster and Timm, 1976). This causes the terminal parts of leaflets to fold perpendicularly to the plane of the rachis, forming a lanceolate tent (Fig. 9C).

The use of broadleaf palms for tent construction appears to be the most common. Foster and Timm (1976) found that nearly two-thirds of the tents observed in Costa Rica were constructed from fronds of the simple bilobed *Asterogyne martiana*. Although broadleaf palms may offer greater protection from rain and sunlight, pinnate leaves may enhance crypticity, if contrasting, white facial and dorsal stripes complement the disruptive pattern of leaflets (Foster and Timm, 1976).

The pteropodid *Cynopterus sphinx* constructs tents from large pinnate fronds of the *Corypha* palm by chewing veins of the leaflets, creating "a characteristic flask-shaped pattern within the blade," which causes the collapse of distal leaflets to form the sides of the tent (Goodwin, 1979). Solitary bats and small groups cling with their toes to the veins of the "roof," hanging against the sides of the shelter where they apparently are less conspicuous from below. The only other pteropodid known to modify plants for shelter is *Cynopterus brachyotis*, which sometimes bites off the center seed string of the *Kitul* palm, leaving a hollow in which to roost (Phillips, 1924).

FIGURE 8. (A) Typical tent roost of *Uroderma bilobatum* constructed from a *Prichardia pacifica* frond. The inset illustrates the manner in which the plications of the frond were chewed. (Drawn from a photo in Barbour, 1932.) (B) Group of *Uroderma bilobatum* roosting under a *Prichardia* tent. (Photo by I. F. Greenbaum.)

FIGURE 9. (A) Palmate palm frond modified into a tent by *Artibeus watsoni*. (B) Two *A. watsoni* hanging from underside of tent. (Photo by M. D. Tuttle.) (C) Tent roost in a pinnate palm, *Scheelea rostrata*, constructed by *A. jamaicensis*. (Photo courtesy of R. M. Timm. With permission from the Association of Tropical Biology.)

Future studies on the ecology of tent-making bats should attempt to determine how much time and how many bats are involved in tent construction. The relative cost appears to be high, considering the number of veins that are severed and the temporary nature of these roosts. Judging from the observations of Barbour (1932), new tents from *Prichardia* fronds are constructed by small groups of *Uroderma bilobatum;* single bats are apparently unable to complete a tent in one night. The number of bites required to sever the veins of a frond may range from as few as 44, in tents formed from the simple bilobed leaves of *Asterogyne,* to as many as 80, in the large palmate leaves of *Prichardia* (Foster and Timm, 1976). These observations suggest that tent making typically involves the cooperation of several bats, but how many individuals are involved and whether both sexes participate remains to be clarified. Although solitary males and groups of female *Uroderma* with their young have been observed in tents (Barbour, 1932), no bats have been observed in the act of tent making. At what age tent making begins and whether it is an innate or learned behavior is unknown. The fact that *Ectophylla alba* constructs tents from plant species having similar leaf forms suggests that at least for this species some specificity exists. There is some latitude in the use of different leaf forms for tent construction among other tent-making species. Barbour (1932) and Foster and Timm (1976) suggest that the initial bites used in tent construction are probably made while bats hover, but direct observations are needed to verify this suggestion and to confirm whether additional bites are made by bats when crawling on or hanging from the leaf.

Because tent roosts constructed from leaves are by nature ephemeral, bats no doubt regularly engage in tent making. Timm and Mortimer (1976) suggested that each colony of *E. alba* may use a series of tents, since they seldom observed bats using the same tent for more than two consecutive days. Similarly, Barbour (1932) observed considerable turnover in the use of tent-roosts by *Uroderma bilobatum*. If a high roost turnover occurs, it would be interesting to determine whether there is any consistent group fidelity and whether bats that share a tent-roost are genetically related.

A unique form of roost construction has been reported for *Mysticina tuberculata,* a monotypic species endemic to New Zealand (Daniel, 1979). This bat uses its teeth and claws to excavate cavities and tunnels in the wood of fallen kauri (*Agthis australis*) trees, where individuals form dense clusters in the largest cavities and rest "head-to-tail like peas in a pod" in adjoining tunnels. The heat generated in these wet tunnels and cavities by a colony of 150–200 bats can result in temperatures of 39°C and a humidity of 100%. Daniel (1979) suggested that this "rodent-like" behavior may have evolved in response to a lack of predators and competition from other mammals.

The mere presence of active bats in sheltered roosts can have a profound affect on the roost environment. The accumulation of bat guano, with its relatively high water-holding capacity, can raise the humidity of roosts, (Verschuren, 1957) as can the accumulation of urine on wooden beams in buildings (Kunz, 1973b). Heat produced as a by-product of metabolism and trapped in crevices and cavities can markedly raise the temperature of a roost during periods of occupancy (Dwyer and Harris, 1972; Voûte, 1972; Kunz, 1974), as well as over longer periods (Henshaw, 1960; Herreid, 1963; Dwyer and Hamilton-Smith, 1965; Tuttle, 1975). Daily and seasonal temperature increments will vary according to the number of bats present, their level of activity, the size and configuration of the roost, and the radiative and conductive properties of the roost substrate. Some bats have successfully exploited otherwise cool caves by selecting sites that facilitate the entrapment of metabolic heat. The importance of selecting roost sites that enhance the energy economy of bats has been thoroughly treated by Dwyer and Hamilton-Smith (1965), Dwyer (1971), Humphrey (1975), and Tuttle (1975). The prolonged use of caves by large numbers of bats can lead to the erosion of ceilings (Dwyer and Hamilton-Smith, 1965; Tuttle, 1975), which may improve the gripping quality of the substrate as well as enhance the entrapment of metabolic heat (Tuttle, 1975). Cavities in trees and buildings may be altered by the scratching action of bats, and this too may improve the gripping quality of the roost substrate (Ryberg, 1947).

The use of a roost by bats for extended periods may lead to temporary and sometimes inalterable changes in the roost environment. If modifications are severe, this may explain why some roosts become temporarily or permanently abandoned. The accumulation of nitrogeneous wastes and feces may create high concentrations of ammonia in some caves (Constantine, 1958b, 1967; Mitchell, 1964; Dwyer and Hamilton-Smith, 1965; Brosset, 1966) and cause physiological stress on its inhabitants or alter species composition. As the concentration of atmospheric ammonia increases in caves, species diversity tends to decrease (Studier, 1966). For example, in Mexico and the southwestern United States *Tadarida brasiliensis* is the only bat remaining when atmospheric ammonia concentrations reach a maximum. The accumulation of guano in tree cavities and similar shelters in buildings may reduce the amount of roosting space (Ryberg, 1947; Medway and Marshall, 1970); similarly, the incrustation of crystallized urea on roost substrates, especially in buildings (Ryberg, 1947) may limit accessibility to preferred roost sites. Odors produced by decomposing guano and urine (Davis, 1944; Hall and Dalquest, 1963; Goodwin, 1970; Constantine, 1967) may attract various predators. The prolonged use of trees by fruit-eating megachiropterans may lead to severe damage and sometimes death of the roost tree due to the accumulation of urine and feces (Ayensu, 1974) and partial or complete defoliation (Nelson, 1965a; Okon, 1974).

2.1.7. Mixed-Species Associations

During nonbreeding periods interspecific associations appear to occur regularly among bats that use internal shelters (e.g., Verschuren, 1957; Goodwin and Greenhall, 1961; Rosevear, 1965; Brosset, 1966, 1974; Villa-R., 1966; Barbour and Davis, 1969; Wallin, 1969; Medway, 1969; Dalquest and Walton, 1970; Kingdon, 1974; Fenton and Kunz, 1977; Bradbury, 1977a; Lekagul and McNeely, 1977). Most of these associations appear to be casual, perhaps resulting from limited numbers of suitable roost sites or from the convergence in requirements for temperature, moisture, and darkness. Most species appear to use separate shelters during the maternity period, although exceptions include species that roost in different parts of the same shelter.

Bats may benefit energetically by roosting in direct contact with other species or in close proximity to large active aggregations, where benefits are derived from an increase in roost temperature. Both situations prevail in Cuban caves, where *Brachyphylla nana* and *Erophylla sezekorni* roost among large aggregations of *Phyllonycteris poeyi,* and where *Mormoops blainvilli* and *Pteronotus* spp. roost separately but in the same parts of the cave as *Phyllonycteris poeyi* (Silva-Taboada and Pine, 1969). Similarly, the roost associations in

Jamaican caves between *Pteronotus parnellii* and *Monophyllus redmani* and *Natalus major* and *N. micropus* may benefit energetically from a mutually warmed environment (Goodwin, 1970). While these and many other roost associations may be casual, the possibility exists that some have or will become obligate.

Evidence in support of the latter suggestion is based on observations by Dwyer (1968), who found maternity colonies of *Miniopterus australis* occupying cool caves at the southern limit of its distribution in New South Wales (Australia), where this bat is invariably associated with large aggregations of *Miniopterus schreibersii*. In other parts of Australia where caves are intrinsically warmer, these two species do not regularly roost together. Similarly, Tuttle (1976a) suggested that the reproductive success of small maternity colonies of *Myotis grisescens* in Florida caves may be augmented when this bat forms colonies with *Myotis austroriparius*. Bearing on this argument are the findings of Tuttle (1975, 1976c) that postnatal growth rates and postflight survival of *Myotis grisescens* can be severely reduced if colony sizes are too small to sufficiently increase the cave temperature. Whether the success or failure of other species populations can be explained on the basis of interspecific associations invites further study. Notwithstanding improved energy economy, another potential benefit that may be derived from interspecific roost associations is reduced predation, resulting either from improved predator surveillance or benefits derived from the so-called "selfish herd" effect (see Hamilton, 1971). Potential disadvantages of interspecific associations may result from misdirected social behavior (Bradbury, 1977a), competition for roost space, increased conspicuousness to predators, and an increased incidence of parasites and disease (see Constantine, 1970).

2.1.8. Synanthropy: A Paradox of Human Influence

Many bats have successfully adapted to a variety of man-made structures for roosts. The exploitation of these structures as substitutes for caves, tree cavities, and other natural roosts supports the view that most bats are highly adaptable and opportunistic in roost selection. Tombs, crypts, ancient ruins, wells, cellars, mine tunnels, storm sewers, basements, and other structures of stone and brick are regularly used by "cave-dwelling" species. The interiors of walls, attics, and the hollow floor spaces of human dwellings, church lofts, barns, schools, and other such structures have commonly become substitutes for natural tree cavities. Buildings, especially of European-style architecture, offer a rich variety of internal roosting places for bats, often more diverse than in their original habitat (Ryberg, 1947; Gaisler, 1963b; Voûte, 1972). Crevices under tile and corrugated metal roofs, expansion joints, and spaces beneath wood shingles and shutters provide alternatives to natural roosts such as rock crevices and exfoliating tree bark. Bridges may also provide suitable roosting places (Davis and Cockrum, 1963; Villa-R, 1966), especially in older-style structures with wood supports, stone bridges with open ends having cavelike chambers, and wooden railway bridges. Coincident with modern road and highway improvements, many of these sites have given way to bridge designs of steel and concrete, which are generally unsuitable for bat roosts.

The exploitation of man-made structures as substitutes for natural roosts (Ryberg, 1947; Gaisler, 1963a, 1963b; Brosset, 1966; Villa-R, 1966; Barbour and Davis, 1969; Wallin, 1969; Fenton, 1970; Sluiter *et al.*, 1971; Gaisler *et al.*, 1979) offers compelling evidence that bats quickly take advantage of newly available structures. Within historical times a variety of species have become predominantly linked to buildings, especially during maternity periods. In northern Europe these include *Rhinolophus hipposideros* (Gaisler, 1963a, 1963b); *Myotis mystacinus* (Nyholm, 1965); *Eptesicus nilssoni*, *Pipistrellus pipistrellus*, *Plecotus auritus*, and *Plecotus austriacus* (Wallin, 1969; Horáček, 1975). In India *Taphozous melanopogon*, *Taphozous perforatus*, and *Megaderma lyra* almost exclusively roost in man-made structures (Brosset, 1962a, 1962c), as does *Tadarida pumila* in parts of Africa (Kingdon, 1974). *Molossus molossus* is the most prevalent house bat in Trinidad (Greenhall and Stell, 1960) and perhaps in the Neotropics. In Panama *Myotis nigricans* is almost invariably found in buildings (Wilson, 1971), and in Paraguay *Myotis nigricans* and *Myotis albescens* are both strongly dependent on roosts in buildings constructed within the last century (Myers, 1977). In North America *Eptesicus fuscus*, *Myotis lucifugus*, and *Myotis yumanensis* have so completely adapted to man-made structures during maternity periods that there are few records from natural roosts (Barbour and Davis, 1969).

The association of bats with man-made structures appears to vary geographically. *Rhinolophus hipposideros* in northern Europe almost always uses buildings in summer; but in southern Europe, where caves are more abundant, this bat is less dependent on human dwellings (Gaisler, 1963b). Gaisler found no direct evidence that the use of buildings has enabled *R. hipposideros* to extend its distribution, but Fenton (1970) suggested that the use of buildings by *Myotis lucifugus* in Canada may have allowed this bat to extend its distribution to otherwise uninhabitable regions. Similarly, *M. velifer* and *Tadarida brasiliensis* both have expanded recent distributions beyond the limit of caves in North America by using man-made structures for maternity roosts (Kunz, 1973b, 1974; Kunz *et al.*, 1980). Davis *et al.* (1962) suggested that the use of buildings by *T. brasiliensis* in Texas allowed populations to increase by as much as 15% above the number before modern building construction. Wilson (1971) suggested that the availability of buildings in Panama may have allowed *M. nigricans* to expand its former distribution and increase in local abundance.

Introduced and cultivated plant species provide suitable roost sites for many foliage-roosting species. Clumped stands of bamboo (*Gigantochloa scortechnii*) maintained in forest reserves in Malaysia provide an abundance of potential roost sites for *Tylonycteris pachypus* and *T. robustula* (Medway and Marshall, 1972).

Similarly, the occurrence of *Myotis frater* in Japan may be due solely to the conservation of bamboo (Wallin, 1969). The introduction and cultivation of banana plants in parts of Africa provides a continuous roost resource for *Pipistrellus nanus* (LaVal and LaVal, 1977) and *M. bocagei* (Brosset, 1976). The introduction and establishment of the neem tree in West Africa at the turn of the last century has provided a favorite roost (and food resource) for *Epomophorus gambianus* (Ayensu, 1974) and other fruit-eating megachiropterans. In the Neotropics *Uroderma bilobatum* has successfully adapted its tent-making habits to an introduced palm, *Prichardia pacifica* (Barbour, 1932).

Although building construction, mining operations, and the introduction of various plant species have undeniably led to the increased numbers of some species, other activities of man have been detrimental to the roosting (and feeding) habits of bats (Stebbings, 1980). Extensive deforestation and forest management have had marked impacts on available roosting sites, especially in tropical regions, where tree-roosting species predominate. The effects of continued deforestation will ultimately force many species into small patches, and forest management will most certainly deprive others of the use of tree cavities. One can expect that as the availability and composition of roost resources change, bat faunas will invariably be altered. Species with less specialized roosting habits (e.g., *Carollia perspicillata* and *Artibeus jamaicensis*) are more likely to sustain marked changes in roost availability. However, because the roost and food resources of most species are intimately linked, the reduction or elimination of only one of these resources will have a severe impact on the continued success of many species.

Increased efforts have been made to promote the extermination of bats that roost in man-made structures. A growing interest in building restoration and home energy conservation, especially in temperate regions, has led to the elimination of many roosts located in such structures (Greenhall, 1982). The use of chemicals as wood preservatives, treatment against wood-boring insects, and the treatment of roost sites with toxic chemical repellents and pesticides has inadvertantly altered roost sites and reduced or eliminated populations of several species (Braaksma, 1980).

The inevitable consequences of forest management, building restoration, deforestation, the increased recreational use of caves, and vandalism has doubtless led to a reduction in the numbers and diversity of roost sites and bats in many regions. Efforts to stem the loss of valuable roosts have been made during the past two decades, especially in Europe (e.g., Daan, 1980) and more recently in North America, to provide protection to bats by encouraging the construction of artificial roosts (Stebbings, 1974; LaVal and LaVal, 1980; Greenhall, 1982) (See Fig. 10) and the use of well-planned cave gating (Tuttle, 1977) to offset the destruction and disturbance of natural roosts and the eviction of bats from man-made structures. Placing caves in land trusts and their purchase by Federal, state,

FIGURE 10. Roost box used to encourage bats that otherwise may have used tree cavities. (Photo by S. C. Bisserôt.)

and local governments and private organizations have helped to restore and protect many valuable roosts, but additional efforts are needed to insure that other roost resources are equally protected.

2.2. Roost Activities and Time Budgets

2.2.1. Return Behavior and Roost Selection

Bats use a combination of spatial familiarity and acoustic and visual cues to locate roosts (Davis, 1966; Griffin, 1970; Williams and Williams, 1970; Findley and Wilson, 1974; Fenton and Kunz, 1977). Daily return behavior involves considerable ritualized behavior once bats have reached the vicinity of a roost, which often includes a generalized reconnaissance of the site and several landing trials before entry. Return rituals have been described for several species, including *Antrozous pallidus* (Vaughan and O'Shea, 1976), *Myotis dasycneme* (Voûte *et al.*, 1974), *M. nattereri* (Laufens, 1973), *Pipistrellus hesperus* (Cross, 1965; Hayward and Cross, 1979), *P. nanus* (O'Shea, 1980), *P. pipistrellus* (Stebbings, 1968; Swift, 1980), *Tadarida macrotis* (Vaughan, 1959), *Thyroptera tricolor* (Findley and Wilson, 1974), *Tylonycteris pachypus*, and *Tylonycteris robustula* (Medway and Marshall, 1972). Voûte *et al.* (1974) suggested that the prolonged return ritual of *M. dasycneme* may be necessary for establishing contacts with conspecifics. Audible "directive calls" emitted by individual *Antrozous pallidus*, as they congregate in the vicinity of their day roost, may provide a focal point to which other bats are attracted (Vaughan and O'Shea, 1976). After one or more bats enter a roost crevice, these bats answer the directives of flying bats, which subsequently leads to the entry of others. The preentry rituals of *Tylonycteris pachypus* and *T. robustula* apparently promote the establishment of day-roosting groups (Medway and Marshall, 1972).

2.2.2. Communication and Social Interactions

Vision probably plays an important communicative role in bats that use external roosts. For bats that seek roosts in the darkness or in the dimly lit interior of a protected shelter, the perception of light may function mostly as a Zeitgeber for synchronizing endogenous rhythms (see Chapter 5). For most bats sensory communication occurs predominately in the acoustic and olfactory modes. The variety of integumentary and facial glands (Quay, 1970) and the increasingly recognized vocal reportoire of bats (Nelson, 1964; Gould, 1971, 1977; Brown, 1976; Fenton *et al.*, 1977c; Barclay *et al.*, 1979; Porter, 1979a, 1979b; Brown and Grinnell, 1980) emphasize the importance of acoustic and olfactory communication during the roosting period.

Scent glands are most notably located on the head, chest, wings, and in the anal region (Quay, 1970), but little is known of how they are used in a social context. Group odors produced from guano and urine deposition may be important in promoting contact between individuals, but much remains to be learned of the olfactory ability of bats (Bhatnagar, 1975). Vaughan and O'Shea (1976) suggested that part of the return ritual of *Antrozous pallidus* may function primarily to confirm their own scent or that of roost mates. Bats that engage in contact clustering invariably exchange individual odors, which may combine to produce group odors; however, individual recognition probably prevails in most species (e.g., Nelson, 1964; Brown, 1976; Kolb, 1977). Hovering behavior observed in some emballonurids appears to involve scent communication from glandular sacs located on the wings (Bradbury and Emmons, 1974; Bradbury, 1977a). Individual and group spacing patterns seen in pteropodids (Nelson, 1965a), rhinolophids (Ransome, 1978), and emballonurids (Bradbury and Emmons, 1974; Bradbury and Vehrencamp, 1976; Bradbury, 1977a) may involve substrate or conspecific marking. Olfaction also appears to be important in mother–infant recognition in several species, including *Pteropus poliocephalus* (Nelson, 1964, 1965a), *Rousettus aegyptiacus* (Kulzer, 1958, 1961), *A. pallidus* (Brown, 1976), *Myotis myotis* (Krátky, 1971; Kolb, 1977), and *Nycticeius humeralis* (Watkins and Shump, 1981).

Many bats rely strongly on acoustic signals in the day roost. Vocalizations may include audible or ultrasonic components given in response to agonistic encounters with conspecifics (Nelson, 1964, 1965a; Bradbury and Emmons, 1974; Porter, 1979a, 1979b) or the approach of intruders (Verschuren, 1957; Nelson, 1965a), or they may serve as spacing- (Nelson, 1964, 1965a; Brown, 1976; Porter, 1979a, 1979b) or contact-promoting signals (Vaughan and O'Shea, 1976; Kolb, 1977; Porter, 1979a, 1979b). Acoustic recognition between mothers and infants has been reported for several species, including *Plecotus townsendii* (Pearson *et al.*, 1952), *Tadarida condylura* (Kulzer, 1962), *Pteropus poliocephalus* (Nelson, 1964, 1965a), *Nycticeius humeralis* (Jones, 1967; Watkins and Shump, 1981), *Eptesicus fuscus* (Davis *et al.*, 1968; Gould, 1971), *Myotis lucifugus* (Gould, 1971), *Desmodus rotundus* (Schmidt, 1972; Gould, 1977), *Myotis myotis* (Kolb, 1977), *Macrotus californicus, Carollia perspicillata, Leptonycteris sanborni* (Gould, 1977), and *Myotis velifer* (Brown and Grinnell, 1980), and these appear to promote contact between individuals beyond the effective range of olfaction.

Whether or not bats vocalize during the roosting period depends on the prevailing environmental conditions, the number of bats present, and the type of social organization. The molossid bat *Otomops wroughtoni* either roosts singly or in small groups and remains silent and motionless during the day (Brosset, 1962c). By contrast, gregarious species such as *Miniopterus schreibersii*

(Brosset, 1962c; Dwyer, 1964), *Tadarida brasiliensis* (Davis *et al.*, 1962; Constantine, 1967), *T. condylura, T. midas, Eidolon helvum* (Rosevear, 1965; Okon, 1974), *Rousettus leschenaulti, Pteropus giganteus* (Brosset, 1962a), and *Dobsonia muluccensis* (Dwyer, 1975), often betray their presence at a roost by constant chatter. On hot days *Myotis myotis* responds with an increased frequency of audible vocalizations and restlessness (DeCoursey and DeCoursey, 1964). Similarly, *Antrozous pallidus* emits irritation buzzes and squabble notes on hot days, which may promote individual spacing (Vaughan and O'Shea, 1976). The frequent audible vocalizations of *Myotis lucifugus* on cool days appear to be associated with attempts by individuals to gain or protect central positions within clusters (Barclay *et al.*, 1979; Burnett and August, 1981).

2.2.3. Time-Activity Budgets

While it may be relatively simple to quantify the amount of time that bats spend in their day roost (because times of entry and departure and periods of occupancy can be recorded or observed directly), placing individual behavior in a temporal context and establishing the appropriate environmental conditions during the roosting period are more problematical. Because most roosting bats are sensitive to human disturbance, prolonged observations are certain to disrupt normal roosting behavior unless experiments are designed to minimize or reduce disturbance. Direct observations making use of ambient light or supplemented with infrared light and night-viewing devices (see Burnett and August, 1981) hold the greatest promise for observing the undisturbed behavior of bats in roosts. Tape recordings of audible sounds (Ransome, 1978) may prove helpful under some situations. If direct observations are made under conditions of ambient light, they should be designed to prevent bats from seeing the observer.

The period immediately following the return of gregarious species is predominated by cluster formation and settling (McCann, 1934; Kulzer, 1961; Dwyer, 1964; Burnett and August, 1981). Among territorial species conflict for roost positions, as observed in *Pteropus poliocephalus* (Nelson, 1965a), *P. giganteus* (Neuweiler, 1969), and *Saccopteryx bilineata* (Bradbury and Emmons, 1974) may involve periods of intense conflict between males and the relocation of displaced individuals. Within an hour or two following returns, a lull in activity becomes evident.

The day-roost period may occasionally be interrupted by bouts of spontaneous activity, including self-grooming (Nelson, 1965a; Bradbury and Emmons, 1974; Wickler and Seibt, 1976; Burnett and August, 1981), allogrooming (Bradbury, 1977a), copulation (McCracken and Bradbury, 1981), and flight. Burnett and August (1981) directly quantified the day-roosting activity of bats in a maternity roost of *Myotis lucifugus*, based on observations of individual focal groups, using a night-viewing device (Fig. 11). Their analysis of five behavioral

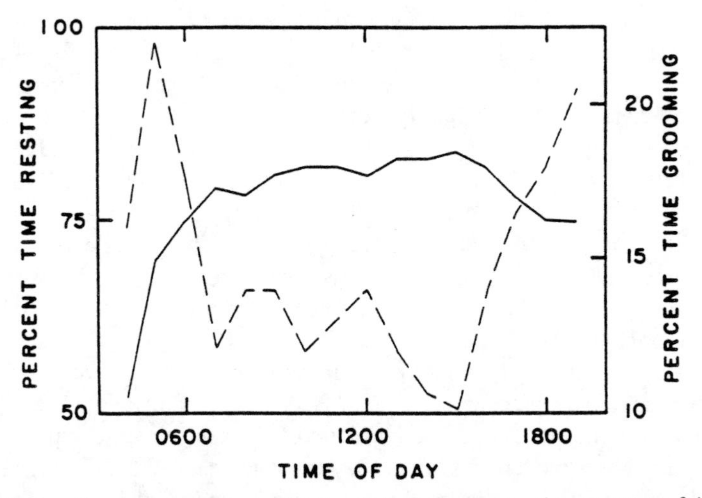

FIGURE 11. Time budget of day-roosting *Myotis lucifugus* showing the percentage of time spent resting (solid line) and that spent grooming (broken line), averaged over a 13-week period. (From Burnett and August, 1981. With permission from the American Society of Mammalogists.)

categories (resting, active, grooming, moving, and flight) indicate that the day-roosting period was predominated by rest (79%). Grooming occurred primarily following the return from feeding and again before the onset of nightly departure. Although grooming accounted for a relatively small percentage (14%) of the day-roosting time budget, it represented more than half of the energy expended during the same period.

Short daytime flights occur most commonly among externally roosting bats, most notably pteropodids (Verschuren, 1957; Rosevear, 1965; Nelson, 1965a; Ayensu, 1974; Kingdon, 1974; Goodwin, 1979) and emballonurids (Bradbury and Emmons, 1974; Bradbury and Vehrencamp, 1976). At certain times bats that roost in protected shelters engage in flight during the day. Daytime flights may involve changes in roost position in response to intruders (Verschuren, 1957; Nelson, 1965a; Constantine, 1967), retreat from sunlight (Ayensu, 1974), practice flights by young bats (Davis and Hitchcock, 1965; Kunz, 1973b, 1974), and occasional feeding flights beneath the tree canopy (Bradbury and Emmons, 1974). Burnett and August (1981) found that spontaneous daytime flights of bats in a maternity roost of *Myotis lucifugus* accounted for less than 1% of the total activity budget.

The timing and rate of defecation is determined by the type of food eaten, the recency of food consumption, and the rate of digestion. Defecation occurs most commonly in the first few hours following the return of bats to the roost (Kunz, 1974; Wickler and Seibt, 1976; Bradbury, 1977b; Ransome, 1978; Rumage, 1979). Ransome (1978) and Rumage (1979), respectively, reported that the peaks of defecation for *Rhinolophus ferrumequinum* and *Myotis lucifugus* occurred within the first 2–3 hr following return to the roost, followed by a lull in mid- to late afternoon and a slight increase shortly before nightly departure. Generally the high rate of defecation in early morning reflects the rapid digestion and food-passage time following the most recent feeding (see Klite, 1965; Nelson, 1965b; Buchler, 1975). The timing and rate of urination closely parallels the pattern of defecation in *R. ferrumequinum* (Ransome, 1978). Seasonal differences in daily timing and rates of defecation and urination will be strongly influenced by energy considerations, water balance, reproductive state, age, physiological condition, and the level of activity (Kunz, 1974; Buchler, 1975; Ransome, 1978).

The timing of some roost activities may be influenced as much by the environment as it is by endogenous factors. Bats that roost in buildings and rock crevices that are subjected to the radiant heat of the sun (directly or indirectly) commonly engage in movements in mid-day to avoid heat stress (Gaisler, 1963a; Licht and Leitner, 1967; Wilson, 1971; O'Farrell and Studier, 1973; Kunz, 1974; Vaughan and O'Shea, 1976). Many bats seek roosts that promote torpor on cool days (and some appear to select roost sites that promote torpor until midday) and facilitate passive arousals from solar heating in the late afternoon (Vaughan and O'Shea, 1976; O'Shea, 1980).

Preemergence behavior may begin as early as two hours before emergence and may include intensive grooming (Neuweiler, 1969; Burnett and August, 1981), increased locomotor activity (Twente, 1955; Dwyer, 1964; Jimbo and Schwassmann, 1967; Wilson, 1971; Voûte, 1972; Voûte *et al.*, 1974; Kunz, 1974; Gaur, 1980), and heightened vocal activity. Marimuthu *et al.* (1978) suggested that the increased vocal activity of *Hipposideros speoris* prior to emergence may be a form of social entrainment.

Because many roost activities, including metabolism, can be affected by the thermal environment of the roost, an analysis of time-activity budgets requires a thorough knowledge of the environmental conditions during the roosting period. Presently there are no published studies where the complexities of the roost environment have been thoroughly examined. In most studies little more than daily, weekly, or monthly averages or extreme temperatures have been reported, and little consideration has been given to where the temperatures are recorded relative to the location of bats within a roost or to the extent of spatial variability of temperatures within a roost. Notable exceptions include studies where hourly or continuous temperatures have been recorded in both occupied and unoccupied roosts (Licht and Lietner, 1967; Dwyer and Harris, 1972; Voûte, 1972; Kunz, 1973b, 1974, 1980; Vaughan and O'Shea, 1976; Humphrey *et al.*, 1977; Burnett and August, 1981). Use of these data, however, is problematical when estimating temperature-dependent metabolic rates (e.g., Studier and O'Farrell, 1976). For example, Burnett and August (1981) found that the estimated energy costs of day roosting in *Myotis lucifugus* was two times greater when they used temperatures of cooler, unoccupied roosts as compared to warmer temperatures of occupied roosts.

Temperatures recorded in crevices and cavities occupied by active bats are typically warmer than unoccupied sites owing to the entrapment of metabolic heat and sometimes because the body surface of bats may be in direct contact with temperature probes (e.g., Voûte, 1972; Kunz, 1974, 1980). Which of these conditions (occupied or unoccupied) more appropriately represents the thermal environment of roosting bats? For clustered bats the answer depends on the position of a bat within the cluster. When bats are densely packed, the body surfaces of centrally located bats are almost entirely subjected to temperatures equivalent to the surface temperatures of adjacent bats. Those roosting on the periphery of a cluster will be subjected to a thermal environment intermediate between conditions in the center of a cluster and that experienced by nearby solitary bats, which in the latter case would be comparable to temperatures in "unoccupied" roosts. Bats roosting on the periphery of clusters in open areas will normally experience cooler temperatures than do peripherally roosting bats confined to crevices. As Burnett and August (1981) found, bats on the periphery of open clusters showed higher levels of activity than those in central positions, probably because of the exposure to cooler temperatures. Future studies that address daily activity budgets should consider quantifying the hemispheric thermal environment of both solitary and clustered bats, using a combination of modeling and empirically derived measures of heat flux via radiation, convection, conduction, and evaporation (see Bakken, 1976).

2.3. Roost Fidelity

Factors affecting roost fidelity include the relative abundance and permanency of roost sites, the proximity and stability of food resources, response to predator pressure, and human disturbance. Roost fidelity may change seasonally, and it can be affected by reproductive condition, sex, age, and social organization (Verschuren, 1957; Humphrey, 1975; Bradbury, 1977a). Bats show little loyalty to foliage roosts that are abundant and temporary, but they typically show a strong attachment to permanent sites such as caves, tree hollows, and man-made structures. Roost fidelity appears to be highest during the maternity period (Humphrey, 1975; Tuttle, 1976a). In temperate regions bats show a high degree of fidelity to hibernacula but lesser degrees of loyalty to transient roosts and swarming sites (LaVal and LaVal, 1980). Although many bats appear to be loyal to certain preferred roosts, there is a growing recognition that most bats establish and maintain familiarity with one or more alternate roosts. Tuttle (1976a) suggested that reports of apparent disloyalty probably result from insufficient knowledge of a species' normal movement patterns.

Fidelity to a home area rather than to a specific roost appears to be a common characteristic among foliage-roosting bats. The foliage-roosting *Artibeus lituratus* and *Vampyrodes caraccioli* seldom use the same roost for more than two consecutive days (Morrison, 1980). *Ectophylla alba* uses freshly cut *Heliconia* leaves for up to two days before moving to other sites (Timm and Mortimer, 1976). Similarly, the tent-making pteropodid *Cynopterus sphynx* makes frequent movements to other foliage roosts (Goodwin, 1979). Findley and Wilson (1974) reported that *Thyroptera tricolor* remained up to 24 hr in unfurled *Heliconia* leaves, and the comparable stage of banana leaves becomes unsuitable as roosts for *Myotis bocagei* and *Pipistrellus nanus* after one or two days (Brosset, 1974; LaVal and LaVal, 1977).

The relationship between roost permanency, roost fidelity, and social organization is complex. For example, *Artibeus jamaicensis* may use a variety of day roosts, including foliage, hollow trees, and caves (Goodwin and Greenhall, 1961; Jimbo and Schwassmann, 1967; Goodwin, 1970; Foster and Timm, 1976; Morrison, 1978a, 1979). Individuals that use foliage roosts seldom remain at one roost for more than a few days (Foster and Timm, 1976; Morrison, 1979), and commuting distances to feeding trees are relatively short (Morrison, 1978a). Commuting distance is nearly two times greater for harem males and females that roost in hollow trees as it is for males that roost in foliage (Morrison, 1978a). For female groups and their harem males, fidelity to a particular tree hollow is higher than it is for bachelor males in foliage roosts. Morrison (1979) suggested that the availability and defensibility of suitable tree holes by the male *A. jamaicensis* probably facilitates strong roost fidelity and harem maintenance. In other areas where foliage roosts and caves are predominantly used as day roosts, the defense of roosts by males may be prohibitive (Morrison, 1979). Porter (1979a) reported male defense of harem groups in the cave-dwelling *Carollia perspicillata*, whose harems were maintained in close proximity to other groups. Similarly, in other species, such as *Myotis adversus* (Dwyer, 1970), *Tylonycteris pachypus*, *T. robustula* (Medway and Marshall, 1972), and *Pipistrellus nanus* (O'Shea, 1980), the high degree of roost fidelity of harem males reflects the defense of roost sites from intrusion by other males.

For some species the relative proximity and stability of food resources appear to have a bearing on roost fidelity. *Desmodus rotundus* may use several roosts when prey populations are widely scattered (Wimsatt, 1969; Turner, 1975), but when prey populations are stable and located near roost sites, fidelity to a single roost appears to be high (Young, 1971). Among temperate insectivorous species fidelity to day roosts is commonly low following the breakup of maternity roosts in late summer and early autumn, and this may, among other factors, be prompted by reduced insect abundance (e.g., Gaisler, 1963a; Stebbings, 1968; Laufens, 1973; Kunz, 1974; Ransome, 1978).

Some bats are likely to change roosts frequently if they are subject to severe predator pressures. Bradbury and Emmons (1974) suggested that frequent changes in roost sites by small groups of *Saccopteryx leptura* may reduce detection by predators. By contrast, its congener *S. bilineata* shows a high degree of roost fidelity, because it relies more on crypsis and on the ability to seek shelter in dark buttress cavities of trees. Similarly, the frequent movements of the male *Artibeus jamaicensis, A. lituratus, Vampyrodes caraccioli* (Morrison, 1978a, 1979, 1980), and other foliage-roosting bats may reduce their vulnerability to predators.

Apparent disloyalty to roost sites may be due as much to human disturbance as it is to other factors (see Nyholm, 1965; Stebbings, 1968; Humphrey and Kunz, 1976; LaVal and LaVal, 1980). When bats are disturbed, they often abandon traditional roosts and take up residence at one or more alternate roosts (Pearson *et al.*, 1952; Sluiter and van Heerdt, 1966; Tuttle, 1976a).

3. NIGHT ROOSTS

Night roosts include places used to ingest food transported from nearby feeding areas, resting places for bats following one or more feeding bouts, feeding perches used by sit-and-wait predators, and calling roosts as part of leks. They may promote digestion and energy conservation, offer retreat from predators, serve as centers for information transfer about the location of food patches, and facilitate social interactions.

Factors governing the selection of night roosts vary widely among species, but roost availability, darkness, shelter from wind, proximity to feeding areas, and reduced risks of predation probably are most important. Bats are opportunistic in their choice of night roosts, but the overriding factor seems to be that they are located in the vicinity of feeding areas so that costly commutes to day roosts can be avoided and risks of predation minimized. Night roosts occur in a variety of places, including areas beneath bridges (Krutzsch, 1955a; Dalquest, 1957; Davis and Cockrum, 1963; Hirshfeld *et al.*, 1977), on rock surfaces (Dalquest, 1947; Nyholm, 1965; Howell, 1979), in rock crevices (Cross, 1965; Hayward and Cross, 1979; Hirshfeld *et al.*, 1977), in caves and mine tunnels (Sanborn and Nicholson, 1950; Dalquest, 1947; Vaughan, 1959; Davis *et al.*, 1968; Kunz, 1974; O'Shea and Vaughan, 1977), in abandoned and occupied buildings (Dalquest, 1947; Krutzsch, 1954; Schowalter *et al.*, 1979; Anthony *et al.*, 1981), in porches, breezeways, and garages (Dalquest, 1947; Vaughan, 1959; Barbour and Davis, 1969), in barns (Orr, 1954; Hoffmeister and Goodpaster, 1954; Kunz, 1974; Anthony *et al.*, 1981), park shelters (Kunz, 1973a), thatch houses (Hall and Dalquest, 1963; O'Shea, 1980), on the walls of buildings (Brosset, 1962b;

Fenton *et al.*, 1977a), on branches in small trees and shrubs (Dalquest, 1947; Nyholm, 1965; Hirshfeld *et al.*, 1977), and on desert plants (Howell, 1979).

3.1. Resting Places

Bats invariably use flight in pursuit of "prey," and most species have evolved nightly activity patterns to minimize the amount of time spent in flight. The timing and duration of nightly "rest" periods vary with species, according to the length of the night, their reproductive condition, prey availability, prevailing temperature, feeding success, food-passage time, and social interactions. One and occasionally two prolonged night-roosting periods are known for some insectivorous species and two nectarivorous bats. Typically, insectivorous bats enter night roosts after an initial feeding period (Krutzsch, 1954; Kunz, 1973a, 1974; Anthony and Kunz, 1977; O'Shea and Vaughan, 1977; Funakoshi and Uchida, 1978; Anthony *et al.*, 1981), often departing to feed or drink one or more times before eventually returning to their day roost. Nectarivorous bats may rest for short intervals ($\approx$ 20 min) in feeding areas and later retreat to protected shelters for prolonged night roosting (Howell, 1979).

Other bats roost for short intervals during the night to consume prey that they have captured in flight or on the ground, only later to engage in an extended rest period. This behavior seems most common among insectivorous species that take relatively large prey. Vaughan (1959) interpreted the intermittent night traffic of *Macrotus californicus* in and out of day roosts as evidence that individuals feeding near the day roost were transporting prey to eat. The intermittent returns and departures of *Antrozous pallidus* at night roosts (Beck and Rudd, 1960; Orr, 1954; O'Shea and Vaughan, 1977) commonly involves the transport of large insects and other arthropod prey. Similar behavior has been reported for *Eptesicus fuscus* (Krutzsch, 1946). In the most northern latitudes, where summer nights are short, bats appear to forgo an extended night-roosting period. In Finland *Myotis mystacinus* and *Myotis daubentoni* each have a single feeding period interrupted by short rest stops (Nyholm, 1965). Similarly, in Sweden *Eptesicus nilssoni* has only one feeding period, interrupted by intermittent rests (Ryberg, 1947), but in Germany this bat has two feeding periods interrupted by a longer night-roosting period (Eisentraut, 1951).

Solitary bats and those that form small colonies typically return to their day roost at night, whereas species that form large aggregations seldom return to their day roost before dawn. For example, at places where *Tadarida brasiliensis* forms small daytime colonies in buildings (Krutzsch, 1955a; Davis *et al.*, 1962) and under bridges (Krutzsch, 1955a; Davis and Cockrum, 1963; Hirshfeld *et al.*, 1977) many, if not most, individuals return to these sites at night. By contrast, few individuals from large colonies return before sunrise (Davis *et al.*, 1962). Thus increased time associated with long-distance commuting in some species

(Williams *et al.*, 1973; Bateman and Vaughan, 1974) may promote the use of night roosts located in proximity to feeding areas.

Several factors interact to influence patterns of night-roost use, and this is especially apparent during the maternity period. Maternity roosts are commonly used as night roosts by lactating females and newly volant young, including *Myotis dasycneme* (Voûte *et al.*, 1974), *M. grisescens* (Tuttle, 1975; LaVal and LaVal, 1980), *M. nattereri* (Laufens, 1973), *M. lucifugus* (Anthony and Kunz, 1977; Anthony *et al.*, 1981), *M. myotis* (Krátky, 1971), *M. velifer* (Kunz, 1973b, 1974), *Antrozous pallidus* (Beck and Rudd, 1960; O'Shea and Vaughan, 1977), *Pipistrellus javanicus* (=*abramus*) (Funakoshi and Uchida, 1978), *P. pipistrellus* (Swift, 1980), and *Rhinolophus ferrumequinum* (Ransome, 1973, 1978). Early returns to maternity roosts at night by terminally pregnant and lactating females may reflect a lower feeding efficiency (during pregnancy) and a need to suckle young (during lactation). Early returns of newly volant young bats to maternity roosts at night probably can be explained by their inefficient foraging and a continued need to nurse when they are first learning to fly (Kunz, 1974).

The selection of some types of night roosts and the duration of occupancy may be influenced directly or indirectly by lunar periodicity. Some desert bats apparently use more protected shelters during brighter lunar periods than during darker ones (Hirshfeld *et al.*, 1977). Fenton *et al.* (1977a) noted that at least three species of insectivorous bats in Africa forego a second feeding period and undergo a prolonged night-roosting period during bright moonlight, in an area where bat hawks feed on bats (Fenton *et al.*, 1977b). The significance of this and similar behavior, termed "lunar phobia" may represent a response to increased risks of predation (Fenton *et al.*, 1977b; Morrison, 1978b, 1980; Chapter 5), but it does not account for the reduced activity of bats during bright moonlight when they are not exposed to severe predation pressure. Anthony *et al.* (1981) found that the early entry of *Myotis lucifugus* into night roosts on bright moonlit nights could best be explained as a response to low insect abundance. In Africa Fenton *et al.* (1977a) also observed reduced insect activity on nights with bright moonlight, but noted that some bats fed more often beneath the tree canopy and less in open areas, suggesting that some insects may fly preferentially in areas sheltered by moonlight.

The most thorough study of night-roosting ecology and behavior has been on *Myotis lucifugus* (Anthony *et al.*, 1981). The amount of time that night roosts, separate from maternity roosts, are occupied by this bat is closely associated with reproduction and prey abundance. The use of night roosts by *M. lucifugus* appears to promote digestion and provides conditions that minimize energy expenditure when efficient foraging is prohibited during cool periods and when insect abundance is low. At these times *M. lucifugus* retreats to small cavities (Fig. 12), where individuals typically form tight clusters, and otherwise remain

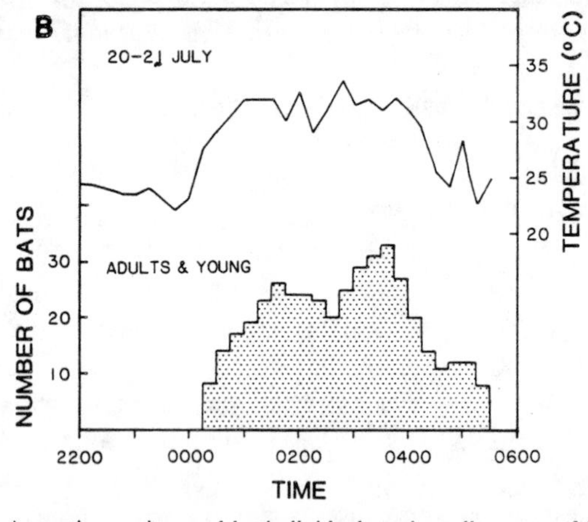

FIGURE 12. (A) A mortice cavity used by individuals and small groups of *Myotis lucifugus* for night roosting. (Photo by T. H. Kunz.) (B) Temperature profile in a mortice cavity denoting the period of night-roost occupancy and the number of bats present contributing to the temperature increase. (Modified from Anthony *et al.*, 1981.)

alert. During pregnancy these separate night roosts are used extensively and bats usually remain throughout a continuous night-roosting period. During lactation separate night roosts are used intermittently, mostly by nonreproductive females at maternity colonies, and individual turnover is high. In late summer, when young are weaned, the number of bats using separate night roosts increases and a single period of occupancy prevails with only moderate individual turnover.

Some bats may enter torpor in night roosts to promote energy economy. For example, *Antrozous pallidus* enters nightly torpor during cool months (O'Shea and Vaughan, 1977) as does *Pipistrellus nanus* in the dry season, when the density of insect prey is low (O'Shea, 1980), but at other times they remain active.

We know little about the social interactions of bats in night roosts and how they may influence times of occupancy, roost composition, and how social interactions in night roosts may affect the social organization of bats in day roosts or during feeding periods. Males of *Pipistrellus nanus* occupy day roosts at night for extended periods during the mating season, defending these roosts from other males and advertising their locations to females (O'Shea, 1980). In species that defend roosts against incursions of conspecifics during the day but not at night, this situation could lead to paternity leaks at night roosts and negate the benefits derived from the vigorous defense of day roosts (Porter, 1979a). Start and Marshall (1976) suggested that the primary role of communal roosting and group foraging in the flower-visiting bat *Eonycteris spelaea* may be to function as a center for information transfer (see Ward and Zahavi, 1973) on the location of widely scattered food sources. Similarly, the periodic night-roosting behavior of *Leptonycteris sanborni* may facilitate information transfer and provide optimum conditions for promoting energy economy, the digestion of food, grooming, and pollen consumption (Howell, 1979). The use of night roosts by *Myotis lucifugus* has no apparent social function (Barclay *et al.*, 1979).

3.2. Feeding Perches

Several members of the Megadermatidae, Nycteridae, and Hipposideridae are sit-and-wait predators that closely integrate night roosting with feeding. Feeding perches are commonly established within a few meters of day roosts and are usually conspicuous owing to the accumulation of culled remains of prey (e.g., Verschuren, 1957; Rosevear, 1965). These bats may hang from feeding perches from one to several hours, acoustically and sometimes visually scanning the immediate environment for prey (Vaughan, 1976, 1977). They make short feeding sallies (<10 sec), returning to a perch, where the hard parts are culled and the palatable parts consumed. The amount of time these bats typically spend in feeding perches exceeds the time spent in flight. *Cardioderma cor* may spend over 95% of its time on feeding perches, allocating the remainder to short flights to nearby perches and in the pursuit of prey (Vaughan, 1976). The amount of time *Hipposideros commersoni* spends in its feeding perch exceeds the time spent flying by a factor of 4.5 to 1 (Vaughan, 1977).

Similar, although less-detailed observations have been made on *Megaderma lyra* and *Megaderma spasma*, both species hunt near their day roost, to which they usually retreat at night to consume their prey (Brosset, 1962b). *Nycteris thebaica* forages in the vicinity of its day roost, feeding in open areas, but consumes its prey in separate feeding roosts (Rosevear, 1965). *Macroderma gigas* usually eats its prey at the place of capture, but it also provisions its young by taking food to the day roost (Douglas, 1967), as does *Vampyrum spectrum* (Vehrencamp *et al.*, 1977).

Whether these bats return to their day roosts at night or use separate feeding perches probably reflects compromises involving risks of predation, commuting costs, and the size of their prey. Solitary sit-and-wait predators are likely to be at risk from predation if and when they consume prey in exposed areas, but this risk should be balanced against the additional costs and risks of predation while commuting and transporting extra baggage to a safer roost.

3.3. Feeding Roosts

Bats that eat fruit, flowers, nectar, pollen, and leaves are almost without exception restricted to the Megachiroptera, the New World Phyllostomidae (Chapter 8), and to the Mystacinidae (Daniels, 1979). Phyllostomids typically transport fruit from fruiting trees to separate feeding roosts, where all or parts of the food items are eaten. By contrast, fruit-eating megachiropterans more commonly consume fruit at fruiting trees.

3.3.1. Fruit-Eating Microchiroptera

Among New World fruit-eating bats, *Carollia perspicillata* (Heithaus and Fleming, 1978), *Artibeus jamaicensis* (Jimbo and Schwassmann, 1967; Morrison, 1978a, 1978b), *A. lituratus*, and *Vampyrodes caraccioli* (Morrison, 1980) are best known for their use of separate feeding roosts. Predator pressure appears to be the leading selective factor promoting the use of these roosts (Fenton and Fleming, 1976; Fenton *et al.*, 1977b; Heithaus and Fleming, 1978; Morrison, 1978a, 1978b, 1980; Howe, 1979). However, the use of certain trees as feeding roosts may represent a compromise between decreased detection by predators, costs associated with traveling greater distances (Morrison, 1978a, 1980), and selection for efficient seed dispersal (Chapter 9). Preferred feeding roosts are often located in trees with a densely leaved crown, downwind from parent trees, and over moist areas (Janzen *et al.*, 1976). *Carollia perspicillata* typically selects feeding roosts within 30–40 m of a fruiting tree, usually in dense foliage, and less than 4 m above the ground (Heithaus and Fleming, 1978). Up to 40–50 trips are made each night between a fruiting tree and feeding roost. If a fruit is too large to transport, it may be eaten in the fruiting tree (Goodwin and Greenhall, 1961).

The feeding roosts of *Artibeus jamaicensis* are often located on the underside of small palms, located at distances ranging from 25 to 400 m from fruiting trees (Jimbo and Schwassmann, 1967; Morrison, 1978a). Caves used as day roosts may also be used as feeding roosts (Goodwin, 1970; Gardner, 1977). *Artibeus jamaicensis* typically makes 10–15 round trips per night between fruiting trees and feeding roosts, spending over 80% of the night in its roost. How much of this time is allocated to actual feeding and how much to other activities remains to be determined. The amount of time spent in feeding roosts is modified on bright, moonlit nights, when bats suspend feeding and return early to their day roost. On dark nights *A. jamaicensis* spends less time occupying feeding roosts and more time searching for newly ripened fruit trees (Morrison, 1978a, 1978b). Similar although less-detailed observations have been reported for *A. lituratus* and *Vampyrodes caraccioli*, both of which fly directly to fruiting trees upon departure from their day roost (Morrison, 1980). Both species may visit two or three trees in the course of a night at distances ranging from 150 to 2300 m from a day roost. These bats transport fruit to feeding roosts located less than 100 m from fruiting trees, but their nightly activity does not appear to be influenced by moonlight.

3.3.2. Fruit-Eating Megachiroptera

While New World fruit-eating bats use separate feeding roosts, most fruit-eating megachiropterans spend the night in the same trees in which they feed. What may account for this difference lies in their contrasting modes of orientation and navigation; phyllostomids rely mostly on echolocation, whereas megachiropteans depend principally on vision (Novick, 1977). Although pteropodids commonly feed in the dark, they apparently require subdued light (including moonlight) for navigation (Gould, 1978) and thus are likely to minimize flight at night. *Rousettus* is capable of echolocating, but apparently it cannot carry food in its mouth while doing so, thus accounting for its feeding and night-roosting activities in the same tree (Lekagul and McNeely, 1977). Low risks of predation may also explain why some megachiropterans night roost and feed in the same trees (Fenton *et al.*, 1977b).

3.4. Calling Roosts

Nocturnal calling roosts are common among male epomophorine bats in Africa, including *Hypsignathus*, *Epomops*, *Micropteropus*, and *Epomophorus* (Brosset, 1966; Kingdon, 1974; Wickler and Seibt, 1976; Bradbury, 1977a, 1977b). One of the most thoroughly studied of these is *H. monstrosus* (Bradbury, 1977b). Upon leaving its day roost, males assemble in calling aggregations known as leks, usually located in riparian forests along streams and rivers. Individual calling roosts are typically 10 m apart and often located in unusually rich food patches, yet there is no evidence that males defend these resources. Most of the time at calling roosts is allocated to overt displays, attacks on other males, loud calling, and grooming. The calling frequency varies during the night, with an early calling session devoted to mate attraction and copulation; a morning session apparently involves territorial establishment.

The calling roosts of the male *Epomophorus wahlbergi* are similar to *Hypsignathus monstrosus*, as described by Wickler and Seibt (1976). Soon after leaving day roosts, male *Epomophorus wahlbergi* fly to nearby trees, where they hang from small branches approximately 2–3 m above the ground. After a period of calling and physical display individuals shift their calling perches to other trees. Similar calling roosts are formed by *Epomops franqueti* (Bradbury, 1977a). At night males establish calling roosts that are usually widely dispersed, located at distances of 100 m or more. Males may have several calling roosts to which they move at regular intervals during the night and steady calling may continue for hours.

Little is known of individual time budgets for these bats. For example, it would be interesting to determine what compromises are made by males between time spent calling and time allocated to foraging. Bradbury (1977b) suggests that the need for foraging time may prohibit most male *Hypsignathus monstrosus* from using both calling sessions in the same night. After departing from calling roosts, most male *Hypsignathus* forage until dawn and then fly directly to day roosts, although some apparently return to the calling assemblage a second time. It would be interesting to know whether the few individuals that return to a calling assemblage a second time in the same night contribute to most of the matings.

4. SUMMARY

The roosting ecology of bats can be viewed as a compromise of opposing selective pressures derived from roost and nonroost origins. The availability of roosts, roost dimensions, energetic considerations, and risks of predation are major determinants of roost use. Roosting habits may vary seasonally, according to sex, reproductive condition, social organization, and food habits. The type of roost, the number of occupants, roost associates, and roost activities are influenced by the manner of flight, the mode of orientation, the dispersion and abundance of food resources, predation risks, social interactions, and energy economy imposed by body size and the physical environment.

Bats have successfully exploited a variety of shelters, including caves, rock and tree crevices, foliage roosts, tree cavities, and man-made structures. Some caves provide spacious chambers that support the largest known mammalian aggregations. Cave topography and structure may enhance the energy economy of roosting bats by the entrapment of metabolic heat and promote population substructuring and the evolution of diverse social systems. Although caves and tree cavities provide protection against most predators and buffers against fluctuations in temperature and adverse weather, tree cavities offer limited space for large aggregations and some caves are unsuitable as roosts, especially when located at considerable distances from profitable feeding areas and when they are too cold or too warm to promote efficient thermoregulation. Some bats have evolved specialized roosting postures, pelage characteristics, and body shapes to accommodate crevice-dwelling habits.

Foliage roosts offer the advantage of being ubiquitous and abundant, but they are relatively temporary and thus require bats to make frequent relocations. Most foliage-roosting bats are solitary or form small groups, and they are mostly distributed in tropical regions. Large aggregations of foliage-roosting bats are exclusively restricted to the Megachiroptera, because most are precluded from using internal shelters owing to their inability to echolocate. Foliage-roosting bats include forms with highly specialized foot pads and wrist pads for roosting on the moist surfaces of leaves, some modify leaves into tentlike structures, and others rely on crypsis. Crypsis may be enhanced by pelage colors that resemble ripe fruits and dead leaves, by countershading, disruptive markings, reticulate wing venation, and motionless postures.

Factors promoting high roost fidelity include roost permanency, morphological specialization, proximity to food resources, the stability of food resources, low risks of predation, microclimatic stability, and complex social organization. As a rule, bats show the highest fidelity toward roosts in tropical caves and the lowest toward foliage roosts. Roost fidelity is highest among females during the maternity period and lowest among solitary males.

The use of man-made structures as substitutes for natural roosts provides convincing evidence that many bats are highly opportunistic in their roost selection. Some bats have become so dependent on man-made structures that there are few recent records from natural shelters; others have extended their former distributions into otherwise uninhabited regions. The introduction and cultivation of certain plant species have increased the availability of roost (and food) resources for some bats. Paradoxically, the adverse consequences of deforestation, forest management, building restoration, the increased recreational and commercial use of caves, and vandalism has led to a decrease in the number and diversity of roosts and bats in some regions.

The daytime activity of bats can be characterized as a period of rest interrupted by periods of spontaneous and rhythmical activity. Bats are most active in their day roost following their return from feeding. This is followed by a lull in activity in midday, with an increase in activity occurring shortly before nightly departure. The amount of time that bats allocate to day-roost activities is influenced by the type of roost, its microclimate, the risks of predation, and the kinds of social interactions among roost mates. In contrast to bats that roost in protected shelters, bats that roost in exposed situations appear to allocate a greater proportion of their time to predator surveillance. Species that form large aggregations are generally more active than solitary bats and those roosting in small groups.

Night roosts may be used by bats to consume food that has been transported from nearby feeding areas, and they may serve as resting places following one or more feeding bouts, as feeding perches for sit-and-wait predators, and as calling and mating roosts for territorial species. The use of night roosts can promote the digestion of food, provide retreat from predators, and serve as centers for information transfer. Selection of night roosts in small protected areas may be important to the energy economy of bats, especially for those species living in cool, temperate environments. The use of night roosts and the duration of occupancy may be influenced by colony size, lunar periodicity, prey abundance, predator pressure, and ambient temperature. The selection of night roosts in or near foraging areas should be important in reducing the risks of predation and the time and energy costs associated with lengthy commutes to day roosts. Some fruit-eating bats may transport fruit to separate feeding roosts, whereas others feed and roost at night in the same trees. Differences between these two strategies may be influenced by differences in predation risks at fruit trees and by contrasting modes of orientation during flight.

ACKNOWLEDGMENTS. I want to thank the Trustees of Boston University for allowing me the freedom during a sabbatical leave to prepare this chapter. I am grateful to Edythe L. P. Anthony, Robert M. R. Barclay, and M. Brock Fenton for criticizing parts of an early draft, and to Sdeuard C. Bisserôt, Christopher D. Burnett, Ira F. Greenbaum, Douglas W. Morrison, Robert M. Timm, Merlin D. Tuttle, and Terry A. Vaughan, who kindly provided copies of original photographs. I want to thank Marylin Massaro and Marie Rutzmoeller (Museum of Comparative Zoology, Harvard University) for processing loans of specimens from which some of the illustrations were prepared, and Peg Esty for preparing the original drawings. Finally, I thank my wife, Margaret, and my children, Pamela and David, for their enduring patience and understanding.

5. REFERENCES

Allen, G. M. 1939. Bats. Harvard University Press, Cambridge, Massachusetts, 368 pp.

Anthony, E. L. P., and T. H. Kunz. 1977. Feeding strategies of the little brown bat, *Myotis lucifugus*, in southern New Hampshire. Ecology, **58**:775–780.

Anthony, E. L. P., M. H. Stack, and T. H. Kunz. 1981. Night roosting and the nocturnal time budget of the little brown bat, *Myotis lucifugus:* Effects of reproductive status, prey density, and environmental conditions. Oecologia, **51**:151–156.

Ayensu, E. S. 1974. Plant and bat interactions in West Africa. Ann. Mo. Bot. Gard., **61**:702–727.

Bakken, G. S. 1976. A heat transfer analysis of animals: Unifying concepts and the application of metabolism chamber data to field ecology. J. Theor. Biol., **60**:337–384.

Barbour, R. W., and W. H. Davis. 1969. Bats of America. University Press of Kentucky, Lexington, 286 pp.

Barbour, T. 1932. A peculiar roosting habit of bats. Q. Rev. Biol., **7**:307–312.

Barclay, R. M. R., M. B. Fenton, and D. W. Thomas. 1979. Social behavior of the little brown bat, *Myotis lucifugus*. II. Vocal communication. Behav. Ecol. Sociobiol., **6**:137–146.

Bateman, G. C., and T. A. Vaughan. 1974. Night activities of mormoopid bats. J. Mammal., **55**:45–65.

Beck, A. J., and R. L. Rudd. 1960. Nursery colonies in the pallid bat. J. Mammal., **41**:266–267.

Bhatnagar, K. P. 1975. Olfaction in *Artibeus jamaicensis* and *Myotis lucifugus* in the context of vision and echolocation. Experientia, **31**:856.

Braaksma, S. 1980. Further details on the distribution and protection of bats in the Netherlands. Pp. 179–183, *in* Proceedings of the Fifth International Bat Research Conference. (D. E. Wilson). Texas Tech Press, Lubbock, 434 pp.

Bradbury, J. W. 1977a. Social organization and communication. Pp. 1–72, *in* Biology of bats. Vol. 3. (W. A. Wimsatt, ed.). Academic Press, New York, 651 pp.

Bradbury, J. W. 1977b. Lek mating behavior in the hammer-headed bat. Z. Tierpsychol., **45**:225–255.

Bradbury, J. W., and L. H. Emmons. 1974. Social organization of some Trinidad bats. I. Emballonuridae. Z. Tierpsychol., **36**:137–183.

Bradbury, J. W., and S. L. Vehrencamp. 1976. Social organization and foraging in emballonurid bats. I. Field studies. Behav. Ecol. Sociobiol., **1**:337–381.

Brosset, A. 1962a. The bats of central and western India. Part I. J. Bombay Nat. Hist. Soc., **59**:1–57.

Brosset, A. 1962b. The bats of central and western India. Part II. J. Bombay Nat. Hist. Soc., **59**:583–624.

Brosset, A. 1962c. The bats of central and western India. Part III. J. Bombay Nat. Hist. Soc., **59**:707–746.

Brosset, A. 1962d. The bats of central and western India. Part IV. J. Bombay Nat. Hist. Soc., **60**:337–355.

Brosset, A. 1966. La biologie des Chiroptères. Masson, Paris, 237 pp.

Brosset, A. 1974. Structure sociale des populations de chauves-souris. J. Psych. Paris, **1**:85–102.

Brosset, A. 1976. Social organization in the African bat, *Myotis boccagei*. Z. Tierpsychol., **42**:50–56.

Brown, P. E. 1976. Vocal communication in the pallid bat, *Antrozous pallidus*. Z. Tierpsychol., **41**:34–54.

Brown, P. E., and A. D. Grinnell. 1980. Echolocation ontogeny in bats. Pp. 355–377, *in* Animal sonar systems. (R. G. Busnel and J. F. Fish, eds.). Plenum Press, New York, 1135 pp.

Buchler, E. R. 1975. Food transit time in *Myotis lucifugus* (Chiroptera: Vespertilionidae). J. Mammal., **56**:252–255.

Burnett, C. D., and P. V. August. 1981. Time and energy budgets for day roosting in a maternity colony of *Myotis lucifugus*. J. Mammal., **62**:758–766.

Chapman, F. M. 1932. A home making bat. Nat. Hist., New York, **32**:555.

Constantine, D. G. 1958a. Ecological observations on lasiurine bats in Georgia. J. Mammal., **39**:64–70.

Constantine, D. G. 1958b. Bleaching of hair pigments in bats by the atmosphere of caves. J. Mammal., **39**:513–520.

Constantine, D. G. 1966. Ecological observations on lasiurine bats in Iowa. J. Mammal., **47**:34–41.

Constantine, D. G. 1967. Rabies transmission by air in bat caves. U.S. Public Health Service, Publication 1617, 51 pp.

Constantine, D. G. 1970. Bats in relation to the health, welfare, and economy of man. Pp. 320–449, *in* Biology of bats. Vol. 2. (W. A. Wimsatt, ed.). Academic Press, New York, 477 pp.

Cross, S. P. 1965. Roosting habits of *Pipistrellus hesperus*. J. Mammal., **46**:270–279.

Daan, S. 1980. Long term changes in bat populations in the Netherlands: A summary. Lutra, **22**:95–105.

Dalquest, W. W., and D. W. Walton. 1970. Diurnal retreats of bats. Pp. 162–187, *in* About bats. (B. H. Slaughter and D. W. Walton, eds.). Southern Methodist University Press, Dallas, 339 pp.

Dalquest, W. W. 1957. Observations on the sharp-nosed bat, *Rhynchiscus nasio* (Maximilian). Tex. J. Sci., **9**:218–226.

Dalquest, W. W., and D. W. Walton. 1970. Diurnal retreats of bats. Pp. 162–187, *in* About bats. (B. H. Slaughter and D. W. Walton, eds.). Southern Methodist University Press, Dallas, 339 pp.

Daniel, M. J. 1979. The New Zealand short-tailed bat, *Mystacina tuberculata:* A review of present knowledge. N. Z. J. Zool., **6**:357–370.

Davis, R. 1966. Homing performance and homing ability in bats. Ecol. Monogr., **36**:201–237.

Davis, R., and E. L. Cockrum. 1963. Bridges utilized as day roosts by bats. J. Mammal., **44**:428–430.

Davis, R. B., C. F. Herreid II, and H. L. Short. 1962. Mexican free-tailed bats in Texas. Ecol. Monogr., **32**:311–346.

Davis, W. B. 1944. Notes on Mexican mammals. J. Mammal., **25**:370–403.

Davis, W. H., and H. B. Hitchcock. 1965. Biology and migration of the bat, *Myotis lucifugus*, in New England. J. Mammal., **46**:296–313.

Davis, W. H., R. W. Barbour, and M. D. Hassell. 1968. Colonial behavior of *Eptesicus fuscus*. J. Mammal., **49**:44–50.

DeCoursey, G., and P. J. DeCoursey. 1964. Adaptive aspects of activity rhythms in bats. Biol. Bull., **126**:14–27.

Dobson, G. E. 1876. On peculiar structures in the feet of certain species of mammals which enable them to walk on smooth perpendicular surfaces. Proc. Zool. Soc. London, **1876**:526–535.

Dobson, G. E. 1877. Protective mimicry among bats. Nature, **15**:354.

Douglas, A. M. 1967. The natural history of the ghost bat *Macroderma gigas* (Microchiroptera, Megadermatidae) in western Australia. West. Aust. Nat., **10**:125–137.

Dwyer, P. D. 1964. Seasonal changes in activity and weight of *Miniopterus schreibersii blepotis* (Chiroptera) in north-eastern New South Wales. Aust. J. Zool., **12**:52–69.

Dwyer, P. D. 1968. The little bent-winged bat—Evolution in progress. Aust. Nat. Hist., **1968**:55–58.

Dwyer, P. D. 1970. Social organization of the bat *Myotis adversus*. Science, **168**:1006–1008.

Dwyer, P. D. 1971. Temperature regulation and cave-dwelling in bats: An evolutionary perspective. Mammalia, **35**:424–455.

Dwyer, P. D. 1975. Notes on *Dobsonia moluccensis* (Chiroptera) in the New Guinea Highlands. Mammalia, **39**:113–118.

Dwyer, P. D., and E. Hamilton-Smith. 1965. Breeding caves and maternity colonies of the bent-winged bat in south-eastern Australia. Helictite, **4**:3–21.

Dwyer, P. D., and J. A. Harris. 1972. Behavioral acclimatization to temperature by pregnant *Miniopterus* (Chiroptera). Physiol. Zool., **45**:14–21.

Eisentraut, M. 1951. Die Ernäheung der Fledermäuse. Zool. Jahrb. Abt. Syst., **79**:118–128.

Fenton, M. B. 1970. Population studies of *Myotis lucifugus* (Chiroptera: Vespertilionidae) in Ontario. Life Sci. Contr. R. Ont. Mus., **77**:1–34.

Fenton, M. B., and T. H. Fleming. 1976. Ecological interactions between bats and nocturnal birds. Biotropica, **8**:104–110.

Fenton, M. B., and T. H. Kunz. 1977. Movements and behavior. Pp. 351–364, *in* Biology of bats of the New World family Phyllostomatidae. Part II. (R. J. Baker, J. K. Jones, Jr., and D. C. Carter, eds.). Spec. Publ. Mus. Texas Tech Univ., Lubbock, **7**:1–364.

Fenton, M. B., N. G. H. Boyle, T. M. Harrison, and D. J. Oxley. 1977a. Activity patterns, habitat use, and prey selection by some African insectivorous bats. Biotropica, **9**:73–85.

Fenton, M. B., D. H. M. Cumming, and D. J. Oxley. 1977b. Prey of bat hawks and availability of bats. Condor, **79**:495–497.

Fenton, M. B., J. J. Belwood, J. H. Fullard, and T. H. Kunz. 1977c. Responses of *Myotis lucifugus* (Chiroptera: Vespertilionidae) to calls of conspecifics and to other sounds. Can. J. Zool., **54**:1443–1448.

Findley, J. S., and D. E. Wilson. 1974. Observations on the Neotropical disk-winged bat, *Thyroptera tricolor*. J. Mammal., **55**:562–571.

Foster, M. S., and R. M. Timm. 1976. Tent-making by *Artibeus jamaicensis* (Chiroptera: Phyllostomatidae) with comments on plants used by bats for tents. Biotropica, **8**:265–269.

Funakoshi, K., and T. A. Uchida. 1978. Studies on the physiological and ecological adaptation of temperate insectivorous bats. III. Annual activity of the Japanese house-dwelling bat, *Pipistrellus abramus*. J. Fac. Agric. Kyushu Univ., **23**:95–115.

Gaisler, J. 1963a. The ecology of lesser horseshoe bat (*Rhinolophus hipposideros hipposideros* Bechstein, 1800 in Czechoslovakia, Part I. Acta Soc. Zool. Bohem., **27**:211–233.

Gaisler, J. 1963b. The ecology of lesser horseshoe bat (*Rhinolophus hipposideros hipposideros* Bechstein, 1800) in Czechoslovakia, II: ecological demands, problems of synanthropy. Acta Soc. Zool. Bohem., **27**:322–327.

Gaisler, J. 1979. Ecology of bats. Pp. 281–342, *in* Ecology of small mammals. (D. M. Stoddard, ed.). Chapman and Hall, London, 386 pp.

Gaisler, J., V. Hanak, J. Dungel. 1979. A contribution to the population ecology of *Nyctalus noctula* (Mammalia: Chiroptera). Acta Sci. Nat. Brno., **13**:1–38.

Gardner, A. L. 1977. Feeding habits. Pp. 293–350, *in* Biology of bats of the New World family Phyllostomatidae. Part II. (R. J. Baker, J. K. Jones, Jr., and D. C. Carter, eds.). Spec. Publ. Mus. Texas Tech Univ., Lubbock, **13**:1–364.

Gaur, B. S. 1980. Roosting ecology of the Indian desert rat-tailed bat, *Rhinopoma kinneari* Wroughton. Pp. 125–128, *in* Proceedings of the Fifth International Bat Research Conference. (D. E. Wilson, A. L. Gardner, eds.). Spec. Publ. Mus. Texas Tech Univ., Lubbock, 434 pp.

Glassman, S. F. 1972. A revision of B. E. Dahlgren's index of American palms. Phanerogamarum Monogr., **6**:1–294.

Goodwin, G. G. 1946. The mammals of Costa Rica. Bull. Am. Mus. Nat. Hist., **87**:271–473.

Goodwin, G. G., and A. M. Greenhall. 1961. A review of the bats of Trinidad and Tobago. Bull. Am. Mus. Nat. Hist., **122**:187–302.

Goodwin, R. E. 1970. The ecology of Jamaican bats. J. Mammal., **51**:571–579.

Goodwin, R. E. 1979. The bats of Timor: Systematics and ecology. Bull. Am. Mus. Nat. Hist., **163**:73–122.

Gould, E. 1971. Studies of maternal–infant communication and development of vocalizations in the bats *Myotis* and *Eptesicus*. Comm. Behav. Biol. A, **5**:263–313.

Gould, E. 1977. Echolocation and communication. Pp. 247–279, *in* Biology of bats of the New World family Phyllostomatidae. Part II. (R. J. Baker, J. K. Jones, Jr., and D. C. Carter, eds.). Spec. Publ. Mus. Texas Tech Univ., Lubbock, **13**:1–364.

Gould, E. 1978. Foraging behavior of Malaysian nectar-feeding bats. Biotropica, **10**:184–193.

Greenhall, A. M. 1982. House bat management. Resource Publication 143, U.S. Fish and Wildlife Service, Washington, D.C., 33 pp.

Greenhall, A. M., and G. Stell. 1960. Bionomics and chemical control of free-tailed house bats (*Molossus*) in Trinidad. Spec. Sci. Rep. Wildl., No. 53, U.S. Fish Wildl. Serv., Washington, D.C. 20 pp.

Griffin, D. R. 1970. Migrations and homing of bats. Pp. 233–264, *in* Biology of bats. Vol. 1. (W. A. Wimsatt, ed.). Academic Press, New York, 406 pp.

Hall, E. R., and W. W. Dalquest. 1963. The mammals of Veracruz. Univ. Kans. Publ. Mus. Nat. Hist., **14**:165–362.

Hamilton, W. D. 1971. Geometry of the selfish herd. J. Theor. Biol., **31**:295–311.

Hamilton, W. J., III, and K. E. F. Watt. 1970. Refuging. Ann. Rev. Ecol. Syst., **1**:263–286.

Hayward, B. J., and S. P. Cross. 1979. The natural history of *Pipistrellus hesperus* (Chiroptera: Vespertilionidae). Office Res. West. N.M., **3**:1–36.

Heithaus, E. R., and T. H. Fleming. 1978. Foraging movements of a frugivorous bat, *Carollia perspicillata* (Phyllostomatidae). Ecology, **48**:127–143.

Henshaw, R. E. 1960. Responses of free-tailed bats to increases in cave temperature. J. Mammal., **41**:396–398.

Herreid, C. F., II, 1963. Temperature regulation of Mexican free-tailed bats in cave habitats. J. Mammal., **44**:560–573.

Hill, J. E. 1969. The generic status of *Glischropus rosseti* Oey, 1951 (Chiroptera: Vespertilionidae). Mammalia, **33**:133–139.

Hirshfeld, J. R., Z. C. Nelson, and W. G. Bradley. 1977. Night roosting behavior in four species of desert bats. Southwest. Nat., **22**:427–433.

Hoffmeister, D. F., and W. W. Goodpaster. 1954. The mammals of the Hauchuca Mountains, southeastern Arizona. University of Illinois Press, Urbana, 152 pp.

Horáček, I. 1975. Notes on the ecology of bats of the genus Plecotus. Geoffroy, 1818 (Mammalia: Chiroptera). Vestn. Cesk. Spol. Zool., **39**:195–210.

Howe, H. F. 1979. Fear and frugivory. Am. Nat., **114**:925–931.

Howell, D. J. 1979. Flock feeding in *Leptonycteris:* Advantages to the bat and to the host plant. Am. Nat., **114**:23–49.

Howell, D. J., and J. Pylka. 1977. Why bats hang upside down: A biomechanical hypothesis. J. Theor. Biol., **69**:625–631.

Humphrey, S. R. 1975. Nursery roosts and community diversity of Nearctic bats. J. Mammal., **56**:321–346.

Humphrey, S. R., and J. B. Cope. 1976. Population ecology of the little brown bat, *Myotis lucifugus*, in Indiana and north-central Kentucky. Spec. Publ. Amer. Soc. Mamm., **4**:1–81.

Humphrey, S. R., and T. H. Kunz. 1976. Ecology of a Pleistocene relict, the western big-eared bat (*Plecotus townsendii*), in the southern Great Plains. J. Mammal., **56**:470–494.

Humphrey, S. R., A. R. Richter, and J. B. Cope. 1977. Summer habitat and ecology of the endangered Indiana bat, *Myotis sodalis*. J. Mammal., **58**:334–346.

Hutchinson, G. E. 1950. Survey of contemporary knowledge of biogeochemistry. 3. The biogeochemistry of vertebrate excretion. Bull. Am. Mus. Nat. Hist., **96**:1–554.

Ingles, L. G. 1953. Observations on Barro Colorado Island mammals. J. Mammal., **34**:266–268.

Janzen, D. H. 1976. Why tropical trees have rotten cores. Biotropica, **8**:110.

Janzen, D. H., G. A. Miller, J. Hackforth-Jones, C. M. Pond, K. Hooper, and D. P. Janos. 1976. Two Costa Rican bat-generated seed shadows of *Andira inerimus* (Leguminosae). Ecology, **57**:1068–1075.

Jepsen, G. L. 1970. Bat origins and evolution. Pp. 1–64, *in* Biology of bats. Vol. 1 (W. A. Wimsatt, ed.). Academic Press, New York, 406 pp.

Jimbo, S., and H. O. Schwassmann. 1967. Feeding behavior and daily emergence pattern of "*Artibeus jamaicensis*" Leach (Chiroptera, Phyllostomidae). Atas Simp. Biota Amazonica, **5**:239–253.

Jones, C. 1967. Growth, development, and wing loading in the evening bat, *Nycticeius humeralis* (Rafinesque). J. Mammal., **48**:1–19.

Jones, C. 1971. The bats of Rio Muni, West Africa. *J. Mammal.*, **52**:121–140.

Jones, C. 1972. Comparative ecology of three pteropid bats in Rio Muni, West Africa. J. Zool., **167**:353–370.

Kingdon, J. 1974. East African mammals. Vol. II. Part A. Insectivores and bats. Academic Press, London, 341 pp.

Klite, P. D. 1965. Intestinal bacterial flora and transit time of three Neotropical bat species. J. Bacteriol., **90**:375–379.

Kolb, A. 1977. Wie erkennen sich Mutter und Junges des Mausohrs, *Myotis myotis*, bei der Rückkehr vom Jagdflug wieder? Z. Tierpsychol., **44**:423–431.

Krátky, J. 1971. Zur Ethologie des Mausohrs (*Myotis myotis* Borkhausen, 1797). Zool. Listy., **20**:131–138.

Krutzsch, P. H. 1946. Some observations on the big brown bat in San Diego County, California. J. Mammal., **27**:240–242.

Krutzsch, P. H. 1954. Notes on the habits of the bat *Myotis californicus*. J. Mammal., **35**:539–545.

Krutzsch, P. H. 1955a. Observations on the Mexican free-tailed bat, *Tadarida mexicana*. J. Mammal., **36**:236–242.

Krutzsch, P. H. 1955b. Observations on the California mastiff bat. J. Mammal., **36**:407–414.

Kulzer, E. 1958. Untersuchungen über die Biologie von Flughunden der Gattung *Rousettus* Gray. Z. Morphol. Ökol. Tiere, **47**:374–402.

Kulzer, E. 1961. Über die Biologie der Nil-Flughunde (*Rousettus aegyptiacus*). Nat. Volk, **91**:219–228.

Kulzer, E. 1962. Über die Jugendentwicklung der Angola-Bulldogfledermaus, *Tadarida condylura*. Säugetierkd. Mitt., **10**:116–124.

Kunz, T. H. 1973a. Resource utilization: Temporal and spatial components of bat activity in central Iowa. J. Mammal., **54**:14–32.

Kunz, T. H. 1973b. Population studies of the cave bat (*Myotis velifer*): Reproduction, growth, and development. Occas. Pap. Mus. Nat. Hist. Univ. Kans., **15**:1–43.

Kunz, T. H. 1974. Feeding ecology of a temperate insectivorous bat (*Myotis velifer*). Ecology, **55**:693–711.

Kunz, T. H. 1980. Daily energy budgets of free-living bats. Pp. 369–392, *in* Fifth International Bat Research Conference. (D. E. Wilson and A. L. Gardner, eds.). Texas Tech Press, Lubbock, 434 pp.

Kunz, T. H., J. R. Choate, and S. B. George. 1980. Distributional records for three species of mammals in Kansas. Trans. Kans. Acad. Sci., **83**:74–77.

Lack, D. 1968. Ecological adaptations for breeding in birds. Methuen, London, 409 pp.

Lang, J., and J. P. Chapin. 1917. The American Museum Congo Expedition collection of bats. Part II. Notes on the distribution and ecology of central African Chiroptera. Bull. Am. Mus. Nat. Hist., **37**:476–496.

Laufens, G. 1973. Beiträge zur Biologie der Fransenfledermäuse (*Myotis nattereri* Koh, 1818). Z. Säugetierkd., **38**:1–14.

LaVal, R. K., and M. L. LaVal. 1977. Reproduction and behavior of the African banana bat, *Pipistrellus nanus*. J. Mammal., **58**:403–410.

LaVal, R. K., and M. L. LaVal. 1980. Ecological studies and management of Missouri bats, with emphasis on cave-dwelling species. Terrestrial Series No. 8, Missouri Department of Conservation, Jefferson City, 93 pp.

Lawrence, B., and A. Novick. 1963. Behavior as a taxonomic clue: Relationships of *Lissonycteris* (Chiroptera). Breviora, **184**:1–16.

Lekagul, B., and J. A. McNeely. 1977. Mammals of Thailand. Association for the Conservation of Wildlife, Bangkok, 758 pp.

Licht, P., and P. Leitner. 1967. Behavioral responses to high temperatures in three species of California bats. J. Mammal., **48**:52–61.

McCann, C. 1934. Notes on the flying-fox (*Pteropus giganteus* Brunn.). J. Bombay Nat. Hist. Soc., **37**:143–149.

McCracken, G. F., and J. W. Bradbury. 1981. Social organization and kinship in the polygynous bat *Phyllostomus hastatus*. Behav. Ecol. Sociobiol., **8**:11–34.

Maeda, K. 1974. Eco-behavior of the large noctule, *Nyctalus lasiopterus* in Sappora, Japan. Mammalia, **38**:461–487.

Marimuthu, G., R. Subbaraj, and M. K. Chandrashekaran. 1978. Social synchronization of the activity rhythm in a cave dwelling insectivorous bat. Naturwissenschaften, **65**:600.

Medway, Lord. 1969. The wild mammals of Malaya. Oxford University Press, London, 127 pp.

Medway, Lord. 1971. Observations of social and reproductive biology of the bent-winged bat *Miniopterus australis*. J. Zool., **165**:261–273.

Medway, Lord, and A. G. Marshall. 1970. Roost site selection among flat-headed bats (*Tylonycteris* spp.). J. Zool., **161**:237–245.

Medway, Lord, and A. G. Marshall. 1972. Roosting association of flat-headed bats, *Tylonycteris* species (Chiroptera: Vespertilionidae) in Malaysia. J. Zool., **168**:463–482.

Mitchell, H. P. 1964. Investigations of the cave atmosphere of a Mexican bat colony. J. Mammal., **45**:568–577.

Morrison, D. W. 1978a. Foraging ecology and energetics of the frugivorous bat *Artibeus jamaicensis*. Ecology, **59**:716–723.

Morrison, D. W. 1978b. Lunar phobia in a neotropical fruit bat, *Artibeus jamaicensis* (Chiroptera: Phyllostomidae). Anim. Behav., **26**:852–856.

Morrison, D. W. 1979. Apparent male defense of tree hollows in the fruit bat, *Artibeus jamaicensis*. J. Mammal., **60**:11–15.

Morrison, D. W. 1980. Foraging and day-roosting dynamics of canopy fruit bats in Panama. J. Mammal., **61**:20–29.

Myers, P. 1977. Patterns of reproduction of four species of vespertilionid bats in Paraguay. Univ. Calif. Publ. Zool., **107**:1–41.

Nelson, J. E. 1964. Vocal communication in Australian flying foxes (Pteropodidae: Megachiroptera). Z. Tierpsychol., **21**:857–870.

Nelson, J. E. 1965a. Behaviour of Australian Pteropodidae (Megachiroptera). Anim. Behav., **8**:544–557.

Nelson, J. E. 1965b. Movements of Australian flying foxes. Aust. J. Zool., **13**:53–73.

Neuweiler, G. 1969. Verhaltensbeobachtungen an einer indischen Flughundkolonie (*Pteropus g. giganteus* Brünn). Z. Tierpsychol., **26**:166–199.

Novick, A. 1977. Acoustic orientation. Pp. 73–287, *in* Biology of bats. Vol. 3. (W. A. Wimsatt, ed.). Academic Press, New York, 651 pp.

Nyholm, E. S. 1965. Zur Okologie von Myotis mystacinus (Leisl.) und M. daubentoni (Leisl.) (Chiroptera). Ann. Zool. Fenn., **2**:77–123.

O'Farrell, M. J., and E. H. Studier. 1973. Reproduction, growth, and development in *Myotis thysanodes* and *M. lucifugus* (Chiroptera: Vespertilionidae). Ecology, **54**:18–30.

Ognev, S. I. 1962. Mammals of eastern Europe and northern Asia: Insectivora and Chiroptera. Vol. I. Translated from Russian. Israel Program for Scientific Translations, Office of Technical Services, U.S. Department of Commerce, Washington, D.C., 487 pp.

Okon, E. E. 1974. Fruit bats at Ife: Their roosting and food preferences. Niger. Field, **39**:33–40.

Orr, R. T. 1954. Natural history of the pallid bat, *Antrozous pallidus* (LeConte). Proc. Calif. Acad. Sci., **28**:165–246.

Osgood, W. H. 1932. Mammals of the Kelly-Roosevelt and Delacour Asiatic expeditions. Field Mus. Nat. Hist., Zool. Ser., **18**:236–238.

O'Shea, T. J. 1980. Roosting, social organization and the annual cycle of a Kenya population of the bat *Pipistrellus nanus*. Z. Tierpsychol., **53**:171–195.

O'Shea, T. J., and T. A. Vaughan. 1977. Nocturnal and seasonal activities of the pallid bat, *Antrozous pallidus*. J. Mammal., **58**:269–284.

Pearson, O. P., M. R. Koford, and A. K. Pearson. 1952. Reproduction of the lump-nosed bat (*Corynorhinus rafinesquei*) in California. J. Mammal., **33**:273–320.

Peterson, R. L. 1965. A review of the flat-headed bats of the family Molossidae from South America and Africa. Life Sci. Contr. R. Ont. Mus., **64**:3–32.

Phillips, W. W. A. 1924. A guide to the mammals of Ceylon. Ceylon J. Sci., **13**:1–63.

Porter, F. L. 1979a. Social behavior in the leaf-nosed bat, *Carollia perspicillata*. 1. Social organization. Z. Tierpsychol., **49**:406–417.

Porter, F. L. 1979b. Social behavior in the leaf-nosed bat, *Carollia perspicillata*. 2. Social communication. Z. Tierpsychol., **50**:1–8.

Quay, W. B. 1970. Integument and derivatives. Pp. 1–56, *in* Biology of bats. Vol. 2. (W. A. Wimsatt, ed.). Academic Press, New York, 447 pp.

Ransome, R. D. 1973. Factors affecting the timing of births of the greater horse-shoe bat (*Rhinolophus ferrumequinum*). Period. Biol., **75**:169–175.

Ransome, R. D. 1978. Daily activity patterns of the greater horseshoe bat, *Rhinolophus ferrumequinum* from April to September. Pp. 259–274, *in* Proceedings of the Fourth International Bat Research Conference. (R. J. Olembo, J. B. Castelino, and F. A. Mutere, eds.). Kenya National Academy of Arts and Sciences, Nairobi, 328 pp.

Rosevear, D. R. 1965. The bats of West Africa. British Museum of Natural History, London, 418 pp.

Rumage, W. T., III, 1979. Seasonal changes in dispersal pattern and energy intake of *Myotis lucifugus*. Unpublished M.A. thesis, Boston University, Boston, 50 pp.

Ryberg, O. 1947. Studies on bats and bat parasites. Bokförlaget Svensk Natur, Stockholm, 330 pp.

Sanborn, C. C. 1949. Extension of range of the African bat, *Myotis bocagei cupreolus* Thomas. J. Mammal., **30**:315.

Sanborn, C. C., and A. J. Nicholson. 1950. Bats from New Caledonia, the Solomon Islands, and New Herbrides. Fieldiana (Zool.), Chicago Nat. Hist. Mus., **31**:313–338.

Schliemann, H., and B. Mags. 1978. *Myzopoda aurita*. Mammal. Species, **116**:1–2.

Schmidt, U. 1972. Die sozialen Laute juveniler Vampirfledermäuse (*Desmodus rotundus*) und ihrer Mütter. Bonn. Zool. Beitr., **23**:310–316.

Schowalter, D. B., J. R. Gunson, and L. D. Harder. 1979. Life history characteristics of little brown bats (*Myotis lucifugus*) in Alberta. Can. Field-Nat., **93**:243–251.

Shump, K. A., Jr., and A. U. Shump. 1980. Comparative insulation in vespertilionid bats. Comp. Biochem. Physiol., **66A**:351–354.

Silva-Taboada, G., and R. H. Pine. 1969. Morphological and behavioral evidence for the relationship between the bat genus *Brachyphylla* and Phyllonycterinae. Biotropica, **1**:10–19.

Sluiter, J. W., and P. F. van Heerdt. 1966. Seasonal habits of the noctule bat (*Nyctalus noctula*). Arch. Neerl. Zool., **16**:423–439.

Sluiter, J. W., P. F. van Heerdt, and A. M. Voûte. 1971. Contribution to the population biology of the pond bat, *Myotis dasycneme*. Decheniana Beih., **14**:1–44.

Start, A. N., and A. G. Marshall. 1976. Nectarivorous bats as pollinators of trees in West Malaysia. Pp. 141–150, *in* Tropical trees: Variation, breeding, and conservation. Linnean Society Symposium Series No. 2. (J. Burley and B. T. Styles, eds.). Academic Press, New York, 243 pp.

Stebbings, R. E. 1968. Measurement, composition and behavior of a large colony of the bat *Pipistrellus pipistrellus*. J. Zool., **156**:15–33.

Stebbings, R. E. 1974. Artificial roosts for bats. J. Devon Trust Nat. Conserv., **6**:114–119.

Stebbings, R. E. 1980. An outline global strategy for the conservation of bats. Pp. 173–178, *in* Fifth International Bat Research Conference. (D. E. Wilson and A. L. Gardner, eds.). Texas Tech Press, Lubbock, 434 pp.

Studier, E. H. 1966. Studies on mechanisms of ammonia tolerance of the guano bat. J. Exp. Zool., **163**:79–86.

Studier, E. H., and M. J. O'Farrell. 1976. Biology of *Myotis thysanodes* and *M. lucifugus* (Chiroptera: Vespertilionidae). III. Metabolism, heart rate, breathing rate, evaporative water loss and general energetics. Comp. Biochem. Physiol., **54A**:423–432.

Suthers, R. A. 1970. Vision, olfaction, taste. Pp. 265–309, *in* Biology of bats. Vol 2. (W. A. Wimsatt, ed.). Academic Press, New York, 477 pp.

Suthers, R. A. 1978. Sensory ecology of mammals. Pp. 253–287, *in* Sensory ecology: Review and perspectives. (M. A. Ali, ed.). Plenum Press, New York, 597 pp.

Swift, S. M. 1980. Activity patterns of pipistrelle bats (*Pipistrellus pipistrellus*) in north-east Scotland. J. Zool., **190**:285–295.

Tate, G. H. H. 1942. Review of the vespertilionine bats, with special attention to genera and species of the Archbold collections. Bull. Am. Mus. Nat. Hist., **80**:221–297.

Timm, R. M., and J. Mortimer. 1976. Selection of roost sites by Honduran white bats, *Ectophylla alba* (Chiroptera: Phyllostomatidae). Ecology, **57**:385–389.

Turner, D. C. 1975. The vampire bat. Johns Hopkins University Press, Baltimore and London, 145 pp.

Tuttle, M. D. 1975. Population ecology of the gray bat (*Myotis grisescens*): Factors influencing early growth and development. Occas. Pap. Mus. Nat. Hist. Univ. Kans., **36**:1–24.

Tuttle, M. D. 1976a. Population ecology of the gray bat (*Myotis grisescens*): Philopatry, timing, and patterns of movement, weight loss during migration, and seasonal adaptive strategies. Occas. Pap. Mus. Nat. Hist. Univ. Kans., **54**:1–38.

Tuttle, M. D. 1976b. Collecting techniques. Pp. 71–88, *in* Biology of bats of the New World Phyllostomatidae. Part I. (R. J. Baker, J. K. Jones, Jr., and D. C. Carter, eds.). Spec. Publ. Mus. Texas Tech Univ., Lubbock, **10**:1–218.

Tuttle, M. D. 1976c. Population ecology of the gray bat (*Myotis grisescens*): Factors influencing growth and survival of newly volant young. Ecology, **57**:587–595.

Tuttle, M. D. 1977. Gating as a means of protecting cave dwelling bats. Pp. 77–82, *in* Proceedings of the National Cave Management Symposium. (T. Aley and D. Rhodes, eds.). Speleobooks, Albuquerque, 106 pp.

Tuttle, M. D. 1979. Status, causes of decline, and management of endangered gray bats. J. Wildl. Manage., **43**:1–17.

Tuttle, M. D., and L. R. Heaney. 1974. Maternity habits of *Myotis leibii* in South Dakota. Bull. S. Calif. Acad. Sci., **73**:80–83.

Tuttle, M. D., and D. E. Stevenson. 1978. Variation in the cave environment and its biological implications. Pp. 108–121, *in* Proceedings of the National Cave Management Symposium. (R. Zuber, J. Chester, S. Gilbert, and D. Rhodes, eds.). Adobe Press, Albuquerque, 140 pp.

Twente, J. W., Jr. 1955. Some aspects of habitat selection and other behavior of cavern dwelling bats. Ecology, **36**:706–732.

Vaughan, T. A. 1959. Functional morphology of three bats: Eumops, Myotis, Macrotus. Publ. Mus. Nat. Hist. Univ. Kans., **12**:1–153.

Vaughan, T. A. 1970a. The skeletal system. Pp. 98–138, *in* Biology of bats. Vol 1. (W. A. Wimsatt, ed.). Academic Press, New York, 406 pp.

Vaughan, T. A. 1970b. The transparent dactylopatagium minus in phyllostomatid bats. J. Mammal., **51**:142–145.

Vaughan, T. A. 1976. Nocturnal behavior of the African false vampire bat (*Cardioderma cor*). J. Mammal., **57**:227–248.

Vaughan, T. A. 1977. Foraging behavior of the giant leaf-nosed bat (*Hipposideros commersoni*). East Afr. Wildl. J., **15**:237–249.

Vaughan, T. A., and T. J. O'Shea. 1976. Roosting ecology of the pallid bat, *Antrozous pallidus*. J. Mammal., **57**:19–42.

Vehrencamp, S. L., F. G. Stiles, and J. W. Bradbury. 1977. Observations on the foraging behavior and avian prey of the Neotropical carnivorous bat, *Vampyrum spectrum*. J. Mammal., **58**:469–478.

Verschuren, J. 1957. Ecologies, biologie, et systématique des Chiroptères. Exploration du Parc

National de la Garamba. No. 7 (Mission H. de Saeger). Institute des Parcs Nationaux du Congo Belge, Bruxelles, 473 pp.

Verschuren, J. 1966. Introduction à l'écologie et à la biologie des Chiroptères. Pp. 25–65, Exploration deu Parc National Albert. No. 2 (Mission F. Bouliere et J. Verschuren). Institut des Parcs Nationaux du Congo Belge, Bruxelles.

Villa-R., B. 1966. Los murcielagos de Mexico. Inst. Biol., Univ. Nac. Auto. Mexico, Mexico City, 491 pp.

Voûte, A. M. 1972. Bijdrage tot de Oecologie van de Merrvleermuis, *Myotis dasycneme* (Boie, 1825). Unpublished doctoral dissertation, Universiteit Utrecht, Utrecht, 159 pp.

Voûte, A. M., J. W. Sluiter, and M. P. Grimm. 1974. The influence of the natural light–dark cycle on the activity rhythm of pond bats (*Myotis dasycneme* Boie 1825) during summer. Oecologia, **17**:221–243.

Walker, E. P. 1975. Mammals of the world. Vol 1. John Hopkins University Press, Baltimore, 644 pp.

Wallin, L. 1969. The Japanese bat fauna. Zool. Bidrag. Uppsala, **37**:223–440.

Ward, P., and A. Zahavi. 1973. The importance of certain assemblages of birds as "information centres" for food finding. Ibis, **115**:517–534.

Watkins, L. C., and K. A. Shump, Jr. 1981. Roosting behavior in the evening bat, *Nycticeius humeralis*. Am. Midl. Nat., **105**:258–268.

Wickler, W., and U. Seibt. 1976. Field studies of the African fruit bat, *Epomophorus wahlbergi* (Sundovall), with special reference to male calling. Z. Tierpsychol, **40**:345–376.

Wickler, W., and D. Uhrig. 1969. Verhalten und Ökologische Nische der Gelflügelfledermaus, *Lavia frons* (Geoffroy) (Chiroptera, Megadermatidae). Z. Tierpsychol., **26**:726–736.

Williams, T. C., and J. M. Williams. 1970. Radio tracking of homing and feeding flights of a Neotropical bat, *Phyllostomus hastatus*. Anim. Behav., **18**:302–309.

Williams, T. C., L. C. Ireland, and J. M. Williams. 1973. High altitude flights of the free-tailed bat, *Tadarida brasiliensis*, observed with radar. J. Mammal., **14**:807–821.

Wilson, D. E. 1971. Ecology of *Myotis nigricans* (Mammalia: Chiroptera) on Barro Colorado Island, Panama Canal Zone. J. Zool., **163**:1–13.

Wilson, D. E., and J. S. Findley. 1972. Randomness in bat homing. Am. Nat., **106**:418–424.

Wilson, E. O. 1975. Sociobiology. Belknap Press, Cambridge, 697 pp.

Wimsatt, W. A. 1969. Transient behavior, nocturnal activity patterns, and feeding efficiency of vampire bats (*Desmodus rotundus*) under natural conditions. J. Mammal., **50**:233–244.

Wimsatt, W. A., and B. Villa-R. 1970. Locomotor adaptations in the disc-winged bat, *Thyroptera tricolor*. Am. J. Anat., **129**:89–119.

Young, A. M. 1971. Foraging of vampire bats (*Desmodus rotundus*) in Atlantic wet lowland Costa Rica. Rev. Biol. Trop., **18**:73–88.

Chapter 2

Ecology of Bat Reproduction

P. A. Racey

Department of Zoology
University of Aberdeen
Aberdeen AB9 2TN, Scotland

1. INTRODUCTION

The *ecology of reproduction* became firmly established as a discrete area of investigation with the publication of Sadlier's monograph in 1969, *The Ecology of Reproduction in Wild and Domestic Animals*. I witnessed the gestation of this book, for Sadlier wrote it in the Wellcome Institute in the London Zoo, where I was working at the time on a doctoral thesis. It stimulated me, as it has stimulated others, to adopt techniques and approaches to the study of wild mammals similar to those long used by students of domestic species.

The purpose of this review is to examine the present state of knowledge on the effects of environmental factors on reproduction in the second largest order of mammals—the bats. The diversity of breeding cycles will be considered initially, and in particular the factors responsible for their timing. In discussing these cycles, only those relatively few studies that reveal something of the interrelationship between environmental factors and reproduction have been selected for review. The effect of environmental factors on specific reproductive events and adaptations is then considered.

The roost environment is clearly important for successful reproduction and is dealt with by Kunz (Chapter 1). Growth and survival, while receiving brief treatment here, is the subject of Tuttle and Stevenson's contribution (Chapter 3).

2. THE TIMING OF BREEDING SEASONS

The patterns of reproduction among the Chiroptera have been reviewed frequently in recent years (Wimsatt, 1969; Carter, 1970; Anciaux de Faveaux, 1973, 1978; Wilson, 1973, 1979; Gustafson, 1979; Krutzsch, 1979; Jerrett, 1979; Oxberry, 1979). Seasonal monestry characterizes all bats of temperate latitudes and is also frequently manifest in tropical insectivorous and frugivorous species. Polyestry is evident in tropical insectivores and frugivores and may be either bimodal, as in Neotropical phyllostomids (e.g., *Artibeus* and *Uroderma;* Fleming, 1971, 1973), or continuous (e.g., *Eonycteris spelaea;* Beck and Lim, 1973). Both these variations of polyestry typically result in the production of two litters a year, and sometimes three, as in the Neotropical insectivore *Myotis nigricans* (Wilson and Findley, 1970). Such polyestry is frequently facilitated by the occurrence of a postpartum estrus (Myers, 1977).

The period of estrus is extended in bats of temperate latitudes and coincides with hibernation, so that several months delay occurs between insemination and fertilization in all genera so far investigated (Wimsatt, 1969; Racey, 1979), with the exception of *Miniopterus,* where fertilization occurs soon after insemination and implantation is delayed (Courrier, 1927).

The reproductive periodicity manifest in many of these bats reflects the seasonal variations in food supply. Bats of temperate latitudes give birth only during the summer months (Oxberry, 1979), when the insect food supply is most abundant (Taylor, 1963), and aseasonal polyestry occurs in those species whose food supply is not thought to fluctuate, such as the sanguivorous vampire *Desmodus rotundus* (see Wilson, 1979).

As more information about the reproductive cycles of tropical bats becomes available, generalizations about the type of reproductive cycle shown by a particular group of bats become more difficult. There is considerable variation within the same family and even within the same genus at the same latitude. For example, *Pipistrellus ceylonicus* is seasonally monestrous but stores spermatozoa at 20° north latitude (Madhavan, 1971), and both *P. mimus* and *P. dormeri* breed throughout the year at the same latitude (Gopalakrishna *et al.*, 1975; Madhavan, 1979), although in both these species there are biannual peaks of pregnancy. *Hipposideros fulvus* was studied by Madhavan *et al.* (1976) at the same latitude and also shows the strict monestrous periodicity seen in *P. ceylonicus*, although the season of birth differs by 4 months.

Tropical bats such as *Molossus molossus* and *M. ater* continue to breed at much the same time each year when maintained in constant photoperiods of 12 hr light and 12 hr dark, suggesting the existence of an endogenous circannual rhythm of reproduction (Häussler *et al.*, 1981). The extent to which the timing of such circannual rhythms can be modified by experimental variation of environmental factors other than photoperiod promises to shed much light on the way in which the observed reproductive seasonality in the Chiroptera has evolved.

2.1. Effect of Variations in Latitude

The timing of breeding varies with latitude in many species of seasonally breeding mammals, suggesting that the reproductive cycle is initiated by changes in photoperiod (Sadlier, 1969; Millar and Glover, 1973). The timing of reproduction also varies with latitude in several species of bats. Thus the breeding season in the polyestrous *Myotis adversus* is longer at 22°30′ south latitude than it is at 27°30′ south latitude. Individual females produce at least two successive litters at both latitudes, and there is evidence that females at lower latitudes may produce three (Dwyer, 1970).

Kitchener (1975) showed that the time of birth in *Chalinolobus gouldii* varied with latitude in Western Australia (which extends from 13 to 35° south latitude). At lower latitudes, where the climate is tropical, births took place from late September to early October; in the central region they occurred in late October and early November; and in southwestern Australia, where the climate is Mediterranean and where Kitchener established that the storage of spermatozoa intervened between copulation and ovulation, births took place in late November and early December. It seems likely that such variations in latitude affect reproduction primarily through the ambient temperature, which affects both the thermal economy of the bats and the availability of food, rather than by a photoperiodic effect on the endocrine control of reproduction.

The breeding season for *Cynopterus sphinx* began about 2 months later at 16°30′ north latitude (Sreenivasan *et al.*, 1974) than further north, around Nagpur (about 20° north latitude) (Moghe, 1956), where rainfall was low and summers were warmer and winters cooler than at more southerly latitudes.

Later commencement of parturition at higher latitudes is well documented for *Myotis lucifugus* (Fenton, 1970; Humphrey and Cope, 1976), and the period over which births occur is shorter in more northern populations, as for *M. lucifugus* in Alberta (Schowalter *et al.*, 1979). Against the trend, however, parturition occurred later in southern Alberta. *M. velifer* gave birth 2–3 weeks sooner in Texas than in Kansas (Kunz, 1973), and in *Eptesicus fuscus* parturition occurred one month earlier in Kansas than in North Dakota (Kunz, 1974a). In general however, later parturition is a consequence of later ovulation associated with later arousal from hibernation at higher latitudes.

Of particular interest is the pattern of reproduction shown by bats in the vicinity of the equator. Anciaux de Faveaux (1977) attempted to define the biological equator from information on the reproductive periodicity of 18 monestrous bat species studied in Africa by himself and others, particularly Brosset

(1968). Above 4°13′ north latitude the timing of reproduction conforms to the boreal pattern, with parturition in March, whereas below 1°30′ south latitude it conforms to the austral pattern, with parturition in October. In the transition zone between the two cycles, some colonies of *Hipposideros* show the boreal cycle and others the austral cycle, even in adjacent cave roosts (Brosset and Saint-Girons, 1980). Brosset (1968) suggested that sexual isolation within these sympatric populations was responsible for the observed differences in the size of individuals of *H. caffer* showing either boreal or austral cycles. Such dimorphism was even evident in the same caves. Such incompatible reproductive rhythms within sympatric populations are rare among vertebrates (White, 1978).

Dimorphic forms of *Hipposideros commersoni*, separable on the basis of forearm length and echolocation frequency, were described at 4° south latitude by Pye (1972), and McWilliam (1982) has found that these two forms have an austral reproductive cycle. McWilliam (1982) has further evidence for reproductive isolation, in that both boreal and austral cycles are evident in the small form in the same cave. Although these observations clearly question Anciaux de Faveaux's definition of the biological equator, they are more important in illustrating the process of speciation. The key to how such reproductive isolation has arisen must be in differences in environmental factors influencing specific reproductive processes, either locally, in populations inhabiting the same or adjacent caves, or in the greater differences (of food supply and photoperiod) likely to be experienced by bats migrating for part of the year to another locality.

2.2. Rainfall and Its Effect on Food Supply

Rainfall is the most important climatic factor affecting plant phenology in the tropics, and Janzen (1973) has shown that insect abundance in and over vegetation is directly related to seasonal patterns of flowering, fruiting, and leaf flush in that vegetation. Many studies of reproduction in seasonally breeding bats have shown that young are produced at a time when the rainfall is highest (Pirlot, 1967) and the food supply most abundant. Births occur just before the onset of peak rains in both the insectivorous *Hipposideros caffer* (Mutere, 1970) and the frugivorous *Eidolon helvum* (Mutere, 1967), and during peak rains in the insectivorous *Otomops martiensseni* (Mutere, 1973a). Clearly the abundance of food for the young, resulting from the rains, is the ultimate or evolutionary factor (Baker, 1938) responsible for the observed reproductive seasonality. It is more difficult to establish the proximate or stimulatory causes responsible for the initiation of spermatogenesis, which in *Otomops* occurs at the end of the wet season, and estrus, which occurs at the end of the dry season.

Marshall and Corbet (1959) and Mutere (1973b) studied the reproductive cycle of *Tadarida pumila* at 0°26′ and 0°6′ north latitude, respectively, where this bat breeds throughout the year. Both studies reveal two clear peaks in the occurrence of pregnant females. Two peaks also occur in rainfall, with maximum precipitation coinciding with the time that most bats are lactating and young are becoming independent. Although Marshall and Corbet (1959) were impressed by the uniformity of the equatorial environment ("rain falls throughout the year, the environment remains lush and insect food is plentiful"), Mutere (1973b) pointed out that the numbers of moths caught in a Malaise trap (at Makerere, some distance from where his bats were collected) varied threefold, with highest catches coinciding with maximum rainfall. Mutere's (1973b) study included a congener *T. condylura*, which has a clear bimodal breeding rhythm with pregnancy peaks coinciding with times of least rainfall, so that bats are lactating and young are weaned at the time of peak rains. This is also the case for *Rousettus aegyptiacus*, studied by Mutere (1968) at 0°22′ north latitude.

An attempt to relate breeding periodicity in bats (principally phyllostomids) to what is known about insect and plant phenology was also made by Fleming *et al.* (1972). They found that most Neotropical frugivores produce two young annually, one toward the end of the dry season and the other in the middle of the wet season. According to Janzen (1967), the dry season throughout much of Central America is the principal flowering season, and the late dry season and early wet season the principal fruiting season. In Panama, and possibly also Costa Rica, frugivorous bats produce young in the middle of this dry season, just after one fruiting season and just before another. Although young produced during the first birth season become independent when fruit levels (determined in another study by Smythe, 1970) are relatively high, those born in the second period become independent when food may be declining in abundance. It would be interesting to compare the survival of young born in either breeding season. Further south, in Columbia (4–6° north latitude), where an adequate food supply is continuously available to frugivores, *Artibeus lituratus*, one of the species reproducing seasonally in Panama (Fleming *et al.*, 1972), gives birth throughout the year (Tamsitt and Valdivieso, 1965). Fleming *et al.* (1972) also showed that most Neotropical insectivorous bats give birth during the wet season, when insect numbers are highest (Janzen and Schoener, 1968; Skutch, 1950).

The degree of intra- and interspecific synchrony in the reproductive cycles of these bats suggests common environmental triggers. In Panama females become pregnant after a period of reproductive quiescence, at or just after the annual peak of rainfall and when day length is decreasing. However, in western Costa Rica, where the dry season begins earlier, the same species becomes pregnant about one month earlier than in Panama, suggesting that decreased rainfall rather than photoperiod may be the proximate factor (Fleming *et al.*, 1972).

The work of Fleming *et al.* (1972) was extended by Bonaccorso (1979), who examined the foraging and reproductive ecology in 35 bat species at Barro Colorado Island in Panama (9°10′ north latitude). Bats were separated into nine feeding guilds on the basis of diet and method of food procurement; species utilizing food sources abundant over most of the year have two litters annually. In the frugivorous guilds births are synchronized within populations and coincide with the two seasonal peaks in fruit abundance. The first pregnancy of the gleaning carnivore guild, which depends on large insects as a primary food source, occurs during relative food scarcity, but births occur as large insects become abundant. The first lactation and the entire second reproductive cycle occur during the months of insect abundance. Bonaccorso's study is important in emphasizing the coincidence of food peaks with lactation rather than with pregnancy.

In one of the few studies of reproductive phenology in South American bats, Taddei (1976) examined 16 species of phyllostomids at 19–22° south latitude, where a rainy season from October to March (with 85% of the annual rainfall) alternates with a dry season occupying the rest of the year. There are two periods of reproduction, the principal one from June to October–November, and the secondary one, continuous with the first, resulting in births in February–March. Although sexually mature males are found throughout the year, the incidence of pregnancies increases from the middle of the cold dry season in June and July and remains above 40% until the middle of the warm rainy season in December and January. The greatest incidence of lactating females occurs during the rainy season in October–November and February–March.

An attempt to quantify insect availability and relate this to reproductive periodicity in order to distinguish different patterns of parental investment was made by Bradbury and Vehrencamp (1977) for four species of emballonurids in Costa Rica. These authors point out in an earlier publication that the scope and intensity of insect sampling was limited (Bradbury and Vehrencamp, 1976). However, the numbers of insects caught were converted to biomass (using an allometric relationship between length and mass) and adjusted for the relative insect richness of the habitat utilized by each bat species in each season and again for the relative area of each habitat, to give a value for food availability in biomass units plotted throughout the year as the bats moved between habitats. The computed values for biomass units varied substantially for similar reproductive states in different bat species, from which Bradbury and Vehrencamp (1977) concluded that the patterns of parental investment differed. Species using less seasonal habitats (*Rhynchonycteris naso* and *Saccopteryx leptura*) are most likely to have more than one parturition a year, and periods of parturition tend to be relatively asynchronous. Those species utilizing highly seasonal food supplies (*S. bilineata* and *Balantiopteryx plicata*) show a single synchronous birth peak annually. However, Bradbury and Vehrencamp (1976: Table 11, p. 378, and Fig. 11, p. 369) show that although *S. bilineata* forages in areas of peak insect abundance during gestation and lactation (January to June), when insects are at a peak in most habitats, the computed biomass units available to this species are inexplicably low and result in resorption or abortion (Bradbury and Vehrencamp, 1977; Bradbury, 1979).

Bradbury and Vehrencamp (1977) consider that where food peaks are synchronized with gestation, the survival of pregnant females will be favored over that of lactating females and dispersing young. Where food peaks coincide with lactation and the dispersal of young, pregnant females are at highest risks during gestation. These authors also maintain that for the two species using seasonal habitats, food peaks are synchronized with lactation and dispersal (although this is not apparent from the data presented), while for those in stable habitats gestation coincides with maximum food levels.

However, mammalian lactation demands far more energy than pregnancy (Migula, 1969), and such an energy demand is exacerbated in smaller species owing to scaling effects (Hanwell and Peaker, 1977). Although many mammals (including bats) have evolved various mechanisms to distribute the energy cost of pregnancy over a longer period, lactation, once initiated, does not lend itself to interruption or modulation, so that adequate food supply during lactation and weaning is the most important selection pressure in the timing of mammalian reproductive cycles.

In Malaysia, although any month may be wettest or driest in any particular year, over most of the country rainfall peaks occur in October–November, with a subsidiary peak in March or April (Medway, 1972). Gould (1978), in summarizing knowledge of the breeding cycles of Malaysian bats, suggests that there seems to be two main birth peaks for insectivorous bats, one between September and November and the other between February and May. Thus *Tylonycteris pachypus* and *T. robustula*, studied by Medway (1972) at Ulu Gombak (3°20′ north latitude), gave birth in April and May after a gestation period of 12–13 weeks, which is very long for bats weighing only 4 and 6 g, respectively. Although insect numbers do not vary dramatically in Malaysia (Ward, 1969), the period of gestation and lactation in *Tylonycteris* coincides with a December–July period of temporary increases in the number of insects (light-trapped at Singapore, 1°40′ north latitude), and Medway considers that the long gestation period may be an adaptation to an irregular food supply. Insects are least abundant from August to October, after the young have reached adult size. It is interesting to note that passerine birds, which are dependent on an adequate supply of insects to rear their young, also nest from January to June in Malaysia (Gibson-Hill, 1952; McClure and Osman, 1968; Ward, 1969).

Lim (1970) attempted to correlate the occurrence of pregnancy in *Cynopterus brachyotis* with rainfall and the relative abundance of fruits and flowers as reflected by their percentage of occurrence in stomachs. The highest pregnancy rates were recorded from April to June at 3° north latitude in the Malaysian lowlands and coincided with the highest incidence of fruits in the diet, and as these diminished, the percentage of floral parts increased. At a similar latitude, but at an altitude of 1524–2012 m, most of the ten species of fruit bats studied by Lim (1973) were pregnant from February to June, with a peak from February to April, several months earlier than in the lowlands, but coinciding with the flowering and fruiting of trees.

It is apparent from the foregoing that although there have been many studies of the reproductive periodicity of the Chiroptera, most authors have attempted to relate their findings to biotic or climatic factors either measured or described subjectively by others. Few have planned investigations to measure variations in rainfall and food supply in the same location and at the same time as bats are sampled; instead, most have used data on food supply obtained at different times and in different places, and since the pattern of insect abundance in the tropics may vary from year to year (Wolda, 1978), extrapolations may lead to false conclusions. One of the few studies to monitor changes in climate and food supply in the same locality as the bats were sampled is that of McWilliam (1982) on a multispecies association of cave bats in coastal Kenya (4°15′ south latitude), including *Coleura afra*, where rainfall, insect availability, and bimodal breeding are convincingly correlated (Fig. 1).

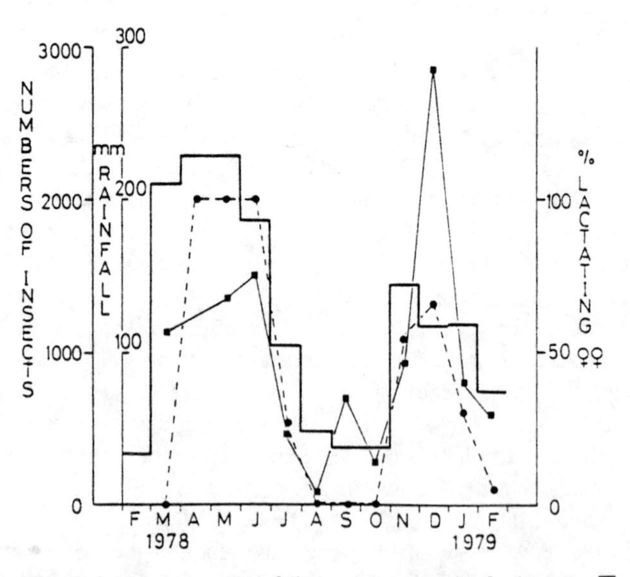

FIGURE 1. The relationship between rainfall (——) (mm/month), food supply (■——■) (numbers of insects caught per night), and reproduction (●---●) (percentage of bats lactating per catch) for *Coleura afra* at Diani, Kenya. A light trap was operated twice a month. The percentage of females lactating per catch refers to the highest value found each month. (From McWilliam, 1982.)

3. ENVIRONMENTAL FACTORS AFFECTING SPECIFIC REPRODUCTIVE EVENTS

3.1. Spermatogenesis and Androgenesis

Spermatogenesis in hibernating bats takes place during summer. The seminiferous tubules then regress and the sperm produced are stored in the cauda epididymidis throughout the following winter (Racey, 1979). Mating is initiated in autumn but may continue throughout the winter, depending on the species, except those in the genus *Miniopterus*, in which the entire male reproductive system regresses before hibernation (Courrier, 1927).

Associated with sperm storage in hibernating vespertilionid and rhinolophid bats is continuing activity of the accessory glands of reproduction and sustained libido. However, differences in the histological appearance of the Leydig cells of New and Old World bats during hibernation have caused much speculation as to whether the exaggerated asynchrony between the endocrine and exocrine functions of the testis is real or apparent (Wimsatt, 1969; Gustafson, 1979; Racey and Tam, 1974).

The Leydig cells of New World species appear involuted during winter, whereas those of Old World species appear active, and involuting only after arousal from hibernation in spring. The involution in the New World species has been supported by a study of their fine structure, which reveals that those cell organelles associated with steroidogenesis are much reduced during winter, and although this does not preclude the possibility of limited hormone secretion, it correlates with the drastic reduction in plasma testosterone levels observed at that time (Gustafson, 1979). Nevertheless, sufficient testosterone is detectable in the blood plasma of both New and Old World species during winter to account for their continued libido and the survival of epididymal spermatozoa (Gustafson and Shemesh, 1976; Racey and Tam, 1974; Racey, 1974a).

The depressed rate of hormone secretion during winter is probably a reflection of the general metabolic rate. Gustafson (1979), in analyzing Racey's (1974a) results, pointed out that the testicular content of testosterone and its plasma levels were higher in noctules fed twice weekly and housed in an open-sided lean-to that reflected natural temperature fluctuations, than in those housed at 5°C, where they were in continuous torpor. These conditions probably represent the two extremes of what occurs in the natural environment. The weight of testes and accessory glands was also higher in bats that hibernated continuously than in those which hibernated intermittently, providing additional evidence that continuous hibernation slowed the regressive changes that occurred in more active animals (Racey, 1974a).

Most studies of New World species were carried out at lower latitudes than similar studies on Old World species (Table I). Favorable environmental condi-

TABLE I
Variations in Latitude of Bat Colonies Studied Histologically

Species in which Leydig cells and accessory glands appear active during hibernation	Locality	Approximate latitude	References
Pipistrellus pipistrellus	England	52° N	Racey and Tam (1974)
Rhinolophus hipposideros	Czechoslovakia	49° N	Gaisler and Titlbach (1964)
P. pipistrellus *R. ferrumequinum* *Eptesicus serotinus* *Myotis (=Vespertilio) capaccinii* *Vespertilio murinus*	France	49° N	Courrier (1927)
Species in which Leydig cells are involuted and accessory glands appear active during hibernation			
Myotis lucifugus	New York State	42° N	Gustafson (1979)
Plecotus townsendii	California	39° N	Pearson *et al.* (1952)
M. lucifugus, M. grisescens	Missouri	38° N	Miller (1939)
Pipistrellus hesperus	Texas	32° N	Krutzsch (1975)
M. velifer	Texas	31° N	Krutzsch (1961, personal communication)

tions during autumn at lower latitudes may allow the complete pituitary–gonad cycle to run its course and result in loss of gonadotrophic stimulation of the Leydig cells before environmental conditions deteriorate sufficiently to necessitate hibernation. Reduced plasma clearance rates (as demonstrated in caribou by Whitehead and West, 1977) or extra gonadal sources of androgen (as demonstrated in *Myotis lucifugus* by Krutzsch and Wells, 1960) could account for the observed, albeit reduced, plasma testosterone levels and androgen-dependent functions such as libido, sperm storage, and active accessory glands during winter. In species studied at higher latitudes in the Old World, hibernation might intervene sooner after the completion of gametogenesis and retard the cycle of pituitary-gonad activity, which would progress to completion only after arousal the following spring.

As in most seasonally breeding mammals, there appears to be a period after the completion of spermatogenesis during which the pituitary–gonad axis is refractory to stimulation by environmental factors. Thus, if *Myotis lucifugus* is aroused from hibernation and maintained at elevated temperatures during the first half of the hibernation period, the testes remain involuted (Gustafson, 1979). If bats experience increased temperatures and food supply in the second half of the hibernation period, spermatogenesis is initiated prematurely (Gustafson, 1979; Racey and Tam, 1974). In the case of the noctule *Nyctalus noctula*, increase in testicular weight is recorded within 2 weeks of arousal and increased food supply, reflecting increased neuroendocrine activity during the second half of hibernation (Racey, 1974a).

In an attempt to investigate whether increased photoperiod was also involved in the resurgence of testicular activity in spring, Racey (1978) aroused adult and immature *Pipistrellus pipistrellus* from hibernation prematurely and maintained them in 8 hr of light followed by 16 hr of darkness (8L:16D) and 16 hr of darkness followed by 8 hr of light (16D:8L) at summer temperatures and with food freely available. There was no significant difference between testis weights in either photoperiod, although spermatogenesis was more advanced in adult animals than in those undergoing puberty in the same conditions. In another experiment pipistrelles maintained in total darkness at summer temperatures and with food freely available had significantly smaller testes than those similarly maintained at 14L:10D. However, 8 weeks in the latter conditions resulted in much larger testes than 16 weeks at either 8L:16D or 16L:8D in a previous year. The most likely interpretation of these results is that total darkness inhibits spermatogenesis (as in rats; see Sadlier, 1969) and that the enhancement of testicular development, compared to the previous year, that resulted from 14L:10D, reflects variations in the ease with which bats adapt to captivity (in particular learning to feed) rather than indicating that such a photoperiod is stimulatory. To date, there is no experimental evidence that increased photoperiod is a proximate cause of spermatogenesis in bats, although further investigation is called for.

However, the period of minimal testicular activity in *Myotis nigricans* at 9° north latitude coincides with the time between the autumn equinox and winter solstice and suggested to Wilson and Findley (1971) that declining day length after the summer solstice is responsible for switching off the pituitary gland. Changes in photoperiod at this latitude are slight (63′ from summer to winter solstice), but such changes are perceptible by some mammals (Goss, 1977). Although the pineal gland is thought to be involved in mediating the effects of photoperiod on the hypothalamus (Reiter and Follett, 1980), Pevet and Racey (1981) could find no evidence of seasonal changes in the ultrastructure of pinealocytes in *Pipistrellus pipistrellus*, and no evidence, therefore, that the pineal gland is involved in the inhibition of reproductive activity as it is in some other mammals (Reiter and Follett, 1980).

An interesting contrast to the situation in temperate-zone hibernators, where spermatogenesis proceeds during the summer, is seen in the rat-tailed bat *Rhi-*

nopoma microphyllum (=kinneari), where this process is initiated as bats migrate to the winter roost and proceeds while the bats are torpid (Anand Kumar, 1965). *Rhinopoma* has abdominal testes, and it is tempting to speculate that spermatogenesis in this species can only progress at decreased body temperature. Carrick and Setchell (1977) provide some evidence that abdominal testes are associated with mammals having a more variable body temperature than those with scrotal testes.

3.2. Estrus and Ovulation

The reproductive cycles of most hibernating bats of temperate latitudes are unique among mammals in that follicles grow in the ovaries during autumn, the females then come into estrus and copulate, and the outstanding follicle survives in the ovary and spermatozoa survive in the uterus until spring, when ovulation and fertilization occur (Wimsatt, 1944b; Wimsatt and Parks, 1966). An exception is *Miniopterus*, where ovulation and fertilization occur before hibernation (Courrier, 1927). It is surprising that no experiments have been undertaken to investigate those factors (such as declining day length or prehibernal fattening) that may be associated with and perhaps responsible for natural estrus in these bats; instead, efforts have concentrated on attempts to induce the premature rupture of the follicles of hibernation.

It has even been suggested that if conditions in temperate regions were to change so that cold winters and consequent food shortage no longer necessitated hibernation in insectivorous bats, then ovulation and development of the young would follow copulation in autumn (Baker and Bird, 1936). This stimulating hypothesis is, however, at variance with most of the subsequent discoveries of the control of ovulation in hibernating bats.

It has long been clear from studies of ovarian morphology (Guthrie and Jeffers, 1938a, 1938b; Wimsatt, 1944b) and subsequently confirmed by Herlant's (1956, 1968, 1970) pituitary cytology, that the ovaries receive continuous stimulation with follicle-stimulating hormone (FSH) during the hibernation period (Richardson, 1979). Attempts to induce ovulation have concentrated on inducing maturation of the oocyte and follicular rupture, either by arousal from hibernation alone or accompanied by the administration of hormones of both pituitary and chorionic origin. Working with *Myotis lucifugus*, Guthrie and Jeffers (1938c) found that spontaneous ovulation correlated with a rise in temperature did not occur before February, whereas ovulation could be induced by means of gonadotrophic hormones as early as December. The fertilization and consequent development of ova did not occur, however, before the time of spontaneous ovulation.

With a related species (*Myotis grisescens*) Guthrie and Smith (1940) and Smith (1951) were unable to induce ovulation by means of human chorionic gonadotrophin (HCG) in August and early September, even though the ovaries contained antral follicles. This hormone has an action similar to that of luteinizing hormone (LH) but is isolated from the urine of pregnant women and is thus easier to obtain than the pituitary hormone. Experiments in late September, before the bats migrated to the hibernaculum, showed that 25% (n = 8) of the injected animals ovulated, and in early October, after migration had occurred, only 20% (n = 20) were induced to ovulate. Thereafter a much larger proportion of the injected animals ovulated and Smith (1951) concluded that the ovary becomes more responsive to gonadotrophin stimulation as winter progresses, corresponding with growth of the outstanding follicle.

Sluiter and Bels (1951) studied follicular growth and spontaneous ovulation in captive bats during the hibernation period. Of four *Myotis myotis* caught on November 18 and confined at 5–12°C, only one, which died 16 days after capture, had ovulated. Of four specimens caught on December 28 and kept first for 12 days at 5–13°C and then for 4–8 days at 20–22°C, three ovulated and the other showed an oocyte in the first meiotic division. Thus increased environmental temperature and resulting arousal from hibernation will cause ovulation as early as January in *M. myotis. Myotis emarginatus,* however, taken from hibernation in early January showed no sign of follicular growth or ovulation when kept at 18–20°C.

In order to test the sensitivity of the ovary to stimulation by gonadotrophins during the first months of hibernation, Herlant (1968) treated *Myotis daubentoni,* maintained at a constant 9°C, with HCG, and seven out of nine females ovulated. Ovulation was also induced in *M. dasycneme* subjected to similar treatment. Although there is no mention of the use of controls, it is nevertheless surprising to find bats ovulating in environmental temperatures equivalent to those of the hibernaculum. Herlant concluded, like Wimsatt (1960), that the obstacle to ovulation in hibernating bats did not appear to originate in the ovary but was probably due to the low pituitary secretion of LH.

Ramaswami and Anand Kumar (1966) studied the effect of exogenous hormones on the semihibernant bat *Rhinopoma microphyllum* and showed that although the breeding season had ended and the bats had apparently entered a period of rest, ovulation could be induced by several pituitary and gonadal hormones.

Skreb (1955) was the first to use low doses of gonadotrophins in a sequential regimen of pregnant mare serum gonadotrophin (PMSG, which has similar biological properties to FSH) and HCG. Although he successfully induced *Nyctalus noctula* to ovulate in the second half of winter, he considered that development of the egg was inhibited in the treated animals. Thus in 35 experimental animals, 31 ovulated to give 32 eggs. In 25 of these the occurrence of two polar bodies, two pronuclei, or cleavage indicated that fertilization had taken place, although cleavage was noted in only seven cases (28%). There was also a high incidence of ovulation in control animals: 35% ovulated (n = 20) and in six fertilization had occurred; of these six, cleavage was noted in five (83%), and one conceptus had reached the uterus.

Racey (1976) achieved a 70% ovulation rate in *Pipistrellus pipistrellus* aroused for a fortnight before injection and treated with low doses of HCG in the first half of hibernation. Although only two out of a total of 48 animals used as controls in these experiments responded to prolonged arousal by ovulating (Racey, 1972), the first polar body had been extruded from the oocyte in the antral follicles of 1/7 of the controls aroused from hibernation for 2 days, 1/8 of those aroused for 9 days, and 8/18 of those aroused for 15 days, indicating that ovulation was increasingly imminent (Racey, 1976).

The difficulty of summarizing these experiments is exacerbated by the fact that several investigators were unaware of the results of previous experiments. Hibernating or torpid animals are generally unresponsive to hormonal stimulation and only those that are aroused will ovulate. Herlant (1956, 1968, 1970) has shown that pituitary cells presumed to secrete LH (but see Richardson, 1979) involute after parturition, and there is a refractory period before they respond to arousal of the bat from hibernation, indicating that the delay in ovulation associated with hibernation is not facultative but involves prolonged and obligate changes in the cycle of hypothalamic and pituitary activity.

Although Herlant (1968) also observed small islets of LH cells showing slight activity during hibernation, prolonged arousal in the first half of hibernation has resulted in ovulation, with subsequent pregnancy in only isolated instances. Thus Zondek (1933) and Caffier and Kolbow (1934) recorded pregnancies in single unspecified bats in December and Sluiter and Bels (1951) recorded an eight-cell pregnancy in *Myotis myotis* collected on November 18 and kept at 5–13°C for 17 days. Parturition has also occurred in November in *Eptesicus fuscus* maintained at 27°C (with constant artificial light in the morning and irregular light in the afternoons). Two females gave birth—one had been in captivity for several years (and had been blinded by enucleation for a year), and the other had been born in captivity during the preceding summer (J. Simmons, personal communication).

Arousal from hibernation followed by the administration of exogenous gonadotrophins is usually required for follicular rupture in the first half of hibernation, although the arousal resulting from the transfer of a hibernating animal to a warm environment for a few minutes before injection is obviously different from arousal associated with a constant warm environment and adequate food supply for several days. Such differences must affect events subsequent to ovulation, particularly the formation and maintenance of the corpus luteum. Comparison of events subsequent to ovulation in previous work is hampered by the fact that there has been no attempt to standardize the conditions of arousal.

Once the refractory period has run its course, ovulation may be easily induced by arousal from hibernation alone. The increasing ease with which ovulation and subsequent pregnancy may be induced by prolonged arousal from hibernation without the assistance of exogenous gonadotrophins during the second half of hibernation is evidence that such arousal is a necessary prerequisite for the resurgence of full hypothalamic and pituitary activity. The interval between arousal and ovulation appears to decrease throughout hibernation. In *Pipistrellus pipistrellus* aroused in January, there is an average delay of 11 days between arousal and ovulation, whereas in animals in which hibernation had been artificially prolonged until May, ovulation occurred immediately. This indicates that LH accumulates in the pituitary gland progressively during hibernation and arousal acts by stimulating the activity of its hypothalamic releasing hormone. The manipulation of environmental conditions has thus resulted in a 19-week variability in the time of birth of captive *P. pipistrellus* (Racey, 1972).

Bats living in cold conditions during the time of ovulation may ovulate and give birth more synchronously than those experiencing a more variable climate. The available evidence from laboratory colonies of *Pipistrellus pipistrellus* and *Nyctalus noctula* support this hypothesis (Racey, 1972) and as a result of a field study on *P. pipistrellus,* Rakhmatulina (1972) considers that a cold spring was responsible for a 40% contraction of the period over which births occurred as compared with the previous year.

Skreb (1954) produced the only evidence implicating light in the control of ovulation in bats. Noctules (*Nyctalus noctula*) were removed from their hibernaculum on March 6 and brought into the laboratory (18–20°C) but not fed. A group of eight animals was kept in complete darkness and after about a fortnight only one had ovulated, whereas all seven animals kept in natural daylight had done so. In general, bringing bats into laboratory conditions results in arousal from hibernation and increased metabolism, with increased temperature rather than light being the most important stimulus. This often results in ovulation, as in the group maintained by Skreb under natural-lighting conditions. Total darkness, on the other hand, may have had an overriding inhibitory effect.

However, these conclusions are not supported by observations of Häussler *et al.* (1981, and personal communication) on Neotropical bats. Harems of *Molossus molossus* and *M. ater* maintained for over 3 years in 12L:12D gave birth to 60 young with fertile F_1 generations. Single females gave birth to a single young a year, in one of two birth periods (September–October or March). A pair of *M. molossus* maintained in total darkness for 1 month at the time of mating gave birth 3 months after they were restored to 12L:12D. These observations suggest that photoperiod is not involved in the timing of estrus and ovulation, but since the time of birth varies only slightly from year to year, they suggest the existence of an endogenous circannual rhythm of reproduction.

3.3. Mating

Mating systems in the Chiroptera have been extensively reviewed by Bradbury (1977), who has himself made major contributions in this field. Bradbury and Emmons (1974) thus found that males of *Saccopteryx bilineata* defend territories in the buttress cavities of large trees during the day, where they perform an elaborate variety of visual, vocal, and olfactory displays. Females are distributed among these territories in harems of from one to eight females to each male, and although harem females appear to be very aggressive to newcomers of their own sex, there was considerable movement of females between the several harems making up a colony. In a subsequent study Bradbury and Vehrencamp (1976) showed that the colony foraging site is partitioned into individual harem territories defended by harem males and containing the individual beats of all current harem females. In this way the subdivision of the roost site is mapped directly onto the foraging dispersion.

The social organization and breeding of the Australian *Pteropus poliocephalus* studied by Nelson (1965a, 1965b) is related to the phenology of eucalyptus trees, the blossom of which provides the main source of food for the bats and increases in abundance during the summer. In spring females arrive at traditional breeding sites and give birth to young. After a month or two, males set up individual territories around females and copulation occurs in late summer when the young have been weaned. Shortly afterward, as the supply of eucalyptus blossom dwindles, breeding camps break up into male and female groups that move nomadically and separately over different ranges.

Humphrey and Cope (1976) suggested the existence of demes of *Myotis lucifugus* and proposed that the swarming behavior evident in this species in eastern North America (Fenton, 1969) brings together bats dispersed throughout the range of a deme, to increase the probability of finding a mate and prevent local inbreeding.

3.4. Delayed Fertilization

Vespertilionid and rhinolophid bats inhabiting temperate latitudes undergo spermatogenesis during summer, and mating is initiated in autumn. Spermatozoa are stored for up to 7 months in the female reproductive tract and retain their fertilizing capacity until spring, when females ovulate and become pregnant (Wimsatt, 1969; Racey, 1979). Male bats also store fertile sperm for prolonged periods (Racey, 1973a), even after females have ovulated, and it seems that the responsibility for sperm storage is shared between the sexes, with different sexes taking greater or lesser shares, depending on the species. Thus the fact that all female *Nyctalus noctula* and *Pipistrellus pipistrellus* taken from hibernation are inseminated suggests that in these species the main burden of sperm storage falls on the female (Racey, 1974b). Copulation during winter, particularly in bats of the genus *Myotis*, is well documented (Wimsatt, 1945; Gilbert and Stebbings, 1958; Stebbings, 1965) and Pearson *et al.* (1952) report that although all females of *Plecotus townsendii* are carrying sperm by the third week in September, copulations can be recorded as late as February. Strelkov (1962) found that an increasing proportion of females of four species (*M. daubentoni*, *M. dasycneme*, *M. mystacinus*, and *Plecotus auritus*) were inseminated as hibernation progressed, suggesting that in these species the responsibility for storing the sperm that will eventually fertilize is more equally shared between males and females.

Early workers observed the coincidence of sperm storage and hibernation and suggested that torpor was necessary for the fertilizing capacity of sperm to be retained (Hartman, 1933). This idea has persisted so that Myers (1977) cites Wimsatt (1960) and McNab (1974) to contend that "hibernation is currently thought to be a *sine qua non* of storage of spermatozoa."

This conclusion arises from the observations of Guthrie (1933) and Wimsatt (1944a), that spermatozoa disappear from the reproductive tracts of bats that are active. Rice (1957) also observed that populations of *Myotis austroriparius* living in peninsular Florida do not experience the cold temperature and paucity of insects prevalent in western Florida, do not hibernate, and do not mate until spring. Clearly, in part of its range the burden of sperm storage in this species falls entirely on the male.

Although there is evidence that repeated arousal and frequent activity during the period of hibernation may be inimical to the storage of spermatozoa, it is now clear that bats are often naturally active for a considerable part of the period of sperm storage. For example, the earliest insemination recorded for *Pipistrellus pipistrellus* in England was on September 23 for a female taken in a mist net while feeding (Racey, 1972). This species does not appear in hibernacula in the same region until the end of November (Racey, 1973b). Insect catches at night in England during October and November can approach those during August and September (Williams, 1939), and it seems likely that foraging and sperm storage coincide for at least 2 months before hibernation begins in *P. pipistrellus*. There are clear species differences in this respect, since in Canada Thomas *et al.* (1979) observed that *Myotis lucifugus* began mating in late August, 15–20 days before the first torpid females were found in their hibernacula.

Captive *Nyctalus noctula* used by Racey (1973a) in experiments that successfully demonstrated the fertilizing capacity of sperm stored by males for up to 7 months, had been fed at least twice weekly throughout the period of sperm storage, and were frequently active. Activity of free-living bats during winter is also widely reported, although there are intraspecific as well as interspecific differences associated with different strategies evolved by both sexes or different species to cope with energy conservation and water loss during winter. Thus Saint-Girons *et al.* (1969) and Ransome (1971) found that *Rhinolophus ferrume-*

quinum studied at 46°–51° north latitude aroused to feed regularly during winter if the environmental temperature rose above the threshold for insect flight. At 33° north latitude, however, Funakoshi and Uchida (1978) found that *R. ferrumequinum* hibernates deeply, almost regardless of changes in ambient temperature, whereas *R. cornutus* arouses occasionally to feed during winter. In contrast, *Miniopterus schreibersii*, also studied by these workers in Japan at 33° north latitude, arouses repeatedly and in large numbers to forage during winter. Daan (1973) found that in three species of *Myotis* there was a tendency for females to arrive at their cave hibernacula earlier in autumn and to leave earlier in spring than males, but that no more than 10% of the bats left the caves after arousal during winter.

At 36° north latitude in Nevada O'Farrell *et al.* (1967) mist-netted *Pipistrellus hesperus* and *Myotis californicus* throughout winter at air temperatures varying from 1 to 31°C. Of these, 48% of *Myotis* were taken at temperatures of 2–6°C, and 11% of *Pipistrellus* at temperatures of 5°C or below. Further north at 42° north latitude Folk (1940) has also recorded winter activity of *M. lucifugus*, and *Eptesicus fuscus* continues to be active throughout winter in Indiana (Whitaker and Mumford, 1972). In contrast to this, species such as *M. grisescens* and *P. subflavus* do not seem able to eliminate hibernation to take advantage of a nearly continuous food supply at the southern limit of their distribution, and they have been termed obligate hibernators by McNab (1974), who noted that most copulations in these species took place in autumn. In contrast, *M. austroriparius* remains active throughout much of the winter in peninsular Florida, and as a result copulation occurs in spring (Rice, 1957).

In recent years sperm storage has been established in tropical bats. In central India (about 21° north latitude) Gopalakrishna and Madhavan (1971) observed that *Pipistrellus ceylonicus* copulates early in June and ovulates in July. Sperm storage has also been reported for *Scotophilus heathi* at 21–25° north latitude (Gopalakrishna and Madhavan, 1978; Krishna and Dominic, 1978), and in Malaysia, Medway (1972) observed that *Tylonycteris pachypus* and *T. robustula* were inseminated in November and December but did not ovulate until early January. In Paraguay, between 22 and 25° south latitude, Myers (1977) reported that *Lasiurus* (= *Dasypterus*) *ega*, *Eptesicus furinalis*, and *Myotis albescens* copulate in May, yet *M. albescens* does not become pregnant until late July, and females of *Lasiurus* and *Eptesicus* not until August. Sperm storage was established in *Chalinolobus gouldii* in the southwest of Western Australia (28–35° south latitude) (Kitchener, 1975), although the extent to which hibernation occurs at this latitude has yet to be determined, since several individuals have been shot in flight in midwinter. It is also of interest to establish whether *Chalinolobus* stores sperm at more tropical latitudes in Australia, and it seems likely that further research on the reproductive cycles of tropical bats will reveal more species storing sperm. Thus Rasweiler (1977) observed that nonpregnant inseminated female *Noctilio albiventris* were captured with high frequency for over 7 weeks. Although the fertilizing capacity of stored sperm remains to be demonstrated in these tropical species, they provide clear evidence that prolonged torpor is not a prerequisite of prolonged sperm survival.

Synchrony of births may have been one of the evolutionary causes of sperm storage, especially in tropical bats (Racey, 1979). Since copulation is seldom synchronized in a population, a period of sperm storage provides an opportunity to bring all females into the same reproductive state. If ovulation then occurs in response to an environmental change experienced by the whole population, a mechanism exists whereby the time of births can be easily synchronized.

3.5. Pregnancy and Lactation

The restriction of foraging in insectivorous bats during the course of evolution to the hours of twilight and darkness (see Chapter 5) means that no food is consumed by pregnant or lactating bats for periods of 16–20 hr. Such prolonged periods of inanition are rare among mammals, particularly during the most energy-demanding phases of their annual cycles. It is hardly surprising that most of the daily food intake in *Myotis lucifugus* and *M. velifer* occurs in the first 2 hr following emergence (Kunz, 1974b; Anthony and Kunz, 1977).

Kunz (1974b) was first to investigate the effects of pregnancy and lactation on foraging in insectivorous bats, and his work on *Myotis velifer* was extended by Anthony and Kunz (1977) to *M. lucifugus*. Two feeding periods were evident in these bats during summer, with a major period of activity after sunset and a reduced secondary period before sunrise. Between these periods bats may make use of night roosts (see Chapter 1). During lactation most bats return to the nursery roost during the night to suckle their young. Pregnant female *M. velifer* emerged from their roosts at dusk sooner than lactating females and increased their daily food consumption throughout most of their gestation, although in the terminal days of pregnancy food consumption was reduced, owing to either increased wing loading and the consequent difficulties of foraging efficiently as the fetal mass increased or to abdominal crowding. Similar results were obtained for the greater horseshoe bat, *Rhinolophus ferrumequinum*, by Ransome (1973), who considered that a phase of heterothermy during late pregnancy was due to the physical bulk of the developing embryo reducing the amount that could be eaten at a time when feeding opportunities were thought to be reduced anyway owing to decreased night length.

Of the daily food consumption of adult female *Myotis velifer*, 80% occurred during the first 2 hr after emergence. Maximum daily levels of food consumption in adult females, amounting to 25–30% of body mass, was reached near the time of weaning, when the energy demands of lactation were highest (Kunz, 1974b). Pregnant female *M. lucifugus* weighing about 10 g consumed an average of 2.5 g insects (2.72 kJ/g prefeeding body mass) each night, lactating females 3.7 g (4.23 kJ), and juveniles 1.8 g (2.47 kJ). Comparing the composition of the diet with the availability of prey indicated that pregnant bats consumed insects 3–10 mm in length in approximate proportions encountered during June, when insect availability was low and unpredictable. However, lactating, postlactating, and nonreproductive females fed more selectively in July, when insects were more abundant (Anthony and Kunz, 1977).

In contrast to the bimodal foraging pattern established for *Myotis velifer* and *M. lucifugus* throughout the summer, *Pipistrellus pipistrellus* changes its pattern of foraging at the time of parturition (Swift, 1980). During pregnancy in May and June, most bats leave the roost once each night soon after dusk and return between midnight and dawn. After parturition in late June, the activity pattern becomes bimodal, as in North American species, with peaks after dusk and before dawn, and the bats return to the nursery roost during the night to feed their young. The average time spent outside the roost by individual bats varies between 2.5 and 5 hr during the summer but does not increase with the progress of pregnancy and lactation. The activity pattern of night-flying insects, as recorded with a suction trap, is also bimodal, with peaks after dusk and before dawn, corresponding with the maximum number of bats foraging during lactation. Energy demands during pregnancy are clearly not high enough for the bats to remain active throughout the hours of darkness in order to exploit the predawn insect peak. *Pipistrellus pipistrellus* frequently change their roost just before parturition (Swift, 1981), and it would be interesting to establish the extent to which this is due to roost factors (see Chapter 1) or to the availability of food. Roer (1968) described similar behavior in *M. myotis*.

Bradbury and Vehrencamp (1976, 1977) found that reproductive females (presumably those that are pregnant and lactating) fed in a group over rich aggregations of aerial insects, while nonreproductive females (which could have been prepubertal) fed in large solitary beats in lower prey densities.

Although young bats of many species retain strong associations with their mothers well past the onset of flight (see Bradbury, 1977) several species of insectivorous bats of temperate latitudes provide an interesting contrast in that mothers leave the roost soon after weaning (Sluiter and van Heerdt, 1966; Stebbings, 1968; Rakhmatulina, 1972; Kunz, 1974a; Swift, 1980). *Myotis thysanodes* is exceptional in that young males, then young females, leave the maternity roost before adult females (O'Farrell and Studier, 1975). Stebbings (1968) and Kunz (1974b) suggested that the early departure of mothers may serve to reduce competition for available food, a hypothesis that is open to testing by measuring feeding rates and relative insect densities. Rakhmatulina (1972) considered that the emigration of females was better explained by the different metabolic adaptations of mothers and young. Higher environmental temperature in the maternity roost together with clustering assists in the maintenance of high body temperature and an optimal growth rate in juvenile bats, whereas adults, in order to accumulate fat, require a lower temperature, which they find in different roost sites. For *Nyctalus noctula*, Sluiter and van Heerdt (1966) suggested that parous females disperse to find mating roosts as soon as the young are weaned.

3.5.1. Effect of Variations in Temperature and Food Supply on the Gestation Period of Bats

Since homeothermy renders most mammals to a large extent independent of temperature, any effects this variable may have on gestation might manifest themselves in those species that do not maintain a constant body temperature. Such is the case with heterothermic bats.

Eisentraut (1937) first drew attention to the effect of variations of environmental temperature on the rate of fetal development in bats. He observed that pregnant *Myotis myotis*, in spite of a more active metabolism associated with pregnancy, were still capable of torpor, which in a low enough environmental temperature could become a state of hibernation lasting for several days. Animals that he kept singly in rooms at different temperatures with available food contained fetuses of different size when killed: the lower the temperature the smaller the fetus. However, since the time of ovulation in hibernating bats bears no relation to the time of copulation (Racey, 1979), the fetuses Eisentraut was comparing may have been at different stages of development at the beginning of his experiment.

A similar effect of environmental temperature on the length of gestation was suggested by Kolb (1950), who related a 3-week delay, compared with the previous year, in the appearance of the first young of a colony of the lesser horseshoe bat, *Rhinolophus hipposideros*, to the occurrence of a 3-week cold spell in May. Eisentraut (1937) suggested that the rate of fetal development was slowed in response to lowered environmental temperature, whereas Kolb considered that it was stopped. Pearson *et al.* (1952) made use of Eisentraut's results to explain the marked variability of their estimates of the gestation period (56–100 days) of the American long-eared bat, *Plecotus townsendii*, examined in different localities and in different years, but considered that it was the preimplantation stages of development that were extended as a result of maternal torpor.

In an attempt to clarify the effect of variations of temperature on the length of gestation in bats, groups of palpably pregnant *Pipistrellus pipistrellus* were maintained by Racey (1973c) at 5, 10, 20, 25, 30, and 35°C, with food freely available, for periods of 14–17 days. They were then returned to an unheated room where environmental conditions approximated those in a natural roost, and their mean date of birth recorded and compared with that of control groups maintained in the same room since capture during early pregnancy. The mean date of birth of bats maintained at 10, 20, and 25°C was not significantly different from that of their respective controls, and the gestation period in these conditions was 44 days. However, the mean date of birth of bats maintained at 5°C was 5 days later than the control mean, indicating that the rate of fetal development had been significantly retarded. Conversely, the mean date of birth of bats maintained at 30 and 35°C occurred 2.0 and 3.3 days, respectively, before the control mean, indicating that the rate of fetal development had been significantly accelerated. When food was withdrawn and palpably pregnant bats maintained at 5, 10, and 11–14°C for 12–14 days, they became torpid. At the end of the experimental period they were returned to the control environment and their mean date of birth was delayed by a period that did not differ significantly from that of the experimental period, indicating that gestation had been halted for the period of torpor. These laboratory experiments suggest that the rate of fetal growth in wild populations of heterothermic bats may vary according to the external conditions of food supply and temperature.

In order to investigate the extent to which gestation length varied naturally from year to year, Racey and Swift (1981) compared the progress of pregnancy in 2 successive years in a colony of *Pipistrellus pipistrellus* near the northern border of their distribution (57° north latitude) and found that gestation length varied by 10 days when the weather conditions and numbers of flying insects differed appreciably from year to year. The gestation length of pipistrelles established by dissection (Deanesly and Warwick, 1939) and as the interval between the arousal of captive bats from hibernation and parturition (Racey, 1973c) was 44 and 45 days respectively. Racey and Swift (1981) recorded intervals of 41 and 51 days between ovulation and parturition in 2 successive years so that a large percentage variation occurs naturally and supports Racey's (1973c) contention that the concept of a fixed gestation period for heterothermic bats is invalid.

At high latitudes low temperatures experienced by pregnant females may result in a delay of fetal development, especially if the food supply is inadequate (Racey and Swift, 1981). Conversely, the high temperatures frequently recorded in maternity roosts (see Chapter 1) may result in the acceleration of fetal development (Racey, 1973c). However, at tropical latitudes temperatures may reach such high levels as to have a converse effect. Thus Krishna (1978) suggests that embryonic development in *Taphozous longimanus* was delayed in midpregnancy at 25°30′ north latitude when environmental temperature was at a peak presumably because the bats aestivated. Such a delay would account for the gestation period lasting for 4 months in this 5-g bat.

It is of interest to examine other features of pregnancy in bats that may be associated with the ability to halt fetal development. Authors frequently compare the lengths of gestation of different species, a comparison confounded by differences in the mass of females and their term fetuses. If the overall rate of growth of the fetus of a particular species can be established, then meaningful comparisons can be made (Hugget and Widdas, 1951). The specific fetal growth

TABLE II
Comparison of Fetal Growth Rates in Bats with Other Mammals[a]

Order	Fetal growth rates
Chiroptera	0.01–0.04
Primates	0.01–0.09
Rodentia	0.04–0.28
Artiodactyla	0.04–0.18
Insectivora	0.06–0.11
Carnivora	0.04–0.21
Perissodactyla	0.06–0.16
Lagomorpha	0.12–0.20

[a]After Frazer and Hugget, 1974.

velocity (Hugget and Widdas, 1951) has been calculated for several species of bats and varies from 0.01 to 0.04 (Racey, 1973c). When these values are compared with those calculated by Frazer and Hugget (1974) for other mammalian orders (Table II), it appears that bats have the slowest rates of fetal growth among mammals. A slow rate of fetal growth is also characteristic of primates, and this was correlated by Payne and Wheeler (1967) with the fact that they produce smaller offspring than nonprimates for the same length of gestation. Comparison of data from *Pipistrellus pipistrellus* with the allometric equation of Leitch *et al.* (1959) relating maternal and neonatal mass indicates that young *Pipistrellus* are also proportionally smaller at birth than those of many other mammals (Racey, 1972). Payne and Wheeler (1967) demonstrated that the stress imposed by pregnancy on a primate mother is less than for a nonprimate of similar mass. Smith (1956) reached a similar conclusion for bats as a result of metabolic studies on *Myotis lucifugus*. Although the metabolic rate increased fivefold during pregnancy, it was always less than the expected metabolic rate for a nonpregnant mammal of the same mass. The bat fetus is thus produced at a very low metabolic cost that is spread over a relatively long period. Kallen and Wimsatt (1962) found that in *M. lucifugus* the circulating blood volume during pregnancy increased by only 17% (as compared with 40% in pregnant humans). These authors also commented on the surprisingly low metabolism of this bat when pregnant.

3.5.2. Thermoregulation and Energy Budgets during Pregnancy and Lactation

Insectivorous bats are unique in the ease with which they can reduce their body temperature (T_b) in the roost in order to save the energy otherwise required to maintain a homeothermic state (Bartholomew, 1972). The extent to which they make use of this capacity during pregnancy and lactation has proved to be difficult to establish, but it is vital to an evaluation of energy budgets (Kunz, 1980; Studier and O'Farrell, 1980).

Studier and O'Farrell (1972) examined the thermoregulatory ability of *Myotis thysanodes* and *M. lucifugus* maintained in the laboratory during pregnancy and lactation and based their generalizations on the response of the majority of individuals tested. They found that the proportion of regulators during pregnancy is significantly greater in midpregnancy than in early or late pregnancy, thus explaining the later result of Studier *et al.* (1973), derived from bomb calorimetry, that energy demand during pregnancy is hyperbolic.

Ransome (1973) came to conclusions similar to those of Studier and O'Farrell (1972) and suggested two phases of heterothermy during pregnancy in the greater horseshoe bat (*Rhinolophus ferrumequinum*): the first occurs in April and May and was considered to be due to limited insect captures during flight resulting from low environmental temperature; the second occurs during June and July when the bulk of the developing embryo reduces the amount that can be eaten and when feeding opportunities are thought to be limited due to short night length. Ransome (1973) considered that horseshoe bats are homeothermic during lactation, in contrast to *Myotis lucifugus* and *M. thysanodes*.

Studier and O'Farrell (1972) considered that females carrying large embryos may be at a selective disadvantage energetically if they regulated at low (16°C) or even moderate (16–20 or 24°C) ambient temperatures (T_a) and stated that "because above a T_a of 20°C for *Myotis lucifugus* and 24°C for *M. thysanodes* there is no measurable means of segregating regulators from conformers, based on differences in metabolism or $T_b - T_a$ differential, there would appear to be no selective advantage to regulation at this time and a selective disadvantage energetically." But the findings of Racey (1973c) suggest that the higher the T_b the faster the rate of fetal development, and this may be a selective advantage if it results in earlier parturition, so that young can achieve adult size before insect numbers decrease. Furthermore, Studier and O'Farrell (1972) ignore the fact that the rate of fetal development of conformers may be retarded compared with that of regulators (Humphrey, 1975).

Studier *et al.* (1973) consider that the loss of homeothermy occurs as an immediate prerequisite to the rapid fetal growth of late pregnancy. This increased energy demand occurs earlier during gestation in *Myotis thysanodes* than in *M. lucifugus*, and such differences between congeners is one of the most significant features to emerge from the work of Studier and his colleagues. Thus the energy invested in late embryo growth in *M. lucifugus* (0.54 kJ/day) is significantly greater than in *M. thysanodes* (0.33 kJ/day). Studier and O'Farrell (1972) found that *M. lucifugus* and, to a lesser extent, *M. thysanodes* regulate T_b during lactation if the energy required to do so is not excessive. Those lactating bats that also regulate, maintain their T_b significantly lower than pregnant or postlactating regulators, and in *M. lucifugus* T_b after lactation is significantly lower than during pregnancy. However, overall, lactation in *M. thysanodes* requires about four times the daily energy investment of pregnancy, a factor similar to that recorded in small homeothermic mammals (Migula, 1969). Because of the increased energy invested in lactation, Studier *et al.* (1973) consider that the regulation of body temperature during this period would be even more harmful than during late pregnancy, and the mother's ability to remain heterothermic postpartum is of even more value as an energy-saving device than during late pregnancy. This argument insofar as it relates to lactation is convincing but has yet to be proven for pregnancy. The fact that Racey (1973c) showed that fetal growth in *Pipistrellus pipistrellus* is accelerated at high T_a's suggests that this process may be slowed in conformers, and it has yet to be established that fetal growth and milk production can occur as efficiently in conformers as in regulators.

Studier and O'Farrell (1980) tabulate the energy demand during various stages of reproduction in *Myotis lucifugus* and *M. thysanodes* and show that it remains the same or decreases from late pregnancy to lactation, respectively. Although these authors admit that their estimates may be minimal, such a pattern of energy demand would be unique among mammals (Migula, 1969; Randolph *et al.*, 1977) and clearly depends on the bat's ability to become heterothermic or conform during lactation. The strongest evidence behind Studier and O'Farrell's (1980) view of the energy costs of reproduction is their finding (1972, 1976) that oxygen consumption of nonregulating and regulating *M. lucifugus* and *M. thysanodes* becomes indistinguishable at T_a's exceeding 20 and 24°C, respectively, and above this level the metabolic energy demand of regulators does not differ significantly from that of nonregulators, so that there should be no selective energetic advantage to differences in thermoregulatory performance.

However, there is little evidence from field studies (Kunz, 1973; Tuttle, 1975, 1976) that temperate-zone bats conform during the last trimester of pregnancy or during lactation, unless they are experiencing extreme conditions, either very high (thermoneutral) temperatures or very low temperatures, which result in a shortage of food. Furthermore, from data accumulated during a field study Anthony and Kunz (1977) estimated that the energy utilized by *Myotis lucifugus* rose from 2.39 kJ g^{-1} day^{-1} during pregnancy to 3.72 kJ g^{-1} day^{-1} during lactation.

In estimating energy demand, Studier and O'Farrell (1980) assume a nightly flight time of 2 hr and consider that to reach the energy intake of 32 kJ/day for lactating *Myotis lucifugus* suggested by Anthony and Kunz (1977), females would have to fly for 6 hr each night, which they consider to be highly improbable. However, Swift (1980), working with *Pipistrellus pipistrellus* at the northern border of its distribution, where night length is short, found that the average time spent by bats outside the roost varied from 2.5 to 5 hr during the summer (and there was no evidence that they were using night roosts).

The ability of bats to maintain higher T_b's during pregnancy was investigated directly by Racey and Swift (1981), using rectal thermistors and recording for 24-hr periods. Pregnant *Pipistrellus pipistrellus* that were fed freely maintained their T_b's within a few degrees of the maximum recorded for each individual, with a mean rectal temperature of 35°C. The temperature of bats that were deprived of food for 24 hr before the start of the experiment fell to 24–25.5°C, with T_a throughout the experiments varying from 22.5 to 23°C. This emphasizes the importance of an adequate food supply in maintaining a high T_b. Similar observations have been made for Neotropical bats by Arata (1972) and Rasweiler (1973).

That pregnancy hormones may be involved in the maintenance of higher T_b's is suggested by a single nonpregnant *Pipistrellus pipistrellus* maintained by Racey and Swift (1981) with free access to food that maintained a lower T_b than did pregnant bats. Racey and Swift also found that plasma progesterone levels and corpus luteum volumes increased during pregnancy to reach a maximum about 6 days before the first births. Similar patterns of plasma progesterone concentration were demonstrated for *Tadarida brasiliensis mexicana* and *Antrozous pallidus* by Jerrett (1979) and Oxberry (1979), respectively. The metabolites of progesterone, particularly etiocholanolone, are pyrogenic (Kappas *et al.*, 1960), and Racey and Swift (1981) suggested that high progesterone levels may be associated with an ability to maintain a high T_b, which is in turn necessary to maintain fetal growth.

3.5.3. Delayed Implantation

Mutere (1967) established that *Eidolon helvum* exhibits a monestrous, seasonal, and synchronized breeding cycle and in Kampala, Uganda (0°20′ north latitude), mates in April to June, although implanted embryos are not evident until October and November. The time of implantation of the embryos coincides with the October–November rainfall peak, which Mutere suggested may stimulate the process, since photoperiod varies little at this latitude. Births take place in February and March, just before the onset of the higher of the two rainfall peaks, which is maintained for 3 months and results in an increased availability of fruits for lactating females and their young.

The occurrence of delayed implantation in *Eidolon helvum* was confirmed by Fayenuwo and Halstead (1974) at Ile Ife, Nigeria (7°24′ north latitude), but implantation coincided exactly with the start of the dry season, and birth with the beginning of the wet season. Fayenuwo and Halstead (1974) suggested that the sudden change to the hot dry season may provide the environmental stimulus to implantation. Since the time of mating at Ile Ife was not recorded, it is not possible to determine whether the overall length of gestation is the same in both localities. The length of delay appears to be 3 months at Kampala.

Delayed implantation has been recorded in *Miniopterus schreibersii* and *M.*

fraterculus, and the lengthening of the delay period with increasing latitude (Table III) has been reviewed by Dwyer (1963), Wimsatt (1969), and Richardson (1977). Two questions emerge from the variations in timing of reproductive events in *Miniopterus*. The first is the stimulus effecting implantation of the blastocyst. At both northern and southern latitudes implantation takes place when day length is 10–12 hr and increasing, suggesting that the winter solstice marks the onset of the environmental stimulus for implantation (Racey, 1981). Experimental work on captive *Miniopterus* in controlled lighting is required to test this hypothesis. In all cases young are born in midsummer, at a time favorable to the lactating females.

The second question is whether the variations in lengths of gestation in *Miniopterus* is a facultative response to local environmental conditions at different latitudes or an adaptation that has become incorporated into the genome. If the latter is the case and populations of *M. schreibersii* at the limits of their range require an obligate delay in implantation that is also fixed in length, then they should be accorded specific status. However, the fact that *M. schreibersii* gave birth in October in one study at Naracoorte, South Australia (Hamilton-Smith, 1972), and in December in another study (Richardson, 1977) suggests a facultative variation in the time of mating or the length of gestation. Attention has also been drawn to histological differences between northern and southern populations of *Miniopterus*. At 45° north latitude the corpus luteum of *M. schreibersii* does not appear to be active during the coincident periods of hibernation and delayed implantation (Peyre and Herlant, 1967). At 30° south latitude implantation occurs during hibernation, and the corpus luteum is well developed throughout the period of delay, although Wallace's (1978) data suggest that it may decrease in size slightly during hibernation. These observations suggested to both Wallace (1978) and Richardson (1977) that differences exist in the endocrine control of implantation in *Miniopterus* at different latitudes. Although histological criteria provide further evidence that hibernation inhibits the secretory activity of endocrine glands (Wimsatt, 1969), such criteria have frequently confounded our understanding of the endocrinology of bats and other mammals (Racey and Tam, 1974). Data on variations in plasma progesterone throughout gestation, as obtained by radioimmunoassay in other vespertilionids (Burns and Easley, 1977; Jerret, 1979; Oxberry, 1979; Racey and Swift, 1981), are needed to resolve this question in *Miniopterus*.

As a result of observing unattached uterine trophoblasts, Menzies (1973) claimed that delayed implantation lasting from 1–2 months occurred in *Hipposideros caffer* and *Rhinolophus landeri* in Nigeria. However, two of the four blastocysts of *H. caffer* illustrated by Menzies were located in implantation cavities, and only part of the trophoblast was unattached in the other two, presumably as a result of shrinkage of the endometrium, which commonly occurs after fixation in Bouin's fluid (Racey and Potts, 1971).

TABLE III
Variations in the Duration of Delayed Implantation with Latitude in *Miniopterus*

Species	Latitude	Duration of gestation (months)	Duration of delay (months)	Time of implantation	Day length (hr)	References
M. schreibersii	25°50′S	8	4	July–August	10–11	van der Merwe (1979)
M. schreibersii	28–31°S	7	2½	August	11–12	Richardson (1977)
M. schreibersii	30°S	7	2½	August	11–12	Wallace (1977)
M. fraterculus	30°S	5	2½	August	11–12	Bernard (1980)
M. schreibersii	45°N	10	5	February	11–12	Courrier (1927)
M. schreibersii	45°N	10	5	February	11–12	Peyre and Herlant (1967)

3.5.4. Delayed Development

Delayed development among bats was first reported in the phyllostomid *Macrotus californicus* by Bradshaw (1961, 1962), who suggested that reduction in body temperature slows the rate of cell division and hence retards embryonic growth during the "winter" period from October to March. Although *Macrotus* leaves its roosts in warm (27°C) abandoned mine tunnels to forage during mild winters, no quantitative data are available on the seasonal availability of food. Burns *et al.* (1972) investigated whether shortage of food led to reduced body temperature during the period of delay, by maintaining pregnant females in environmental temperatures of 24°C and a 14-hr photoperiod, with food freely available. Although the bats gained weight, the gestation period was not altered as a result, demonstrating that the period of delay is obligate.

Bleier (1975) confirmed delayed development in *Macrotus* and showed that for 4½ months of the 8-month gestation only very slow growth of the embryo occurs. He also observed bats feeding each night during winter and accumulating considerable quantities of fat. However, Burns *et al.* (1972) found that plasma thyroxine levels fell during early and midpregnancy, suggesting that body temperature might also fall. Bradshaw recorded rectal temperatures as low as 25.7°C, and the regular occurrence of adaptive hypothermia in *Macrotus* would explain how food eaten during winter results in the accumulation of fat. The period of delay is also associated with reduced progesterone (Burns and Easley, 1977) and estrogen levels (Burns and Wallace, 1975), suggesting that the corpus luteum acts as a clock for the timing of the events during pregnancy.

Delayed embryonic development occurs in one of the two reproductive cycles of *Artibeus jamaicensis* (Fleming, 1971). In Panama mating occurs first at the end of the dry season, and births occurring 4 months later coincide with a peak in the abundance of small fruits. Blastocysts resulting from conceptions at postpartum estrus implant but then enter diapause for 3 months before developing normally for 4 months. They are born at the end of the dry season just prior to a peak in the availability of large fruits. If diapause did not occur, these young would be born when the availability of fruits is lowest.

Delayed development also occurs in *Miniopterus australis* at 3°50′ north latitude (Medway, 1971). This 5–6 g bat is monestrous, mates in November and December, and gives birth 4½–5 months later. Medway considered that the prolonged gestation period is an adaptation to a perenially low and variable food resource, with gestation and birth coinciding with the period of limited seasonal increase in the abundance of insect prey. Although there are no quantitative data on food availability, it is significant that the peak nesting season of terrestrial birds (Gibson-Hill, 1952), many of which rear their young on an insect diet, coincides with gestation and birth in *M. australis*. At 28–31° south latitude Richardson (1977) also recorded a gestation period of 4½ months for *M. aus-*

tralis, but at an intermediate latitude of 15°15′ south Baker and Bird (1936) report a gestation period of 110 days (although the data these authors present suggest a slightly longer gestation of 120 days).

Ramakrishna and Rao (1977) investigated the reproductive cycles of two populations of *Rhinolophus rouxi*. In specimens collected at Bangalore (12°58′ north latitude), tubal ova were observed on November 22, early uterine morulae on December 6–8, unimplanted uterine blastocysts on December 12–14, and early implantating blastocysts on December 21. At Khandala (18°52′ north latitude), however, ovulation and fertilization occurred during the first week of January. Late uterine morulae were found in bats collected between January 7 and 23, unimplantated blastocysts between February 1 and 12, and early implanting blastocysts on February 23. Although it is not clear what techniques were used to establish the stage of pregnancy, differences appear to have been established in the duration of the earlier stages of pregnancy. Ramakrishna and Rao (1977) concluded that embryos remained free in the uterus of *R. rouxi* for 40–45 days at Khandala, and although implantation took place sooner after fertilization at Bangalore, subsequent development was retarded. In both populations the young are born just before the monsoon, which is 4 weeks earlier at Bangalore than at Khandala, and since the gestation period is the same in both localities (150 ± 8 days), it is clear that the postimplantation rate of fetal growth varies.

Krishna (1978) and Krishna and Dominic (in press) showed that *Taphozous longimanus* breeds twice annually in quick succession at Varanasi (25°30′ north latitude). The first pregnancy of the year commences between mid-January and early February. Parturition begins in the last week of April, but pregnant females near term were collected up to the middle of May. The length of the first pregnancy was estimated as 105 ± 5 days and the length of the second, resulting from a postpartum estrus and terminating in August, is 86 ± 5 days. A similar pattern was established by Krishna (1978) in *Cynopterus sphinx*, which also breeds twice annually in quick succession. The first pregnancy is initiated in mid- or late October, terminates in March, and has a duration of 150 ± 5 days. The second pregnancy of the year begins in early April, terminates in July or early August, and lasts for 121 ± 5 days. Krishna suggested that retarded embryonic development at particular stages of one pregnancy was responsible for the intraspecific difference in gestation lengths, but it is possible that different rates of fetal growth throughout pregnancy are responsible.

The first pregnancy of the year in *Eptesicus furinalis* and *Myotis albescens* studied by Myers (1977) in Paraguay at 22–25° south latitude occurred during the cooler winter months and was thought to last longer than subsequent pregnancies occurring during the summer, and there is some evidence that a period of retarded embryonic development also occurs in the funnel-eared bat, *Natalus stramineus* (Mitchell, 1965; Wimsatt, 1975).

3.5.5. Lactation

It is suggested by Ben Shaul's (1962) review of milk composition in wild mammals that mammals that nurse their young on a scheduled basis produce milk of a higher energy content than those that nurse continuously or on demand, and Jenness and Studier (1976), in drawing attention to the high calculated energy content of bat milk, imply that bats also nurse their young on a scheduled basis. Apart from observations that lactating bats return to the nursery roost during the night to suckle their young (O'Farrell and Studier, 1973; Anthony and Kunz, 1977; Swift, 1980), evidence for scheduled suckling is lacking. The high energy content of bat milk also agrees with Blaxter's (1961) prediction that small mammals would produce relatively more milk energy than would large mammals. Blaxter (1961, 1971) and Hanwell and Peaker (1977) point out that milk yield, energy output in milk, and mammary weight are scaled similarly to metabolic rate, and that small mammals face a relatively short period of intense metabolic demand, while large mammals face a longer period of lesser demand. It is not surprising, therefore, to find that lactation in insectivorous bats is of relatively short duration, most species being weaned between 4 and 8 weeks of age (Kleiman, 1969; Bogan, 1972; Kunz, 1973; O'Farrell and Studier, 1973). In Neotropical frugivorous phyllostomids the estimated duration of lactation is 2–4 months (Fleming *et al.*, 1972; Jenness and Studier, 1976). Pteropodids take longer, from 15 to 20 weeks (see Bradbury, 1977; Chapter 3). Brosset (1969) reports that in *Hipposideros commersoni (=gigas)* studied in Gabon, lactation lasts for 5 months, and in Kenya, McWilliam (1982) found that weaning occurred in this species at 16 weeks, with one young individual voiding fecal pellets containing insect remains after 12 weeks. In captive *Desmodus rotundus*, Schmidt and Manske (1973) found that lactation extended between 3 and 9 months.

The relationship between the energy content of different diets and the energy demands of lactation have been little investigated and would clearly be a profitable line of inquiry, particularly in some tropical bats that switch seasonally between diets of fruit and pollen and nectar (Heithaus *et al.*, 1975), nursing one young while feeding primarily on fruit and the next while on a diet consisting mostly of nectar and pollen (Humphrey and Bonnaccorso, 1979).

Jenness and Studier (1976), using data from Schmidt and Manske (1973) and McNab (1973), estimate the energy required by neonatal and adult vampires at different T_a's and emphasize the extent to which the energy demand on the mother is dependent on whether or not the neonates regulate their T_b. They conclude that vampires nursing homeothermic young would need to feed more than once a night.

In general, female bats depart from their maternity roosts at dusk to feed but return during the night to suckle young (see Chapter 1). Young are left by their mothers in the maternity roost, where they cluster to varying extents. In *Myotis thysanodes*, O'Farrell and Studier (1975) observed that several guardian females remain with the young bats during the night, suckling and retrieving, but in most species mothers recognize, retrieve, and suckle their own young (see Bradbury, 1977; Brown, 1976; O'Farrell and Studier, 1973).

In species forming very large maternity colonies, such as *Tadarida brasiliensis mexicana*, where a nursery colony may comprise many millions of bats, the recognition of young presents greater problems, and female *Tadarida* nurse young without apparent reference to family bonds (Davis *et al.*, 1962). Communal suckling also appears to be the case in colonies of up to 100,000 *Miniopterus schreibersii* studied by Brosset (1962, 1966) in India, who reports that females alight on a mass of young and suckle the first two to seize their nipples. Suckling in these colonies may be even more indiscriminate, since Brosset also records adult females with milk in their stomachs. Strong roost fidelity (see Chapter 1) may result in a particular area of the roost consisting of related bats, so this behavior may be justified in terms of kin selection. Studies on mice have shown that communally suckled young nursed longer and grew faster than those suckled by single mothers (Sayler and Salmon, 1971). Similar investigations are needed in laboratory colonies of *Tadarida* and *Miniopterus*.

Schmidt and Manske (1973) observed that after about 3 months of age, captive vampires regurgitated small amounts of blood that were eaten by the young. Häussler (personal communication) has observed that captive *Molossus ater* carry food to their young, and A. N. McWilliam (personal communication) notes that lactating *Coleura* return to the maternity roost with cheek pouches and stomachs full of masticated insects. These observations raise the possibility that feeding nonflying young with adult food is more widespread in bats than has formerly been suspected. Such behavior would significantly decrease the energy requirements of the lactating female.

3.6. Environmental Factors Affecting the Growth and Survival of Young

Since most species of temperate-zone bats hibernate to avoid the shortage of insect food during winter, the processes of fat storage, mating, gestation, lactation, growth and development of young, and, for some species, migration have all to be accomplished in approximately half a year. Humphrey (1975) considers that bats of temperate latitudes have met this challenge by colonizing maternity roosts that have a high survival value, by having few young, and by bearing them at a relatively advanced state of development. These adaptations have generally resulted in an exceptionally low rate of preweaning mortality (see Chapter 3). Herreid (1967), for example, estimated such mortality to be 1.3% in *Tadarida brasiliensis mexicana*, equivalent to an annual survival of 0.92, and O'Farrell and Studier (1973) estimated neonatal mortality as 1% for *Myotis thysanodes* and 2% for *M. lucifugus*. Christian's (1956) estimate of preweaning mortality in *Eptesicus fuscus* was considerably higher, at 7%, and similar to that of Humphrey *et al.* (1977) in *M. sodalis*.

Foster *et al.* (1978) suggest that the habit of *Myotis austroriparius* of roosting in caves over water serves to exclude predators and maintain this species' requirement for high humidity. Such a roosting habit also contributes to the high preweaning mortality in this species, since it precludes the retrieval of fallen young, which drown. Of the young, 12% die before weaning, mostly during the first week of life (Foster *et al.*, 1978). Congeners such as *M. keenii, M. lucifugus*, and *M. thysanodes*, which typically roost in drier places during summer, will retrieve fallen young (Brandon, 1961; O'Farrell and Studier, 1973). Because *M. austroriparius* enters hibernation only for brief periods and is active for a greater part of the year than are many North American species, Foster *et al.* (1978) speculate that it is at greater risk from predators and accidents than less active species. This risk factor together with a high preweaning mortality may have resulted in selection for twinning, since *M. austroriparius* is the only Nearctic member of its genus that normally has twins.

Humphrey (1975) considers that colonial roosting confers physiological advantages, mainly thermal, ensuring optimum growth of embryos and young, and a high degree of homeothermy has also been demonstrated in the cave-dwelling microchiropterans *Tadarida brasiliensis, Miniopterus schreibersii*, and *Myotis grisescens* by Herreid and Schmidt-Neilsen (1966), Dwyer (1964), and Tuttle (1975), respectively. Humphrey (1975) does not consider the availability of food in the vicinity of the roost to be a factor limiting the growth of young, except at higher latitudes, although both Stebbings (1968) and Kunz (1974b) suggested that local competition for food explained why postlactating female *Pipistrellus pipistrellus* and *Myotis velifer* left their maternity roosts as soon as the young were weaned. Emigration of females before young was confirmed for *Myotis lucifugus* by O'Farrell and Studier (1975), who found that the situation was reversed for *Myotis thysanodes*, with young males and then young females leaving before adult females.

Although there have been numerous studies on the growth of young bats, few have considered ecological factors (see Chapter 3). Unusually cool summer weather resulted in a slower growth of young *Myotis sodalis* roosting in trees in eastern Indiana (Humphrey *et al.*, 1977). This resulted in young bats being exposed to freezing weather in the maternity roost, delayed the first flights by 2 weeks and migration by 3 weeks, and may have affected fat storage and subsequent survival during hibernation.

Tuttle (1975) considers that cave-dwelling bats of the temperate zone thermoregulate during lactation, despite the energetic cost, as a condition of the normal growth of young in cold environments. He has also shown that through clustering behavior and the selection of roost sites bats can markedly alter the microclimate to which young are exposed, and as a result growth rates of young *Myotis grisescens* can vary by 50% from colony to colony. Growth rates increase with colony size, which in turn determines roost temperature. Tuttle (1975) also found that growth and survival are significantly affected by the distance that newly volant *M. grisescens* travel to a foraging area (Tuttle, 1976). Pagels and Jones (1974) also consider that the comparatively rapid increase in body weight of young *Tadarida brasiliensis cynocephala* is due to the energy conserved as a result of clustering.

In a ten-year study conducted in the German Eifel, Roer (1973) found a much higher mortality rate among the young of the mouse-eared bat *Myotis myotis* when the weather in the central European area was dominated by repeated cold fronts. For example, in 1961 preweaning mortality amounted to 43%, and in 1971 to 29%; in contrast, only a 2% mortality was observed in the warm summer of 1964. As a result of the cold weather, the temperature in the maternity roosts fell to 8–10°C, causing heavily pregnant females to abort and lactating females to desert their young in search of more sheltered maternity roosts.

3.7. Puberty and Subsequent Fertility and Fecundity

The ages at which bats attain puberty have been tabulated by Tuttle and Stevenson (Chapter 3). Although considerable inter- and intraspecific variations are apparent in the age of sexual maturity of the vespertilionid and rhinolophid bats, Racey (1974b) suggested that some females of all vespertilionid species for which information is available attain sexual maturity in their first autumn. Spermatogenesis also takes place during the first uninterrupted 6-month period of activity after birth. This contrasts with the situation in rhinolophids, in which the prepubertal period is extended until the second or third autumn, as in the greater horseshoe bat in Liguria (Dinale, 1964). The most extreme example of an extended prepubertal period occurs in *Rhinolophus ferrumequinum* at the northern border of its distribution in Dorset, United Kingdom (about 51° north latitude), where the age of females at the birth of their first young varies from 3 to 7 years, with a mean age of 5 years (R. E. Stebbings, personal communication). Dinale (1968) considers that the difference in the time of onset of sexual maturity among different species of rhinolophids is correlated with body size according to the formula $a = b/k^2$, where a is the age of sexual maturity of the smaller species, b the age of the larger one, and k the ratio between the condylobasal length of the skull in the two species. Dinale (1968) also commented on the extent of variation between individuals, and it seems likely that a similar relationship between body size and sexual maturity is obtained within species. Schowalter and Gunson (1979) confirmed variations in the times of puberty in the vespertilionid *Eptesicus fuscus* by observing that yearling females have a lower pregnancy rate than do older females.

The effect of nutrition on the onset of puberty in mammals has been reviewed by Sadlier (1969), and Frisch (1980) has postulated that a particular ratio of fat to lean mass is normally necessary for puberty and the maintenance of female reproductive activity in the human and the rat. Jones and Ward (1976), on the other hand, propose that in the weaverbird *Quelea quelea* breeding is controlled by the state of protein reserves. Although the control mechanism may thus vary among vertebrates, it seems likely that the intrasexual variation in the age of puberty in bats is due to nutritional factors, possibly resulting from the late birth of young and their failure to achieve threshold body weight in their first autumn.

Once puberty is achieved, reproductive rates among vespertilionids frequently approach 100% (Humphrey, 1975; Humphrey and Cope, 1970). In his study on *Pipistrellus pipistrellus*, Rakhmatulina (1972) recorded 100% pregnant females, although Deanesly and Warwick (1939) found that 11% ($n = 164$), and Hurka (1966) recorded that 28% ($n = 37$), of the females of this species were not pregnant during the summers of study. In *Rhinolophus ferrumequinum* at about 51° north latitude some females produce young only in alternate years until the age of 9 or 10, when they appear to become reproductively senescent (R. E. Stebbings, personal communication), and Gaisler (1966) reported that 7% of the adult female *R. hipposideros* examined during this study did not reproduce. It seems likely that such impaired reproductive performance may be due to poor nutritional status, which was also thought to cause pregnancy failure in the tropical emballonurid *Saccopteryx bilineata* (Bradbury and Vehrencamp, 1976, 1977).

Among mammals in general, litter size is the dominant factor determining the energy requirement of breeding females (Millar, 1977). This may explain why a small litter size is such a remarkably constant feature of chiropteran reproduction. Of the 39 species of bats in the United States, 26 are monotocous, and Myers (1978) considers this to be an adaptation for flight. Twins may occur very occasionally in such species, and Schowalter *et al.* (1979) found two sets of twins in 312 pregnancies in *Myotis lucifugus*. In only a few species twins or triplets are the rule (Kunz, 1971; Foster *et al.*, 1978; Myers, 1978), and in these, reproductive losses appear to be inversely related to litter size (Foster *et al.*, 1978). In species where litter size varies, twins are generally smaller at birth than singletons (Wimsatt, 1945; Foster *et al.*, 1978).

The twinning rate varies with maternal age and geographical distribution. Yearling female *Antrozous pallidus* have one young, whereas older bats have twins (Davis, 1969). Populations of *Eptesicus fuscus* in western North America have one young, and eastern populations two (Cockrum, 1955; Brenner, 1968; Kunz, 1974a), and in *Lasiurus borealis* smaller litter sizes are recorded in Iowa than slightly further south, in Kansas (Kunz, 1971).

Twin births occur more commonly in *Pipistrellus pipistrellus* in Scotland (Deanesly and Warwick, 1939; Swift, 1981) than in England (Racey, 1972), and a single instance of triplets has been recorded in Russia (Rakhmatulina, 1972).

Yoshiyuki *et al.* (1970) have also shown variation in the number of embryos *P. javanicus* from two to four. Twins occur frequently in noctules *Nyctalus noctula* in continental Europe (Bels, 1952; Panyutin, 1963), and triplets have also been recorded (Ryberg, 1947). In the United Kingdom there is only a single record of a wild noctule apparently nursing twins (Jenyns, 1846), whereas in captivity four sets of twins have been recorded (Kleiman and Racey, 1969; Racey, 1972). It may be that the rapid increase in the quantity of food provided for captive noctules in spring (Kleiman and Racey, 1969) increased the ovulation rate, as occurs in "flushed" sheep (see Sadlier, 1969). However, critical studies of the effect of food supply on the ovulation rate and litter size are lacking, although Humphrey (1975) revealed a highly significant inverse correlation between litter size and the size of maternity colonies of Nearctic bats and suggested that this may be related to demands on the available food supply. Similarly, Mills *et al.* (1975) found that the number of young reared by female *Eptesicus fuscus* was significantly reduced at high population densities.

Little data is available on factors affecting the relationship between ovulation rate and litter size in bats. In *Eptesicus fuscus* and *Pipistrellus subflavus* two to seven ova are shed and most implant, although resorption reduces the number of developing embryos so that singletons or twin births result (Wimsatt, 1945). Kunz's (1974a) estimate for prenatal mortality in *E. fuscus* was 34%. Similarly two to nine ova are shed from each ovary in *Chalinolobus gouldii*, although the litter size in this species is two (Kitchener, 1975). Whether such prenatal losses are always a reflection of uterine capacity or factors in the external environment remains to be established. However, in *Myotis lucifugus*, where the ovulation rate normally corresponds with litter size (Wimsatt, 1945), Schowalter *et al.* (1979) found that three out of eight nulliparous females caught during adverse weather conditions were resorbing embryos.

In *Eptesicus furinalis*, which breeds twice a year in Paraguay, the ovulation rate in the first breeding period varied from three to five, although resorption reduced the number of embryos to two at birth. In the second breeding period only one egg was ovulated and no resorptions were recorded (Myers, 1977). The first pregnancy coincides with periods of cold weather and a presumed reduction in the availability of food, so that the associated high ovulation rate and litter size would not appear to be the result of improved maternal nutrition. Nevertheless, such a disparity in fertility between successive pregnancies provides an ideal opportunity for identifying the factors responsible. Thus Rakhmatulina (1972) reported a marked reduction in the twinning rate in *Pipistrellus pipistrellus* in successive years, associated with a cold and protracted spring during the second year of study.

4. SUMMARY

The timing of birth varies with latitude in several species of bats, although this effect is thought to operate through variations in the ambient temperature and food supply rather than through changes in photoperiod, particularly since there is no clear experimental evidence that photoperiod is involved in the timing of gametogenesis. The only suggested effect of photoperiod on reproduction in bats is in *Miniopterus*, which delays implantation to varying extents over a wide range of latitudes but where this event coincides with a similar and increasing photoperiod.

Although few studies have measured climatic variables and food supply at the same time as investigating bat reproductive cycles, an adequate food supply during lactation and weaning appears to be the single most important evolutionary factor in the timing of such cycles. Synchrony between food abundance and lactation is achieved by delays of fertilization, implantation, or development. The dependence on an adequate food supply during lactation makes rainfall the most important climatic factor affecting bat reproductive cycles in the tropics, because of its effect on the phenology of insects and plants. In temperate latitudes, where rainfall is less seasonal, ambient temperature is more important in affecting reproductive phenology and the availability of food. Increased temperature and food supply will thus induce arousal from hibernation and premature ovulation and spermatogenesis in hibernating bats.

The widespread occurrence of foraging activity during winter indicates that retention of the fertilizing capacity of sperm stored by male and female hibernating bats does not depend on low body temperature, and this is further supported by the discovery of sperm storage in several species of tropical bats.

The rate of fetal development is accelerated experimentally if heterothermic bats are fed freely in a hot environment, and slowed in a cold one. When deprived of food in a cold environment, bats become torpid and fetal development is halted for the duration of torpor. Similar variations in gestation length in response to adverse conditions has also been shown in the natural environment. The heterothermic ability of bats complicates the estimation of energy budgets during pregnancy and lactation.

ACKNOWLEDGMENTS. This review was prepared during the tenure of Natural Environmental Research Council grant GR3/3381. I am grateful to my colleagues A. G. Marshall, R. E. Stebbings, S. M. Swift, A. N. McWilliam, M. van der Merwe, and D. W. Thomas, who commented on an earlier version, and particularly to T. H. Kunz, who made substantial improvements, and Miss Susan Johnston, who typed the manuscript.

5. REFERENCES

Anand Kumar, T. C. 1965. Reproduction in the rat-tailed bat *Rhinopoma kinneari*. J. Zool., **147**:147–155.

Anciaux de Faveaux, M. 1973. Essai de synthèse sur la reproduction de Chiroptères d'Afrique (Region faunistique Ethiopienne). Period. Biol., **75**:195–199.

Anciaux de Faveaux, M. 1977. Definition de l'équateur biologique en fonction de la reproduction de Chiroptères d'Afrique centrale. Ann. Soc. R. Zool. Belg., **107**:79–89.

Anciaux de Faveaux, M. 1978. Les cycles annuels de reproduction chez les Chiroptères cavernicoles du Shaba (S-E Zaïre) et du Rwanda. Mammalia, **42**:453–490.

Anthony, E. L. P., and T. H. Kunz. 1977. Feeding strategies of the little brown bat, *Myotis lucifugus*, in southern New Hampshire. Ecology, **58**:775–786.

Arata, A. A. 1972. Thermoregulation in Colombian *Artibeus lituratus* (Chiroptera). Mammalia, **36**:86–92.

Baker, J. R. 1938. Evolution of breeding seasons. Pp. 161–177, *in* Evolution. Essays on aspects of evolutionary biology. Presented to Professor E. S. Goodrich. (G. R. de Beer, ed.). Clarendon Press, Oxford, 350 pp.

Baker, J. R., and T. F. Bird. 1936. The seasons in a tropical rain forest (New Hebrides). Part 4. Insectivorous bats (Vespertilionidae and Rhinolophidae). J. Linn. Soc. London, Zool., **40**:143–161.

Bartholomew, G. A. 1972. Aspects of timing and periodicity of heterothermy. Pp. 663–680, *in* Hibernation and hypothermia, perspectives and challenges (F. E. South *et al.*, eds.). Elsevier, New York, 743 pp.

Beck, A. J., and Boo-Liat Lim. 1973. Reproductive biology of *Eonycteris spelaea*, Dobson (Megachiroptera) in West Malaysia. Acta Trop., **30**:251–260.

Bels, L. 1952. Fifteen years of bat banding in the Netherlands. Publties natuurh. Genoot. Limburg, **5**:1–99.

Ben Shaul, D. M. 1962. The composition of the milk of wild animals. Int. Zoo. Yearb., **4**:333–342.

Bernard, R. T. F. 1980. Reproductive cycles of *Miniopterus schreibersii natalensis* (Kuhl, 1819) and *Miniopterus fraterculus* Thomas and Schwann 1906. Ann. Transvaal Mus., **32**:55–64.

Blaxter, K. L. 1961. Lactation and the growth of young. Pp. 305–361, *in* Milk: The mammary gland and its secretion. Vol. II (S. K. Kon and A. T. Cowie, eds.). Academic Press, London, 423 pp.

Blaxter, K. L. 1971. The comparative biology of lactation. Pp. 51–69, *in* Lactation (I. R. Falconer, ed.). Butterworths, London, 467 pp.

Bleier, W. J. 1975. Early embryology and implantation in the California leaf-nosed bat *Macrotus californicus*. Anat. Rec., **182**:237–254.

Bogan, M. A. 1972. Observations on parturition and development in the hoary bat *Lasiurus cinereus*. J. Mammal., **53**:611–614.

Bonaccorso, F. J. 1979. Foraging and reproductive ecology in a Panamanian bat community. Bull. Fla. State Mus. Biol. Ser., **24**:359–408.

Bradbury, J. W. 1977. Social organization and communication. Pp. 1–72, *in* Biology of bats. Vol. 3. (W. A. Wimsatt, ed.). Academic Press, New York, 651 pp.

Bradbury, J. W. 1979. Behavioural aspects of reproduction in Chiroptera. J. Reprod. Fertil., **56**:431–438.

Bradbury, J. W., and L. H. Emmons. 1974. Social organization of some Trinidad bats. I. Emballonuridae. Z. Tierpsychol., **36**:137–183.

Bradbury, J. W., and S. L. Vehrencamp. 1976. Social organization and foraging in emballonurid bats. I. Field studies. Behav. Ecol. Sociobiol., **1**:337–381.

Bradbury, J. W., and S. L. Vehrencamp. 1977. Social organization and foraging in emballonurid bats. IV. Parental investment patterns. Behav. Ecol. Sociobiol., **2**:19–29.

Bradshaw, G. V. R. 1961. Le cycle de reproduction de *Macrotus californicus* (Chiroptera, Phyllostomatidae). Mammalia, **25**:117–119.

Bradshaw, G. V. R. 1962. Reproductive cycle of the California leaf-nosed bat, *Macrotus californicus*. Science, **136**:645–646.

Brandon, R. A. 1961. Observations of young keen bats. J. Mammal., **42**:400–401.

Brenner, F. J. 1968. A three-year study of two breeding colonies of the big brown bat, *Eptesicus fuscus*. J. Mammal., **49**:775–778.

Brosset, A. 1962. The bats of central and western India. Part III. J. Bombay Nat. Hist. Soc., **59**:707–746.

Brosset, A. 1966. La biologie des chiroptères. Masson, Paris, 240 pp.

Brosset, A. 1968. La permutation du cycle sexuel saisonnier chez le Chiroptère *Hipposideros caffer*, au voisinage de l'équateur. Biol. Gabon., **4**:325–341.

Brosset, A. 1969. Recherches sur la biologie des Chiroptères troglophiles dans le nord-est du Gabon. Biol. Gabon., **5**:93–116.

Brosset, A. and H. Saint-Girons. 1980. Cycles de reproduction des microchiroptères troglophiles du nord-est du Gabon. Mammalia, **44**:225–232.

Brown, P. E. 1976. Vocal communication in the pallid bat, *Antrozous pallidus*. Z. Tierpsychol., **41**:34–54.

Burns, J. M., and R. G. Easley. 1977. Hormonal control of delayed development in the California leaf-nosed bat, *Macrotus californicus*. III. Changes in plasma progesterone during pregnancy. Gen. Comp. Endocrinol., **32**:163–166.

Burns, J. M., and W. E. Wallace. 1975. Hormonal control of delayed development in *Macrotus waterhousii*. II. Radioimmunoassay of plasma estrone and estradiol 17-β during pregnancy. Gen. Comp. Endocrinol., **25**:529–533.

Burns, J. M., R. J. Baker, and W. J. Bleier. 1972. Hormonal control of "delayed development" in *Macrotus waterhousii*. I. Changes in plasma thyroxine during pregnancy and lactation. Gen. Comp. Endocrinol., **18**:54–58.

Caffier, P., and H. Kolbow. 1934. Anatomisch-physiologische Genitalstudien an Fledermäusen zur Klarung der therapeutischen Sexualhormonwirkung. Z. Geburtschilfe Gynaekol., **108**:185–235.

Carrick, F. N., and B. P. Setchell. 1977. The evolution of the scrotum. Pp. 165–170, *in* Reproduction and evolution. (J. H. Calaby and C. H. Tyndale-Briscoe, eds.). Australian Academy of Sciences.

Carter, D. C. 1970. Chiropteran reproduction. Pp. 233–246, *in* About bats. (B. H. Slaughter and D. W. Walton, eds.). Southern Methodist University Press, Dallas, 339 pp.

Christian, J. J. 1956. The natural history of a summer aggregation of the big brown bat, *Eptesicus fuscus fuscus*. Am. Midl. Nat., **55**:66–95.

Cockrum, E. L. 1955. Reproduction in North American bats. Trans. Kans. Acad. Sci., **58**:487–511.

Courrier, R. 1927. Etude sur le déterminisme des caractères sexuels secondaires chez quelques mammifères à l'activité testiculaire périodique. Arch. Biol., **37**:173–334.

Daan, S. 1973. Activity during natural hibernation in three species of vespertilionid bats. Neth. J. Zool., **23**:1–71.

Davis, R. 1969. Growth and development of young pallid bats, *Antrozous pallidus*. J. Mammal., **50**:729–736.

Davis, R. B., C. F. Herreid II, and H. L. Short. 1962. Mexican free-tailed bats in Texas. Ecol. Monogr., **32**:311–346.

Deanesly, R., and T. Warwick. 1939. Observations on pregnancy in the common bat (*Pipistrellus pipistrellus*). Proc. Zool. Soc. London A., **109**:57–60.

Dinale, G. 1964. Studi sui Chirotteri italiani. II. Il raggiungimento della maturità sessuale in *Rhinolophus ferrumequinum* Schreber. Atti Soc. Ital. Sci. Nat. Mus. Civ. Stor. Nat. Milano, **103**:141–153.

Dinale, G. 1968. Studi sui Chirotteri italiani. VII. Sul raggiungimento della maturità sessuale nei Chirotteri europei ed in particolare nei Rhinolophidae. Arch. Zool. Ital., **54**:51–71.

Dwyer, P. D. 1963. Reproduction and distribution in *Miniopterus* (Chiroptera). Aust. J. Sci., **25**:435–436.

Dwyer, P. D. 1964. Seasonal changes in activity and weight of *Miniopterus schreibersii blepotis* (Chiroptera) in north-eastern New South Wales. Aust. J. Zool., **12**:52–69.

Dwyer, P. D. 1970. Latitude and breeding season in a polyestrus species of *Myotis*. J. Mammal., **51**:405–410.

Eisentraut, M. 1937. Die Wirkung niedriger Temperaturen auf die Embryonentwicklung bei Fledermäusen. Biol. Zentralbl., **57**:59–74.

Fayenuwo, J. O., and L. B. Halstead. 1974. Breeding cycle of straw-colored fruit bat, *Eidolon helvum*, at Ile Ife, Nigeria J. Mammal., **55**:453–454.

Fenton, M. B. 1969. Summer activity of *Myotis lucifugus* (Chiroptera: Vespertilionidae) at hibernacula in Ontario and Quebec. Can. J. Zool., **47**:597–602.

Fenton, M. B. 1970. Population studies of *Myotis lucifugus* (Chiroptera: Vespertilionidae) in Ontario. Life Sci. Contrib. R. Ont. Mus., **77**:1–34.

Fleming, T. H. 1971. *Artibeus jamaicensis:* delayed embryonic development in a Neotropical bat. Science, **171**:402–404.

Fleming, T. H. 1973. The reproductive cycles of three species of opossums and other mammals in the Panama Canal Zone. J. Mammal., **54**:439–455.

Fleming, T. H., E. T. Hooper, and D. E. Wilson. 1972. Three central American bat communities: structure, reproductive cycles and movement patterns. Ecology, **53**:555–569.

Folk, G. E. 1940. Shift of population among hibernating bats. J. Mammal., **21**:306–315.

Foster, G. W., S. R. Humphrey, and P. P. Humphrey. 1978. Survival rate of young southeastern brown bats, *Myotis austroriparius*, in Florida. J. Mammal., **59**:299–304.

Frazer, J. F. D., and A. St. G. Hugget. 1974. Species variations in the foetal growth rate of eutherian mammals. J. Zool., **174**:481–509.

Frisch, R. 1980. Pubertal adipose tissue: is it necessary for normal sexual maturation? Evidence from the rat and human female. Fed. Proc. Fed. Am. Soc. Exp. Biol., **39**:2395–2400.

Funakoshi, K., and T. A. Uchida. 1978. Studies on the physiological and ecological adaptation of temperate insectivorous bats. III. Hibernation and winter activity in some cave dwelling bats. Jap. J. Ecol., 28:237–261.

Gaisler, J. 1966. Reproduction in the lesser horseshoe bat (*Rhinolophus hipposideros hipposideros* Bechstein, 1800). Bijdr. Dierkd., **36**:45–64.

Gaisler, J., and M. Titlbach. 1964. The male sexual cycle in the lesser horseshoe bat (*Rhinolophus hipposideros hipposideros* Bechstein, 1800). Acta Soc. Zool. Bohem., **28**:268–277.

Gibson-Hill, C. A. 1952. The apparent breeding seasons of land birds in North Borneo and Malaya. Bull. Raffles Mus., **24**:270–294.

Gilbert, O., and R. E. Stebbings. 1958. Winter roosts of bats in West Suffolk. Proc. Zool. Soc. London, **131**:329–333.

Gopalakrishna, A., and A. Madhavan. 1971. Survival of spermatozoa in the female genital tract of the Indian vespertilionid bat, *Pipistrellus ceylonicus chrysothrix* (Wroughton). Proc. Indian Acad. Sci. B, **73**:43–49.

Gopalakrishna, A., and A. Madhavan. 1978. Viability of inseminated spermatozoa in the Indian vespertilionid bat *Scotophilus heathi* (Horsefield). Indian J. Exp. Biol., **16**:852–854.

Gopalakrishna, A., R. S. Thakur, and A. Madhavan. 1975. Breeding biology of the southern dwarf pipistrelle *Pipistrellus mimus mimus* (Wroughton) from Maharashtra, India. Pp. 225–240, *in* Dr. B. S. Chauhan. Commem. Vol. Zool. Soc. India.

Goss, R. J. 1977. Photoperiodic control of antler cycles in deer. IV. Effects of constant light:dark ratios on circannual rhythms. J. Exp. Zool., **201**:379–382.

Gould, E. 1978. Rediscovery of *Hipposideros ridleyi* and seasonal reproduction in Malaysian bats. Biotropica, **10**:30–32.

Gustafson, A. W. 1979. Male reproductive patterns in hibernating bats. J. Reprod. Fertil., **56**:317–331.

Gustafson, A. W., and M. Shemesh. 1976. Changes in plasma testosterone levels during the annual reproductive cycle of the hibernating bat *Myotis lucifugus lucifugus* with a survey of plasma testosterone levels in adult male vertebrates. Biol. Reprod., **15**:9–24.

Guthrie, M. J. 1933. Notes on the seasonal movements and habits of some cave bats. J. Mammal., **14**:1–19.

Guthrie, M. J., and K. R. Jeffers. 1938a. Growth of follicles in the ovaries of the bat *Myotis lucifugus lucifugus*. Anat. Rec., **71**:477–496.

Guthrie, M. J., and K. R. Jeffers. 1938b. A cytological study of the ovaries of the bats *Myotis lucifugus lucifugus* and *Myotis grisescens*. J. Morphol., **62**:523–557.

Guthrie, M. J., and K. R. Jeffers. 1938c. The ovaries of the bat *Myotis lucifugus lucifugus* after injection of hypophyseal extract. Anat. Rec., **72**:11–36.

Guthrie, M. J., and E. W. Smith. 1940. The pregnancy urine factor as a stimulus for ovulation in the bat and augmentation of its effect. Am. J. Anat. Suppl. 2, **76**: Abstr. **144**:66.

Hamilton-Smith, E. 1972. The bat population of the Naracoorte Caves area. Proc. Nat. Conf. Aust. Speleol. Fed., **8**:66–75.

Hanwell, A., and M. Peaker. 1977. Physiological effects of lactation on the mother. Symp. Zool. Soc. London, **41**:297–312.

Hartman, C. G. 1933. On the survival of spermatozoa in the female genital tract of the bat. Q. Rev. Biol., **8**:185–193.

Häussler, U., E. Möller, and U. Schmidt. 1981. Zur Haltung und Jugendtwicklung von *Molossus molossus* (Chiroptera). Z. Säugetierk. **46**:337–351.

Heithaus, E. R., T. H. Fleming, and P. A. Opler. 1975. Foraging patterns and resource utilization in seven species of bats in a seasonal tropical forest. Ecology, **56**:841–854.

Herlant, M. 1956. Corrélations hypophysogénitales chez la femelle de la chauve-souris, *Myotis myotis* (Borkhausen). Arch. Biol., Liège, **67**:89–180.

Herlant, M. 1968. Cycle sexuel chez les chiroptères des regions temperées. Pp. 111–126, *in* Cycles génitaux saisonniers des mammifères sauvages. (R. Canivenc, ed.). Masson, Paris.

Herlant, M. 1970. La localisation cytologique des activités gonadotropes au niveau du lobe antérieur de l'hypophyse. Pp. 181–194, *in* Ovo-implantation, human gonadotrophins and prolactin. (P. Hubinont, F. Leroy, C. Robyn, and P. Leleux, eds.). Karger, Basel.

Herreid, C. F., II. 1967. Mortality statistics of young bats. Ecology, **48**:310–312.

Herreid, C. F., II, and K. Schmidt-Nielsen. 1966. Oxygen consumption, temperature and water loss in bats from different environments. Am. J. Physiol., **211**:1108–1112.

Hugget, A. St. G., and W. F. Widdas. 1951. The relationship between mammalian foetal weight and conception age. J. Physiol., **114**:306–317.

Humphrey, S. R. 1975. Nursery roosts and community diversity of Nearctic bats. J. Mammal., **56**:321–346.

Humphrey, S. R., and F. J. Bonaccorso. 1979. Population and community ecology. Pp. 409–441, *in* Biology of bats of the New World family Phyllostomatidae. Part III. (R. J. Baker, J. K. Jones, Jr., and D. C. Carter, eds.) Spec. Publ. Mus. Texas Tech Univ., Lubbock, **10**:1–441.

Humphrey, S. R., and J. B. Cope. 1970. Population samples of the evening bat *Nycticeius humeralis*. J. Mammal., **51**:339–401.

Humphrey, S. R., and J. B. Cope. 1976. Population ecology of the little brown bat *Myotis lucifugus* in Indiana and North Central Kentucky. Spec. Publ. Amer. Soc. Mamm., **4**:1–81.

Humphrey, S. R., A. R. Richter, and J. B. Cope. 1977. Summer habitat and ecology of the endangered Indiana bat, *Myotis sodalis*. J. Mammal., **58**:334–346.

Hurka, L. 1966. Beitrag zur Bionomie, Ökologie, und zur Biometrik der Zwergfledermaus (*Pipistrellus pipistrellus* Schreber, 1774) (Mammalia: Chiroptera) nach dem Beobachtungen in Westbohmen. Acta Soc. Zool. Bohem. **30**:228–246.

Janzen, D. H. 1967. Synchronization of sexual reproduction of trees within the dry season in Central America. Evolution, **21**:620–637.

Janzen, D. H. 1973. Sweep samples of tropical foliage insects: Effects of seasons, vegetation types, elevation, time of day and insularity. Ecology, **54**:687–708.

Janzen, D. H., and T. W. Schoener. 1968. Differences in insect abundance and diversity between wetter and drier sites during a tropical dry season. Ecology, **49**:96–110.

Jenness, R., and E. H. Studier. 1976. Lactation and milk. Pp. 201–218, *in* Biology of bats of the New World family Phyllostomatidae, Part I. (R. J. Baker, J. K. Jones, Jr., and D. C. Carter, eds.). Spec. Publ. Mus. Texas Tech Univ., Lubbock, **10**:1–218.

Jenyns, L. 1846. Observations in natural history. Van Voorst, London.

Jerrett, D. P. 1979. Female reproductive patterns in nonhibernating bats. J. Reprod. Fertil., **56**:369–378.

Jones, P. J., and P. Ward. 1976. The level of reserve protein as the proximate factor controlling the timing of breeding and clutch-size in the red-billed Quelea *Quelea quelea*. Ibis, **118**:547–574.

Kallen, F. C., and W. A. Wimsatt. 1962. Circulating red cell volume in little brown bats in summer. Am. J. Physiol., **203**:1182–1184.

Kappas, A., W. Soybel, P. Glickman, and D. K. Fukushima. 1960. Fever-producing steroids of endogenous origin in man. Arch. Intern. Med., **105**:701–708.

Kitchener, D. J. 1975. Reproduction in female Gould's wattled bat *Chalinolobus gouldii* (Gray) (Vespertilionidae), in Western Australia. Aust. J. Zool., **23**:29–42.

Kleiman, D. G. 1969. Maternal care, growth rate and development in the noctule (*Nyctalus noctula*), pipistrelle (*Pipistrellus pipistrellus*), and serotine (*Eptesicus serotinus*) bats. J. Zool., **157**:187–211.

Kleiman, D. G., and P. A. Racey. 1969. Observations on noctule bats (*Nyctalus noctula*) breeding in captivity. Lynx, **10**:65–77.

Kolb, A. 1950. Beitrage zur Biologie einheimischer Fledermause. Zool. Jahrb. Abt. Syst. Ökol. Geogr. Tiere, **78**:547–572.

Krishna, A. 1978. Aspects of reproduction in some Indian bats. Unpublished Ph.D. dissertation, Banaras Hindu University, Varanasi.

Krishna, A., C. J. Dominic. 1978. Storage of spermatozoa in the female genital tract of the Indian vespertilionid bat, *Scotophilus heathi* Horsefield. J. Reprod. Fertil., **54**:319–321.

Krishna, A., and C. J. Dominic. 1982. Differential rates of fetal growth in two successive pregnancies in the emballonurid bat, *Taphozous longimanus* Hardwicke. Biol. Reprod., (in press).

Krutzsch, P. H. 1961. The reproductive cycle in the male vespertilionid bat *Myotis velifer*. Anat. Rec., **139**:309.

Krutzsch, P. H. 1975. Reproduction of the canyon bat *Pipistrellus hesperus*, in southwestern United States. Am. J. Anat., **143**:163–200.

Krutzsch, P. H. 1979. Male reproductive patterns in nonhibernating bats. J. Reprod. Fertil., **56**:333–344.

Krutzsch, P. H., and W. W. Wells. 1960. Androgenic activity in the interscapular brown adipose tissue of the male hibernating bat (*Myotis lucifugus*). Proc. Soc. Exp. Biol. Med., **105**:578–581.

Kunz, T. H. 1971. Reproduction of some vespertilionid bats in central Iowa. Am. Midl. Nat., **86**:477–486.

Kunz, T. H. 1973. Population studies of the cave bat (*Myotis velifer*): Reproduction, growth and development. Occas. Pap. Mus. Nat. Hist. Univ. Kans., **15**:1–43.

Kunz, T. H. 1974a. Reproduction, growth and mortality of the vespertilionid bat, *Eptesicus fuscus*, in Kansas. J. Mammal., **55**:1–13.

Kunz, T. H. 1974b. Feeding ecology of a temperate insectivorous bat (*Myotis velifer*). Ecology, **55**:693–711.

Kunz, T. H. 1980. Daily energy budgets of free-living bats. Pp. 369–392, *in* Proceedings of the Fifth International Bat Research Conference. (D. E. Wilson and A. L. Gardner, eds.). Texas Tech Press, Lubbock, 434 pp.

Leitch, I., F. E. Hytten, and W. Z. Billewicz. 1959. The maternal and neonatal weights of some Mammalia. Proc. Zool. Soc. London, **133**:11–28.

Lim, Boo-Liat. 1970. Food habits and breeding cycle of the Malaysian fruit-eating bat, *Cynopterus brachyotis*. J. Mammal., **51**:174–177.

Lim, Boo-Liat. 1973. Breeding pattern, food habits and parasitic infestation of bats in Gunong Brinchang. Malay. Nat. J., **26**:6–13.

McClure, H. E., and H. B. H. Osman. 1968. Nesting of birds in a coconut–mangrove habitat in Selangor. Malay. Nat. J., **22**:18–25.

NcNab, B. K. 1973. Energetics and the distribution of vampires. J. Mammal., **54**:131–144.

McNab, B. K. 1974. The behavior of temperate cave bats in a subtropical environment. Ecology, **55**:943–958.

McWilliam, A. N. 1982. Adaptive responses to seasonality in four species of Microchiroptera in coastal Kenya. Unpublished Ph.D. dissertation, University of Aberdeen, Aberdeen, 290 pp.

Madhavan, A. 1971. Breeding habits in the Indian vespertilionid bat, *Pipistrellus ceylonicus chrysothrix* (Wroughton). Mammalia, **35**:283–306.

Madhavan, A. 1979. Breeding habits and associated phenomena in some Indian bats. Part V—*Pipistrellus dormeri*—Vespertilionidae. J. Bombay Nat. Hist. Soc., **75**:426–433.

Madhavan, A., D. R. Patil, and A. Gopalakrishna. 1976. Breeding habits and associated phenomena in some Indian bats. Part IV—*Hipposideros fulvus fulvus* (Gray)—Hipposideridae. J. Bombay Nat. Hist. Soc., **75**:96–103.

Marshall, A. J., and P. S. Corbet. 1959. The breeding biology of equatorial vertebrates: reproduction of the bat *Chaerephon hindei* Thomas at latitude 0°26′N. Proc. Zool. Soc. London, **132**:607–616.

Medway, Lord. 1971. Observations of social and reproductive biology of the bent-winged bat *Miniopterus australis* in northern Borneo. J. Zool., **165**:261–273.

Medway, Lord. 1972. Reproductive cycles of the flat-headed bats *Tylonycteris pachypus* and *T. robustula* (Chiroptera: Vespertilionidae) in a humid equatorial environment. J. Linn. Soc. London, Zool., **51**:33–61.

Menzies, J. I. 1973. A study of leaf-nosed bats (*Hipposideros caffer* and *Rhinolophus landeri*) in a cave in Northern Nigeria. J. Mammal., **54**:930–945.

Migula, P. 1969. Bioenergetics of pregnancy and lactation in European common vole. Acta Theriol., **14**:167–179.

Millar, J. S. 1977. Adaptive features of mammalian reproduction. Evolution, **31**:370–386.

Millar, R. P., and T. D. Glover. 1973. Regulation of seasonal sexual activity in an ascrotal mammal, the rock hyrax, *Procavia capensis*. J. Reprod. Fertil., Suppl., **19**:203–220.

Miller, R. E. 1939. The reproductive cycle in male bats of the species *Myotis lucifugus lucifugus* and *Myotis grisescens*. J. Morphol., **64**:267–295.

Mills, R. S., G. W. Barrett, and M. P. Farrell. 1975. Population dynamics of the big brown bat (*Eptesicus fuscus*) in southwestern Ohio. J. Mammal., **56**:591–604.

Mitchell, G. C. 1965. A natural history of the funnel-eared bat *Natalus stramineus*. Unpublished Masters thesis, University of Arizona.

Moghe, M. A. 1956. On the development and placentation of a megachiropteran bat—*Cynopterus sphinx gangeticus* (Anderson). Proc. Nat. Inst. Sci. India B, **22**:48–55.

Mutere, F. A. 1967. The breeding biology of equatorial vertebrates: Reproduction in the fruit bat, *Eidolon helvum*, at latitude 0°20′N. J. Zool., **153**:153–161.

Mutere, F. A. 1968. The breeding biology of the fruit bat *Rousettus aegyptiacus* E. Geoffroy living at 0°22′S. Acta Trop., **25**:97–108.

Mutere, F. A. 1970. The breeding biology of equatorial vertebrates: reproduction in the insectivorous bat, *Hipposideros caffer* living at 0°27′N. Bijdr. Dierkd., **40**:56–58.

Mutere, F. A. 1973a. A comparative study of reproduction in two populations of the insectivorous bats, *Otomops martiensseni*, at latitudes 1°5′S and 2°30′S. J. Zool., **171**:79–92.

Mutere, F. A. 1973b. Reproduction in two species of equatorial free-tailed bats (Molossidae). East Afr. Wildl. J., **11**:271–280.

Myers, P. 1977. Patterns of reproduction of four species of vespertilionid bats in Paraguay. Univ. Calif. Publ. Zool., **107**:1–41.

Myers, P. 1978. Sexual dimorphism in size of vespertilionid bats. Am. Nat., **112**:701–711.

Nelson, J. E. 1965a. Behaviour of Australian Pteropodidae (Megachiroptera). Anim. Behav., **13**:544–557.

Nelson, J. E. 1965b. Movements of Australian flying foxes (Pteropodidae: Megachiroptera). Aust. J. Zool., **13**:53–73.

O'Farrell, M. J., and E. H. Studier. 1973. Reproduction, growth and development in *Myotis thysanodes* and *M. lucifugus* (Chiroptera: Vespertilionidae). Ecology, **54**:18–30.

O'Farrell, M. J., and E. H. Studier. 1975. Population structure and emergence activity in *Myotis thysanodes* and *M. lucifugus* (Chiroptera: Vespertilionidae) in northeastern New Mexico. Am. Midl. Nat., **93**:368–376.

O'Farrell, M. J., W. G. Bradley, and G. W. Jones. 1967. Fall and winter bat activity at a desert spring in southern Nevada. Southwest. Nat., **12**:163–171.

Oxberry, B. A. 1979. Female reproductive patterns in hibernating bats. J. Reprod. Fertil., **56**:359–367.

Pagels, J. F., and C. Jones. 1974. Growth and development of the free-tailed bat, *Tadarida brasiliensis cynocephala* (LeConte). Southwest. Nat., **19**:267–276.

Panyutin, K. K. 1963. Reproduction in the common noctule. Uchen. Zap. Mosk. Obl. Pedagog. Inst., **126**:63–66.

Payne, P. R., and E. F. Wheeler. 1967. Comparative nutrition in pregnancy. Nature, **215**:1134–1136.

Pearson, O. P., M. R. Koford, and A. K. Pearson. 1952. Reproduction of the lump-nosed bat (*Corynorhinus rafinesquei*) in California. J. Mammal., **33**:273–320.

Pevet, P., and P. A. Racey. 1981. The pineal gland of nocturnal mammals. (1) The ultrastructure of the pineal gland in the pipistrelle bat (*Pipistrellus pipistrellus* L.): Presence of two populations of pinealocytes. Cell Tissue Res. **216**:253–271.

Peyre, A., and M. Herlant. 1967. Ovo-implantation différée et déterminisme hormonal chez le Minioptère *Miniopterus schreibersi* K. (Chiroptère). C. R. Séances. Soc. Biol. Paris, **161**:1779–1782.

Pirlot, P. 1967. Periadicité de la reproduction chez les Chiroptères neotropicaux. Mammalia, **31**:361–366.

Pye, J. D. 1972. Bimodal distribution of constant frequencies in some hipposiderid bats (Mammalia: Hipposideridae). J. Zool., **166**:323–335.

Racey, P. A. 1972. Aspects of reproduction in some heterothermic bats. Unpublished Ph.D. dissertation, University of London, London, 377 pp.

Racey, P. A. 1973a. The viability of spermatozoa after prolonged storage by male and female European bats. Period. Biol., **75**:201–205.

Racey, P. A. 1973b. The time of onset of hibernation in Pipistrelle bats, *Pipistrellus pipistrellus*. J. Zool., **171**:465–467.

Racey, P. A. 1973c. Environmental factors affecting the length of gestation in heterothermic bats. J. Reprod. Fertil., Suppl., **19**:175–189.

Racey, P. A. 1974a. The reproductive cycle in male noctule bats *Nyctalus noctula*. J. Reprod. Fertil., **41**:169–182.

Racey, P. A. 1974b. Ageing and assessment of reproductive status of Pipistrelle bats, *Pipistrellus pipistrellus*. J. Zool., **173**:264–271.

Racey, P. A. 1976. Induction of ovulation in the pipistrelle bat, *Pipistrellus pipistrellus*. J. Reprod. Fertil., **46**:481–483.

Racey, P. A. 1978. The effect of photoperiod on the initiation of spermatogenesis in pipistrelle bats, *Pipistrellus pipistrellus*. Pp. 255–258, *in* Proceeding of the Fourth International Bat Research Conference. (R. J. Olembo, J. B. Castelino, and F. A. Mutere, eds.). Kenya National Academy of Arts and Sciences. 328 pp.

Racey, P. A. 1979. The prolonged storage and survival of spermatozoa in Chiroptera. J. Reprod. Fertil., **56**:391–402.

Racey, P. A. 1981. Environmental factors affecting the length of gestation in mammals. Pp. 199–213, *in* Environmental factors in mammal reproduction. (D. Gilmore and B. Cook, eds.). MacMillan, London, 320 pp.

Racey, P. A., and D. M. Potts. 1971. A light and electron microscope study of early development in the bat *Pipistrellus pipistrellus*. Micron, **2**:322–348.

Racey, P. A., and S. M. Swift. 1981. Variations in gestation length in a colony of pipistrelle bats (*Pipistrellus pipistrellus*) from year to year. J. Reprod. Fertil., **61**:123–129.

Racey, P. A., and W. H. Tam. 1974. Reproduction in male *Pipistrellus pipistrellus* (Mammalia: Chiroptera). J. Zool., **172**:101–122.

Rakhmatulina, I. K. 1972. The breeding, growth and development of pipistrelles in Azerbaidzhan. Sov. J. Ecol., **2**:131–136.

Ramakrishna, P. A., and K. V. B. Rao. 1977. Reproductive adaptations in the Indian rhinolophid bat *Rhinolophus rouxi* (Temminck). Curr. Sci., **46**:270–271.

Ramaswami, L. S., and T. C. Anand Kumar. 1966. Effect of exogenous hormones on the reproductive structures of the female bat *Rhinopoma* during the non-breeding season. Acta Anat., **63**:101–123.

Randolph, P. A., J. C. Randolph, K. Mattingly, and M. M. Foster. 1977. Energy costs of reproduction in the cotton rat, *Sigmodon hispidus*. Ecology, **58**:31–45.

Ransome, R. D. 1971. The effect of ambient temperature on the arousal frequency of the hibernating greater horseshoe bat, *Rhinolophus ferrumequinum* in relation to site selection and the hibernation site. J. Zool., **164**:353–371.

Ransome, R. D. 1973. Factors affecting the timing of births of the greater horseshoe bat (*Rhinolophus ferrumequinum*). Period. Biol., **75**:169–175.

Rasweiler, J. J., IV. 1973. Care and management of the long tongued bat, *Glossophaga soricina* (Chiroptera: Phyllostomatidae), in the laboratory, with observations on estivation induced by food deprivation. J. Mammal., **54**:391–404.

Rasweiler, J. J., IV. 1977. Preimplantation development, fate of the zona pellucida, and observations on the glycogen-rich oviduct of the little bulldog bat, *Noctilio albiventris*. Am. J. Anat., **150**:269–300.

Reiter, R. J., and B. K. Follett. 1980. Seasonal reproduction in higher vertebrates. Progress in reproductive biology 5, Karger, Basel, 221 pp.

Rice, D. W. 1957. Life history and ecology of *Myotis austroriparius* in Florida. J. Mammal., **38**:15–32.

Richardson, B. A. 1979. The anterior pituitary and reproduction in bats. J. Reprod. Fertil., **56**:379–389.

Richardson, E. G. 1977. The biology and evolution of the reproductive cycle of *Miniopterus schreibersii* and *M. australis* (Chiroptera: Vespertilionidae). J. Zool., **183**:353–375.

Roer, H. 1968. Zur Frage den Wochenstuben—Quartiertreue weiblicher Mausohren (*Myotis myotis*). Bonn. Zool. Beitr., **19**:85–96.

Roer, H. 1973. Uber die Ursachen hoher Jugendmortalitat beim Mausohr, *Myotis myotis* (Chiroptera, Mamm.). Bonn. Zool. Beitr., **24**:332–341.

Ryberg, O. 1947. Studies on bats and bat parasites. Bokforlaget Svensk Natur., Stockholm, 330 pp.

Sadlier, R. M. F. S. 1969. The ecology of reproduction in wild and domestic animals. Methuen, London, 321 pp.

Saint-Girons, H., A. Brosset, and M. C. Saint-Girons. 1969. Contribution à la connaissance du cycle annuel de la chauve-souris *Rhinolophus ferrumequinum* (Schreber, 1774). Mammalia, **33:** 357–470.

Sayler, A., and M. Salmon. 1971. An ethological analysis of communal nursing by the house mouse (*Mus musculus*). Behaviour, **40:**62–85.

Schmidt, U., and U. Manske. 1973. Die Jugendentwicklung der Vampirfledermause (*Desmodus rotundus*). Z. Säugetierk., **38:**14–33.

Schowalter, D. B., and J. R. Gunson. 1979. Reproductive biology of the big brown bat (*Eptesicus fuscus*) in Alberta. Can. Field Nat., **93:**48–54.

Schowalter, D. B., J. R. Gunson, and L. D. Harder. 1979. Life history characteristics of little brown bats (*Myotis lucifugus*) in Alberta. Can. Field-Nat., **93:**243–251.

Skreb, N. 1954. Experimentelle Untersuchungen uber die ausseren Ovulationsfaktoren bei der Fledermausen *Nyctalus noctula*. Naturwissenschaften, **41:**484.

Skreb, N. 1955. Influence des gonadotrophines sur les noctules (*Nyctalus noctula*) en hibernation. C. R. Séances Soc. Biol. Paris, **149:**71–74.

Skutch, A. F. 1950. The nesting seasons of Central American birds in relation to climate and food supply. Ibis, **92:**185–222.

Sluiter, J. W., and L. Bels. 1951. Follicular growth and spontaneous ovulation in captive bats during the hibernation period. Proc. K. Ned. Akad. Wet. Ser. C., **54:**585–593.

Sluiter, J. W., and P. F. van Heerdt. 1966. Seasonal habits of the noctule bat (*Nyctalus noctula*). Arch. Neerl. Zool., **16:**423–439.

Smith, E. W. 1951. Seasonal responses of follicles in the ovaries of the bat *Myotis grisescens* to pregnancy urine factor. Endocrinology, **49:**67–72.

Smith, E. W. 1956. Pregnancy in the little brown bat. Am. J. Physiol., **185:**61–64.

Smythe, N. 1970. Relationships between fruiting seasons and seed dispersal methods in a neotropical forest. Am. Nat., **104:**25–35.

Sreenivasan, M. A., H. R. Bhat, and G. Geevarghese. 1974. Observations on the reproductive cycle of *Cynopterus sphinx sphinx* Vahl, 1797 (Chiroptera: Pteropidae). J. Mammal., **55:**200–202.

Stebbings, R. E. 1965. Observations during sixteen years on winter roosts of bats in west Suffolk. Proc. Zool. Soc. London, **144:**137–143.

Stebbings, R. E. 1968. Measurements, composition and behaviour of a large colony of the bat *Pipistrellus pipistrellus*. J. Zool., **156:**15–33.

Strelkov, P. 1962. The peculiarities of reproduction in bats (Vespertilionidae) near the northern border of their distribution. Pp. 306–311, *in* International Symposium on Methods in Mammalogical Investigation, Brno., 1960. Praha.

Studier, E. H., and M. J. O'Farrell. 1972. Biology of *Myotis thysanodes* and *M. lucifugus* (Chiroptera: Vespertilionidae)—I. Thermoregulation. Comp. Biochem. Physiol., **41A:**567–595.

Studier, E. H., and M. J. O'Farrell. 1976. Biology of *Myotis thysanodes* and *M. lucifigus* (Chiroptera: Vespertilionidae)—III. Metabolism, heart rate, breathing rate, evaporative water loss and general energetics. Comp. Biochem. Physiol., **54A:**423–432.

Studier, E. H., and M. J. O'Farrell. 1980. Physiological ecology of *Myotis*. Pp. 415–424, *in* Proceedings of the Fifth International Bat Research Conference. (D. E. Wilson and A. L. Gardner, eds.). Texas Tech Press, Lubbock, 434 pp.

Studier, E. H., V. L. Lysengen, and M. J. O'Farrell. 1973. Biology of *Myotis thysanodes* and *M. lucifigus* (Chiroptera: Vespertilionidae)—II. Bioenergetics of pregnancy and lactation. Comp. Biochem. Physiol., **44A:**467–471.

Swift, S. M. 1980. Activity patterns of pipistrelle bats (*Pipistrellus pipistrellus*) in north-east Scotland. J. Zool., **190:**285–295.

Swift, S. M. 1981. Foraging, colonial and maternal behaviour of bats in Scotland. Unpublished Ph.D. dissertation, University of Aberdeen, Aberdeen, 208 pp.

Taddei, V. A. 1976. The reproduction of some Phyllostomidae (Chiroptera) from the northwestern region of the State of Sao Paulo. Bol. Zool. Univ. Sao Paulo, **1**:313–330.

Tamsitt, J. R., and D. Valdivieso. 1965. Reproduction of the female big fruit-eating bat, *Artibeus lituratus palmarum*, in Colombia. Caribb. J. Sci., **5**:157–166.

Taylor, L. R. 1963. Analysis of the effect of temperature on insects in flight. J. Anim. Ecol., **32**:99–117.

Thomas, D. W., M. B. Fenton, and R. M. R. Barclay. 1979. Social behaviour of the little brown bat *Myotis lucifugus*. 1. Mating behaviour. Behav. Ecol. Sociobiol., **6**:129–136.

Tuttle, M. D. 1975. Population ecology of the gray bat (*Myotis grisescens*): Factors influencing early growth and development. Occas. Pap. Mus. Nat. Hist. Univ. Kans., **36**:1–24.

Tuttle, M. D. 1976. Population ecology of the gray bat (*Myotis grisescens*): Factors influencing growth and survival of newly volant young. Ecology, **57**:587–595.

van der Merwe, M. 1979. Foetal growth curves and seasonal breeding in the Natal clinging bat *Miniopterus schreibersi natalensis*. S. Afr. J. Zool., **14**:17–21.

Wallace, G. I. 1978. A histological study of the early stages of pregnancy in the bent-winged bat (*Miniopterus schreibersii*) in north-eastern New South Wales, Australia (30°27′S). J. Zool., **185**:519–537.

Ward, P. 1969. The annual cycle of the yellow-vented bulbul *Pycnonotus goiavier* in a humid equatorial environment. J. Zool., **157**:25–45.

Whitaker, J. O., and R. E. Mumford. 1972. Notes on occurrence and reproduction of bats in Indiana. Proc. Indiana Acad. Sci., **81**:376–383.

White, M. J. D. 1978. Modes of speciation. W. H. Freeman, San Francisco, 455 pp.

Whitehead, P. E., and N. O. West. 1977. Metabolic clearance and production rates of testosterone at different times of the year in male caribou and reindeer. Can. J. Zool., **55**:1692–1697.

Williams, C. B. 1939. An analysis of four years captures of insects in a light trap. Part 1. Trans. R. Entomol. Soc. London, **89**:79–132.

Wilson, D. E. 1973. Reproduction in Neotropical bats. Period. Biol., **75**:215–217.

Wilson, D. E. 1979. Reproductive patterns. Pp. 317–378, *in* Biology of bats of the New World family Phyllostomatidae. Part III. (R. J. Baker, J. K. Jones, Jr., and D. C. Carter, eds.). Spec. Publ. Mus. Texas Tech Univ., Lubbock, **16**:1–441.

Wilson, D. E., and Findley, J. S. 1970. Reproductive cycle of a Neotropical insectivorous bat *Myotis nigricans*. Nature, **225**:1155.

Wilson, D. E., and Findley, J. S. 1971. Spermatogenesis in some Neotropical species of *Myotis*. J. Mammal., **52**:420–426.

Wimsatt, W. A. 1944a. Further studies on the survival of spermatozoa in the female reproductive tract of the bat. Anat. Rec., **88**:193–204.

Wimsatt, W. A. 1944b. Growth of the ovarian follicle and ovulation in *Myotis lucifugus*. Am. J. Anat., **74**:129–173.

Wimsatt, W. A. 1945. Notes on breeding behavior, pregnancy and parturition in some vespertilionid bats of the eastern United States. J. Mammal., **26**:23–33.

Wimsatt, W. A. 1960. Some problems of reproduction in relation to hibernation in bats. Bull. Mus. Comp. Zool., **124**:249–267.

Wimsatt, W. A. 1969. Some interrelations of reproduction and hibernation in mammals. Symp. Soc. Exp. Biol., **23**:511–549.

Wimsatt, W. A. 1975. Some comparative aspects of implantation. Biol. Reprod., **12**:1–40.

Wimsatt, W. A., and H. F. Parks. 1966. Ultrastructure of the surviving follicle of hibernation and of the ovum–follicle cell relationship in the vespertilionid bat *Myotis lucifugus*. Symp. Zool. Soc. London, **15**:419–454.

Wolda, H. 1978. Seasonal fluctuations in rainfall, food and abundance of tropical insects. J. Anim. Ecol., **47**:369–381.

Yoshiyuki, M., M. Iijima, and Y. Ogawara. 1970. The embryo size in a population of *Pipistrellus abramus* in Saitama, Japan. J. Mammal. Soc. Jpn., **5**:74–75.

Zondek, B. 1933. Action of folliclin and prolan on the reproductive organs of the bat during hibernation. Lancet, **225**:1256–1257.

Chapter 3

Growth and Survival of Bats

Merlin D. Tuttle and Diane Stevenson

Vertebrate Division
Milwaukee Public Museum
Milwaukee, Wisconsin 53233

1. INTRODUCTION

The literature on bat growth and survival is extensive. Nevertheless, few studies have made or permit within or among species comparisons of performance under varied combinations of biological and physiological parameters. In particular, surprisingly few studies have taken advantage of the fact that many species occupy wide ranges of environmental conditions. Growth and survival frequently have been treated as species-specific constants. Such treatment neglects many important determinants of success and likely has inhibited meaningful hypothesis testing and further investigation. Despite these limitations, existing studies demonstrate impressive variation. In fact, available literature indicates that few if any details of growth and survival are constant over time, even for individual colonies within a single species.

Detailed descriptions of growth and developmental stages in bats are already available (Orr, 1970). Krátky (1970), Fenton (1970b), Kunz (1973), and Pagels and Jones (1974) provide additional observations on the development of dentition and pelage, and Buchler (1980), and Brown and Grinnell (1980) review and report on the ontogeny of echolocation. Further specifics of growth and development and maternal care in phyllostomid bats are provided by Kleiman and Davis (1979). A number of survival studies are available and will be cited along with a discussion of their methods of analysis. Where possible, we will not restate previously available details of chiropteran growth and survival, but we will emphasize within and among species comparisons of adaptive responses to different environments. Our discussion will focus on variation and its probable causes, in an attempt to stimulate future investigations of a comparative nature.

2. PRENATAL GROWTH AND DEVELOPMENT

2.1. Length of Gestation

For most mammals the length of gestation is "fixed" and "species characteristic," with minimal variation (Weir and Rowlands, 1973). In contrast, many bats vary prenatal growth rates in response to environmental conditions (see Chapter 2). In temperate species development is delayed when bats fall into torpor (Eisentraut, 1936; Pearson *et al.*, 1952; Racey, 1969, 1973; Ransome, 1973), but in subtropical and tropical species delays may occur in the absence of torpor through alternative mechanisms (Mutere, 1967; Fleming, 1971; Burns *et al.*, 1972). As a consequence, lengths of gestation vary widely, both among (44 days–11 months) and within (56–100 days) species (Carter, 1970; Orr, 1970).

The two most important factors determining rates of prenatal development, and in turn length of gestation, appear to be temperature and food availability. Pearson *et al.* (1952) reported lengths of gestation ranging from 56 to 100 days in *Plecotus townsendii* and noted that length varied greatly among colonies and even among years for a single colony. They attributed this variation to periods of protracted cold, during which the bats became torpid. Myers (1977) suggested that the second gestation period in the polyestrous *Eptesicus furinalis* may be shorter than the first, and attributed the difference to lower ambient temperatures.

Racey (1969, 1973) found that the length of gestation in *Pipistrellus pipistrellus* could be "profoundly affected by environmental conditions." Pregnant pipistrelles deprived of food and subjected to temperatures of 11–14°C for 13 days entered torpor and showed mean parturition times 14½ days longer than for controls that were fed and kept at 18–26°C (the approximate range of their natural environment). However, pipistrelles provided with all the food they could eat remained homeothermic and did not differ from controls in mean lengths of pregnancy at temperatures ranging 10–25°C.

Ransome (1973) reported similar effects of environmental conditions on the length of gestation in *Rhinolophus ferrumequinum*. He described two phases of slowed embryonic development. The first occurred in early spring, when bats became torpid during adverse climatic conditions, and the second occurred in late pregnancy, when physical bulk of the embryo and relatively short nights appeared to force a reduction in food intake. Mean birth dates in his bats varied between July 6 and 9 in two years, apparently due "to factors affecting length of gestation, rather than its onset." He concluded that the primary effect of climatic temperature was its reduction of insect food resources.

Low temperature probably exerts a dual impact by simultaneously reducing prey abundance and increasing the energetic costs of homeothermy. A captive colony of pregnant and lactating *Myotis lucifugus* remained homeothermic but required approximately three times as much food to maintain constant weight at 21–26°C as at 33°C (Stones, 1965; Stones and Wiebers, 1965, 1967). Theoretically, rates of prenatal development should be proportional to food availability when temperature is constant or to temperature when food is constant. Deviations from these predictions would be expected to the extent that energy is conserved through behavioral thermoregulation (see Tuttle, 1975).

Behavioral thermoregulation may be used by bats either to increase or to reduce rates of embryonic development. Dwyer (1964) and Dwyer and Harris (1972) reported pronounced tendencies for pregnant *Miniopterus schreibersii*, believed to be near term, to segregate into relatively cool peripheral roosts. There they dropped into torpor while other bats from the same colonies remained active. Dwyer and Harris interpreted this behavior as a means of delaying parturition until a more appropriate time. Tuttle (unpublished data) made similar observations in maternity colonies of *Myotis grisescens*.

Such apparent heterothermic regulation of the onset of parturition in *Miniopterus schreibersii* and *Myotis grisescens* is probably the same behavior that Ransome (1973) classified as phase II heterothermy in *Rhinolophus ferrumequinum*. Whether or not such behavior is merely a response to stress from lack of sufficient food, as suggested by Ransome, or an adaptive behavior used by near-term pregnant females to facilitate synchrony of parturition, as believed by Dwyer and Harris and by Tuttle, remains to be determined. If the latter, then females nearing term early should enter torpor much more readily than those that are late (assuming equivalent conditions). Clearly, further investigations of the behavior of near-term pregnant females early and late in the parturition cycle and of the performance of pregnant bats under strictly controlled conditions of both food and temperature are needed.

2.2. Time and Synchrony of Parturition

2.2.1. Temperate Bats

Time and synchrony of parturition and the number of litters produced annually vary geographically (Orr, 1970). For temperate-zone bats reproductive cycles are forced into a relatively short time span that is tightly controlled by the rigors of climate (Wilson, 1979). Only a single litter per year is possible, and the timing of parturition is highly synchronous. As expected, parturition times in widely distributed species characteristically are later toward the latitudinal extremes (Dwyer and Hamilton-Smith, 1965; Dwyer, 1970; Kunz, 1974a; LaVal and LaVal, 1979; Schowalter *et al.*, 1979).

Most temperate-zone bats form maternity colonies that share thermoregulatory costs during the period of postnatal development (Dwyer, 1971; Kunz, 1973; Humphrey, 1975; Tuttle, 1975). The energetic advantages are large (Herreid, 1963, 1967; Tuttle, 1975; Trune and Slobodchikoff, 1976) and inversely proportional to the ambient temperature and the body size of the bat (Dwyer, 1971; Tuttle, 1975). Therefore small species and those occupying the coolest roosts are likely to be most strongly selected for synchrony of parturition. Young born either early or late would face increased thermoregulatory costs.

A variety of other factors, such as the seasonality of food resources, the social facilitation of foraging, and predator avoidance (Hoogland and Sherman, 1976; Snapp, 1976; Bradbury and Vehrencamp, 1977b), also may select for synchrony, even in the absence of thermoregulatory constraints. Nevertheless, temperate bats do seem to fit predictions based on thermoregulatory costs, at least at the extremes of body size and roost temperature. *Myotis grisescens* rears young in the coldest roosts (as low as 13.0°C) reported for any North American bat (Barbour and Davis, 1969; Tuttle, 1975) and exhibits extreme synchrony of parturition (Tuttle, 1975). At the opposite extreme *M. lucifugus* occupies the warmest roosts (often as high as 55°C) and shows the least synchrony known for a bat of similar size and distribution (Barbour and Davis, 1969; O'Farrell and Studier, 1973; Schowalter and Gunson, 1979). As expected, *Eumops perotis*, the largest North American bat and a user of relatively warm roosts, shows the least synchrony of parturition (Barbour and Davis, 1969).

Rakhmatulina (1972) provides the only evidence of weather effects on synchrony of parturition. Parturition in a single colony of *Pipistrellus pipistrellus* lasted almost 25 days in 1967 but only 15 days in 1969. He stated that "this was undoubtedly due to the cold and protracted spring of 1969." Assuming that the roost (in a building) was cooler in 1969, selection should have favored greater synchrony in that year. Perhaps this is an example of the extent to which synchrony can be achieved through behavioral thermoregulation of gestation (see Section 2.1).

Although climatic conditions are known to affect both the time and synchrony of parturition, additional important factors may await discovery. Pearson *et al.* (1952) found differences in the mean dates of parturition in colonies of *Plecotus townsendii* that could not be correlated with weather. In 1959 the average birth date for one colony was 4 days later than in 1949, whereas at another it was 7 days earlier than in 1949. Both colony locations were colder in the spring of 1949. O'Farrell and Studier (1973) found highly synchronous versus asynchronous parturition patterns in colonies of *Myotis thysanodes* and *M. lucifugus*, respectively, even though both occupied a single building. The former species often rears young in relatively cool cave roosts, where strong selection may have favored synchronous parturition in the past.

2.2.2. Subtropical and Tropical Bats

In subtropical and tropical regions longer seasons and moderate climates permit a much wider range of options. Many bats in these environments are polyestrous, mostly bimodally or seasonally, and both asynchronous and synchronous patterns of parturition are well documented (Wilson and Findley, 1970; Wilson, 1979).

Artibeus literatus in Colombia (Wilson, 1979), *Taphozous longimanus* in India (Brosset, 1962a), and *Myotis mystacinus* in Malaysia (Lim, 1973) apparently give birth continuously and asynchronously year-round. Closer examination, however, may show that more births occur in some periods than in others. In Costa Rica *Molossus sinaloae* gives birth continuously but shows bimodal peaks in the annual cycle (LaVal and Fitch, 1977). An especially interesting pattern is found in *Pipistrellus dormeri* in India. This bat breeds continuously and asynchronously, year-round, producing litters of both one and two young each, with twins much more common from November through early February and singles most common from late February through October (Madhavan, 1978). *Eptesicus furinalis* in Paraguay produces only two litters annually. The average number of embryos in the first litter is 1.9. The second litter invariably consists of one young (Myers, 1977). *Desmodus rotundus*, from Mexico, south at least to Trinidad, breeds continuously and asynchronously but apparently produces only two litters of one young each annually (Fleming *et al.*, 1972; Wilson, 1979).

Some bats bear young continuously and asynchronously but during only half of the annual cycle. In Panama (Wilson and Findley, 1970; Wilson, 1971) and Paraguay (Myers, 1977) *Myotis nigricans* fits this pattern for 6 months, followed by intervals of varying length in which no young are born. *Myotis albescens* in Paraguay (Myers, 1977) and *M. adversus* in southern Queensland (Dwyer, 1970) show similar cycles. Both species produce at least two and sometimes three litters of one annually.

At the opposite extreme a wide variety of tropical bats produce only one young per year in a highly synchronous period of parturition. Examples include the following: Pteropodidae, *Pteropus geddiei* (Baker and Baker, 1936), *P. giganteus* (Marshall, 1947; Brosset, 1962a), *P. poliocephalus* (Nelson, 1965a); Emballonuridae, *Saccopteryx bilineata*, *Balantiopteryx plicata* (Bradbury and Vehrencamp, 1977b), *Taphozous nudiventris* (=*kachensis*), *T. melanopogan*, *T. perforatus* (Brosset, 1962a); Megadermatidae, *Megaderma lyra;* Rhinolophidae, *Rhinolophus lepidus;* Hipposideridae, *Hipposideros bicolor*, *H. speoris* (Brosset, 1962b); Phyllostomidae, *Brachphylla cavernarum* (Nellis and Ehle, 1977); and Vespertilionidae, *Tylonycteris pachypus*, *T. robustula* (Medway, 1972), and *Miniopterus schreibersii* (Brosset, 1962c). Other bats follow an intermediate pattern. *Dobsonia moluccensis* gives birth asynchronously to a single young annually, but nearly all births are restricted to a definite 3-month season (Dwyer, 1975).

The above patterns are likely quite flexible and should not be assumed to be species specific. Contrasting patterns occur among closely related bats living in similar environments (Brosset, 1962a; Bradbury and Vehrencamp, 1977b) and among populations of single species living in different environments (Dwyer, 1970; Wilson, 1979). The widely distributed Neotropical bat, *Artibeus lituratus*, provides an excellent example. Near the equator, in South America, it is a continuous, asynchronous breeder, but in Panama and Costa Rica parturition is discontinuous and bimodal. Farther north, in Mexico, it produces only one young per year in an asynchronous pattern in the period January–July (see Wilson, 1979).

Seasonality of food resources is probably the most important single factor in determining the reproductive patterns of tropical bats (Wilson and Findley, 1970; Fleming *et al.*, 1972; Medway, 1972; Lim, 1973; Heithaus *et al.*, 1975; Bradbury and Vehrencamp, 1976a, 1976b; LaVal and Fitch, 1977; Bonaccorso, 1979; Humphrey and Bonaccorso, 1979). Tropical forests are complex and highly seasonal, with different resources often on differing annual cycles (Fogden, 1972; Bonaccorso, 1979; Humphrey and Bonaccorso, 1979). Wet, moist, and dry forests frequently occur in close proximity but differ in phenology of fruit production. Wet and moist forests tend to provide peaks in both wet and dry seasons, while dry forests generally provide only one. Nectar production is highest during the dry season (Heithaus *et al.*, 1975). The biomass of large insects in a Panamanian moist forest was much higher early in the wet season, declining to an annual low just prior to the next wet season, but the biomass of small insects remained high year-round (Smythe, 1974).

Even this highly simplistic picture of tropical patterns of food availability is apparently adequate to explain many birth patterns of tropical bats. Vampires, *Desmodus rotundus*, prey on stable populations of large vertebrates and predictably show continuous, asynchronous parturition (Wilson, 1979). Many species of both insectivorous and frugivorous bats show annual cycles that at least roughly correspond to known food peaks in their habitat types (Bonaccorso, 1979). Births in polyestrous species are usually timed so that the weaning of young from the first birth peak coincides with the beginning of the rainy season and maximum food abundance. Although these are basic patterns, specific strategies can be far more complex. Species can switch seasonally between or among food types, thereby unexpectedly prolonging reproductive periods, as has been reported for some phyllostomid bats (Bonaccorso, 1979; Humphrey and Bonaccorso, 1979). Many other complicating alternatives, such as local migration among forest types (Bonaccorso, 1979), make generalization difficult.

Perhaps the most exciting areas for further investigation involve the many species that fail to meet expectations based on the above broad generalizations. What unique seasonal strategy, for example, permits *Taphozous longimanus* to give birth continuously, while *T. nudiventris*, *T. melanopogan*, and *T. perforatus*, occupying similar ranges in India, are highly synchronous breeders, giving birth to a single young annually (see Brosset, 1962a)? Why do the latter three species produce only one young annually, and why should they be highly synchronous? Why does *Megaderma lyra*, a species that feeds upon a wide variety of vertebrates, from birds and mammals to reptiles and amphibians, give birth in a brief, highly synchronous period once per year (see Brosset, 1962b), despite food supplies that seem to be relatively stable? This species forms large maternity colonies, while *T. longimanus* is unique among the four species of *Taphozous* listed above in not forming large colonies.

In their excellent comparative studies of Neotropical emballonurids, Bradbury and Vehrencamp (1976a, 1976b, 1977a, 1977b) concluded that food was a limiting resource and that the seasonality of food supplies set the baseline for mortality. Two stable-habitat species produced up to two asynchronous litters of one each per year, whereas two seasonal-habitat species produced a single young per year and showed high synchrony of parturition. Clearly, strategies of adult versus juvenile survival and the selection of subunits of overall habitat may themselves account for considerable variation, even between closely related sympatric species.

Many more comparative studies are needed, comparing both strategies of related species in similar environments and those used by a single species in different environments. Such studies of the strikingly contrasting patterns reviewed here are essential to any further progress in understanding the complexities of bat survival and reproductive strategies, particularly in the tropics.

2.3. Developmental State at Birth

Factors determining the developmental state of young bats at birth remain largely uninvestigated. Members of the Megachiroptera are smaller relative to adults, but more advanced developmentally at birth when compared to the Microchiroptera (Orr, 1970). Within the Microchiroptera members of the family Phyllostomidae are most advanced at birth (Kleiman and Davis, 1979), while members of the Emballonuridae appear to be largest relative to adult size (Davis, 1944). Considerable variation in the relative sizes of the young exists in the Vespertilionidae, but smaller bats generally seem to produce relatively larger young at birth. They are born with very large feet and thumbs, but beyond their ability to cling tenaciously to roosts, all are highly dependent upon their mothers (Orr, 1970). Relative to other mammals, all bats are enormous at birth (Leitch *et al.*, 1959); they weigh approximately 15–30% of the mass of the postpartum mother (Wimsatt, 1960; Orr, 1970; Kleiman and Davis, 1979).

Neonate-to-adult mass and/or forearm-length ratios are available for several species. Literature on these ratios in the families Phyllostomidae and Vespertilionidae is reviewed by Kleiman and Davis (1979), and there are additional relevant reports (Smith, 1956; Wimsatt, 1960; Dwyer, 1963; Rakhmatulina, 1972; Kunz, 1973; Ransome, 1973; Noll, 1979). Unfortunately, most are difficult to evaluate. Methods are not consistent among authors. Some studies most likely include at least day-old young. Many do not match neonates with their own mothers, and most do not consider whether or not the young or adults have fed prior to weighing. Additional complicating factors include failure to clearly state whether or not masses of neonates or near-term fetuses include the placenta, and the fact that many births for newly captured mothers are premature. The inclusion of data from near-term fetuses greatly increases variability, owing to subjectivity in deciding which are near term.

Despite these problems, perusal of the available literature does indicate that considerable intraspecies variability may exist. For example, Smith (1956) weighed 19 neonate *Myotis lucifugus* that were found in their natural roosts with wet umbilical cords still attached and found that their masses were 25.0–31.9% of their mothers' masses. Other even larger intraspecies ranges have been reported (see Kleiman and Davis, 1979), but these may result from biases in choosing "near-term" fetuses or from an inappropriate choice of adult masses for comparison.

Comparative studies, either within or among species, are unavailable but essential to a better understanding of the growth and survival of bats. In general, the size and developmental state of mammals at birth are affected by maternal size, age, nutritional and hormonal condition, litter size, and length of gestation (Everitt, 1968). Wimsatt (1960) noted that total fetal burdens between species bearing one versus two young per litter did not differ appreciably, but that individual twins were smaller. Twins in *Myotis austroriparius* and *Pipistrellus subflavus*, for example, each weigh approximately 15–16% of postpartum females. Foster *et al.* (1978) reported that twin neonates of *M. austroriparius* are born in a very reduced state of development compared to neonates of other nontwinning *Myotis*, but they made no intraspecies comparison.

Litter size often varies geographically or seasonally within species. Yet few comparisons of the relative size and state of development of singles versus twins within individual species have been made. Maeda (1972) reported that single *Nyctalus lasiopterus* were better developed at birth than twins, but Ransome (1973) found no difference in the state of development in newborn *Rhinolophus ferrumequinum*. Size comparisons were not made. Myers (1977) found a significant negative correlation between the size and number of embryos in *Eptesicus furinalis*. Such information, combined with a knowledge of how litter size varies in different environments, both among localities and at single localities during years of climatic extremes, is badly needed.

Additionally, comparisons are needed on relative neonate sizes and states of development within species occupying wide latitudinal or climatic extremes but producing only a single young. Does neonate size vary with roost temperature or latitude? Do larger mothers produce proportionately larger neonates? Is the age of the mother important? Do stressful climatic and feeding conditions only lengthen gestation or might they also result in smaller or less developed young at birth?

2.4. Litter Size

Most bats produce only one young per litter. In the Megachiroptera only two species, *Epomops dobsoni* and *Pteropus rufus*, are known to produce twins, and even in the Microchiroptera litter sizes exceeding one are uncommon (Carter, 1970). Although twins sometimes occur in the Megadermatidae (Brosset, 1962b), Rhinolophidae (Ransome, 1973), Phyllostomidae (Barlow and Tamsitt, 1968), and possibly the Molossidae (Krutzsch, 1955a), most litters of more than a single young are restricted to the Vespertilionidae.

Within the Vespertilionidae most species produce only one young per litter, but a wide variety commonly bear twins. These include *Chalinolobus dwyeri* (Dwyer, 1966b), *Eptesicus fuscus* (Kunz, 1974a), *E. furinalis* (Myers, 1977), *Lasionycteris noctivagans* (Kunz, 1982), *Lasiurus cinereus* (Barbour and Davis, 1969; Mumford, 1969; Sealy, 1978), *Myotis austroriparius* (Rice, 1957), *M. leibii* (Tuttle and Heaney, 1974), *Nyctalus lasiopterus* (Maeda, 1974), *Pipistrellus ceylonicus*, *P. coromandra* (Brosset, 1962c), *P. dormeri* (Madhavan, 1978), *P. nanus* (LaVal and LaVal, 1977), *P. subflavus* (Barbour and Davis, 1969), *Scotophilus heathi*, *S. temmincki* (Brosset, 1962c), *Tylonycteris pachypus*, and *T. robustula* (Medway, 1972). Others, and probably some of the above, produce mostly twins but range from one to three young per litter; these are *Antrozous pallidus* (Orr, 1954), *Nyctalus noctula* (Gaisler *et al.*, 1979), *Nycticeius humeralis* (Easterla and Watkins, 1970), and *P. pipistrellus* (Rakhmatulina, 1972). *Lasiurus borealis* (Kunz, 1971; Mumford, 1973), *L. ega* (Myers, 1977), *L. intermedius*, and *L. seminolus* (Barbour and Davis, 1969) produce mostly litters of three or four, and litters of up to five are possible (Mumford, 1973).

Litter size varies within species according to the mother's age (Davis, 1969; Gaisler *et al.*, 1979), geographically (Stebbings, 1968; Kleiman and Racey, 1969; Kunz, 1974a; Gaisler *et al.*, 1979), seasonally (Myers, 1977; Madhaven, 1978), and even at a single locality in years of differing climatic conditions (Rakhmatulina, 1972). *Eptesicus fuscus* is reported to bear twins more often east than west of the Great Plains in the United States (Barbour and Davis, 1969), but litter size ranges between one and two in both areas (Christian, 1956; Kunz, 1974a; Schowalter and Gunson, 1979), with a single litter of four reported from Georgia (Harper, 1929). In England *Nyctalus noctula* produces mostly one young per litter, rarely two, whereas on the mainland of Europe litters of two are more common, especially among older adults, with rare litters of three (Kleiman and Racey, 1969; Gaisler *et al.*, 1979). Similarly, *Pipistrellus pipistrellus* in England rarely produces more than one young per litter (Stebbings, 1968), but in Russia litters often contain two and rarely three (Rakhmatulina, 1972).

Seasonal climatic differences apparently have a major impact upon litter size in both tropical and temperate environments. In a tropical environment in India *Pipistrellus dormeri* produces mostly twins during the rainy season and more singles thereafter (Madhavan, 1978), when food resources probably are reduced. In a subtropical environment in Paraguay *Eptesicus furinalis* breeds twice annually (Myers, 1977). In the first reproductive period some embryos are lost between fertilization and parturition, but most litters contain two young (range one to two). The second litter is produced as food resources apparently are declining and always consists of only one young. In a less predictable temperate environment in Russia *P. pipistrellus* produces litters ranging from 25.0 to 88.5% twins or triplets, with the smallest litters observed in years of unusally long and cold spring weather (Rakhmatulina, 1972).

In an environment that fluctuates unpredictably early fertilization of several ova, with subsequent resorption of some in response to unfavorable seasons, may permit bats to maximize the number of embryos brought to term (Myers, 1977). *Eptesicus fuscus* often produces four or more ova but normally bears only one or two young (Wimsatt, 1945). Embryo resorption is well known in temperate bats (Pearson *et al.*, 1952; Barbour and Davis, 1969; Myers, 1977), and resorption and abortion apparently occur in response to lean food seasons, even in tropical environments (Bradbury and Vehrencamp, 1977b).

Several possible reasons why bats might vary in the number and size of litters have been suggested by Myers (1977). These include differences in the availability of food relied upon by different species, the evolutionary history of populations of each species, and their present associations with other populations in contiguous but different environments. In general, species that depend upon highly seasonal food supplies (Bradbury and Vehrencamp, 1977b) or live in exposed roosts that are vulnerable to adverse weather conditions or predators (Humphrey, 1975) probably suffer increased mortality and should therefore increase reproductive output.

Despite the fact that many species occupy wide ranges of environmental conditions and roost types, investigators rarely have taken advantage of this circumstance to investigate either intraspecies plasticity or the probable impact of environmental parameters (Myers, 1977). This is particularly true regarding litter size, where we are unable to find a single comparative study. Future studies should compare intraspecies variability in litter size among habitat and roost-type extremes as well as within habitats between years of climatic extremes. Laboratory studies in which food and/or temperature are varied during pregnancy would also be useful.

3. POSTNATAL GROWTH AND DEVELOPMENT

Very little is known about bat growth and development. Even superficial studies are lacking for 14 of 19 chiropteran families, and less than 3% of the known species have been investigated. Of the few studies that are available (Table I), most were conducted in captivity, and 80% are restricted to a single family, the Vespertilionidae. Among these, most are noncomparative, relatively superficial, and limited to temperate-zone bats. They report details of ontogeny but most fail to mention environmental conditions known to have major impacts upon growth and development. Vague, graphical presentations of data and a lack of adequate knowledge of adult proportions for most species further complicates

TABLE I
Growth Rates[a]

Species	Forearm (mm/day)	Weight (g/day)	References
Megachiroptera			
Pteropodidae			
Hypsignathus monstrosus	0.7	1.3	Bradbury, 1977b[b]
Rousettus sp.	1.5		Kulzer, 1969[b]
Microchiroptera			
Rhinolophidae			
Rhinolophus hipposideros	1.0		Rybář, 1971
Phyllostomidae			
Carollia perspicillata	0.8	0.3	Kleiman and Davis, 1979[b]
Vespertilionidae			
Antrozous pallidus	1.5	0.3–0.4	Davis, 1969
Eptesicus fuscus	0.8–1.4	0.3–0.47	Davis *et al.*, 1968; Kunz, 1974; Burnett and Kunz, 1982
E. serotinus	1.2	0.9	Kleiman, 1969[b]
Lasiurus cinereus	0.4		Bogan, 1972[b]
Miniopterus schreibersii	0.7	0.3	Dwyer, 1963
Myotis adversus	0.6		Dwyer, 1970
M. grisescens		0.2–0.4	Tuttle, 1975
M. lucifugus	1.6	0.34	Kunz and Anthony, 1982
M. macrodactylus	0.9		Maeda, 1976[b]
M. myotis	2.0		Krátky, 1970[b]
M. thysanodes	1.5	0.3	O'Farrell and Studier, 1973
M. velifer	1.6	0.4	Kunz, 1973
Nyctalus lasiopterus	0.9–1.6	0.6–1.0	Maeda, 1972[b]
N. noctula	0.9	0.6	Kleiman, 1969[b]
Nycticeius humeralis	0.7	0.1	Jones, 1967[b]
Pipistrellus pipistrellus	0.7–0.8	0.1–0.42	Kleiman, 1969[b]; Rakhmatulina, 1972
Plecotus (=*Corynorhinus*) *townsendii*	1.2		Pearson *et al.*, 1952
Molossidae			
Tadarida brasiliensis			
T. b. cynocephala	0.7–0.9	0.4	Pagels and Jones, 1974
T. b. mexicana	1.0		Short, 1961

[a]Many growth rates are extrapolations from reported growth curves. Calculations of rates are for the first 14 days only.
[b]Captive study.

interpretation. Finally, intraspecies plasticity of growth and development has been thoroughly investigated for only a single species.

One of the few generalizations that presently seems justified is that megachiropterans tend to develop more slowly than microchiropterans (Orr, 1970; Bradbury, 1977a). Also, it appears that postnatal growth curves of bats more closely resemble birds than other mammals (Pierson and Kunz, 1982). In this respect *Myotis lucifugus* may be typical of many bats. Most growth is completed relatively rapidly, prior to initiation of flight (Burnett and Kunz, 1982; Kunz and Anthony, 1982). Weight peaks prior to flying and tends to decrease thereafter until after weaning is complete. Available observations do demonstrate that bat growth and development is highly variable and in need of much more comparative study. Our purpose is to identify environmental factors likely to influence growth and development and to suggest areas in need of further investigation.

3.1. Preflight

Factors that determine preflight success can be divided into two general categories, those at the roost, which affect young bats directly, and those away from roosts, which exert their effect indirectly through their influence on mother bats. At the roost the most important growth-determining factor for most bats, especially subtropical and temperate species, appears to be temperature. In the only comparative study of the effect of temperature on growth, the postnatal preflight growth rates of *Myotis grisescens* occupying similar roosts were found to be proportional to temperature when colony size was constant and proportional to colony size when temperature was constant (Tuttle, 1975). LaVal and LaVal (1980) tested and supported these conclusions. Low temperatures are known to greatly increase metabolic requirements (Stones, 1965; McNab, 1973), and the effect of colony size is consistent with the known ability of larger colonies to raise roost temperatures (Cagle, 1950; Rice, 1957; Henshaw, 1960; Davis *et al.*, 1962; Dwyer, 1963; Dwyer and Hamilton-Smith, 1965; Constantine, 1967; Kunz, 1973; Tuttle, 1975) and to further reduce thermoregulatory costs by clustering (Herreid, 1963, 1967a; Trune and Slobodchikoff, 1976). Temperature may be the single most important factor in determining roost choice for many temperate-zone bats (Harmata, 1973) and is of major importance in limiting the distribution of at least some species (see Chapter 4). Its potential impact on growing bats, even in the tropics, needs further investigation.

Since the growth rates of young bats are highly sensitive to temperature, in species that form maternity colonies, those born either early or late in the parturition period may suffer reduced growth rates. Thermoregulatory costs would be increased owing to a reduced ability to cluster with other bats. The possibility of individual variation within a single colony, based on time of birth, was suggested by Dwyer (1963) and Herreid (1967a). Kunz (1973) tested this hypothesis by comparing early postnatal growth rates of *Myotis velifer* from two cohorts marked as neonates early (June 17) and late (June 27) in the parturition period but found no significant differences in either size or mass. Selection for birth synchrony is probably strongest in colonial species that occupy relatively cool roosts, such as temperate-zone caves.

Although energetic losses to parasites must be considerable for colonial bats, no studies have attempted to estimate these costs or their effects on growth rates. Colonial bats are more heavily parasitized than solitary species (Krutzsch, 1955b), and juveniles of the former are sometimes heavily infested (Stebbings, 1966b). Watkins (1972) noted that ectoparasites at maternity roosts of *Nycticeius humeralis* tended to reach peak abundance when newborn young were in the roost, and Tuttle (unpublished observations) observed late-born young *Myotis grisescens* that were apparently killed by the exceptionally large numbers of mites that attacked them when other colony members left to forage. A wide variety of bat parasites is known (Constantine, 1970; Ubelaker, 1970; Ubelaker *et al.*, 1977; Webb and Loomis, 1977; see Chapter 10), and investigations of their impact upon bat growth, survival, and overall behavior are badly needed. At least for the more common roost-dwelling ectoparasites, such studies seem quite feasible.

Away from the maternity roost, factors such as night length and weather exert indirect effects on the growth and development of young bats by their impact on maternal ability to feed. The growth of young birds is enhanced by increased day length (Woodward and Snyder, 1978), and Ransome (1973) reported probable converse effects of short nights on the prenatal development of *Rhinolophus ferrumequinum*. The fact that adverse weather can affect maternal feeding efficiency is well known. Strong wind (Stebbings, 1966b; Maeda, 1974), fog (Pye, 1971), severe drought (Nellis and Ehle, 1977), or low temperature (Stebbings, 1966b; Roer, 1973) can seriously reduce or prevent feeding. Surprisingly, bats can and do forage in light to moderate rain (Stebbings, 1966b; Kunz, 1973; Maeda, 1974) and sometimes even in heavy rain during thunderstorms (Pye, 1971).

The impact of adverse maternal feeding conditions on the growth of young bats remains unstudied, yet available reports indicate that maternal failure can have extreme consequences. The apparent starvation of young *Myotis myotis* was attributed to inadequate feeding by their mothers during a period of prolonged cold (Roer, 1973), and young *Brachyphylla cavernarum* suffered heavy mortality during a severe drought (Nellis and Ehle, 1977), probably for the same reason. The young of species that roost in relatively exposed roosts, such as in foliage or hollow trees, are especially vulnerable to adverse weather. In addition to receiving reduced nourishment, they may simultaneously face the stress of increased thermoregulatory costs. Krzanowski (1956) reported that young *Nyctalus leisleri* in a "nest box" were very late in developing in an exceptionally cold spring.

Additional factors requiring future investigation involve diet and genetics. Some tropical bats switch seasonally between fruit versus pollen and nectar diets (Heithaus *et al.*, 1975), nursing one young while feeding primarily on fruit and the next while on a diet consisting mostly of nectar and pollen (Humphrey and Bonaccorso, 1979). Does diet affect parental investment, juvenile growth, and survival in these bats? Bats at latitudinal extremes face extremely short growing seasons. Have they adjusted genetically? If so, what happens in species such as *Eptesicus fuscus* and *Lasiurus cinereus*, which have extremely wide latitudinal ranges (Hall, 1981). Under controlled laboratory conditions, would the young of mothers from latitudinal extremes grow at similar rates? Lack (1968) emphasized that differences in growth rates in some birds may be adaptations to particular environments.

3.2. Postflight

3.2.1. First Flight to Weaning

Since more than 70% of all reports of times till first flight and weaning are restricted to temperate-zone vespertilionid bats, and since little documentation of intraspecies variability exists, interpretation of available observations remains difficult. Intraspecies variability under varied environmental conditions is high. *Myotis lucifugus* first flew in 14–15 days in one study (O'Farrell and Studier, 1973), but in 21–30 days in others (Griffin, 1940; Burnett and Kunz, 1982). In *M. grisescens* time till first flight varied among years within a colony and between colonies in the same year (Tuttle, 1975). Known-age young were compared between a colony of 600 young at 13.9°C ambient temperature and a colony of 2200 young at 16.4°C. Significantly increased growth rates in the latter resulted in mean attainment of first flight at 24 days, compared to 33 days in the former. In areas where growing seasons are short, such differences likely affect subsequent survival.

The postflight period is clearly a time of stress. Just prior to first flight fat levels are actually 55% higher than adult levels (Pierson and Kunz, 1982). Fat deposition as a developmental strategy is an apparent response to the stress of learning to fly and to hunt for an unpredictable food source (Kunz, 1973; 1974b; Pierson and Kunz, 1982). Loss of weight between the time of first flight and weaning is common in the vespertilionid bats thus far studied (Short, 1961; Dwyer, 1963; Davis, 1969; Kleiman, 1969; Krátky, 1970; Kunz, 1973; Tuttle, 1976b) and probably occurs in most bats. Success during the postflight period is undoubtedly affected by a bat's state of development at birth, the length of its preflight growth period, and the amount of time allowed till weaning.

Large pteropodids first fly at 9–12 weeks of age and are weaned at 15–20 weeks (Marshall, 1947; Kulzer, 1958; Neuweiler, 1962; Nelson, 1965a). Emballonurids produce enormous young that often first fly at two weeks but sometimes require 5 weeks. Weaning occurs at 4–8 weeks (Brosset, 1962a; Al-Robaae, 1968; Bradbury and Emmons, 1974). In the Rhinolophidae *Rhinolophus rouxi* first fly in approximately 6 weeks and are weaned in roughly 8 weeks (Sreenivasan *et al.*, 1973). Weaning in *Hipposideros commersoni* apparently occurs at a record 20 weeks (Brosset, 1969). Most phyllostomids probably fly in 2½–4 weeks (Kleiman and Davis, 1979), but captive *Desmodus rotundus* required 8–10 weeks to fly and nursed for 3–9 months (Schmidt and Manske, 1973). Such prolonged nursing may have resulted from unnatural conditions of captivity (see Kunz, 1973). Weaning times under natural conditions are unknown for phyllostomids. Vespertilionids first fly in 2 weeks to 2 months, but mostly in 3–4 weeks. Time from birth till weaning ranges from 5 to 8 weeks (Griffin, 1940; Orr, 1954; Brosset, 1962c; Dwyer, 1963; Davis *et al.*, 1968; Kleiman, 1969; Fenton, 1970a; Kunz, 1971, 1973, 1974a; Bogan, 1972; Maeda, 1972, 1976; Rakhmatulina, 1972; O'Farrell and Studier, 1973; Bradbury, 1977a; Jones, 1977; Burnett and Kunz, 1982).

The interval from first flight to weaning is a "grace period" during which young bats must perfect orientation and flight skills, learn to catch prey (Kunz, 1974b; Buchler, 1980), and memorize locations of sometimes distant patches of food resources (Bradbury, 1977a). This period is critical and deserves much more attention. Studies of variation within species occupying contrasting habitats or within a single colony in years of climatic contrasts would be useful. Why, for example, does *Taphozous melanopogan* apparently wean its young within only 2 weeks of first flight (Brosset, 1962a), while other emballonurids thus far studied allow 4–6 weeks (Bradbury, 1977a)?

Growth success in the period between first flight and weaning may be highly variable, even among colonies of a single species. Highly significant differences in the growth patterns of *Myotis grisescens* during the first 3 weeks of flight were related to the distances they were forced to travel to reach appropriate foraging habitat (Tuttle, 1976b). The mean masses of newly volant young were inversely proportional to the distance between the roosts and bodies of water where they foraged. Over a 2-week period mean masses in the two colonies farthest from feeding habitats continually declined, while the young in the closest colonies gained. During the first 3 weeks of flight even forearm growth was affected, with significant growth occurring only in the least stressed colonies. The length of lactation ranged from 58 to 76 days among colonies and was extended by approximately 2 weeks in the most stressed colony (Tuttle, unpublished data). Apparently the time of weaning is at least partially determined by the state of development of young bats and can vary considerably within a single species.

3.2.2. Attainment of Sexual Maturity

Many bats superficially become nearly indistinguishable from adults within as little as 3–4 months, however attainment of sexual maturity often comes much later (Orr, 1970). Ages of reproductive maturity are highly variable among taxa, between males and females of a single species, and sometimes even within a single sex of one species (Table II). In the Megachiroptera reproductive maturity in males is achieved in from 15 months to more than 2 years. Females thus far studied require a similar period, with the exception of *Hypsignathus monstrosus* and *Rousettus leschenaulti*, which require only 5–6 months. In the Microchiroptera both sexes of the families Rhinopomatidae, Emballonuridae, Megadermatidae, and Rhinolophidae usually reach sexual maturity only after a period of 1–2 years. Two exceptions are known in these families. Male *Taphozous melanopogon* (Emballonuridae), and probably females as well, reach maturity in as little as 6 months, and some female *Rhinolophus hipposideros* (Rhinolophidae) are sexually mature in as little as 3 months (though males and some females require 15 months).

Little is known about sexual maturity in members of the Phyllostomidae. *Macrotus californicus* males reach sexual maturity at about 16 months, while females apparently require only 3–4 months. Maturity in female *Artibeus jamaicensis* is reached in 8–11 months, and about 16 months seem to be required in *Phyllostomus hastatus*.

In the Vespertilionidae the age of reproductive maturity is highly variable, ranging approximately 3–16 months in both sexes. Males rarely attain sexual maturity ahead of females, but in nearly half of the species studied females mature ahead of males. In a nearly equal number of species, both first reproduce at the same age. Within a number of species the age of sexual maturity of either sex varies by up to 12 months. Polyestrous species often attain sexual maturity soon after they achieve morphological maturity.

Ages of sexual maturity are now available for many bats in a wide variety of taxa (Table II). These bats represent species of contrasting social organization, living in highly varied environments. Yet, surprisingly little attention has been devoted to the adaptive significance of these varied developmental and reproductive strategies. Many latitudinally and ecologically widespread species offer excellent opportunities for comparative study.

4. SURVIVAL

4.1. Survival Analyses and Results

Bat survival analyses have fallen into three general categories: (1) analyses utilizing demographic information such as age structure and birth rate, (2) studies

TABLE II
Ages at Which Sexual Maturity of Bats Is Reached

	Age at sexual maturity[a]		
Species	♂♂	♀♀	References
Megachiroptera			
Pteropodidae			
Dobsonia moluccensis	>2 years		Dwyer, 1975
Hypsignathus monstrosus	18 months	6 months	Bradbury, 1977b
Pteropus ornatus	~16 months	~16 months	Asdell, 1964
P. poliocephalus	~16 months	~16 months	Nelson, 1965b
P. scapulatus	~16 months	~16 months	Nelson, 1965b
Rousettus leschenaulti	15 months	5 months	Gopalakrishna and Choudhari, 1977
Microchiroptera			
Rhinopomatidae			
Rhinopoma microphyllum (=*kinneari*)	~16 months	~16 months	Anand Kumar, 1965
Emballonuridae			
Balantiopteryx plicata		1 year	Bradbury and Vehrencamp, 1977b
Rhynchonycteris naso		>1 year	Bradbury and Vehrencamp, 1977b
Saccopteryx bilineata		1 year	Bradbury and Vehrencamp, 1977b
S. leptura		>1 year	Bradbury and Vehrencamp, 1977b
Taphozous melanopogon	6 months		Brosset, 1962d
Megadermatidae			
Cardioderma cor	~16 months	~16 months	Matthews, 1941
M. lyra	15 months	19 months	Ramakrishna, 1951; Brosset, 1962b
Hipposideridae			
Hipposideros bicolor	~1 year	~1 year	Brosset, 1962b
H. fulvus	18–19 months	18–19 months	Madhavan and Gopalakrishna, 1978
H. speoris	~16 months	~16 months	Brosset, 1962b
Rhinolophidae			
Rhinolophus euryale	2¼ years	15 months–2¼ years	Dinale, 1968
R. ferrumequinum	15 months–4¼ years	15 months–3¼ years	Asdell, 1964; Dinale, 1964

(continued)

TABLE II (*Continued*)

Species	Age at sexual maturity[a] ♂♂	♀♀	References
R. hipposideros	15 months	3–15 months	Sluiter, 1960; Gaisler and Titlbach, 1964; Gaisler, 1965
R. rouxi		>1 year	Brosset, 1962b
Phyllostomidae			
Artibeus jamaicensis		8–11 months	Fleming *et al.*, 1972
Macrotus californicus	~16 months	3–4 months	Bradshaw, 1961
Phyllostomus hastatus		~16 months	McCracken and Bradbury, 1981
Vespertilionidae			
Eptesicus fuscus	~4 months	~4–16 months	Christian, 1956
E. furinalis	<1 year	<1 year	Myers, 1977
Lasiurus ega	<1 year	<1 year	Myers, 1977
Miniopterus australis		~16 months	Dwyer, 1968
M. schreibersii	~16 months	~16 months	Brosset, 1962c; Dwyer, 1963
Myotis austroriparius	~10–15 months	~3–10 months	Rice, 1957
M. emarginatus		3 months	Sluiter and Bouman, 1951
M. grisescens	~16 months	~16 months	Guthrie, 1933; Miller, 1939
M. lucifugus	~16 months	~4 months	Miller, 1939
M. myotis	15 months	3 months	Sluiter and Bouman, 1951; Sluiter, 1961

(*continued*)

of unaged cohorts, and (3) studies of cohorts of known age. Of the three categories, type 2 has been predominant.

Studies of chiropterans are beset by most of the usual obstacles to valid analysis presented by mammals, plus others more or less unique to the order. The consensus of bat researchers, regardless of approach, is that bats are unusually long-lived for small mammals and that even the best survival estimates are underestimates. Beyond this area of agreement, however, results are varied.

The student of chiropteran survival analyses will be considerably frustrated in attempting to compare the results of various studies. A number of reasons for this difficulty of comparison will be listed at the beginning of the section on

TABLE II (*Continued*)

	Age at sexual maturity[a]		
Species	♂♂	♀♀	References
M. mystacinus		3–15 months	Sluiter, 1954
M. nigricans	2.5–4 months	3–4 months	Wilson, 1971; Myers, 1977
M. velifer	~16 months	~4 months	Hayward, 1970; Kunz, 1973
Nyctalus lasiopterus	~16 months	~4 months	Maeda, 1974
N. noctula	3–15 months	3 months	Kleiman and Racey, 1969; Gaisler *et al.*, 1979
Pipistrellus pipistrellus	~16 months	~4 months	Asdell, 1964; Racey, 1974
Plecotus auritus		~16 months	Stebbings, 1966b
P. rafinesquii	~16 months		Jones and Suttkus, 1975
P. townsendii	~4–18 months	~4 months	Pearson *et al.*, 1952
Scotophilus temmincki	~9 months	~9 months	Gopalakrishna, 1947
Tylonycteris pachypus	3 months	6 months	Medway, 1972
T. robustula	3 months	6 months	Medway, 1972
Molossidae			
Tadarida brasiliensis			
T. b. cynocephala	~16 months		Sherman, 1937
T. b. mexicana	18–22 months	9 months	Short, 1961

[a]Reported in months and years. Many estimates represent the time when active sperm are first visible in males or the time of first estrus in females and do not necessarily imply active breeding.

unaged-cohort analysis, and apply as well to analyses of known-age cohorts. The task of interpretation and comparison is also needlessly confounded by the frequent lack of clear statements of how data were tallied and indications of sample size. Variance estimates on the survival rates usually have not been provided. Granted, the underlying data of many existing studies may not have warranted rigorous statistical treatment; however, the common application of regression and life-table techniques to chiropteran recapture data in these studies implies some degree of reliability. Comparisons within and among species have and will be made based on these presentations. It would seem advisable in future studies to either not imply sufficient reliability or to take care to present the data in such a way as to make comparisons maximally informative and valid.

4.1.1. Demographic Analyses

In one of the early studies of Chiroptera, that attempted to analyze survival, Rice (1957) calculated a birth rate for *Myotis austroriparius* in Florida based on observations of females and young. He then proposed an annual survival rate of 46%, the rate necessary to maintain a stable population given that birth rate. Davis *et al.* (1962) placed Texas *Tadarida brasiliensis* populations into relative age classes by tooth wear, assumed a constant mortality rate for all age classes of bats after the initiation of flight, and assumed (as had Rice) that the population was stable. They then constructed a table of age compositions with a series of hypothetical survival percentages ranging from 50 to 90% and matched the results with the age composition observed in their tooth-wear classes. The most reasonable match was an adult survival rate of 70–80%, with a maximum age of 15 years.

Stebbings (1966b), working with a very small population of *Plecotus* (two species) in England, established that there were approximately nine births each year and that the population remained constant. He then calculated the life expectation for members of the colony to be 4 years, and the survival rate 0.75. Defining life-span as the number of years after which only 10 of an initial total of 1000 individuals are still alive (see Sluiter *et al.*, 1956b), his *Plecotus* would have a life span of 16 years. Dwyer (1966a) used estimates of the age distribution, birth rate, and population size of *Miniopterus schreibersii* in Australia to obtain a juvenile survival rate of 0.75 after 1 year. The life-span was estimated to be 15 years, and life expectancy after the first year 3.5 years.

Relatively few studies have provided estimates of preflight mortality. From a tally of young *Tadarida brasiliensis* that had fallen from the roost, Herreid (1967b) estimated a loss of only 1.3%. Estimates of fallen young were 1% for *Myotis thysanodes* and 2% for *M. lucifugus* in studies by O'Farrell and Studier (1973). In a more recent study Foster *et al.* (1978) found that 11.8% of young *M. austroriparius* died prior to weaning age (approximately 1 month) and that 75% of the preweaning deaths occurred in the first week of life. These last figures were obtained by censusing carcasses below a roost, using tooth-development classes as an aging technique. Based on dead young under the roost, Rakhmatulina (1972) estimated the postnatal mortality of *Pipistrellus pipistrellus* to be 3% in 1967 and 1% in 1969. Cockrum (1955) found the average number of embryos in 41 adult female *Lasiurus borealis* to be 3.00, but the average number of preflight young found with 26 females (2.65) was significantly different. Similarly, Kunz (1971) reported an average prepartum litter size of 3.24 and an average postweaning litter size of 1.41 for *L. borealis*. As summarized by Foster *et al.* (1978), prenatal mortality for this species was 8.6%, and mortality from birth to weaning accounted for 25.7%, for a total loss of 34.3%. Humphrey and Kunz (1976) estimated a 6.5% natality loss and a 4% preweaning loss for

Plecotus townsendii. Christian (1956), from observations of the ratios of newly flying young to adult female *Eptesicus fuscus*, estimated a loss of 7% between birth and weaning. As pointed out by Foster *et al*. (1978), rates of preweaning mortality appear to be related to litter size. In contrast, Bradbury and Emmons (1974) found no mortality during postnatal growth prior to weaning for *Saccopteryx bilineata* in Trinidad.

A number of problems are apparent in, although not necessarily unique to, demographic analyses. When the population dealt with is small, the data are of questionable reliability in extrapolations to the species at large or for comparisons among species. On the other hand, when the population is large, estimations of sex ratio, birth rate, and population size are extremely difficult and subject to error. Furthermore, the techniques allow no examination of the assumptions upon which they are based, such as constant adult survival rate and stable population size. Carcass counts, of course, are biased by undetected removal, as recognized by the authors.

Finally, most analyses depend heavily on error-prone classification schemes for the construction of age distributions. A wide variety of age-determination methods have been used, including sexual characters, dental cementum lines, pelage, long-bone length, epiphyseal closure, and tooth wear. Baagøe (1977) and Racey (1974) provide reviews of the literature and discuss the advantages and difficulties of the various techniques. It is obvious, however, that any such method that depends on determining a stage of growth will be unreliable to the extent that growth is not constant. As Pucek and Lowe (1975) point out in their discussion of aging techniques for small mammals, it is "important to know the errors and confidence intervals of any age estimates based upon growth and development, since these processes are quite variable among themselves." Difficulties kept in mind, it is of interest to note the wide range of survival-rate-values obtained in these studies on various species.

4.1.2. Unaged-Cohort Analyses

In cohort analysis a group of individuals is banded and a recapture history recorded over a period of time. Since many species of bats congregate in hibernation sites during the winter but are difficult to find or capture in summer roosts, the vast majority of bat survival studies have been conducted at winter hibernacula. Unfortunately, if the initial banding of the cohort is done at the hibernation site, there is no reliable means of determining either the ages of the individuals banded or whether summer colonies are experiencing equal survival rates. As a result, the survival estimates obtained are the average survival rate at the prevailing age structure of the population rather than age-specific estimates, and pooled population estimates rather than local subgrouping estimates.

Table III provides a summary of the results of most of the available analyses of unaged cohorts. It is apparent that even within species a wide range of survival values has been obtained. Unfortunately, it is difficult to compare the results of one study with another owing to differences in such factors as the duration of the study, sample size, sampling methods and bias, calculation techniques and basic assumptions. For example, Bezem *et al.* (1960) combined data from several hibernacula and from multiple years of banding and assumed that the age dis-

TABLE III
Mean Annual Rate of Survival of Bats in Unknown-Age Cohorts, Generally Excluding the First Year after Marking

Species	Sex	Duration of study (years)	Percentage survival (%)	References
Rhinolophidae				
Rhinolophus hipposideros	F and M	5	65.7	Bezem *et al.*, 1960
Vespertilionidae				
Myotis dasycneme	F and M	5	67.0–70.0	Sluiter *et al.*, 1971
M. dasycneme	F and M	5	66.7	Bezem *et al.*, 1960
M. mystacinus	F	5	76.7	Sluiter *et al.*, 1956b
M. mystacinus	M	8	80.3	Sluiter *et al.*, 1956b
M. mystacinus	F and M	7	75.2	Bezem *et al.*, 1960
M. emarginatus	F and M	6	70.1	Bezem *et al.*, 1960
M. daubentoni	F and M	4	80.0	Bezem *et al.*, 1960
M. myotis	F and M	4	63.7	Bezem *et al.*, 1960
M. lucifugus	F	13	85.7	Humphrey and Cope, 1976
M. lucifigus	M	11	77.1	Humphrey and Cope, 1976
M. lucifugus	F	10	70.8 ± 2.2	Keen and Hitchcock, 1980
M. lucifugus	M	10	81.6 ± 1.0	Keen and Hitchcock, 1980
M. lucifigus	F	4	63.6–84.6	Brenner, 1974
M. lucifugus	M	4	54.7–85.5	Brenner, 1974
M. sodalis	F	13	4.1–75.9	Humphrey and Cope, 1977
M. sodalis	M	13	36.3–69.9	Humphrey and Cope, 1977
Eptesicus fuscus	F and M	9	66.7–77.4	Beer, 1955
Pipistrellus subflavus	F	8	43.8–73.7	Davis, 1966
P. subflavus	M	14	38.4–98.1	Davis, 1966

tribution, sex ratio, and probability of survival were constant throughout the study (although they allowed for a recognized decrease in population size over the years). Their estimate of survival probability was obtained from the "fitting of straight regression lines to the points obtained by plotting the logarithm of the fraction of recapture against time."

To support their assumption of sex independence, Bezem *et al.* tested the regression values of males versus females. They found no difference in slope; they therefore concluded that no difference in survival probability existed, contrary to the findings of Eisentraut (1950), whom they cited as having found a lower female survival. Tests were also done to check the validity of pooling data from different hibernating caves. Again, no differences in slope (survival) were obtained, but there were significant differences in level among the regression lines, which they interpreted as indicating a difference in the rates of recapture. One such difference in level also had been observed in the sex comparisons, but it was attributed to sampling effects. The assumption of independence of survival from age was supported only by "the evidence of the recapture data."

Davis (1966), on the other hand, treated the sexes separately and explicitly denied, at least for *Pipistrellus subflavus*, the validity of the assumption made by Bezem *et al.* (1960) (and also by Sluiter *et al.*, 1956b) that the probability of survival is independent of age. He also calculated a "percent captured of those surviving" estimate, using data available from his bandings year to year, and used those estimates for his curves and life tables. This corrects for some of the error introduced by failing to recapture individuals despite their continued survival. In contrast to the assumption of constant survival probability, Davis found that his survival percentage curves approximated exponential decay functions. However, even this function only approaches an adequate description of the general results. Survival during the first 2 years was rather low, but survival during the third and fourth years was exceptionally high. Survival rates then decreased beyond 4 years and fell sharply as they approached maximum lifespan. The ranges of figures provided in Table III for his study reflect the variation after the first year; in both cases the low figure is for the last-year class and the high figure is from the third year.

In his discussion of the shape of the survival curves Davis (1966) stated that an "exponential decay curve would apply to an animal population if all losses were due to predation or disease acting equally upon all age groups," but that experience and aging are important in determining survival rates. He tested this hypothesis by fitting a regression line to the middle (3–11) years and calculating confidence bands. Two points for younger and older bats fell outside the 99% bands, and three outside the 95% band, leading him to conclude that survival rate was not independent of age. In sum, Davis was critical of the use of constant survival rates and suspected that even more gradual change existed than would be reflected by fitting three exponential decay functions to the curve (for young, middle-aged, and old individuals).

Beer (1955) also reported a "general trend towards an increased survival rate with an increase in age" for *Eptesicus fuscus*, reflected in Table III. He provided two values for bats beyond the first year after banding, one (the lower figure) for the second year, and one for subsequent years. However, he attempted to compare his values with those found in Europe for *Myotis myotis* by combining his results into a single overall survival rate of 60% per year. Beer calculated his life tables from actual recapture data as opposed to the corrected values used by Davis (1966), perhaps making the latter study a more reasonable approximation to reality. Both studies suffer from small sample size (245 recaptures after banding for Beer; unreported but somewhat larger for Davis) due to the difficulty of finding large numbers of their species for banding and recapture. In addition, the populations included in Beer's study declined by about 60% over 5 years, further confounding interpretation of his results. Another study of *Eptesicus fuscus* (Goehring, 1972), also on unaged cohorts, gave a similar survival rate of 60% in the first year and "higher" thereafter.

Sluiter *et al.* (1971) banded both sexes of *Myotis dasycneme* at summer sites, in addition to winter hibernacula, and so gathered some information on known-age individuals. In their analysis, however, they excluded all males (small sample; they assume survival independent of sex) and all juvenile females, which were never seen after banding. They tallied their recapture data, counting the first recapture as though it were the time of banding, thus making their results equivalent to an unaged cohort. This also resulted in a reduction from their total of 579 recaptures to a considerably smaller final sample size for the study.

The *Myotis lucifugus* cohort groupings of Humphrey and Cope (1976) were composites. To counteract the downward bias of a low sampling success, irregular visits to the recapture sites, the extermination of colonies, bandings late in the study, and unrecorded existence for periods or total escape from recapture, Humphrey and Cope grouped cohorts from a variety of sites and years of banding by sex and season of banding. They then selected the highest value for the year class (number of years since banding) as representative for the cohort. They considered mean rates for the groups to be less realistic than the highest values. Since "no nonlinear patterns were evident" in the points beyond the first year, the survival rate for each curve after year one was approximated with a linear regression line; that is, they assumed minimum survival rates to be constant after 1 year of age, as did Bezem *et al.* (1960) and Sluiter *et al.* (1956b).

One result of Humphrey and Cope's (1976) use of only the highest values to represent the cohort is that "no probability statements may be made about the resultant regression equations. These data do not support statistical tests of the constancy of survival rates within a cohort type or of the equality of the survival rates of different cohort types." Points used in the regression analyses were based on mean cohort sample sizes of 143 (for males) and 291 (for females) at banding.

Numbers of recaptures are not given and are presumably much smaller. Life tables were produced using the values derived from the regression equations, rather than using actual recaptures or a Davis (1966) type of correction. This, of course, results in a constant estimate of survival after the first year in the life table, reflecting their assumption of linearity.

In their later paper on *Myotis sodalis* survival, Humphrey and Cope (1977) selected and analyzed cohorts in the same manner as their 1976 *M. lucifugus* paper. Although these data were supposedly from "healthy, relatively undisturbed populations," their estimates of winter populations varied from 1000 to 3200 to one cave, and from 30 to 200 in the other. In this study, however, they discard the assumption of constant adult survival and draw lines of different slope through various parts of the curves. The result is similar to that postulated by Beer (1955) and Davis (1966), with higher survival during the "middle-age" years than later. As they state, "the adult period of life is characterized by two survival phases. The first is a high and apparently constant rate from 1 to 6 years after marking" (females have an annual survival rate of 75.9%, males 69.6%). These are the higher values of the ranges given in Table I for the study. The second phase, after 6 years, is proposed to have lower and again constant rates of 66.0% for the females up to 10 years and 36.3% for males. The very low rate (4.1%) for females (Table III) appears after 10 years following marking. They state that this may be attributable to sampling error, since a very small number of animals remained alive.

Keen and Hitchcock (1980) provide a careful analysis of data from the bandings of unaged *Myotis lucifugus*, using the stochastic technique of Cormack (1964), which assumes age-independent survival but allows variability from year to year. The technique avoids many of the assumptions and pitfalls of the technique of Bezem *et al.* (Sluiter *et al.*, 1956b) and of the most commonly used regression method (e.g., Davis, 1966; Humphrey and Cope, 1976). The discussion of analytical methods presented by Keen and Hitchcock and their comparison of their results calculated with the three techniques are valuable and relevant to this discussion. Their estimates using Cormack's technique appear in Table III.

Establishment of realistic survival rates based on recaptures of mammals is clearly problematic. The above discussion has touched upon difficulties in the assumptions, design, and analysis of many existing studies on bat populations. There are, moreover, particular difficulties inherent in the interpretation of cohort data when the individuals are unaged.

4.1.3. Known-Age Cohort Analyses

Three papers are available that provide survival estimates based on known-age individuals of *Eptesicus fuscus*. All three involved both banding and recapture of bats at summer maternity colonies. Davis (1967) obtained his data when a colony he had banded was exterminated 10 months following banding. He recovered the following proportions of his original bandings: 60.5% of adult females, 34.8% of juvenile females, 51.3% of adult males, 8.3% of juvenile males (76 bands recovered). Brenner (1968) conducted a 3-year study. He did not provide exact figures on recovery rates, but the graphed data indicate that returns for adult females ranged from approximately 50% (banding year to the next year) to less than 10% (banding year to the second year). He found that "mortality or disappearance rate was higher in juveniles than adults." Mills *et al.* (1975) studied the species for 4 years. They presented survivorship curves for 110 juvenile females from one summer colony and 142 juvenile females from another, calculated using the actual number of recaptures. Survival rates for the first, second, and third years, respectively, for the two colonies (which declined during their study) were 31.9, 71.3, and 28.0%, and 10.5, 70.0, and 57.1%.

Humphrey and Cope (1970) conducted a similar study of a maternity colony of *Nycticeius humeralis*. They calculated minimum survivorship in the first and second year, respectively, to be 23–32 and 60% for females banded as juveniles and 55 and 50% for females banded as adults. Pearson *et al.* (1952) provided 3-year recapture results on known-age individuals of *Plecotus townsendii* banded both in summer and winter. Of summer-banded juvenile females at one colony, 46% returned the next year and only 54% were ever seen there. The same figures for adult females were 73 and 77%, respectively. For another cohort at the same site, 1-year returns were 40% for juveniles and 70% for adults. At a second site, 1-year returns were 38% for juveniles and 81% for adults. Pearson *et al.* (1952) made no judgment as to whether the yearling loss was due to dispersal or death, but said that it "appears likely that much of the loss in the first year occurs before the bats arrive at the winter quarters."

Unfortunately, the results of these summer band-recapture studies are difficult to compare with studies done in hibernacula. The species involved roost primarily in man-made structures in the summer, and the degree of site loyalty expected is considerably lower than that for winter. Although female bats of many species tend to return to the site of their natality while in late pregnancy and during nursing, the activities of the researchers themselves could increase disloyalty. Nursery colonies tend to be particularly sensitive to disturbance. Pearson *et al.* (1952) noted an occurrence where adults moved their young to a roost 1.3 miles away following their banding activities, and Mills *et al.* (1975) noted that their two colonies decreased by 88 and 60% during the 3 years of their return visits.

Humphrey and Cope (1976) present survival data for a number of known-age cohorts (also selected and analyzed in the same manner) in addition to the unaged cohorts discussed in the previous section. It is unclear from their discussion whether the recaptures on these summer-banded cohorts were made in the summer colonies, the winter hibernacula, or a combination of both. If any of the recaptures were at the summer sites, the results are subject to the same sources of error discussed above. If, on the other hand, the recaptures comprising the raw data for regression analysis were solely from winter sites, the samples must have been small; Table XXIV of Humphrey (1971; the basis of the 1976 Humphrey and Cope paper) lists only 320 total recaptures of maternity-banded *Myotis lucifugus* in hibernacula, out of at least 21,532 summer bandings (Table 9, Humphrey and Cope, 1976).

Difficulties set aside, Humphrey and Cope (1976) produced the following values as their best estimates of the percentage survival in known-age cohorts (first year, followed by average for subsequent years): juvenile females, 20.4–47.2, 71.2; juvenile males, 27.9, 74.7; adult females, 48.7, 71.7; adult males, 30.2, 74.4. Comparing these with the values for unaged cohorts (females, 31.3, 85.7; males, 36.5, 77.1), the authors point out the existence of discrepancies that are unexplainable. They conclude that even though life tables on juvenile individuals should give the best demographic picture, shorter recapture histories and less intensive recapture efforts in their study make those results less reliable than the tables presented for their winter-banded cohorts of unknown age.

Pearson *et al.* (1952) were able to distinguish adult and young *Plecotus townsendii* at the hibernacula, and so present some winter-recovery data. In one age grouping of females, 63% of the adults and 44% of the young were recovered during two subsequent winters. In a small sample of males the recovery rate for juveniles was half that for adults.

Stevenson and Tuttle (1981) present data on known-age cohorts analyzed, using two different analysis methods. Banding data from adult *Myotis grisescens* (juveniles and yearlings excluded) were analyzed both with the commonly used regression technique found in Humphrey and Cope (1976) and with nonparametric trend tests. Results were found to differ considerably depending on the method, and the trend tests were felt to better reflect actual survival as indicated by census estimates. Regression results ranged from 57.3 to 76.7% for bats banded as adult females, and from 57.0 to 82.0% for adult males. The wide range represents the differing survival of bats from different summer-banding locations, one of which underwent drastically increased mortality at one point during the study. A comparison among banding locations was possible because bats were banded at summer maternity caves and recovered at common hibernating caves. *Myotis grisescens* is extremely loyal to both summer and winter sites (Tuttle, 1976a).

Stevenson and Tuttle (1981) also present known-age cohort survival data for juvenile *Myotis grisescens* in the same manner as the data already discussed. Bats banded as juvenile males were found to have first half-year survival rates ranging from 6.0% for one summer colony to 51.7% for another (regression values). Female juvenile figures ranged from 13.9 to 58.6%. The subsequent annual survival rate from regression ranged from 65.5 to 85.1% for males and 61.9 to 83.7% for females. Again, many of the values obtained by regression were suspect, and relative survival among colonies was much better represented by the nonparametric methods. Those tests, however, provide no survival rate figures for direct comparison with other studies.

One problem that remains intractable in existing chiropteran research is the fact that even in studies of this third category, the term *known age* really only properly applies to cohorts banded as juveniles. Summer-banded cohorts of adults in most of the above studies excluded juveniles but included any bats of other ages from yearling on up. One study excluded yearlings as well. In terms of traditional life-table/survival analysis, then, these too are essentially unaged cohorts.

A second problem is that the cohort analyses, aged or unaged, do not take into account mortality from the time of birth to the time of banding. As pointed out by Humphrey and Cope (1977), this precludes "real" fixing of the curves of the *y* axis and causes overestimation of mean life expectancy when the common regression techniques are used. Foster *et al.* (1978) applied their preweaning mortality data to this problem and estimated that the lack of that information in their prior curves for the species placed the "set-point of the survival curve 11.8% too high on the *y* axis."

A third difficulty pointed out in the aged-cohort analyses presented by Humphrey and Cope (1976) is the fact that not only is the survival rate much lower the first year following banding than in subsequent years for juvenile cohorts (where it is expected) and unaged cohorts (which contain some juveniles), but it is also markedly lower in adult cohorts. Survival rates in subsequent years in all cohort types appear much more constant and greater than the first-year rate, even to visual inspection of the curves, whether or not the assumption of constant adult survival is accepted. Humphrey and Cope (1976) state that this "suggests that some individuals responded to our banding procedure in a manner that actually or apparently reduced survival during the first year." They cite some possible causes of increased mortality due to banding and consider the possibility of band loss or permanent movement to other sites but feel that none of these seems to adequately explain the phenomenon.

In summary, the pitfalls and difficulties awaiting the student of chiropteran survival are numerous. As pointed out by Keen and Hitchcock (1980) for *Myotis lucifugus*, "An understanding of the major question of age-specific mortality patterns is not likely to emerge from a refinement of winter-banding techniques or analysis, but will come from a statistically rigorous study of survival of *M. lucifugus* marked as newborn animals." The statement is equally true for other bat species as well and would provide an excellent start in the right direction. But no study will totally escape the problems involved in sampling chiropteran populations. Many of the authors cited discuss this issue as it relates to their studies and attempt to deal with it. Stevenson and Tuttle (1981) provide data on sampling bias in *M. grisescens*. Bats segregated by sex and age in hibernaculum clusters to a degree that was highly significant statistically. Learned avoidance behavior was observed. Differential recovery effort and success are also important factors. Clearly, the implication of the term *statistically rigorous* includes careful design to control bias throughout any study.

One interesting type of information that has emerged from the many banding studies and which provides insight into the life-span of the Chiroptera even without statistical treatment, however, is the longevity record. The data in Table IV demonstrate the extreme longevity of bats, a factor that is both a blessing and a curse for research on their survival.

TABLE IV
Longevity Records of Chiroptera

Species	Sex	Age (minimum years)	References
Megachiroptera			
Pteropodidae			
Eidolon helvum		21.8	Sacher and Jones, 1972
Pteropus giganteus	F	17.2	Flower, 1931
Rousettus leachi	F	19.8	Flower, 1931
Microchiroptera			
Rhinolophidae			
Rhinolophus ferrumequinum		26.0	Sluiter *et al.*, 1956a
Rhinolophus hipposideros		18.0	Stebbings, 1977
Phyllostomatidae			
Artibeus jamaicensis	F	7.0	Wilson and Tyson, 1970
Desmodus rotundus		18.0	Lord *et al.*, 1976
Macrotus californicus	F	10.4	Cockrum, 1973
Vespertilionidae			
Myotis bechsteini		7.0	Stebbings, 1977
M. brandti		13.0	Stebbings, 1977
M. dasycneme	F	15.5	Heerdt and Sluiter, 1958
M. daubentoni		18.0	Stebbings, 1977
M. emarginatus	M	14.5	Heerdt and Sluiter, 1958
M. evotis	M	22.0	Cross, 1977, personal communication
M. grisescens	F	16.5	Stevenson and Tuttle, 1981
M. keenii		18.5	Hall *et al.*, 1957
M. leibii (=*subulatus*)	F	12.0	Paradiso and Greenhall, 1967
M. lucifugus	M	30.0	Keen and Hitchcock, 1980
M. myotis		18.0	Stebbings, 1977
M. mystacinus	M	13.5	Heerdt and Sluiter, 1958
M. nattereri	F	14.5	Heerdt and Sluiter, 1960
M. nigricans		7.0	Wilson and Tyson, 1970

(*continued*)

TABLE IV (*Continued*)

Species	Sex	Age (minimum years)	References
M. sodalis	M	13.8	Paradiso and Greenhall, 1967
M. thysanodes	M	18.3	Cross, 1977
M. velifer	F	11.3	Cockrum, 1973
M. (=*Pizonyx*) *vivesi*		10.0	Orr, unpublished data
M. volans	M	21.0	Cross, 1977, personal communication
M. yumanensis	F	8.8	Cockrum, 1973
Antrozous pallidus	M	9.1	Cockrum, 1973
Barbastella barbastellus		18.0	Stebbings, 1977
Eptesicus fuscus	M	19.0	Paradiso and Greenhall, 1967
E. serotinus		6.0	Stebbings, 1977
Lasiurus cinereus	M	2.1	Cockrum, 1973
Miniopterus schreibersii	F	15.5	Aellen, 1978
Pipistrellus pipistrellus		11.0	Stebbings, 1977
P. subflavus	M	14.8	Walley and Jarvis, 1971
Plecotus auritus	M	13.8	Poots, 1977
P. austriacus		12.0	Stebbings, 1977
P. rafinesquii	F	10.1	Paradiso and Greenhall, 1967
P. townsendii	F	16.4	Paradiso and Greenhall, 1967
Molossidae			
Tadarida brasiliensis	F	7.2	Cockrum, 1973

4.2. Survival Determinants

4.2.1. Juvenile

Clearly a wide variety of previously discussed growth-determining factors, including maternal condition during gestation, the time of parturition, the state of development at birth, litter size, roost type and temperature, parasites, adverse weather, and time till first flight and weaning, all have direct bearing on juvenile survival as well. Most factors that affect juvenile survival also affect adults; they simply have a greater impact upon juveniles. To avoid redundancy we will reserve discussion of many such factors for Section 4.2.2. on adult survival. In this section we will simply document unique juvenile problems, high juvenile vulnerability, and factors that cause juvenile survivorship to vary among and within colonies and species.

Maternal stress during pregnancy often leads to prenatal mortality through the resorption or abortion of embryos (Wimsatt, 1945; Pearson *et al.*, 1952; Barbour and Davis, 1969; Kunz, 1971; Bradbury and Vehrencamp, 1977b; Myers, 1977) and may also cause abandonment and starvation of postnatal young

(Rakhmatulina, 1972; Roer, 1973; Nellis and Ehle, 1977). The extent of such mortality varies between years, even in a single colony, when climatic conditions vary (Rakhmatulina, 1972).

Postnatal preflight and newly volant young are vulnerable to falls from the roost. When possible, adult bats often retrieve fallen young (Davis *et al.*, 1968; Barbour and Davis, 1969; Kunz, 1973; O'Farrell and Studier, 1973), but such young are susceptible to numerous arthropods and predators, drowning, or abandonment (Herreid, 1967b; Kunz, 1971, 1973). In species that roost over water, most falling young likely drown (Humphrey, 1975; Foster *et al.*, 1978). This is an especially serious source of mortality when human disturbance at roosts causes large numbers of panicked mothers to drop their young (Tuttle, 1979). Frequency of falling likely varies considerably among localities owing to simple differences in roost configuration and texture.

Both prevolant and newly volant bats are especially susceptible to predators. Flightless young, if not protected by physical barriers, may be taken directly from roosts by a wide variety of predators that range from cockroaches (Wilson, 1971) and snakes (Rice, 1957; Kunz, 1973; Lemke, 1978) to birds and mammals (Allen, 1939; Gillette and Kimbrough, 1970; Mumford, 1973). Newly volant young are unskilled flyers and often concentrate near the roost during early practice flights (Kunz, 1974b; Bradbury, 1977a; Buchler, 1980). At such times these slow-flying poorly maneuverable bats are especially vulnerable. Kunz (1974a) found a bias toward juveniles among *Eptesicus fuscus* caught by barn owls.

Roost selection is perhaps the single most important factor in determining the survival of juvenile bats. Suitable roosting sites are in limited supply (Fenton, 1970a; Dwyer, 1971; Humphrey, 1975; Hayward and Cross, 1979), and some bats are forced to use marginally suitable roosts (Sluiter and van Heerdt, 1966). Such displacement undoubtedly leads to increased mortality. Colonies of *Nycticeius humeralis* occupy a variety of roosts in buildings, some in spacious attics where dangerous practice flights of newly volant young can be made in a protected enclosure, and some in small enclosed roosts (Watkins, 1972). Watkins found a greatly increased mortality in the marginal roosts, where protected practice was impossible. The young of species that characteristically roost in relatively exposed places, such as foliage, tree hollows, and rock cavities, are especially vulnerable to both weather and predation (Humphrey, 1975).

Competition for roosts involves roost location as well as suitability (see Chapter 1). Roosts must be near enough to adequate foraging habitat (Pearson *et al.*, 1952; Tuttle, 1976b). Tuttle showed that the mortality of newly volant *Myotis grisescens* was proportional to the distance between roosts and the nearest over-water foraging area. Growth success and survival were greatly reduced where young were forced to travel much more than 1 km in a single direction.

A further survival factor may involve the extent of maternal care. Species that rear young in protected sites usually leave their young at the roost during foraging, while those occupying exposed roosts normally move their young to more protected places (Bradbury, 1977a). Mothers of several species apparently teach their young to hunt (see Bradbury, 1977a and O'Shea and Vaughan, 1977), while others may not (Buchler, 1980).

4.2.2. Adult

A wide variety of factors have been shown to influence bat survival. They include adverse weather (Campbell, 1931; Davis, 1970; Ryberg, 1947; Stebbings, 1966b; Rakhmatulina, 1972; Mumford, 1973; Roer, 1973; Nellis and Ehle, 1977; Hayward and Cross, 1979) flooding (Hall, 1962; Jegla, 1963; DeBlase *et al.*, 1965; Humphrey, 1978; Tuttle, 1979), predation (Fenton, 1970a; Gillette and Kimbrough, 1970; Laurie, 1971; Wilson, 1971; Watkins, 1972; Krzanowski, 1973; Mumford, 1973; Kunz, 1973, 1974a; Thomas, 1974; Fenton and Fleming, 1976; Fenton *et al.*, 1977; Petretti, 1977; Lemke, 1978; Black *et al.*, 1979; Hayward and Cross, 1979; Humphrey and Bonaccorso, 1979; Ruprecht, 1979), disease and parasites (Allen, 1939; Constantine, 1970; Fenton, 1970a; Ubelaker, 1970; Lim, 1973; Kunz, 1976; Reisen *et al.*, 1976; Ubelaker *et al.*, 1977; Webb and Loomis, 1977; Hayward and Cross, 1979), tooth wear and decay (Christian, 1956; Phillips and Jones, 1970), length of nightly (Tuttle, 1976b) and migratory travel (Constantine, 1967; Paradiso and Greenhall, 1967; Cockrum, 1973; Tuttle and Stevenson, 1977; Zinn and Baker, 1979), accidents (Gillette and Kimbrough, 1970; O'Farrell and Miller, 1972; Hayward and Cross, 1979; Zinn and Baker, 1979), loss of habitat (Humphrey *et al.*, 1977; Humphrey, 1978; Tuttle, 1979; LaVal and LaVal, 1980), human harvesting for food (Brosset, 1966; Wodzicki and Felton, 1975; Halstead, 1977; Funmilayo, 1978; Bruner and Pratt, 1979), human disturbance of roosts (Henshaw, 1972; Mohr, 1952, 1953, 1972, 1975; Stebbings, 1966a; Hamilton-Smith, 1974; Gaisler and Bauerova, 1977; Humphrey, 1978; Tuttle, 1979; LaVal and LaVal, 1980; Rabinowitz and Tuttle, 1980), and pesticide and other poisoning (Cockrum, 1970; Jefferies, 1972; Mohr, 1972; Luckens and Davis, 1964; Reidinger, 1972, 1976; Clark and Prouty, 1976; Geluso *et al.*, 1976; Kunz *et al.*, 1977; Clark *et al.*, 1978; Clark, 1979; Tuttle, 1979; Clark *et al.*, 1980). Additionally, Foster *et al.* (1978) suggest that longer annual cycles of activity should expose bats to more hazards, resulting in higher rates of mortality. Hibernating bats, however, continue to be exposed to a wide variety of mortality factors (Ryberg, 1947; Twente, 1955; Fenton, 1970a; Humphrey, 1978), and Herreid (1964) found no evidence that tropical bats that do not hibernate have reduced maximum life-spans.

Many authors have suggested differential survival between the sexes (see

Tuttle and Stevenson, 1977), however no consistent differences are documented. In many cases small sample sizes, sampling bias, or other biases in the treatment of data make interpretation difficult. In a comparative study of survival in *Myotis grisescens,* Stevenson and Tuttle (1981) found small survival differences favoring either sex in different summer colonies, but no significant differences nor consistent trend. The sexes of several species apparently differ greatly in their habitat choice and/or migratory tendencies (Findley and Jones, 1964; Easterla and Watkins, 1970; Baker, 1978; Hayward and Cross, 1979; Zinn and Baker, 1979), and survival differences in these certainly might be expected. In other species migratory tendencies may differ greatly by sex, but only for members of certain colonies (Tuttle, 1976a). In such species opposite survival trends might be expected between sexes of different colonies. In bats that occupy foliage roosts litter size may be an especially important sex-related survival determinant. Jackson (1961) believed that "falling to the ground under the heavy burden of offspring" was the single most important cause of mortality in *Lasiurus cinereus;* and *L. borealis* mothers with two or more large young attached are frequently found grounded following storms (Tuttle, personal observation). Eisentraut (1947) believed that the rigors of pregnancy increased mortality in female *M. myotis,* and Keen and Hitchcock (1980) included higher metabolic rates during summer and greater exposure to ectoparasites as factors likely to increase female over male mortality in *M. lucifugus.*

The survivorship of bats in general is directly related to the roost types used, the nearness to foraging habitat and hibernating caves, local predator populations, annual weather cycles, and human disturbance factors. Especially in temperate zones the effects of these factors are highly variable and often unpredictable. Interactions may be complex (Foster *et al.*, 1978), and survival variation may be high among colonies of a single species (Stevenson and Tuttle, 1981), even ranging widely at a single location (Roer, 1973).

The impact of predation is likely much greater than generally recognized. Low reproductive rates in most bats greatly increase the importance of seemingly low predation rates (Twente, 1955). Many bats exhibit energetically costly behavior, apparently solely for the purpose of predator avoidance (Jimbo and Schwassmann, 1967; Fenton *et al.*, 1977; Morrison, 1978a, 1978b; Cope *et al.*, 1978; Tuttle, 1979; LaVal and LaVal, 1980). Rupprecht (1979) recently reviewed the importance of bats in owl diets and concluded that the role of owls in chiropteran mortality has been underestimated. Most would-be predators are likely frightened away by human activity, thereby limiting observations of predation. Using a night-vision scope, we observed predation by screech owls at entrances to caves occupied by *Myotis grisescens* only when we hid in a blind. Certainly, additional comparative observations are needed and may be more feasible than generally suspected.

4.3. Survival Strategies

Chiropteran survival strategies, though little known, are undoubtedly complex and highly variable, both within and among species. Tuttle (1975, 1976a, 1976b) compared colonies of *Myotis grisescens* living under a wide variety of conditions and concluded that the most important limiting factors in his study colonies were roost temperature and distance from maternity caves to foraging areas and hibernating caves. He developed a simple word model that stated that suboptimal conditions in any of these limiting parameters must be compensated for by lowered stress in one or both of the other two. He predicted that the greatest population sizes and/or growth rates would be found when all three factors affecting one colony approached optimal conditions (Tuttle, 1976a).

No violations of these predictions were found. Maternity colonies stressed by low roost temperatures or excessive distance from roosts to foraging areas were invariably located near hibernating caves, and colonies that traveled farthest to reach hibernating caves all utilized exceptionally warm roosts in excellent foraging habitat. The largest colonies were found, as expected, in warm caves adjacent to both foraging habitat and hibernation caves (Tuttle, 1976a). Each colony had its own unique set of trade-offs. Some suffered a heavy mortality of newly volant young (Tuttle, 1976b) but avoided further losses from migratory stress (Tuttle, 1976a), while others, for example, those migrating from Florida to northern Alabama and Tennessee, likely lost few newly volant young but undoubtedly lost many later, owing to the stress of long migration (Tuttle, 1976a).

Bradbury and Vehrencamp (1977b) compared survival strategies among four species of emballonurids, two species dependent upon highly seasonal food supplies and two dependent on relatively stable supplies. Each pair shared demographic features in common. The stable-habitat species produced two litters per year, exhibited delayed first reproduction in females, had high adult survival rates, and favored adult over juvenile survival. In contrast, the seasonal-habitat species produced a single young annually, had lower adult survival rates, exhibited early first reproduction in females, and favored juvenile over adult survival. Classical patterns of *r* and *K* selection were apparently followed, with *K* specialists in the most stable habitats.

5. SUMMARY

Plasticity and adaptability appear to be the most consistent features of chiropteran growth and survival. Length of gestation varies with the environment and is probably most strongly affected by temperature and food availability.

Time and synchrony of parturition vary greatly as well, both among and within species. Thermoregulatory constraints and the seasonality of food resources are likely the most important determinants. Temperate species are uniformly monestrous and parturition is highly seasonal and generally synchronous. Tropical species, however, are more variable, ranging from monestrous to polyestrous, bimodal to continous, and asynchronous to synchronous. Even in the tropics the seasonality of food resources may have a strong determining influence.

Developmental state at birth may vary considerably, depending upon litter size and the nutritional state of the mother, but very little information is available. Most bats have only one young per litter, some produce twins, and a few species bear up to five in a litter. Litter size often varies within species, geographically, seasonally, with the age of the mother, or with climatic conditions.

The success of preflight young is affected by such factors as roost temperature, the feeding success of nursing females, and probably parasites. Postflight to weaning growth and success vary greatly within species according to environmental conditions. Age of sexual maturity is also variable, but determining factors remain largely unstudied. The weaning period is one of considerable stress and apparent high mortality.

The wide range of observed strategies, even within single species, makes comparative survival estimates of particular interest in assessing the possible adaptive significance of one strategy versus another. Unfortunately, all three of the most widely used types of survival analyses suffer from serious difficulties and statistical limitations. The aging of individuals for study and sampling bias are two of the more problematical areas. Many factors determine the success and survival of individuals and populations. Although little is currently known of survival strategies in bats, the few available studies indicate that they are complex and highly adaptable to local environmental conditions. Future comparative investigations in all areas of bat growth and survival will be needed to provide a better understanding of this diverse group.

6. REFERENCES

Aellen, V. 1978. Les chauves-souris du Canton de Neuchâtel Suisse (Mammalia, Chiroptera). Bull. Soc. Neuchâtel. Sci. Nat., **101**:5–26.

al-Robaae, K. 1968. Notes on the biology of the tomb bat, *Taphozous nudiventris magnus* v. Wettstein, in Iraq. Säugetierkd. Mitt., **16**:21–26.

Allen, G. M. 1939. Bats. Harvard University Press, Cambridge, 368 pp.

Anand Kumar, T. C. 1965. Reproduction in the rat-tailed bat *Rhinopoma kinneari*. J. Zool., **147**:147–155.

Asdell, S. A. 1964. Chiroptera. Pp. 64–122, *in* Patterns of mammalian reproduction. Cornell University Press, Ithaca, 670 pp.

Baagøe, H. J. 1977. Age determination in bats. Vidensk. Medd. Dan. Naturhist. Foren. **140**:53–93.

Baker, J. R., and Z. Baker. 1936. The seasons in a tropical rain forest (New Hebrides). Part 3. Fruit bats (Pteropidae). J. Linn. Soc. London, Zool., **40**:123–141.

Baker, R. R. 1978. The evolutionary ecology of migration. Holmes and Meier, New York, 1012 pp.

Barbour, R. W., and W. H. Davis. 1969. Bats of America. University Press of Kentucky, Lexington, 286 pp.

Barlow, J. C., and J. R. Tamsitt. 1968. Twinning in American leaf-nosed bats (Chiroptera: Phyllostomatidae). Can. J. Zool., **46**:290–292.

Beer, J. R. 1955. Survival and movements of banded big brown bats. J. Mammal., **36**:242–248.

Bezem, J. J., J. W. Sluiter, and P. F. van Heerdt. 1960. Population statistics of five species of the bat genus *Myotis* and one of the genus *Rhinolophus*, hibernating in the caves of S. Limburg. Arch. Neerl. Zool., **13**:511–539.

Black, H. L., G. Howard, and R. Sternstedt. 1979. Observations on the feeding behavior of the bat hawk (*Macheiramphus alcinus*). Biotropica, **11**:18–21.

Bogan, M. A. 1972. Observations on parturition and development in the hoary bat, *Lasiurus cinereus*. J. Mammal., **53**:611–614.

Bonaccorso, F. J. 1979. Foraging and reproductive ecology in a Panamanian bat community. Bull. Fla. State Mus. Biol. Ser., **24**:359–408.

Bradbury, J. W. 1977a. Social organization and communication. Pp. 1–72, *in* Biology of bats. Vol. 3. (W. A. Wimsatt, ed.). Academic Press, New York, 651 pp.

Bradbury, J. W. 1977b. Lek mating behavior in the hammer-headed bat. Z. Tierpsychol., **45**:225–255.

Bradbury, J. W., and L. Emmons. 1974. Social organization in some Trinidad bats. I. Emballonuridae. Z. Tierpsychol., **36**:137–183.

Bradbury, J. W., and S. L. Vehrencamp. 1976a. Social organization and foraging in emballonurid bats. I. Field studies. Behav. Ecol. Sociobiol., **1**:337–382.

Bradbury, J. W., and S. L. Vehrencamp. 1976b. Social organization and foraging in emballonurid bats. II. A model for the determination of group size. Behav. Ecol. Sociobiol., **1**:383–404.

Bradbury, J. W., and S. L. Vehrencamp. 1977a. Social organization and foraging in emballonurid bats. III. Mating systems. Behav. Ecol. Sociobiol., **2**:1–17.

Bradbury, J. W., and S. L. Vehrencamp. 1977b. Social organization and foraging in emballonurid bats. IV. Parental investment patterns. Behav. Ecol. Sociobiol., **2**:19–29.

Bradshaw, G. V. R. 1961. A life history study of the California leaf-nosed bat, *Macrotus californicus*. Unpublished Ph.D. dissertation, University of Arizona, Tucson, 89 pp.

Brenner, F. J. 1968. A three-year study of two breeding colonies of the big brown bat, *Eptesicus fuscus*. J. Mammal., **49**:775–778.

Brenner, F. J. 1974. A five-year study of a hibernating colony of *Myotis lucifugus*. Ohio J. Sci., **74**:239–244.

Brosset, A. 1962a. The bats of central and western India. Part I. J. Bombay Nat. Hist., Soc., **59**:1–57.

Brosset, A. 1962b. The bats of central and western India. Part II. J. Bombay Nat. Hist. Soc., **59**:583–624.

Brosset, A. 1962c. The bats of central and western India. Part III. J. Bombay Nat. Hist. Soc., **59**:706–746.

Brosset, A. 1962d. La reproduction des Chiroptères de l'ouest et du centre de l'Inde. Mammalia, **26**:166–213.

Brosset, A. 1966. La biologie des chiroptères. Masson, Paris, 237 pp.

Brosset, A. 1969. Recherches sur la biologie des Chiroptères troglophiles dans le nord-est du Gabon. Biol. Gabon., **5**:93–116.

Brown, P. E., and A. D. Grinnell. 1980. Echolocation ontogeny in bats. Pp. 355–377, *in* Animal sonar systems. NATO Advanced Study Institutes Series A. Vol. 28 (R.-G. Busnel and J. F. Fish, eds.). Plenum Press, New York, 1135 pp.

Bruner, P. L., and H. D. Pratt. 1979. Notes on the status and natural history of Micronesian bats. Elepaio J. Hawaii Audubon Soc., **40**:1–4.

Buchler, E. R. 1980. The development of flight, foraging and echolocation in the little brown bat (*Myotis lucifugus*). Behav. Ecol. Sociobiol., **6**:211–218.

Burnett, C. D., and T. H. Kunz. 1982. Growth rates and age-estimation in *Eptesicus fuscus* and comparison with *Myotis lucifugus*. J. Mammal., **63**:33–41.

Burns, J. M., R. J. Baker, and W. J. Bleier. 1972. Hormonal control of "delayed development" in *Macrotus waterhousii*. I. Changes in plasma thyroxin during pregnancy and lactation. Gen. Comp. Endocrinol., **18**:54–58.

Cagle, F. R. 1950. A Texas colony of bats, *Tadarida mexicana*. J. Mammal., **31**:400–402.

Campbell, B. 1931. Bats on the Colorado desert. J. Mammal., **12**:430.

Carter, D. C. 1970. Chiropteran reproduction. Pp. 233–246, *in* About bats. (B. H. Slaughter and D. W. Walton, eds.). Southern Methodist University Press, Dallas, 339 pp.

Christian, J. J. 1956. The natural history of a summer aggregation of the big brown bat, *Eptesicus fuscus fuscus*. Am. Midl. Nat., **55**:66–95.

Clark, D. R., Jr. 1979. Lead concentrations: Bats vs. terrestrial small mammals collected near a major highway. Env. Sci. Technol., **13**:338–341.

Clark, D. R., Jr., and R. M. Prouty. 1976. Organochlorine residues in three bat species from four localities in Maryland and West Virginia, 1973. Pestic. Monit. J., **10**:44–53.

Clark, D. R., Jr., R. K. LaVal, and D. M. Swinford. 1978. Dieldrin-induced mortality in an endangered species, the gray bat (*Myotis grisescens*). Science, **199**:1357–1359.

Clark, D. R., Jr., R. K. LaVal, and A. J. Krynitsky. 1980. Dieldrin and heptachlor residues in dead gray bats, Franklin County, Missouri—1976 versus 1977. Pestic. Monit. J., **13**:137–140.

Cockrum, E. L. 1955. Reproduction in North American bats. Trans. Kans. Acad. Sci., **58**:487–511.

Cockrum, E. L. 1970. Insecticides and guano bats. Ecology, **51**:761–762.

Cockrum, E. L. 1973. Additional longevity records for American bats. J. Ariz. Acad. Sci., **8**:108–110.

Constantine, D. G. 1967. Activity patterns of the Mexican free-tailed bat. Univ. N.M. Publ. Biol., **7**:7–79.

Constantine, D. G. 1970. Bats in relation to the health, welfare, and economy of man. Pp. 319–449, *in* Biology of bats. Vol. 2. (W. A. Wimsatt, ed.). Academic Press, New York, 477 pp.

Cope, J. B., A. R. Richter, and D. A. Seerley. 1978. A survey of the bats in the Big Blue Lake project area in Indiana. Unpublished Final Report, U.S. Army Corps of Engineers, Louisville, 51 pp.

Cormack, R. M. 1964. Estimates of survival from the sighting of marked animals. Biometrika, **54**:429–438.

Cross, S. P. 1977. A survey of the bats of Oregon Caves National Monument. Unpublished Final Report, National Park Service, Cave Junction, Oregon, 40 pp.

Davis, R. 1969. Growth and development of young pallid bats, *Antrozous pallidus*. J. Mammal., **50**:729–736.

Davis, R. B., C. F. Herreid, II, and H. L. Short. 1962. Mexican free-tailed bats in Texas. Ecol. Monogr., **32**:311–346.

Davis, W. B. 1944. Notes on Mexican mammals. J. Mammal., **25**:370–403.

Davis, W. H. 1966. Population dynamics of the bat *Pipistrellus subflavus*. J. Mammal., **47**:383–396.

Davis, W. H. 1967. Survival of *Eptesicus fuscus*. Bat Res. News, **8**:18–19.

Davis, W. H. 1970. Hibernation: Ecology and physiological ecology. Pp. 265–300, *in* Biology of bats. Vol. 1. (W. A. Wimsatt, ed.). Academic Press, New York, 406 pp.

Davis, W. H., R. W. Barbour, and M. D. Hassell. 1968. Colonial behavior of *Eptesicus fuscus*. J. Mammal., **49**:44–50.

DeBlase, A. F., S. R. Humphrey, and K. S. Drury. 1965. Cave flooding and mortality in bats in Wind Cave, Kentucky. J. Mammal., **46**:96.

Dinale, G. 1964. Studi sui Chirotteri italiani. II. Il raggiungimento della maturità sessuale in *Rhinolophus ferrumequinum* Schreber. Atti Soc. Ital. Sci. Nat. Mus. Civ. Stor. Nat. Milano, **103**:141–153.

Dinale, G. 1968. Studi sui Chirotteri italiani. VII. Sul raggiungimento della maturità sessuale hei Chirotteri europei: ed in particolare nei Rhinolophidae. Arch. Zool. Ital., **53**:51–71.

Dwyer, P. D. 1963. The breeding biology of *Miniopterus schreibersi blepotis* (Temminck) (Chiroptera) in north-eastern New South Wales. Aust. J. Zool., **11**:219–240.

Dwyer, P. D. 1964. Seasonal changes in activity and weight of *Miniopterus schreibersii blepotis* (Chiroptera) in north-eastern New South Wales. Aust. J. Zool., **12**:52–69.

Dwyer, P. D. 1966a. The population pattern of *Miniopterus schreibersii* (Chiroptera) in north-eastern New South Wales. Aust. J. Zool., **14**:1073–1137.

Dwyer, P. D. 1966b. Observations on *Chalinolobus dwyeri* (Chiroptera: Vespertilionidae) in Australia. J. Mammal., **47**:716–718.

Dwyer, P. D. 1968. The biology, origin, and adaptation of *Miniopterus australis* (Chiroptera) in New South Wales. Aust. J. Zool., **16**:49–68.

Dwyer, P. D. 1970. Latitude and breeding season in a polyestrous species of *Myotis*. J. Mammal., **51**:405–410.

Dwyer, P. D. 1971. Temperature regulation and cave-dwelling in bats: An evolutionary perspective. Mammalia, **35**:424–453.

Dwyer, P. D. 1975. Notes on *Dobsonia moluccensis* (Chiroptera) in the New Guinea highlands. Mammalia, **39**:113–118.

Dwyer, P. D., and E. Hamilton-Smith. 1965. Breeding caves and maternity colonies of the bent-winged bat in south-eastern Australia. Helictite, **4**:3–21.

Dwyer, P. D., and J. A. Harris. 1972. Behavioral acclimatization to temperature by pregnant *Miniopterus* (Chiroptera). Physiol. Zool., **45**:14–21.

Easterla, D. A., and L. Watkins. 1970. Nursery colonies of evening bats (*Nycticeius humeralis*) in northwestern Missouri and southwestern Iowa. Trans. Mo. Acad. Sci., **4**:110–117.

Eisentraut, M. 1936. Ergebnisse der Fledermausberingung nach dreijähringer Versuchzeit. Z. Morphol. Ökol. Tiere, **31**:1–26.

Eisentraut, M. 1947. Die mit Hilfe der Beringungsmethode erzielten Ergebnisse über Lebensdauer und jährliche Verlustziffern bei *Myotis myotis* Borkh. Experientia, **3**:157–160.

Eisentraut, M. 1950. Beobachtungen über Lebensdauer und jährliche Verlustziffern bei Fledermäusen, ins besondere bei *Myotis myotis*. Zool. Jahrb. Abt. Syst. Ökol. Geogr. Tiere, **78**:297–300.

Everitt, G. C. 1968. Prenatal development of uniparous animals, with particular reference to the influence of maternal nutrition in sheep. Pp. 131–157, *in* Growth and development of mammals (G. A. Lodge and G. E. Lamming, eds.). Plenum Press, New York, 527 pp.

Fenton, M. B. 1970a. Population studies of *Myotis lucifugus* (Chiroptera: Vespertilionidae) in Ontario. Life Sci. Contrib. R. Ont. Mus., **7**:1–34.

Fenton, M. B. 1970b. The deciduous dentition and its replacement in *Myotis lucifugus* (Chiroptera: Vespertilionidae). Can. J. Zool., **48**:817–820.

Fenton, M. B., and T. H. Fleming. 1976. Ecological interactions between bats and nocturnal birds. Biotropica, **8**:104–110.

Fenton, M. B., D. H. M. Cumming, and D. J. Oxley. 1977. Prey of bat hawks and availability of bats. Condor, **79**:495–497.

Findley, J. S., and C. Jones. 1964. Seasonal distribution of the hoary bat. J. Mammal., **45**:461–470.

Fleming, T. H. 1971. *Artibeus jamaicencis:* Delayed embryonic development in a Neotropical bat. Science, **171**:402–404.

Fleming, T. H., E. T. Hooper, and D. E. Wilson. 1972. Three Central American bat communities: Structure, reproductive cycles, and movement patterns. Ecology, **53**:555–569.

Flower, S. S. 1931. Contributions to our knowledge of the duration of life in vertebrate animals—V. Mammals. Proc. Zool. Soc. London, **1931**:145–234.

Fogden, M. P. L. 1972. The seasonality and population dynamics of equatorial forest birds in Sarawak. Ibis, **114**:307–343.

Foster, G. W., S. R. Humphrey, and P. P. Humphrey. 1978. Survival rate of young southeastern brown bats, *Myotis austroriparius*, in Florida. J. Mammal., **59**:299–304.

Funmilayo, D. 1978. Fruit bats for meat: Are too many taken? Oryx, J. Fauna Preserv. Soc., **14**:377–378.

Gaisler, J. 1965. The female sexual cycle and reproduction in the lesser horseshoe bat (*Rhinolophus hipposideros hipposideros* Bechstein, 1800). Acta Soc. Zool. Bohem., **29**:336–352.

Gaisler, J., and Z. Bauerova. 1977. The bat community of Kvêtnice (Czechoslovakia) during thirty years. Lynx, **19**:17–28.

Gaisler, J., and M. Titlbach. 1964. The male sexual cycle in the lesser horseshoe bat (*Rhinolophus hipposideros hipposideros* Bechstein, 1800). Acta Soc. Zool. Bohem., **28**:268–277.

Gaisler, J., V. Hanak, and J. Dungel. 1979. A contribution to the population ecology of *Nyctalus noctula* (Mammalia: Chiroptera). Acta Sci. Nat. Brno., **13**:1–38.

Geluso, K. N., J. S. Altenbach, and D. E. Wilson. 1976. Bat mortality: Pesticide poisoning and migratory stress. Science, **194**:184–186.

Gillette, D. D., and J. D. Kimbrough. 1970. Chiropteran mortality. Pp. 262–283, *in* About bats. (B. H. Slaughter and D. W. Walton, eds.). Southern Methodist University Press, Dallas, 339 pp.

Goehring, H. H. 1972. Twenty-year study of *Eptesicus fuscus* in Minnesota. J. Mammal., **53**:201–207.

Gopalakrishna, A. 1947. Studies on the embryology of Microchiroptera. I. Reproduction and breeding seasons in the South Indian vespertilionid bat—*Scotophilus wroughtoni*, (C. Thomas). Proc. Indian Acad. Sci., **26**:219–232.

Gopalakrishna, A., and P. N. Choudhari. 1977. Breeding habits and associated phenomena in some Indian bats. Part I. *Rousettus leschenaulti* (Desmarest)—Megachiroptera. J. Bombay Nat. Hist. Soc., **74**:1–16.

Griffin, D. R. 1940. Notes on the life-histories of New England cave bats. J. Mammal., **21**:181–187.

Guthrie, M. J. 1933. Notes on the seasonal movements and habits of some cave bats. J. Mammal., **14**:1–19.

Hall, E. R. 1981. The mammals of North America. Vol. 1. 2nd ed. John Wiley and Sons, New York, 600 + 90 pp.

Hall, J. S. 1962. A life history and taxonomic study of the Indiana bat, *Myotis sodalis*. Sci. Pub., Reading Pub. Mus. Art Gallery, **12**:1–68.

Hall, J. S., R. J. Cloutier, and D. R. Griffin. 1957. Longevity records and notes on tooth wear of bats. J. Mammal., **38**:407–409.

Halstead, L. B. 1977. Fruit bats—An example of wildlife management. Niger. Field, **42**:50–56.

Hamilton-Smith, E. 1974. The present knowledge of Australian Chiroptera. J. Aust. Mammal Soc., **1**:95–108.

Harmata, W. 1973. The thermopreferendum of some species of bats (Chiroptera) in natural conditions. Zesz. Nauk. Univ. Jagiellon. Pr. Zool., **19**:127–141.

Harper, F. 1929. Mammal notes from Randolph County, Georgia. J. Mammal., **10**:84–85.

Hayward, B. J. 1970. The natural history of the cave bat *Myotis velifer*. West. N.M. Univ. Res. Sci., **1**:1–74.

Hayward, B. J., and S. P. Cross. 1979. The natural history of *Pipistrellus hesperus* (Chiroptera: Vespertilionidae). West. N.M. Univ. Res. Sci., **3**:1–36.

Heithaus, E. R., T. H. Fleming, and P. A. Opler. 1975. Foraging patterns and resource utilization in seven species of bats in a seasonal tropical forest. Ecology, **56**:841–854.

Henshaw, R. E. 1960. Responses of free-tailed bats to increases in cave temperature. J. Mammal., **41**:396–398.

Henshaw, R. E. 1972. Niche specificity and adaptability in cave bats. Bull. Nat. Speleol. Soc., **34**:61–70.

Herreid, C. F., II. 1963. Temperature regulation and metabolism in Mexican free-tailed bats. Science, **142**:1573–1574.

Herreid, C. F., II. 1964. Bat longevity and metabolic rate. Exp. Gerontol., **1**:1–9.

Herreid, C. F., II. 1967a. Temperature regulation, temperature preference and tolerance, and metabolism of young and adult free-tailed bats. Physiol. Zool., **40**:1–22.

Herreid, C. F., II. 1967b. Mortality statistics of young bats. Ecology, **48**:310–312.

Hoogland, J. L., and P. W. Sherman. 1976. Advantages and disadvantages of bank swallow (*Riparia riparia*) coloniality. Ecol. Monogr., **46**:33–58.

Humphrey, S. R. 1971. Population ecology of the little brown bat, *Myotis lucifugus*, in Indiana and north-central Kentucky. Unpublished Ph.D. dissertation, Oklahoma State University, 138 pp.

Humphrey, S. R. 1975. Nursery roosts and community diversity of Nearctic bats. J. Mammal., **56**:321–346.

Humphrey, S. R. 1978. Status, winter habitat, and management of the endangered Indiana bat, *Myotis sodalis*. Fla. Sci., **41**:65–76.

Humphrey, S. R., and F. J. Bonaccorso. 1979. Population and community ecology. Pp. 409–437, *in* Biology of bats of the New World family Phyllostomatidae. Part III. (R. J. Baker, J. K. Jones, Jr., and D. C. Carter, eds.). Spec. Publ. Mus. Texas Tech Univ., **16**:1–441.

Humphrey, S. R., and J. B. Cope. 1970. Population samples of the evening bat, *Nycticeius humeralis*. J. Mammal., **51**:399–401.

Humphrey, S. R., and J. B. Cope. 1976. Population ecology of the little brown bat, *Myotis lucifigus*, in Indiana and north-central Kentucky. Spec. Publ. Amer. Soc. Mamm., **4**:1–81.

Humphrey, S. R., and J. B. Cope. 1977. Survival rates of the endangered Indiana bat, *Myotis sodalis*. J. Mammal., **58**:32–36.

Humphrey, S. R., and T. H. Kunz. 1976. Ecology of a Pleistocene relict, the western big-eared bat (*Plecotus townsendii*) in the southern Great Plains. J. Mammal., **57**:470–494.

Humphrey, S. R., A. R. Richter, and J. B. Cope. 1977. Summer habitat and ecology of the endangered Indiana bat. *Myotis sodalis*. J. Mammal., **58**:334–346.

Jackson, H. H. T. 1961. Mammals of Wisconsin. University of Wisconsin Press, Madison, 504 pp.

Jefferies, D. J. 1972. Organochlorine insecticide residues in British bats and their significance. J. Zool., **166**:245–263.

Jegla, T. C. 1963. A recent deposit of *Myotis lucifugus* in Mammoth Cave. J. Mammal., **44**:121–122.

Jimbo, S., and H. O. Schwassmann. 1967. Feeding behavior and the daily emergence pattern of "*Artibeus jamaicensis*" Leach (Chiroptera, Phyllostomatidae). Atas Simp. Biota Amazonica, **5**:239–253.

Jones, C. 1967. Growth, development, and wing loading in the evening bat, *Nycticeius humeralis* (Rafinesque). J. Mammal., **48**:1–19.

Jones, C. 1977. *Plecotus rafinesquii*. Mammal. Species, **69**:1–4.

Jones, C., and R. D. Suttkus. 1975. Notes on the natural history of *Plecotus rafinesquii*. Occas. Pap. Mus. Zool. La. State Univ., **47**:1–14.

Keen, R., and H. B. Hitchcock. 1980. Survival and longevity of the little brown bat (*Myotis lucifugus*) in southeastern Ontario. J. Mammal., **61**:1–7.

Kleiman, D. G. 1969. Maternal care, growth rate and development in the noctule (*Nyctalus noctula*), pipistrelle (*Pipistrellus pipistrellus*) and serotine (*Eptesicus serotinus*) bats. J. Zool., **157**:187–211.

Kleiman, D. G., and T. M. Davis. 1979. Ontogeny and maternal care. Pp. 387–402, *in* Biology of bats of the New World family Phyllostomatidae. Part III. (R. J. Baker, J. K. Jones, Jr., and D. C. Carter, eds.). Spec. Publ. Mus. Texas Tech Univ., **16**:1–441.

Kleiman, D. G., and P. A. Racey. 1969. Observations of noctule bats (*Nyctalus noctula*) breeding in captivity. Lynx, **10**:65–77.

Krátky, J. 1970. Postnatale Entwicklung des Grossmausohrs, *Myotis myotis* (Borkhausen, 1797). Acta Soc. Zool. Bohem., **33**:202–218.

Krutzsch, P. H. 1955a. Observations on the California mastiff bat. J. Mammal., **36**:407–414.

Krutzsch, P. H. 1955b. Ectoparasites from some species of bats from western North America. J. Mammal., **36**:457–458.

Krzanowksi, A. 1956. The bats of Pulawy: List of species and biological observations. Acta Theriol., **1**:87–108.

Krzanowski, A. 1973. Numerical comparison of Vespertilionidae and Rhinolophidae (Chiroptera: Mammalia) in the owl pellets. Acta Zool., Cracov, **18**:133–140.

Kulzer, E. 1958. Untersuchen über die Biologie von Flughunden der Gattung *Rousettus*. Z. Morphol. Oekol. Tiere, **47**:374–402.

Kulzer, E. 1969. African fruit-eating bats: Part II. Afr. Wildl. **23**:129–138.

Kunz, T. H. 1971. Reproduction of some vespertilionid bats in central Iowa. Am. Midl. Nat., **86**:477–486.

Kunz, T. H. 1973. Population studies of the cave bat (*Myotis velifer*): Reproduction, growth, and development. Occas. Pap. Mus. Nat. Hist. Univ. Kans., **15**:1–43.

Kunz, T. H. 1974a. Reproduction, growth, and mortality of the vespertilionid bat, *Eptesicus fuscus*, in Kansas. J. Mammal., **55**:1–13.

Kunz, T. H. 1974b. Feeding ecology of a temperate insectivorous bat (*Myotis velifer*). Ecology, **55**:693–711.

Kunz, T. H. 1976. Observations on the winter ecology of the bat fly *Trichobius corynorhini* Cockerell (Diptera: Streblidae). J. Med. Entomol., **12**:631–636.

Kunz, T. H. 1982. *Lasionycteris noctivagans*. Mammal. Species, **173**:1–5.

Kunz, T. H., and E. L. P. Anthony. 1982. Age estimation and postnatal growth rates in the bat *Myotis lucifugus*. J. Mammal., **63**:23–32.

Kunz, T. H., E. L. P. Anthony, and W. T. Rumage III. 1977. Mortality of little brown bats following multiple pesticide applications. J. Wildl. Manage., **41**:476–483.

Lack, D. 1968. Ecological adaptations for breeding in birds. Methuen, London, 409 pp.

Laurie, W. A. 1971. The food of the barn owl in the Serengeti National Park, Tanzania. J. East Afr. Nat. Hist. Soc. Nat. Mus., **28**:1–4.

LaVal, R. K., and H. S. Fitch. 1977. Structure, movement, and reproduction in three Costa Rican bat communities. Occas. Pap. Mus. Nat. Hist. Univ. Kans., **69**:1–28.

LaVal, R. K., and M. L. LaVal. 1977. Reproduction and behavior of the African banana bat, *Pipistrellus nanus*. J. Mammal., 58:403–410.

LaVal, R. K., and M. L. LaVal. 1979. Notes on reproduction, behavior and abundance of the red bat, *Lasiurus borealis*. J. Mammal., **60**:209–212.

LaVal, R. K., and M. L. LaVal. 1980. Ecological studies and management of Missouri bats, with emphasis on cave-dwelling species. Mo. Dep. Conserv. Terrest. Ser., **8**:1–53.

Leitch, I., F. E. Hytten, and W. Z. Billewicz. 1959. The maternal and neonatal weights of some Mammalia. Proc. Zool. Soc. London, **133**:11–28.

Lemke, T. O. 1978. Predation upon bats by *Epicrates cenchris cenchris* in Colombia. Herp. Rev. **9**:47.

Lim, B-L. 1973. Breeding pattern, food habits and parasitic infestation of bats in Gunong Brinchang. Malay. Nat. J., **26**:6–13.

Lord, R. D., F. Murdali, and L. Lazavo. 1976. Age composition of vampire bats (*Desmodus rotundus*) in northern Argentina and southern Brazil. J. Mammal., **57**:573–575.

Luckens, M. M., and W. H. Davis. 1964. Bats: Sensitivity to DDT. Science, **146**:948.

McCracken, G. F., and J. W. Bradbury. 1981. Social organization and kinship in the polygynous bat, *Phyllostomus hastatus*. Behav. Ecol. Sociobiol., **8**:11–34.

McNab, B. K. 1973. Energetics and the distribution of vampires. J. Mammal., **54**:131–144.

Madhavan, A. 1978. Breeding habits and associated phenomena in some Indian bats. Part V—*Pipistrellus dormeri* (Dobson)—Vespertilionidae. J. Bombay Nat. Hist. Soc., **75**:426–433.

Madhavan, A., and A. Gopalakrishna. 1978. Breeding habits and associated phenomena in some Indian bats. IV. *Hipposideros fulvus fulvus* (Gray)—Hipposideridae. J. Bombay Nat. Hist. Soc., **75**:96–103.

Maeda, K. 1972. Growth and development of the large noctule, *Nyctalus lasiopterus* Schreber. Mammalia, **36**:269–278.

Maeda, K. 1974. Eco-ethologie de la grande noctule, *Nyctalus lasiopterus*, á Sapporo, Japan. Mammalia, **38**:461–487.

Maeda, K. 1976. Growth and development of the Japanese large-footed bat, *Myotis macrodactylus*. I. External characters and breeding habits. J. Growth, **15**:29–40.

Marshall, A. J. 1947. The breeding cycle of an equatorial bat (*Pteropus giganteus*) of Ceylon. Proc. Linn. Soc. London, **159**:103–111.

Matthews, L. H. 1941. Notes on the genitalia and reproduction of some African bats. Proc. Zool. Soc. London, **111**:289–346.

Medway, Lord. 1972. Reproductive cycles of the flat-headed bats *Tyloncteris pachypus* and *T. robustula* (Chiroptera: Vespertilionidae) in a humid equatorial environment. J. Linn. Soc. London, Zool., **51**:33–61.

Miller, R. E. 1939. The reproductive cycle in male bats of the species *Myotis lucifugus* and *Myotis grisescens*. J. Morphol., **64**:267–295.

Mills, R. W., G. W. Barrett, and M. P. Farrell. 1975. Population dynamics of the big brown bat (*Eptesicus fuscus*) in southwestern Ohio. J. Mammal., **56**:591–604.

Mohr, C. E. 1952. A survey of bat banding in North America, 1932–1951. Bull. Nat. Speleol. Soc., **14**:3–13.

Mohr, C. E. 1953. Possible causes of an apparent decline in wintering populations of cave bats. Nat. Speleol. Soc. News, **11**:4–5.

Mohr, C. E. 1972. The status of threatened species of cave-dwelling bats. Bull. Nat. Speleol. Soc., **34**:33–47.

Mohr, C. E. 1975. The protection of threatened cave bats. Pp. 57–62, *in* National Cave Management Symposium Proceedings, 1976. Speleobooks, Albuquerque, 146 pp.

Morrison, D. W. 1978a. Foraging ecology and energetics of the frugivorous bat *Artibeus jamaicensis*. Ecology, **59**:716–723.

Morrison, D. W. 1978b. Lunar phobia in a Neotropical fruit bat, *Artibeus jamaicensis* (Chiroptera: Phyllostomidae). Anim. Behav., **26**:852–855.

Mumford, R. E. 1969. The hoary bat in Indiana. Proc. Indiana Acad. Sci., **78**:497–501.

Mumford, R. E. 1973. Natural history of the red bat (*Lasiurus borealis*) in Indiana. Period. Biol., **75**:155–158.

Mutere, F. A. 1967. The breeding biology of equatorial vertebrates: Reproduction in the fruit bat. *Eidolon helvum*, at latitude 0°20′ N. J. Zool., **153**:153–161.

Myers, P. 1977. Patterns of reproduction of four species of vespertilionid bats in Paraguay. Univ. Calif. Publ. Zool., **107**:1–41.

Nellis, D. W., and C. P. Ehle. 1977. Observations on the behavior of *Brachyphylla cavernarum* (Chiroptera) in the Virgin Islands. Mammalia, **41**:403–409.

Nelson, J. E. 1965a. Behavior of Australian Pteropodidae (Megachiroptera). Anim. Behav., **13**:544–557.

Nelson, J. E. 1965b. Movements of Australian flying foxes (Pteropodidae: Megachiroptera). Z. Tierpsychol., **27**:857–870.

Neuweiler, G. 1962. Das Verhalten Indischer Flughunde (*Pteropus giganteus*). Naturwissenschaften, **49**:614–615.

Noll, U. G. 1979. Postnatal growth and development of thermogenesis in *Rousettus aegyptiacus*. Comp. Biochem. Physiol., **63A**:89–93.

O'Farrell, M. J., and B. W. Miller. 1972. Pipistrelle bats attracted to vocalizing females and to a blacklight insect trap. Am. Midl. Nat., **88**:462–463.

O'Farrell, M. J., and E. H. Studier. 1973. Reproduction, growth, and development in *Myotis thysanodes* and *M. lucifugus* (Chiroptera: Vespertilionidae). Ecology, **54**:18–30.

Orr, R. T. 1954. Natural history of the pallid bat, *Antrozous pallidus* (LeConte). Proc. Calif. Acad. Sci., **28**:165–246.

Orr, R. T. 1970. Development: Prenatal and postnatal. Pp. 217–231, *in* Biology of bats. Vol. 1. (W. A. Wimsatt, ed.). Academic Press, New York, 406 pp.

O'Shea, T. J., and T. A. Vaughan. 1977. Nocturnal and seasonal activities of the pallid bat, *Antrozous pallidus*. J. Mammal., **58**:269–284.

Pagels, J. F., and C. Jones. 1974. Growth and development of the free-tailed bat, *Tadarida brasiliensis cynocephala* (LeConte). Southwest. Nat., **19**:267–276.

Paradiso, J. L., and A. M. Greenhall. 1967. Longevity records for American bats. Am. Midl. Nat., **78**:251–252.

Pearson, O. P., M. R. Koford, and A. K. Pearson. 1952. Reproduction of the lump-nosed bat (*Corynorhinus rafinesquei*) in California. J. Mammal., **33**:273–320.

Petretti, F. 1977. Season food of the barn owl (*Tyto alba*) in an area of central Italy. Gerfaut Rev. Sci. Belge Ornithol., **67**:225–233.

Phillips, C. J., and J. K. Jones, Jr. 1970. Dental abnormalities in North American bats. II. Possible endogenous factors in dental caries in *Phyllostomus hastatus*. Univ. Kans. Sci. Bull., **48**:791–799.

Pierson, E. D., and T. H. Kunz. 1982. Energetics of pregnancy and lactation in *Myotis lucifugus* (Chiroptera: Vespertilionidae). Unpublished manuscript.

Poots, L. 1977. Maximum age of some bats according to ring marking data. Eesti Loodus Dorpat., **7**:435–438.

Pucek, Z., and V. P. W. Lowe. 1975. Age criteria in small mammals. Pp. 55–72, *in* Small mammals: Their productivity and population dynamics. (F. B. Golley, K. Petrusewicz, and L. Ryszkowski, eds.). Cambridge University Press, Cambridge, 451 pp.

Pye, J. D. 1971. Bats and fog. Nature, **229**:572–574.

Rabinowitz, A., and M. D. Tuttle. 1980. Status of summer colonies of the endangered gray bat in Kentucky. J. Wildl. Manage., **44**:955–960.

Racey, P. A. 1969. Diagnosis of pregnancy and experimental extension of gestation in the pipistrelle bat, *Pipstrellus pipistrellus*. J. Reprod. Fertil., **19**:465–474.

Racey, P. A. 1973. Environmental factors affecting the length of gestation in heterothermic bats. J. Reprod. Fertil., Suppl., **19**:175–189.

Racey, P. A. 1974. Aging and assessment of reproductive status of Pipistrelle bats, *Pipistrellus pipistrellus*. J. Zool., **173**:264–271.

Rakhmatulina, I. K. 1972. The breeding, growth and development of pipistrelles in Azerbaidzhan. Sov. J. Ecol., **2**:131–136.

Ramakrishna, P. A. 1951. Studies on the reproduction in bats. I. Some aspects of the reproduction in the Oriental vampires, *Lyroderma lyra lyra* (Geoffroy) and *Megaderma spasma* (Linn.). J. Mysore Univ., **12**:107–118.

Ransome, R. D. 1973. Factors affecting the timing of births of the greater horse-shoe bat (*Rhinolophus ferrumequinum*). Period. Biol., **75**:169–175.

Reidinger, R. F., Jr. 1972. Factors influencing Arizona bat population levels. Unpublished Ph.D. dissertation, University of Arizona, Tucson, 172 pp.

Reidinger, R. F., Jr. 1976. Organochlorine residues in adults of six southwestern bat species. J. Wildl. Manage., **40**:677–680.

Reisen, W. K., M. L. Kennedy, and N. T. Reisen. 1976. Winter ecology of ectoparasites collected from hibernating *Myotis velifer* (Allen) in southwestern Oklahoma (Chiroptera: Vespertilionidae). J. Parasitol., **62**:628–635.

Rice, D. W. 1957. Life history and ecology of *Myotis austroriparius* in Florida. J. Mammal., **38**:15–32.

Roer, H. 1973. Uber die Ursachen hoher Jugendmortalitaet beim Mausohr, *Myotis myotis* (Chiroptera, Mamm.). Bonn. Zool. Beitr., **24**:332–341.

Rupprecht, A. L. 1979. Bats (Chiroptera) as constituents of the food of barn owls *Tyto alba* in Poland. Ibis, **121**:489–494.

Rybář, P. 1971. On the problems of practical use of the ossification of bones as age criterion in the bats (Microchiroptera). Pr. studie-Přír. Pardubice, **3**:97–121.

Ryberg, O. 1947. Studies on bats and bat parasites. Bokförlaget Svensk Natur., Stockholm, 330 pp.

Sacher, G. A., and M. L. Jones. 1972. Life spans: Mammals. Pp. 229–230, *in* Biology data book. 2nd ed. (P. L. Altman and D. S. Dittmer, eds.). Federation of the American Society for Experimental Biology. Bethesda.

Schmidt, U., and U. Manske. 1973. Die Jugendentwicklung der Vampirfledermaüse (*Desmodus rotundus*). Z. Säugetierk., **38**:14–33.

Schowalter, D. B., and J. R. Gunson. 1979. Reproductive biology of the big brown bat (*Eptesicus fuscus*) in Alberta. Can. Field-Nat., **93**:48–54.

Schowalter, D. B., J. R. Gunson, and L. D. Harder. 1979. Life history characteristics of little brown bats (*Myotis lucifugus*) in Alberta. Can. Field-Nat., **93**:243–251.

Sealy, S. G. 1978. Litter size and nursery sites of the hoary bat near Delta, Manitoba. Blue Jay, **36**:51–52.

Sherman, H. B. 1937. Breeding habits of the free-tailed bat. J. Mammal., **18**:176–187.

Short, H. L. 1961. Growth and development of Mexican free-tailed bats. Southwest. Nat., **6**:156–163.

Sluiter, J. W. 1954. Sexual maturity in bats of the genus *Myotis*. II. Females of *M. mystacinus* and *M. emarginatus*. Proc. K. Ned. Akad. Wet., **57**:696–700.

Sluiter, J. W. 1960. Reproductive rate of the bat *Rhinolophus hipposideros*. Proc. K. Ned. Akad. Wet., **63**:383–393.

Sluiter, J. W. 1961. Sexual maturity in males of the bat *Myotis myotis*. Proc. K. Ned. Akad. Wet., **64**:243–249.

Sluiter, J. W., and M. Bouman. 1951. Sexual maturity in bats of the genus *Myotis*. I. Size and histology of the reproductive organs during hibernation in connection with age and wear of the teeth in female *Myotis myotis* and *Myotis emarginatus*. Proc. K. Ned. Akad. Wet., **54**:595–601.

Sluiter, J. W., and P. F. van Heerdt. 1966. Seasonal habits of the noctule bat (*Nyctalus noctula*). Arch. Neerl. Zool., **16**:423–439.

Sluiter, J. W., P. F. van Heerdt, and J. J. Bezem. 1956a. Paramètres de population chez le Grand Rhinolophe fer-à-cheval (*Rhinolophus ferrum-equinum* Schreber), estimés par la méthode de reprises après baguages. Mammalia, **35**:254–272.

Sluiter, J. W., P. F. van Heerdt, and J. J. Bezem. 1956b. Population statistics of the bat *Myotis mystacinus*, based on the marking–recapture method. Arch. Neerl. Zool., **12:**63–88.

Sluiter, J. W., P. F. van Heerdt, and A. M. Voûte. 1971. Contribution to the population biology of the pond bat, *Myotis dasycneme* (Boie, 1825). Decheniana, **18:**1–44.

Smith, E. 1956. Pregnancy in the little brown bat. Am. J. Physiol., **185:**61–64.

Smythe, N. 1974. Insect sampling. Pp. 147–157, *in* Smithsonian Tropical Research Institute Environmental Monitoring Program. Smithsonian Institution Environmental Science Program, Panama City, 409 pp.

Snapp, B. D. 1976. Colonial breeding in the barn swallow (*Hirundo rustica*) and its adaptive significance. Condor, **78:**471–480.

Sreenivasan, M. A., H. R. Bhat, and G. Geevarghese. 1973. Breeding cycle of *Rhinolophus rouxi* Temminck, 1835 (Chiroptera: Rhinolophidae), in India. J. Mammal., **54:**1013–1017.

Stebbings, R. E. 1966a. Bats under stress. Stud. Speleol., **1:**168–173.

Stebbings, R. E. 1966b. A population study of bats of the genus *Plecotus*. J. Zool., **150:**53–75.

Stebbings, R. E. 1968. Measurements, composition and behavior of a large colony of the bat *Pipistrellus pipistrellus*. J. Zool., **156:**15–33.

Stebbings, R. E. 1977. Order Chiroptera. Pp. 68–128, *in* Handbook of British mammals. (G. B. Corbet and H. N. Southern, eds.). Blackwells Scientific, Oxford, 520 pp.

Stevenson, D. E., and M. D. Tuttle. 1981. Survivorship in the endangered gray bat (*Myotis grisescens*). J. Mammal., 62:244–257.

Stones, R. C. 1965. Laboratory care of little brown bats at thermal neutrality. J. Mammal., **46:**681–682.

Stones, R. C., and J. E. Wiebers. 1965. Seasonal changes in the food consumption of little brown bats held in captivity at a "neutral" temperature of 92°F. J. Mammal., **46:**18–22.

Stones, R. C., and J. E. Wiebers. 1967. Temperature regulation in the little brown bat, *Myotis lucifugus*. Pp. 97–109, *in* Mammalian hibernation III. (K. C. Fisher, A. R. Dawe, C. P. Lyman, E. Schonbaum, and F. E. South, eds.). American Elsevier, New York, 535 pp.

Thomas, M. E. 1974. Bats as a food source for *Boa constrictor*. J. Herpetol., **8:**188.

Trune, D. R., and C. N. Slobodchikoff. 1976. Social effects of roosting on the metabolism of the pallid bat (*Antrozous pallidus*). J. Mammal., **57:**656–663.

Tuttle, M. D. 1975. Population ecology of the gray bat (*Myotis grisescens*): Factors influencing early growth and development. Occas. Pap. Mus. Nat. Hist. Univ. Kans. **36:**1–24.

Tuttle, M. D. 1976a. Population ecology of the gray bat (*Myotis grisescens*): Philopatry, timing and patterns of movement, weight loss during migration, and seasonal adaptive strategies. Occas. Pap. Mus. Nat. Hist. Univ. Kans. **54:**1–38.

Tuttle, M. D. 1976b. Population ecology of the gray bat (*Myotis grisescens*): Factors influencing growth and survival of newly volant young. Ecology, **57:**587–595.

Tuttle, M. D. 1979. Status, causes of decline, and management of endangered gray bats. J. Wildl. Manage., **43:**1–17.

Tuttle, M. D., and L. R. Heaney. 1974. Maternity habits of *Myotis leibii* in South Dakota. Bull. S. Calif. Acad. Sci., **73:**80–83.

Tuttle, M. D., and D. E. Stevenson. 1977. An analysis of migration as a mortality factor in the gray bat based on public recoveries of banded bats. Am. Midl. Nat., **97:**235–240.

Twente, J. W. Jr. 1955. Aspects of a population study of cavern-dwelling bats. J. Mammal., **36:**379–390.

Ubelaker, J. E. 1970. Some observations on ecto- and endoparasites of Chiroptera. Pp. 247–261, *in* About bats. (B. H. Slaughter and D. W. Walton, eds.). Southern Methodist University Press, Dallas, 339 pp.

Ubelaker, J. E., R. D. Specian, and D. W. Duszynski. 1977. Endoparasites. Pp. 7–56, *in* Biology of bats of the New World family Phyllostomatidae. Part II. (R. J. Baker, J. K. Jones, Jr., and D. C. Carter, eds.). Spec. Publ. Mus. Texas Tech Univ., **13**:1–364.

van Heerdt, P. F., and J. W. Sluiter. 1957. The results of bat banding in the Netherlands in 1956. Nat. Hist. Maandbl., **46**:13–16.

van Heerdt, P. F., and J. W. Sluiter. 1958. The results of bat banding in the Netherlands in 1957. Nat. Hist. Maandbl., **47**:38–41.

van Heerdt, P. F., and J. W. Sluiter. 1960. The results of bat banding in the Netherlands in 1959. Nat. Hist. Maandbl., **49**:42–44.

Walley, H. D., and W. L. Jarvis. 1971. Longevity record for *Pipistrellus subflavus*. Bat Res. News, **12**:17–18.

Watkins, L. C. 1972. *Nycticeius humeralis*. Mammal. Species, **23**:1–4.

Webb, J. P., Jr., and R. B. Loomis. 1977. Ectoparasites. Pp. 57–119, *in* Biology of bats of the New World family Phyllostomatidae. Part II. (R. J. Baker, J. K. Jones, Jr., and D. C. Carter, eds.). Spec. Publ. Mus. Texas Tech Univ., **13**:1–364.

Weir, B. J., and I. W. Rowlands. 1973. Reproductive strategies of mammals. Ann. Rev. Ecol. Syst., **4**:139–163.

Wilson, D. E. 1971. Ecology of *Myotis nigricans* (Mammalia: Chiroptera) on Barro Colorado Island, Panama Canal Zone. J. Zool., **162**:1–13.

Wilson, D. E. 1979. Reproductive patterns. Pp. 317–378, *in* Biology of bats of the New World family Phyllostomatidae. Part III. (R. J. Baker, J. K. Jones, Jr., and D. C. Carter, eds.). Spec. Publ. Mus. Texas Tech Univ., **16**:1–441.

Wilson, D. E., and J. S. Findley. 1970. Reproductive cycle of a Neotropical insectivorous bat, *Myotis nigricans*. Nature, **225**:1155.

Wilson, D. E., and E. L. Tyson. 1970. Longevity records for *Artibeus jamaicensis* and *Myotis nigricans*. J. Mammal., **51**:203.

Wimsatt, W. A. 1945. Notes on breeding behavior, pregnancy, and parturition in some vespertilionid bats of the eastern United States. J. Mammal., **26**:23–33.

Wimsatt, W. A. 1960. An analysis of parturition in Chiroptera, including new observations on *Myotis l. lucifugus*. J. Mammal., **41**:183–200.

Wodzicki, K., and H. Felten. 1975. The peka or fruit bat (*Pteropus tonganus*) of Niue Island, South Pacific. Pac. Sci., **29**:131–138.

Woodward, A. E., and R. L. Snyder. 1978. Effect of photoperiod on early growth and feed conversion in the chukar partridge. Poult. Sci., **57**:341–348.

Zinn, T. L., and W. W. Baker. 1979. Seasonal migration of the hoary bat, *Lasiurus cinereus*, through Florida. J. Mammal., **60**:634–635.

Chapter 4

Evolutionary Alternatives in the Physiological Ecology of Bats

Brian K. McNab

Department of Zoology
University of Florida
Gainesville, Florida 32611

1. INTRODUCTION

Bats are often thought to have poor temperature regulation, to feed on insects, and to enter hibernation. This combination of characteristics is typical only of temperate species. In the last two decades intensive work in the tropics, especially in the New World, has shown the great ecological diversity existing among bats, which in turn permits us to place temperate species in their appropriate ecological and evolutionary perspective and to examine some of the physiological and ecological alternatives available to bats when living in a physically benign environment.

1.1. The Significance of Physiology to the Ecology of Bats

All biologically important activities require energy, but when body size in animals is large, energy stores in the form of fat are of sufficient size, relative to daily energy expenditures, so that day-to-day variations in food availability may be relatively unimportant. However, in most bats fat stores are small relative to demand, especially when the high cost of flight is considered. No mammal, large or small, can escape the necessity of a long-term balance in their energy budget. The question, then, is how such a balance is attained, and not whether it occurs.

One of the most important factors influencing energy expenditure is the thermal regime in the environment. Bats are found in nearly every climate from cold temperate to tropical. Some (few) species, such as the Eurasian *Eptesicus nilssonii* and to a lesser extent the North American *Myotis lucifugus*, even encroach on subarctic environments. In tropical environments the thermal demands placed on bats are normally quite modest, especially if the bats are cave-dwelling species. Many bats in tropical rain forests, in fact, roost in trees, where they find shelter from rain and insolation (see Chapter 1). The only tropical bats to face a rigorous thermal environment are those species that face high roost temperatures in arid regions. Many temperate environments, however, are so seasonally cold that the twofold problem of facing temperatures well below freezing and the absence of significant amounts of available food requires a radical adjustment of energetics to permit bats to live under these conditions.

The solutions to the problems encountered by bats in a temperate environment can be seen in the energetics of temperature regulation. Endothermy can be used only when adequate quantities of food are continuously available. Some food supplies may well be continuously available in many tropical environments, but this condition is highly unlikely in temperate regions, where (judging from birds and mammals other than bats) the only foods that permit continuous homeothermy include seeds, grass, vertebrates, or insect pupae, none of which are used by bats in temperate regions, and most of which are not used by bats anywhere. If a bat were to be a continuous endotherm in a temperate environment, it could only do so by evading the coldest periods through migration to equatorial environments where adequate food supplies exist; that is, bats could behave like the majority of temperate and polar birds; otherwise, temperate bats have to give up temperature regulation and enter seasonal torpor (hibernation) to await the return of periods when adequate food supplies are again available. If migration and hibernation are both acceptable solutions to seasonally harsh conditions, the question remains as to why some species use migration and others use hibernation.

Energetics also has significance for the population biology of mammals as a result of the dependence of fecundity, gestation period, and rate of postnatal growth upon the rate of metabolism (McNab, 1980b). In general, mammals that have high rates of metabolism also have populations that grow rapidly and show large amplitudes in their population fluctuations. Mammals that have high rates of metabolism, in short, tend to be *r* selected, while those that have low rates tend to be *K* selected. This analysis has not been applied to bats.

Water balance also has significance for the ecology of bats. Difficulties in balancing a water budget may arise from living in arid environments with high temperatures and low humidities, or they may result from having a diet high in protein and salt, which would increase the demands for renal water conservation.

1.2. The Significance of Bats for Physiological Ecology

If the concerns of physiology are important for the ecology of bats, bats have been equally important for the demonstration of rules that appear to exist in physiological ecology. Bats are unparalleled subjects for these studies because the order Chiroptera demonstrates the greatest ecological diversity found within any mammalian order, with the possible exception of the order Rodentia. Chiroptera contains species that range in body mass from about 2 g (*Craseonycteris;* Lekagul and McNeely, 1977) to flying foxes that weigh over 1 kg (*Pteropus*). No mammalian order has a greater range in food habits than do bats: some species feed on flying and terrestrial insects, and others feed on terrestrial vertebrates, fish, copepods, shrimp, fruit, nectar, pollen, and on the blood of mammals and birds. Only grazing, browsing, and seed-eating habits are ignored by bats.

This diversity has demonstrated (1) the importance of body size in setting the limits for endothermic temperature regulation, (2) the influence of food habits on the rates of energy expenditure and temperature regulation in endotherms, and (3) the importance of climate, through its action on the availability of food resources, to the physiological and behavioral characteristics of endotherms in general and of bats in particular. As a consequence, such work has relaxed the intellectual grip placed on our view of bats, and thereby on endotherms, produced in part by a temperate parochialism that hitherto has obscured the rules that govern endothermy on an isolated planet.

2. THE ENERGETICS OF BATS

2.1. Factors Determining the Energy Expenditure of Bats

2.1.1. Body Mass and Food Habits

Many factors influence the energy expenditure of bats, the principal one being body mass. The importance of this factor was first described by Kleiber (1932), who pointed out that the total basal rate of metabolism in mammals is proportional to $m^{0.75}$. As a consequence, mass-specific basal rates in mammals are proportional to $m^{-0.25}$ (Fig. 1). This proportionality can be seen within the Pteropodidae, Molossidae, and Vespertilionidae, within the subfamilies Glossophaginae, Stenoderminae, and Phyllostominae of the Phyllostomidae, and within the genera *Phyllostomus* and *Noctilio*. The importance of body size can be shown by the observation that the basal rate varies in Fig. 1 by a factor of 5.8 (= 3.05/0.53), while that predicted by a 157 (= 598/3.8)-fold variation in mass is 3.5 (= $157^{0.25}$), or 60% of the total variation. Nevertheless, an examination of Fig. 1 clearly shows that there is great residual variation in the basal rate independent of the influence of body mass.

Much of this residual variation is correlated with food habits: bats that feed on fruit, nectar, or meat tend to have basal rates equal to or greater than the values expected from the Kleiber relation. Those species that feed on insects tend to have the lowest basal rates, the only apparent exception being *Saccopteryx leptura* (Emballonuridae), the value for which is based on three measurements

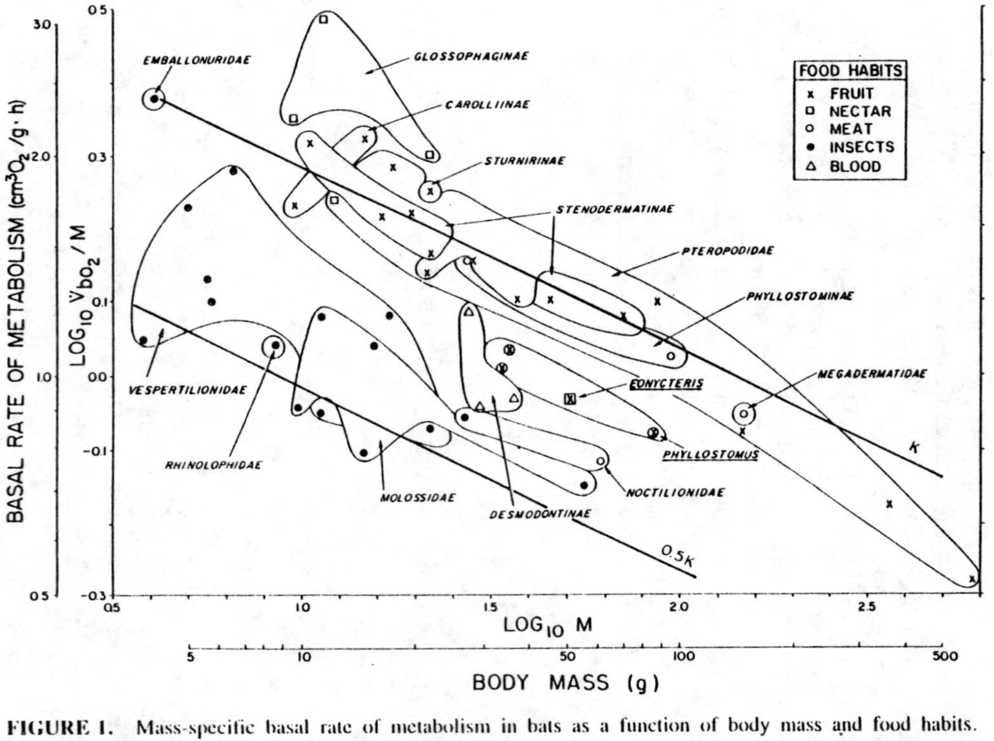

FIGURE 1. Mass-specific basal rate of metabolism in bats as a function of body mass and food habits. Kleiber's (1932) relation is indicated by the linear curve labeled *K*. (Data derived from McNab, 1969; Noll, 1979; Kulzer and Storf, 1980, unpublished observations.)

and is therefore suspect. Bats that feed on blood and others that have a mixed diet, such as *Phyllostomus* (Phyllostomidae) and *Eonycteris* (Pteropodidae), have intermediate basal rates. Although there is an appreciable variation in the basal rate at a given mass within a family (e.g., Vespertilionidae and Molossidae), there seems to be no evidence of the influence of taxonomy *per se* on basal rate. That is, if a bat is insectivorous, it makes no difference to its basal rate whether the bat is a vespertilionid, rhinolophid, molossid, or noctilionid, which is not to say, however, that we understand why some vespertilionids have basal rates that are three times those of other vespertilionids.

The fish-eating bat *Noctilio leporinus* (Noctilionidae), like its insectivorous relative *N. albiventris* (=*labialis*), has a low basal rate, unlike the carnivorous species belonging to the genera *Tonatia*, *Chrotopterus*, and *Macroderma* (Fig. 1), all of which have high basal rates. The problem with a phylogenetic interpretation of this low rate is that *N. leporinus* also feeds heavily on insects; it may well be that the low basal rate found in this bat is produced by a partial (or seasonal?) dependence on insects. In a way, then, this bat, like *Phyllostomus* and *Eonycteris*, may have a low basal rate in relation to a mixed diet.

The variation of basal rate with body size means that temperature regulation depends on body size. The influence of body size on the ability of bats to maintain body temperature independently of environmental temperature is shown within the phyllostomid subfamily Stenoderminae (Fig. 2). Within this series fruit bats weighing over 19 g maintain body temperatures independently of ambient temperature down (at least) to 14°C, and generally to temperatures below 10°C. Species weighing 16 g or less have body temperatures that vary with

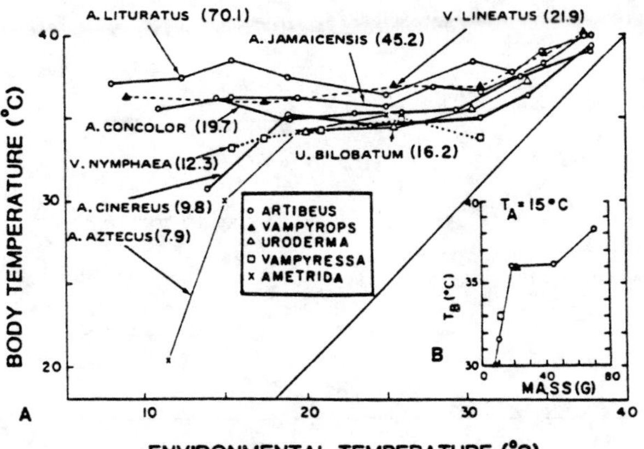

FIGURE 2. (A) Body temperature as a function of environmental temperature and body mass in stenodermine bats (Phyllostomidae). (B) Body temperature in stenodermine bats in relation to body mass at 15°C. (Data from McNab, 1969.)

ambient temperature below 20°C; in fact, the fall in body temperature in these smaller species is nearly proportional to the reduction in body size, especially at masses below 20 g (Fig. 2B). Thus it is not surprising that small stenodermines, such as species belonging to the genera *Vampyressa* and *Ametrida*, as well as the smallest species of the genus *Artibeus*, tend to be limited in distribution to the climatically stable lowland tropics. Such correlations between thermoregulation and distribution occur in other bat groups, for example, *Rhinophylla* in the Carolliinae (Phyllostomidae).

Food habits, through their influence on energetics, also affect temperature regulation. Insectivorous bats weighing 11–18 g have lower body temperatures than do frugivorous or nectarivorous species of similar size at all ambient temperatures (Fig. 3); compare, for example, *Macrotis* and *Synconycteris*, *Molossus* and *Carollia*, and *Histiotus* and *Anoura*. Furthermore, insectivorous bats generally permit body temperature to fall more with a decrease in ambient temperature than is found in species with other food habits; colder temperatures are required for an appreciable decrease in body temperature in frugivorous species. The low body temperatures and limited capacities for temperature regulation in insectivorous species is exactly what should be expected from their low rates of metabolism and small size (McNab, 1970, 1974a). It can be seen in Fig. 3 that the effect of body size on temperature regulation also occurs in insectivorous bats (e.g., compare *Histiotus* and *Macrotis*).

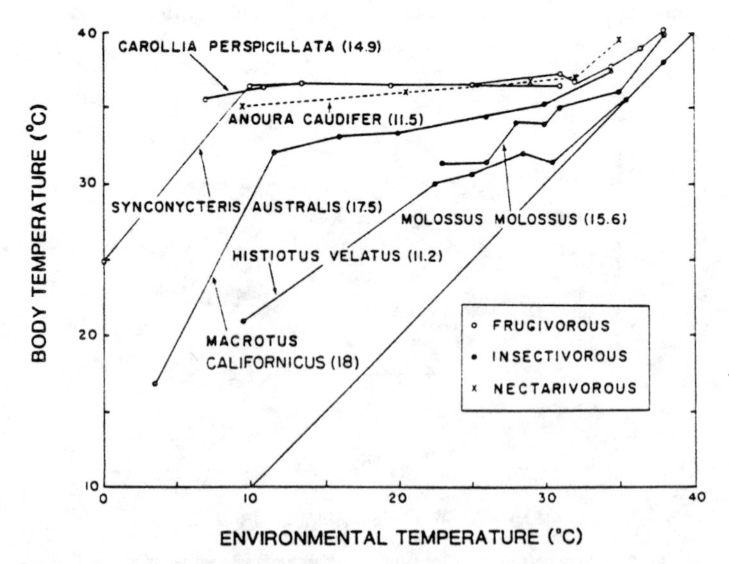

FIGURE 3. Body temperature as a function of environmental temperature and food habits in bats weighing 11–18 g. (Data from McNab, 1969.)

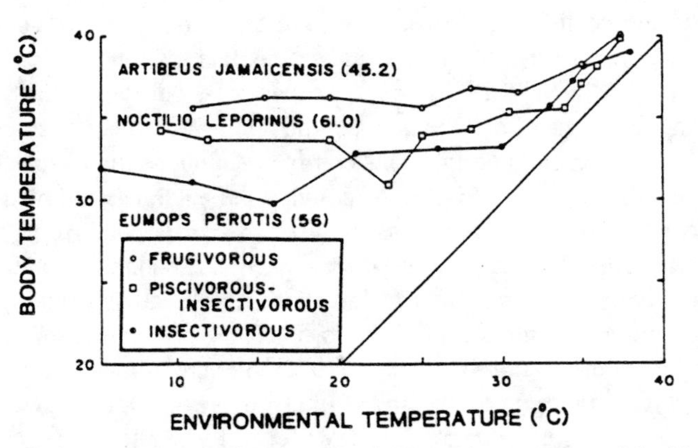

FIGURE 4. Body temperature as a function of environmental temperature and food habits in bats weighing 45–61 g. (Data from McNab, 1969.)

It is obvious from Fig. 3 that most bats in the mass range of 11–18 g, irrespective of food habits, show a decrease in body temperature with a fall in ambient temperature, whereas bats that weigh 45–61 g maintain higher body temperatures at low environmental temperatures (Fig. 4). These larger species also show the influence of food habits: the frugivorous *Artibeus* has the highest temperature, and the insectivorous *Eumops* the lowest temperature. The inter-

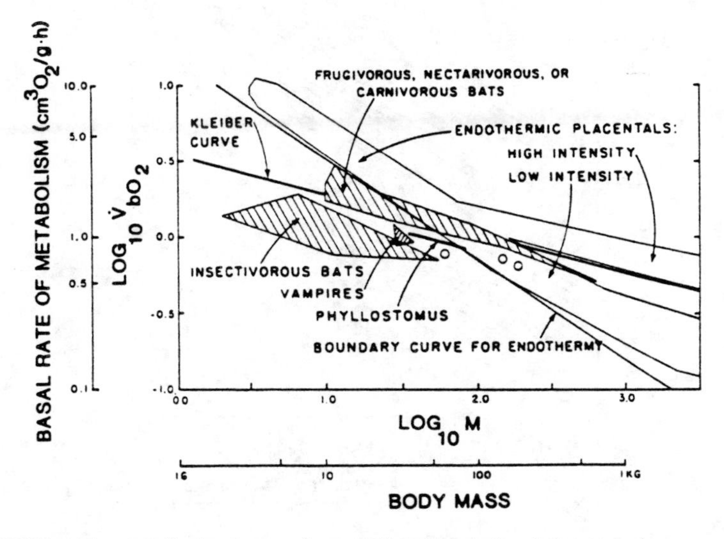

FIGURE 5. Mass-specific basal rate of metabolism in bats in relation to body mass and food habits. Kleiber's relation and the boundary relation for endothermy (see text) are indicated. Data on the basal rates of tree-shrews are indicated by circles and are derived from Bradley and Hudson (1974), Whittow and Gould (1976), and Weigold (1979).

mediate body temperature in *Noctilio leporinus* reflects its low basal rate and (probably) its extensive use of insects, as well as fish, as food.

Recent work (McNab, 1982) has described in detail the interactions that exist between the scaling of metabolism and the scaling of temperature regulation with respect to mass in endotherms. This complex scaling relation in placentals is shown with data from bats in Fig. 5. Although most mammals follow a Kleiber curve at large masses, at smaller masses most mammals conform to another, steeper (in mass-specific units) curve that represents the boundary conditions for effective endothermy. If a mammal follows a Kleiber relation at masses below the boundary curve, it enters into daily torpor. In other words, any mammal found to the left of the boundary curve in Fig. 5 will enter torpor, and the further it is to the left of this curve, the more likely it is to do so. Obviously, all insectivorous bats will enter torpor. The tendency for small nectarivorous and frugivorous bats to have basal rates higher than those expected from the Kleiber relation is easily interpreted as an adjustment to the cost of temperature regulation at a small body size. Nevertheless, this adjustment is not adequate to ensure continuous endothermy in these bats, because the rates are generally below the

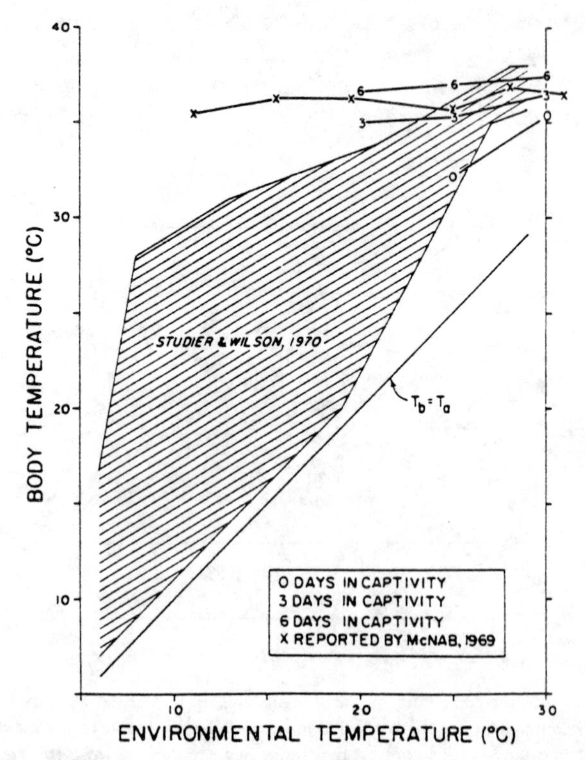

FIGURE 6. The relation of body temperature to environmental temperature in *Artibeus jamaicensis* (Studier and Wilson, 1979). (Additional data from McNab, 1969, and Studier and Wilson, 1970.)

boundary curve (Fig. 5). Large frugivores and carnivores, however, are found well to the right of this boundary and may not normally enter torpor. The best measure in energetics of the capacity for endothermic regulation, therefore, is not the Kleiber relation but, rather, the boundary curve. At masses less than 40 g mammals can have basal rates greater than those described by the Kleiber relation and still be required to enter torpor.

A cautionary note should be given on measurements that demonstrate the thermoregulatory capacities of bats (or other mammals): the performance of individuals belonging to a particular species depends on their nutritional state. This difficulty is readily apparent when trying to study the thermoregulation of insectivorous bats (e.g., *Noctilio albiventris;* McNab, 1969) where it is often difficult to feed them quantities of food sufficient to maintain their mass in the laboratory. Such bats physically deteriorate, with the consequent loss of the ability to thermoregulate. This problem can be solved by measuring temperature regulation only in freshly caught individuals. Frugivorous, carnivorous, and sanguinivorous bats, however, are generally easy to feed in captivity and thus have been thought to be free from this problem. These bats, however, raise another question: are they in as good a condition in the field as they are in the laboratory, where they are usually given a surplus of food? This question was raised by the conflict in data reported by McNab (1969) and by Studier and Wilson (1970). I had found generally excellent temperature regulation in *Artibeus jamaicensis, A. lituratus, Carollia perspicillata,* and *Glossophaga soricina,* whereas Studier and Wilson found poor temperature regulation in these species. This conflict was resolved by Studier and Wilson (1979). They showed that *A. jamaicensis* taken directly from the field showed poor temperature regulation, but the longer that they were maintained in captivity, the higher they regulated body temperature and the more independent it was from ambient temperature (Fig. 6). Equally, the longer these bats were kept in captivity the more their rates of metabolism reflected a pattern typical of endotherms. Thus, fruit bats seem to be able to regulate body temperature precisely, if they are in a good nutritional state, but this state is subject to appreciable variations in the field, apparently due to variations in the availability of food. The poor temperature regulation of field *Artibeus,* then, is not due to physiological incompetence but is a means of reducing the rate at which energy stores are used, thereby increasing survivorship.

There is direct evidence that the nutritional state of bats in the tropics, as measured by their fat deposits, is seasonally variable (Fig. 7); this variability occurs in *Artibeus jamaicensis,* although the annual variation in fat deposits is much greater in certain insectivorous species, such as *Tadarida brasiliensis.* An extensive examination of the seasonal variation in the nutritional states of bats, as indicated by their ability to maintain a temperature differential with the environment and by the sizes of fat stores in individuals taken directly from the field, would be very interesting, especially in relation to climate, food habits, and body

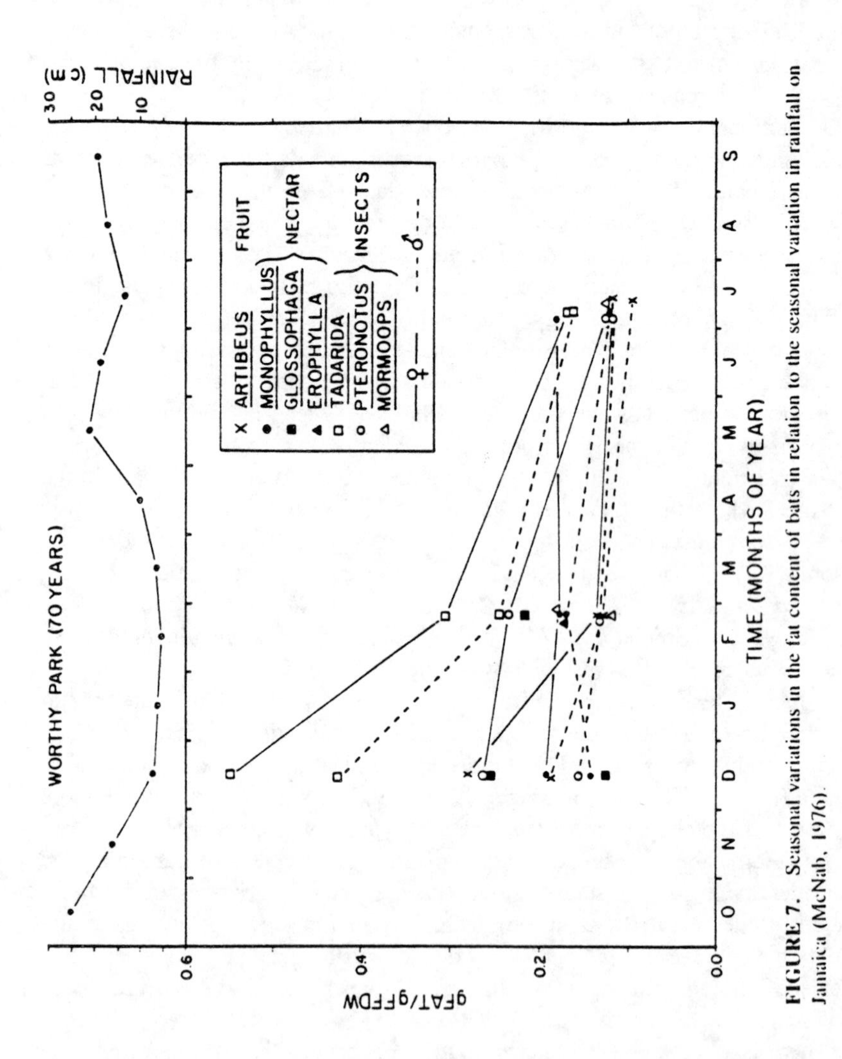

FIGURE 7. Seasonal variations in the fat content of bats in relation to the seasonal variation in rainfall on Jamaica (McNab, 1976).

size. Such work should be pursued on the mainland tropical lowlands, where the greatest ecological diversity in bats is found.

2.1.2. Thermal Conductance

An important aspect of thermal biology concerns the control of heat exchange with the environment. Most bats have standard thermal conductances (Fig. 8), which presumably reflect a small mass. Most estimates of thermal conductance in endotherms depend upon the maintenance of a temperature differential and the solution to the equation

$$C = \dot{V}_{O_2}/\Delta T$$

where C is thermal conductance (usually in $cm^3\ O_2\ g^{-1}\ h^{-1}\ {}^\circ C^{-1}$), $\dot{V}_{O_2}$ is the rate of metabolism ($cm^3\ O_2\ g^{-1}\ h^{-1}$), and ΔT is the temperature differential (°C) (McNab, 1980a). It is difficult to know whether climate has an appreciable effect on conductance in bats, because although temperate bats might be expected to have lower conductances (to conserve heat), temperate species tend to abandon temperature regulation in winter. It could be argued that temperate bats should have high thermal conductances to minimize the equilibrial temperature differential maintained by a torpid bat in winter (see McNab, 1974a). Recently Shump and Shump (1980) have measured insulation directly in several vespertilionids from North America. They found that lasiurines, such as *Lasiurus*

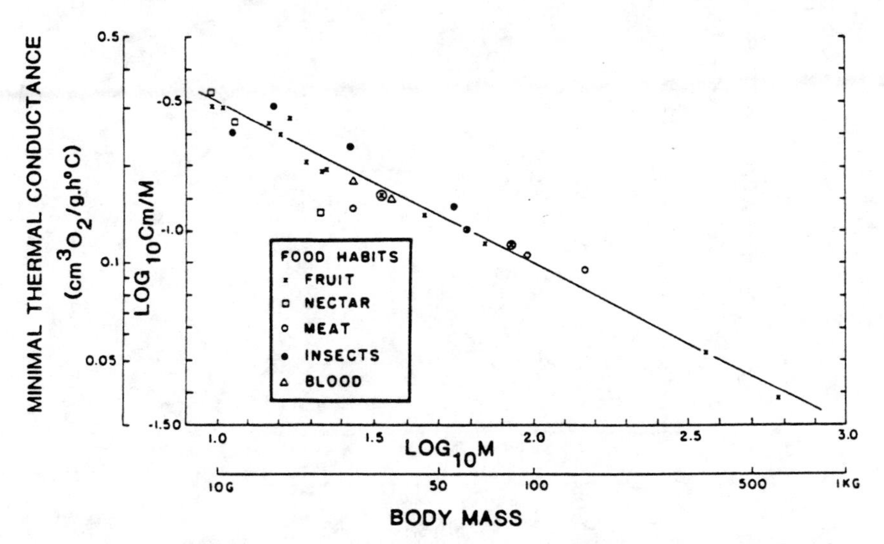

FIGURE 8. Minimal mass-specific thermal conductance in relation to body mass. (Data from McNab, 1969. Standard curve from McNab and Morrison, 1963.)

borealis and *L. cinereus*, which are solitary tree bats, have better insulation in summer than do cave bats, for example, *Myotis lucifugus*, *M. keenii*, and *Eptesicus fuscus*, in winter. These cave bats have an insulation in winter that is about 26% higher than the one they have in summer. The generally gregarious habits and protected roosts of *Myotis* and *Eptesicus* reduce the necessity to have high insulation; however, the roosts of lasiurines may often expose them to temperatures below freezing (Reite and Davis, 1966), at which times temperature regulation is required.

2.1.3. Maintenance of a Temperature Differential with the Environment

One of the physiological consequences of the interaction of body mass, minimal thermal conductance, and the basal rate of metabolism in endotherms is the variation found in the temperature differential that they maintain at the lower

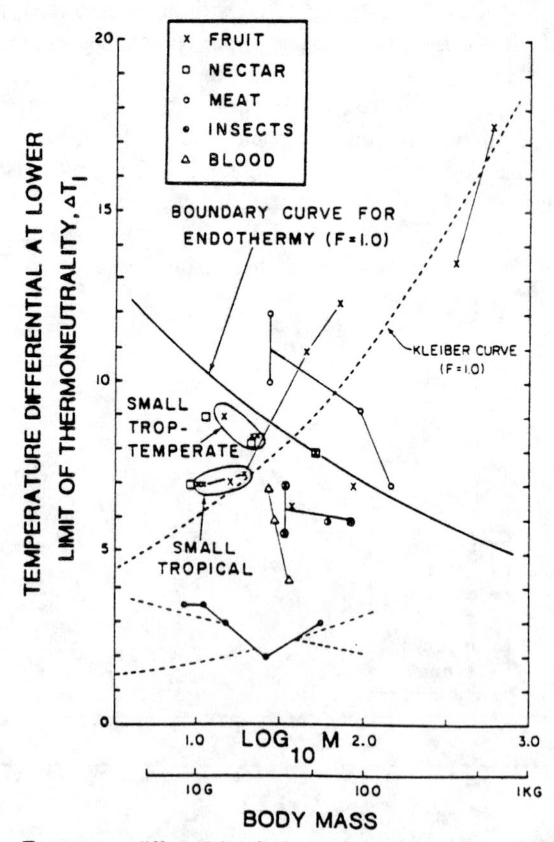

FIGURE 9. Temperature differential at the lower limit of thermoneutrality in bats as a function of mass, climate, and food habits. For a calculation of the curves see text. (Data principally from McNab, 1969.)

limit of thermoneutrality (Fig. 9). At first glance this differential ΔT_l, appears to be chaotically determined, but a closer inspection shows various patterns: (1) at a given mass the smallest differentials are found in insectivorous species, while the largest differentials are found in carnivorous and frugivorous species, as to be expected from the basal rate. (2) Larger frugivorous and insectivorous bats have differentials that increase with mass, as is to be expected from the relation

$$\Delta T_l = 3.42Fm^{0.25}$$

which implies that the basal rate follows a Kleiber relation and that total thermal conductance is proportional to $m^{0.50}$, and where F reflects ecologically significant variations in the basal rate and thermal conductance independent of mass (McNab, 1974b). (3) At small masses fruit bats that marginally enter temperate environments (such as *Carollia* and *Sturnira*) have larger differentials than similarly sized fruit bats that are restricted to the lowland tropics (such as *Ametrida*, *Vampyressa*, *Rhinophylla*, and small *Artibeus*). (4) In carnivorous, sanguinivorous, mixed-diet frugivorous, and small insectivorous and frugivorous bats ΔT_l increases with a decrease in mass, as is to be expected from

$$\Delta T_l = 15.56Fm^{-0.17}$$

which is derived from assuming that the basal rate follows the boundary relation for endothermy (McNab, 1982). Some bats (e.g., insectivores) may show the influence of both the Kleiber relation (at large masses) and the boundary relation (at small masses) (Fig. 9).

2.2. Ecological Significance of Energetics for Bats

2.2.1. Endothermy

A significant factor coupling mammals energetically to the environment is the price that must be paid for endothermy. It is generally agreed that (1) endothermy is expensive in terms of the amount of energy that must be expended to maintain body temperature independently of the environmental temperature, and (2) this cost is greater at a small size. The view that the cost of endothermy is greatest at small masses has been expressed in many papers (e.g., Pearson, 1948; Morrison *et al.*, 1959; Lasiewski, 1963; Tracy, 1977), but what is being said must be examined with care. In general, the absolute cost of endothermy, whether it is measured within or below thermoneutrality, is greater in larger than in smaller individuals (Fig. 10), because the total rate of metabolism is proportional to $m^{0.75}$ in thermoneutrality and to $m^{0.50}$ at temperatures below thermoneutrality. The scaling of the total rate of metabolism with mass is also positive when an

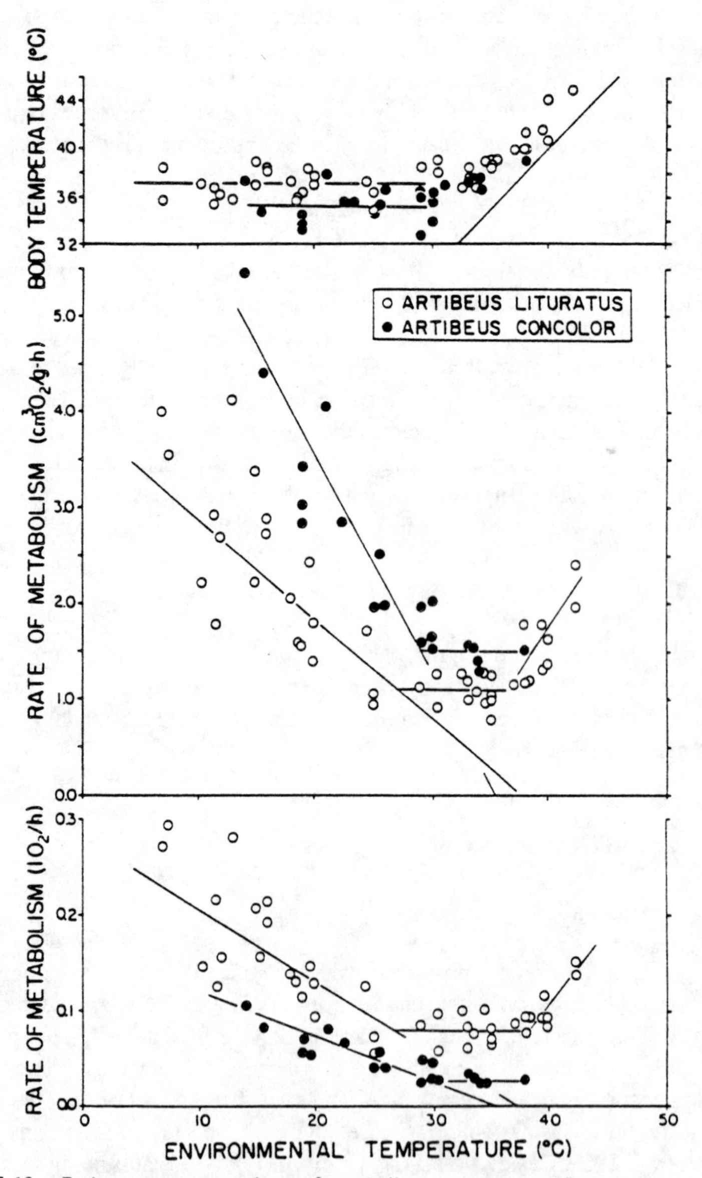

FIGURE 10. Body temperature, total rate of metabolism, and mass-specific rate of metabolism in relation to environmental temperature in *Artibeus lituratus* (mean mass, 70.1 g) and *A. concolor* (mean mass, 19.7 g). (Data from McNab, 1969.)

endotherm follows the "boundary" curve in Fig. 5, which on a mass-specific basis is proportional to $m^{-0.67}$, but proportional to $m^{0.33}$ on a total basis (McNab, 1982). A small size is associated with a high rate of metabolism only when rate is expressed on a mass-specific basis. In spite of their convenience, there is no unique biological significance to mass-specific rates of metabolism (McNab, in preparation). Why, then, do smaller species have lower body temperatures and poorer temperature regulation than larger species (Figs. 2 and 3)?

The factor that often requires small individuals and species to lower body temperature is that they can miss fewer meals without starving than large individuals and species. That is, the time period t during which starvation can be tolerated (without lowering the rate of metabolism and body temperature) is equal to the ratio of the size of the food store in the body (generally fat) divided by the rate at which the store is used to maintain body temperature (Calder, 1974; McNab, 1980c). If the fat store is proportional to the body mass (i.e., $m^{1.0}$) and if the rate of fat usage is proportional to $m^{0.75}$, t is proportional to $m^{0.25}$. In other words, a large mass reduces the immediate impact on survivorship of missing a meal, because with an increase in mass, a fat store increases more than does the rate at which the store is used. This suggests that there is a physiological advantage to a large body size in environments where missing a meal may be common. This relation may constitute a physiological basis for the occurrence of Bergmann's rule (namely, that subspecies of homeotherms living in cold climates tend to have larger body masses than subspecies of the same species living in warm climates) in some mammals. Bats, however, are limited in body size by the type of food they use; an increase in body size may be denied them. Under these circumstances, tolerance of starvation can be increased by reducing the rate at which the stores are used. Consequently, temperature regulation is often relaxed or abandoned at small masses.

2.2.2. Relationship between Energetics and Life-Span

One apparent consequence of the diversity in thermal behavior found in bats is that some species have much longer maximal life-spans than do other mammals of the same size. This view, which was advocated by Bourlière (1958), has been disputed by Herreid (1964). The easiest way to demonstrate the relation existing between these factors would be to plot life-span as a function of the rate of metabolism, but such a relation is complicated by both life-span and rate of metabolism being functions of body mass. However, when the residual variation in life-span about the mean curve on body mass is compared to the residual variation in the basal rate, it can be shown that life-span in mammals falls with an increase in the rate of metabolism independently of the direct influence of body mass (McNab, 1983). This transformation is impossible for bats at the present time, because there are few data on mean life-spans and because it is unclear

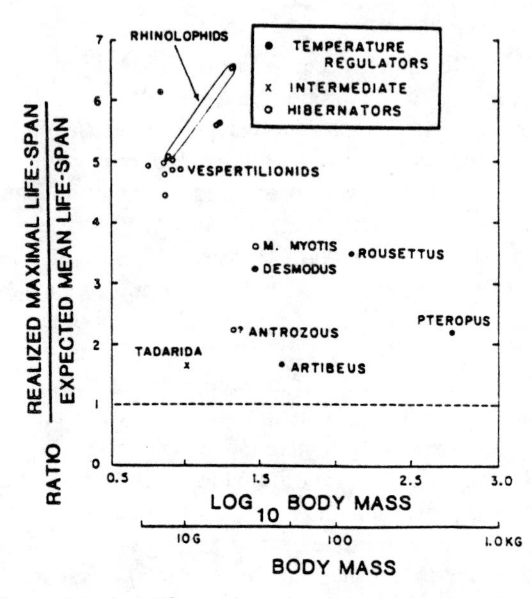

FIGURE 11. Observed maximal life-span in various bats expressed in relation to the mean life-span expected from the relation l (days) = $645.7g^{0.23}$ (derived from Western, 1979) as a function of body mass. (Data on maximal life-spans from Herreid, 1964.)

what the ecologically relevant rate of metabolism would be in species that enter daily or seasonal torpor.

The few available data on maximal life-spans in bats are derived mainly from northern temperate species (see Chapter 3). These data have been corrected for body size by expressing them relative to the mean curve for life-span (Western, 1979). Several points can be seen (Fig. 11): (1) all bats have life-spans greater than expected, in part simply because most data on bats are maximal, not mean, life-spans. (2) Most small vespertilionids and rhinolophids have maximal life-spans that are 4.5–6 times the mean values expected from mass, this large factor reflecting the propensity of these bats to enter torpor. (3) Larger (tropical) phyllostomids and pteropodids have maximal life-spans that are 1.5–3.5 times the mean life-spans expected from mass; these bats rarely, if ever, enter torpor. (4) Some insectivorous bats, such as *Tadarida brasiliensis* and *Antrozous pallidus*, have low life-span ratios (1.5–2.2), suggesting either that adequate information on their longevity is not available (*Antrozous*), or that these bats do not go into the deep torpor characteristic of true hibernators (*Tadarida*). Although the available data are inadequate for a complete analysis of the factors that determine longevity, it can be tentatively concluded that the tendency for small, insectivorous bats to enter torpor and hibernation is a factor directly leading to a longer life-span when compared to other mammals of the same mass.

2.2.3. Adjusting Energy Expenditure in Tropical Environments

A limitation in the amount of energy available to an organism in the environment may well require it to stagger the various energy-demanding activities throughout the year. Thus, there is a seasonal pattern to reproduction in bats that reflects their food habits (McNab, 1969; Mares and Wilson, 1971). Recent studies have reinforced and elaborated this suggestion (see Chapter 2). For

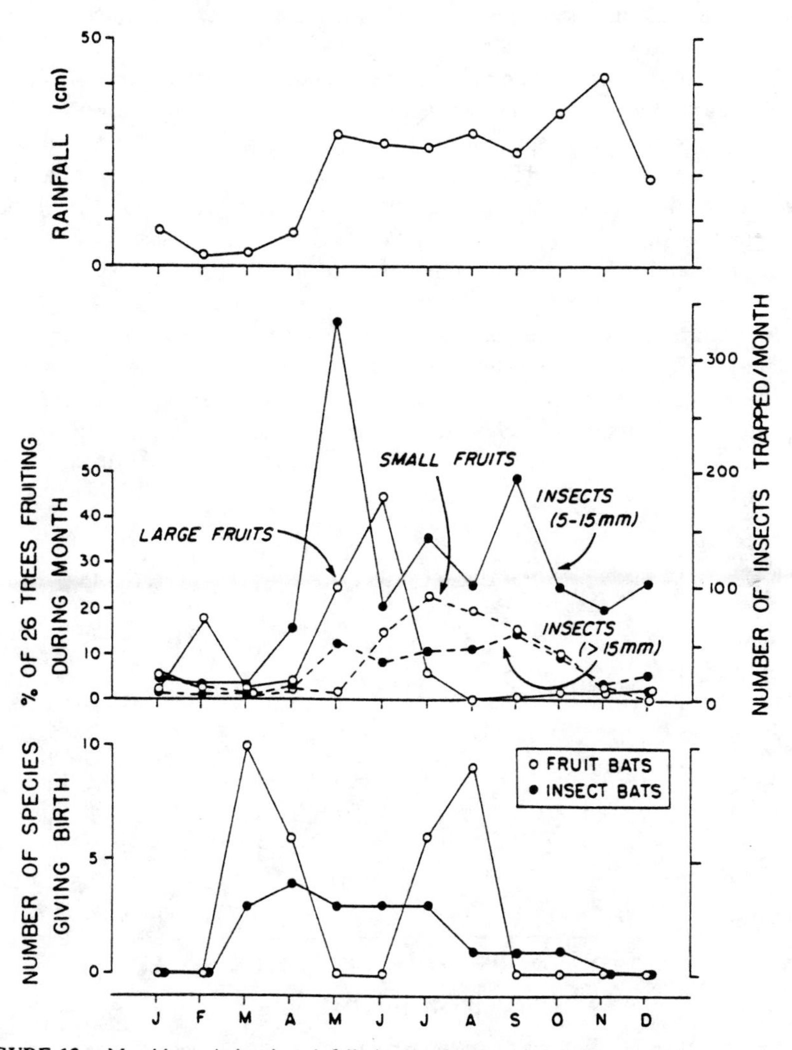

FIGURE 12. Monthly variation in rainfall, in the abundance of fruits and insects, and in the numbers of insectivorous and frugivorous bats giving birth on Barro Colorado Island, Panama. (Data from Fleming *et al.*, 1972; Wilson, 1977.)

example, Fleming *et al.* (1972) showed in Central America the following: (1) insectivorous bats are often seasonally monestrous; (2) many frugivorous bats are seasonally polyestrous; (3) one insectivorous species, *Myotis nigricans*, has an extended reproductive season, with some individuals being polyestrous; and (4) the vampire *Desmodus rotundus* is polyestrous, with year-round reproductive activity. From the viewpoint of energetics, then, the length of the reproductive season in tropical bats appears to be related to the dependability of food resources, with the use of fruit, or a mixture of fruit and insects, permitting a longer period than the use of insects alone (*M. nigricans* excepted; why?), and the use of blood permitting the longest period of reproduction.

The timing of reproduction may also be crucial. Many insectivorous bats are pregnant during the dry season and give birth at the beginning of the wet season, so that the high cost of lactation is borne at a time when the abundance of flying insects is the greatest (Fig. 12). Frugivorous bats have two peaks of birth, one during the peak in abundance of large fruits and the other at the time of the greatest abundance of small fruits (Fig. 12). Humphrey and Bonaccorso (1979) showed that seasonal peaks in lactating *Artibeus jamaicensis* correlate with the number of trees producing fruits (Fig. 13).

As with most organisms, bats adjust their behavior to maximize survival. Thus some species selectively shift their food during the dry season from insects to fruit or to nectar and pollen (Fleming *et al.*, 1972; Heithaus *et al.*, 1975; Humphrey and Bonaccorso, 1979) or migrate to follow the availability of selected flowers (e.g., *Leptonycteris, Choeronycteris;* Humphrey and Bonaccorso,

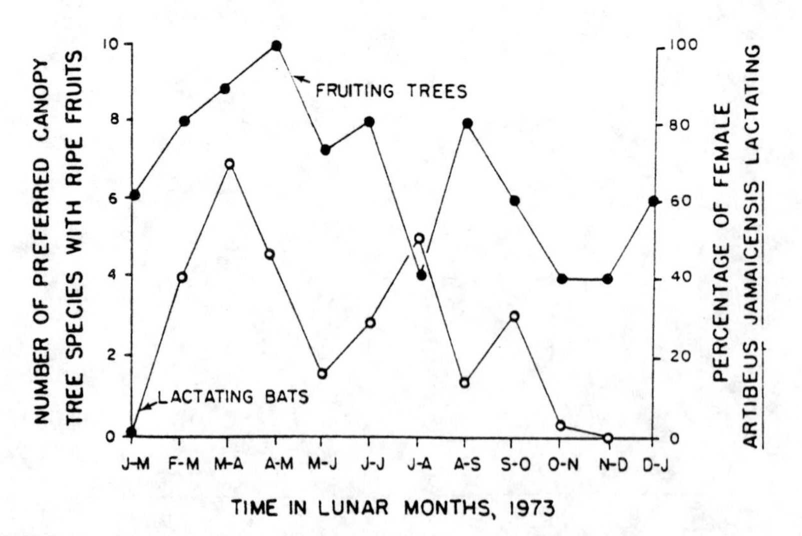

FIGURE 13. Monthly variation in the percentage of female *Artibeus jamaicensis* lactating and in the number of preferred canopy trees with ripe fruits on Barro Colorado Island, Panama (Humphrey and Bonaccorso, 1979).

1979). The problem faced by most insectivorous bats is as much due to the inflexibility of their food habits as it is to feeding on insects. This conclusion has a taxonomic component in the sense that some insectivorous bats, such as vespertilionids, natalids, rhinolophids, and molossids, are strictly insectivorous, while others, especially phyllostomines, are opportunistically insectivorous. The influence of insectivorous habits is further complicated by the generally small size of such bats, which (as stated) reduces their capacity to store fat and to withstand starvation.

2.2.4. Adjusting Energy Expenditure in Temperate Environments

Special consideration should be given to adjustments that bats make to temperate environments because of the extreme environmental conditions that may be faced. The harshest environmental conditions are faced during temperate winters, so that is an appropriate time with which to begin the discussion.

Temperate winters are generally characterized by cold temperatures and a shortage of food. This combination is exceedingly difficult for an endotherm, because cold temperatures demand an increased rate of heat production and thus an increased rate of food consumption. Few foods used by bats are available during a temperate winter. In fact, the only food used by most temperate bats is flying insects. Pollen and nectar feeders only marginally enter the warm temperate zone (e.g., *Leptonycteris* and *Choeronycteris* in southwestern United States), as is the case in fruit eaters (e.g., *Sturnira* and to a lesser extent *Carollia* in Uruguay, southern Brasil, and northern Argentina).

Given the dependence of temperate bats on flying insects as food, it is impossible for bats to remain year-long in temperate climates and maintain endothermic temperature regulation. Therefore only two alternatives are permitted: temperate bats must migrate to warm climates if they are to retain endothermy, or they must enter torpor for extended periods if they are to remain in a temperate environment.

2.2.4a. Migration. Compared to temperate birds, few temperate bats migrate to the tropics or to transequatorial temperate regions to evade a harsh temperate winter (see Baker, 1978). In temperate North America the principal migratory species are either tropical bats that reach their northern limits to distribution in warm temperate regions (e.g., *Tadarida brasiliensis, Leptonycteris nivalis, L. sanborni,* and *Choeronycteris mexicana*), or they are north temperate tree bats (e.g., *Lasiurus cinereus, L. borealis,* and *Lasionycteris noctivagans*), which potentially face very low ambient temperatures in tree roosts. A similar pattern is found in Eurasian bats (Strelkov, 1969). Many temperate vespertilionids and most *Myotis* seasonally move between breeding colonies and hibernating roosts, but these movements usually do not involve long latitudinal

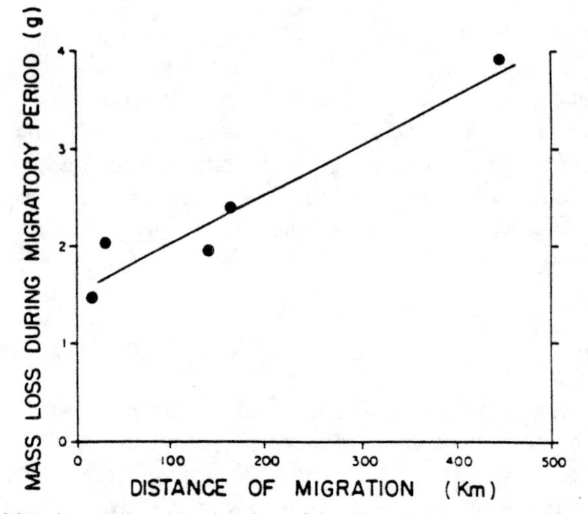

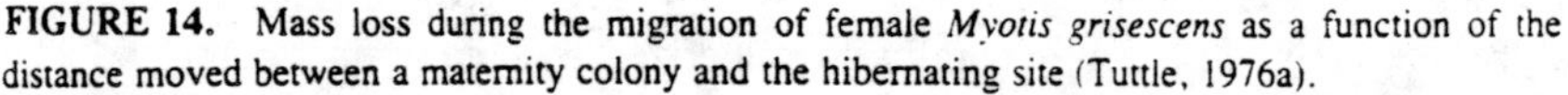

FIGURE 14. Mass loss during the migration of female *Myotis grisescens* as a function of the distance moved between a maternity colony and the hibernating site (Tuttle, 1976a).

movements, except in individuals at the northern limits of distribution, which often may not encounter hibernacula locally.

There are few detailed studies of migration in bats; the only species in which individually banded bats have been shown to make such migratory movements in North America is *Tadarida brasiliensis*, and even then migration is known to occur only in populations that breed in southwestern United States and northern Mexico (Villa-R. and Cockrum, 1962). The seasonal pattern in fat storage has been examined in this species both in migratory (O'Shea, 1976) and nonmigratory populations (Pagels, 1975). There is no reason to believe, however, that fat storage for migration is different from fat storage for local movements and hibernation, so it is convenient to examine the seasonal pattern in fat storage for migration and hibernation at the same time in the next section.

The use of fat for migration in bats is clearly shown in data collected by Tuttle (1976a) on *Myotis grisescens*. The decrease of body mass in females collected during hibernation is directly proportional to the distance between the breeding caves (where the bats were originally caught) and the hibernacula (Fig. 14). Although some of the weight loss may be water, much of the loss represents a loss of body fat. If, however, the bats had had enough time after migration to become rehydrated, then the decrease in mass would have been due to a reduction in fat content alone. Krulin and Sealander (1972) showed that female *M. grisescens* deposit fat having a caloric value of 9.2 kcal/g. Tuttle's data indicate that this bat loses about 0.5 g per 100 km, which translates into an expenditure of about 4.6 kcal per 100 km. Thomas and Suthers (1972) and Thomas (1975) showed that the cost of flight is similar in bats to that in birds. Tucker (1970) estimated that the metabolism of flying vertebrates weighing about 15 g is about

4.9 kcal per 100 km, which is remarkably similar to the estimate derived from Tuttle's measurements.

2.2.4b. Hibernation. Hibernation is the condition in which animals spend much or all of winter in an inactive physiological state (torpor), which in endotherms is characterized by the fall of body temperature to ambient levels, as long as ambient temperature does not fall to or below freezing. Such a state greatly reduces energy expenditure; it is used because an active state cannot be sustained due to limited food supplies.

Due to low rates of metabolism and small body masses (see McNab, 1974a), the temperature differential maintained with the environment by torpid bats is very small, in the order of a few tenths of a degree Celsius. The size of this differential varies to some extent seasonally (Henshaw and Folk, 1966), being somewhat smaller during "deep" hibernation and somewhat greater during early and late hibernation, but in all cases the differentials are within 2°C (Fig. 15). The only time that torpid temperate-zone bats maintain an appreciable temperature differential with the environment is when ambient temperatures approach freezing (Hock, 1951; Henshaw and Folk, 1966); such differentials are produced by an increase in the rate of metabolism (Hock, 1951; McNab, 1974a).

The presence of small temperature differentials in temperate vespertilionids is radically different from the pattern found in the tropical free-tailed bat *Tadarida brasiliensis* at the temperate limits of its distribution in the nonmigratory populations. These bats maintain large differentials in winter (presumably as long as they have a food or fat supply), this differential attaining 24–31°C (Fig. 15). Although this bat can enter torpor, there is no evidence that it enters a "true" hibernation characterized by long periods (i.e., weeks or months) of continuous torpor.

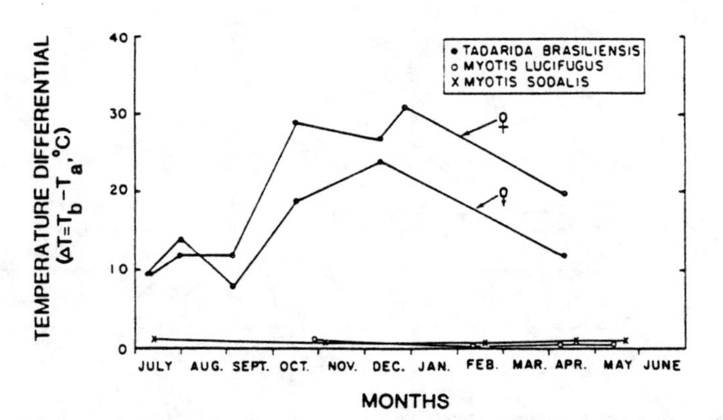

FIGURE 15. Body temperature maintained in *Tadarida brasiliensis* at 6–7°C and in *Myotis lucifugus* and *M. sodalis* at 5°C throughout the year. (Data from Henshaw and Folk, 1966; Pagels, 1975.)

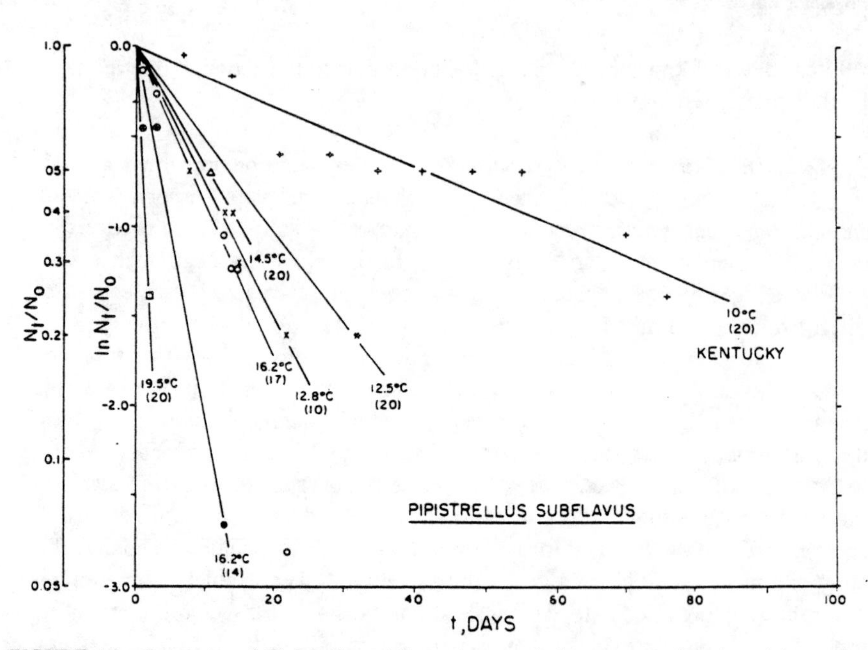

FIGURE 16. Proportion of *Pipistrellus subflavus* that remain in torpor as a function of time and ambient temperature (McNab, 1974a).

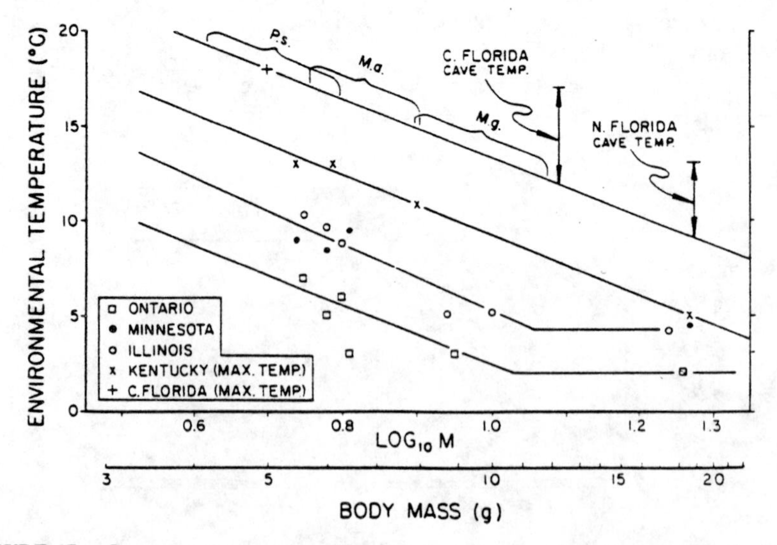

FIGURE 17. Cave temperatures selected by bats as a function of the mass of individual bats (McNab, 1974a).

No mammal remains continuously in torpor during the hibernal period. [For a detailed examination of movement during this period in three European *Myotis*, see Daan (1973).] The length of a period of torpor varies with the ambient temperature, body size, and species (Daan, 1973; McNab, 1974a); it also varies with the time of the year (Daan, 1973). Torpid periods are maximal at low temperatures (as long as the temperature does not approach or fall below freezing) and at small body sizes. For example, 50% of the *Pipistrellus subflavus* occurring in caves move in 1.1 days at 20°C, whereas 50% move in 43.9 days at 10°C (Fig. 16). Furthermore, 50% of *Rhinolophus ferrumequinum* move in 0.9 days at 10°C, the principal differences between these species being that *P. subflavus* weighs 6 g and is a solitary hibernator, whereas *R. ferrumequinum* weighs 25 g and is a clustering hibernator.

The tendency of large bats to have higher levels of activity, that is, shorter periods of continuous torpor, during hibernal periods can be compensated for by the selection of cooler microclimates in caves (Fig. 17). It is therefore clear why *Eptesicus fuscus* chooses the coldest microclimates to hibernate and how the small, solitary *Pipistrellus subflavus* can tolerate high temperatures, but it is unclear why *Pipistrellus* does not choose as low a temperature as *Eptesicus*, thereby saving even more energy than it does at warmer temperatures. It is tempting to suggest that the temperature chosen is, in fact, a compromise between those low temperatures that diminish energy expenditure and those higher temperatures that permit something else to occur. What the "something else" might be is unclear, but it may reflect the sensitivity of spermatozoa to a rapid fall in temperature (Mann, 1954) and that small bats cool more rapidly while entering torpor than do large bats. This balance weighs in favor of low temperatures in larger species and in favor of high temperatures in small species.

One implication of the curves illustrated in Fig. 17 is that there is a boundary curve that describes the maximal temperatures that are compatible with torpor and hibernation (see the Florida curve). Given the warm winter cave temperatures of subtropical climates, the only species capable of using hibernation would have to be small and solitary. Thus in Florida the only bat to hibernate is *Pipistrellus subflavus*. Furthermore, *Myotis grisescens*, which is both large and forms clusters, is presently found only in northern Florida in summer. Most of this population migrates northward to Alabama and Tennessee for winter (Tuttle, 1976a), presumably to find caves that are sufficiently cold for successful hibernation to occur (McNab, 1974a).

Clustering habits, as seen, also modify the thermal behavior of cave bats (McNab, 1974a): for example, *Myotis lucifugus* has a rate constant for movement that is about 12.6 times that of *M. emarginatus* at 10°C, even though both bats weigh about 9 g; *M. lucifugus* tends to hibernate in clusters, while *M. emarginatus* tends to be a solitary hibernator. Clustering bats tend to maintain larger temperature differentials with the environment than do solitary bats, to select cooler microclimates than do solitary bats (Beer and Richards, 1956), and to have smaller clusters in warmer caves than in cooler caves (Twente, 1955; Hall, 1962). Bats also cluster less in mild than in cold weather (Hooper and Hooper, 1956).

It can be concluded that the temperature, season, clustering behavior, and mass all influence the activity of bats during the hibernal period. But some of the variation in activity cannot be explained by this analysis. For example, *Myotis leibii* is small and solitary but selects cold temperatures in caves (Fenton, 1972). Such examples do not deny the analysis given above but simply point out that much work remains to be done on the factors that determine activity in bats (and other mammals) during hibernation.

Even though the rate of metabolism is greatly decreased by inactivity and entrance into torpor, a period of hibernation lasting 6 months or more requires significant amounts of energy. This energy in bats is provided in the form of body fat and is accumulated over relatively short periods at the end of summer (Fig. 18). For example, female *Myotis thysanodes* increased their fat content by ninefold (from 0.23 to 2.06 g per bat) in a period of 11 days (Ewing *et al.*, 1970), and female *Miniopterus* increased body fat by 3.36 g per bat in 32 days (Shimoizumi, 1959). In warm temperate climates some feeding may occur in winter during warmer periods, thereby reducing the drain on the limited energy reserves.

What permits bats to store large amounts of energy as fat over short periods of time? There are several possible answers: bats may have a longer feeding period during late summer; they may collect food more "efficiently" during this season; or they may be more "efficient" in converting food into fat than is

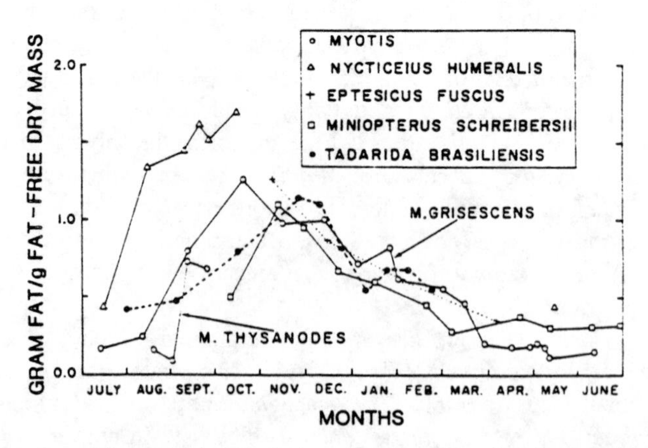

FIGURE 18. Seasonal variation in body fat in various bats. (Data from Beer and Richards, 1956; Shimoizumi, 1959; Baker *et al.*, 1968; Ewing *et al.*, 1970; Weber and Findley, 1970; Krulin and Sealander, 1972.)

normally the case. Several of these factors may contribute to the accumulation of fat, one of the most interesting being originally suggested by Twente (1955) and followed by Krzanowski (1961). They suggested that bats in early fall may accumulate fat by feeding outside of caves, then entering cool to cold caves for roosting, wherein the resting rates of metabolism fall and the majority of the harvested energy is converted into body fat. The calculations of Ewing *et al.* (1970) on the autumnal energy budgets of two North American *Myotis* support this view, namely, that entrance into daily torpor contributes to the accumulation of body fat.

The use of hibernation by temperate bats is closely coupled to their reproductive behavior (see Chapter 2). In these species copulation occurs during "swarming" in the fall at the mouths of caves or during the hibernal period in caves (Thomas *et al.*, 1979). Depending on the species, females either store spermatozoa (most vespertilionids) or have delayed implantation (e.g., *Miniopterus*, *Macrotis*). In both behaviors females remain torpid, or at least reduce the number of arousals, to prevent ovulation or implantation. In fact, there is much evidence that hibernating females are less active than males of the same species (Cross, 1965; McNab, 1974a; Tuttle, 1976a); this reduction in activity is accomplished by females hibernating as solitary individuals or in groups smaller than those of males (Davis and Hitchcock, 1964; Fenton, 1970, 1972), or by the selection of colder microclimates (Beer and Richards, 1956). At the southern limits of distribution some temperate bats face warm cave microclimates that, given the size of the bats or their clustering behavior, do not permit extended periods of torpor. Thus, female *Myotis grisescens*, as noted, migrate northward to hibernate in northern Alabama (at a minimal distance of 440 km), central Tennessee (510 km), and northern North Carolina (660 km).

Almost all observations on the hibernation of temperate bats have been made on cave bats (Davis, 1970). Little is known of the physiology and ecology of tree bats. In summer most tree bats, at least in North America, roost among leaves or in Spanish moss (in the South), while a few (especially *Nycticeius*) roost in tree hollows. These roosting sites offer much less protection from low temperatures than is the case in most caves and, from the viewpoint of hibernation, do not provide stable temperatures even in winter. It is therefore not surprising that tree bats are often migratory, or that they have heavy insulation and are able to withstand moderate levels of supercooling (Reite and Davis, 1966). It is of considerable interest that the tree bat *Lasiurus borealis* remains in torpor at higher temperatures than some (most?) cave bats. Such a difference would permit tree bats to remain in torpor even though air temperatures showed a temporary increase. For example, *L. borealis* may remain in torpor at temperatures up to 20°C (Davis and Reite, 1967); in winter this bat is seen feeding only at air temperatures greater than 19°C (Davis and Lidicker, 1956). Much more information is required to determine the extent to which tree bats differ from cave bats in terms of physiology, ecology, and behavior.

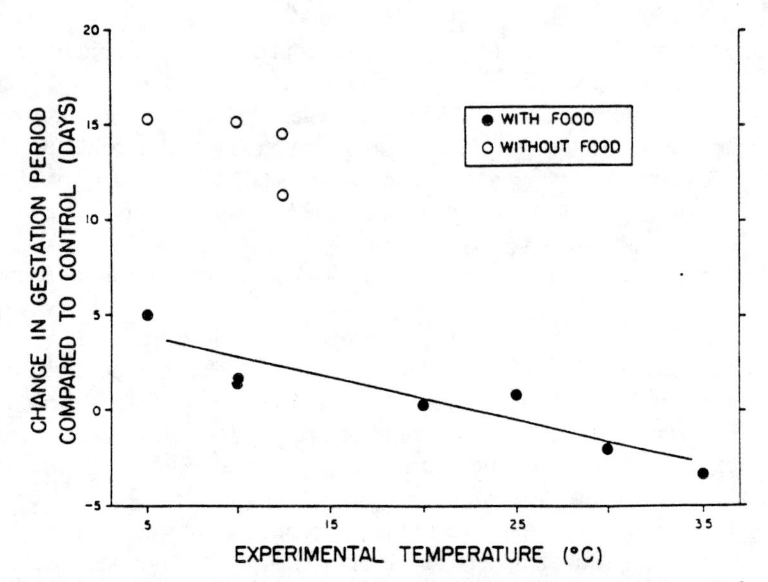

FIGURE 19. Variation in the length of the gestation period in *Pipistrellus pipistrellus* as a function of ambient temperature, with and without food (Racey, 1973). The absolute gestation periods at each temperature were compared to a control group housed in a room whose ambient temperature varied between 10 and 25°C. This technique permitted a correction to be made among the various groves of bats, which may have started the experiment at different stages of pregnancy. Notice that by feeding the bats the gestation period is shortened, presumably because the bats were able to maintain a higher body temperature and consequently higher fetal growth rates.

2.2.4c. Maternity Roosts. During the summer many bats use roosting sites that differ from those used in winter. For example, *Myotis sodalis*, which hibernates in large clusters in caves, leaves caves to give birth and rear young (Humphrey *et al.*, 1977), while others like *Myotis grisescens* (Tuttle, 1975, 1976a) and *Miniopterus schreibersii* (Dwyer, 1963, 1964; Dwyer and Harris, 1972) spend winter and summer in caves, although even in these cases there usually is a seasonal shift from one cave to another, or from one section of a cave to another. Sometimes a fraction of the hibernating population remains at the hibernating cave in summer, but these individuals are usually males, an observation that emphasizes the relationship of seasonal movement to reproduction.

The length of the gestation period and the period of postnatal growth and development depends in mammals on the complex interaction existing among body mass, the rate of metabolism, and the level of body temperature (McNab, 1980b). As has been seen, most temperate insectivorous bats show only modest levels of thermoregulation, at least as solitary individuals. Therefore these species could reduce the time from birth to independence if they chose roosts with high ambient temperatures during the reproductive period. Racey (1973), for example, has shown under laboratory conditions that the gestation period of

Pipistrellus pipistrellus increases at low ambient temperatures and is shortened at high temperatures (Fig. 19), presumably because of the dependence of the gestation period and intrauterine rates of development on the rate of metabolism and body temperature. Similarly, the postnatal growth rate of *Myotis grisescens* in the field increases with cave temperature (Fig. 20). A detailed examination of the dependent nature of gestation period, postnatal development, and survival in bats are found in Racey (Chapter 2) and in Tuttle and Stevenson (Chapter 3).

It is important for the gestation period and the period of postnatal development to be short in temperate environments, because the time over which adequate food supplies are dependably found is short. Not only must gestation, development, and independence be accomplished before autumn, but young bats must also store adequate amounts of fat to survive the hibernal period. It is therefore necessary for young bats in a temperate environment to expend little energy on maintenance *per se*, a condition that can be accomplished through the use of warm maternity roosts. Humphrey *et al.* (1977) have shown that during cool summers young *Myotis sodalis* attain a free-flying stage about 2 weeks late, and the last migrants depart maternity colonies 3 weeks late, which may expose them to freezing temperatures and require an emergency entrance into torpor. This delay leaves young bats little opportunity to accumulate fat reserves. Winter survivorship in first-year *M. grisescens* was lower in lean than in fat individuals (Tuttle, 1976b).

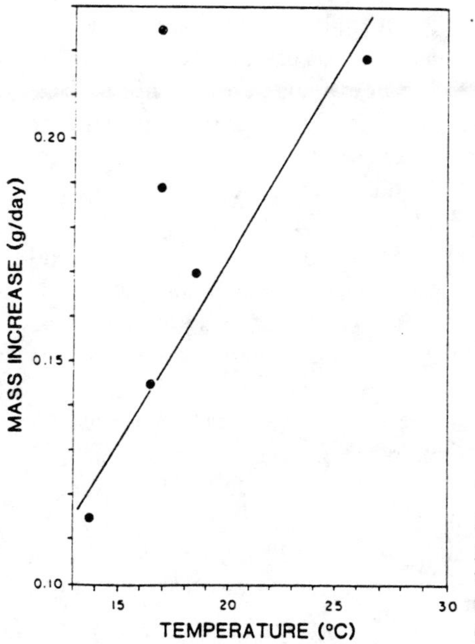

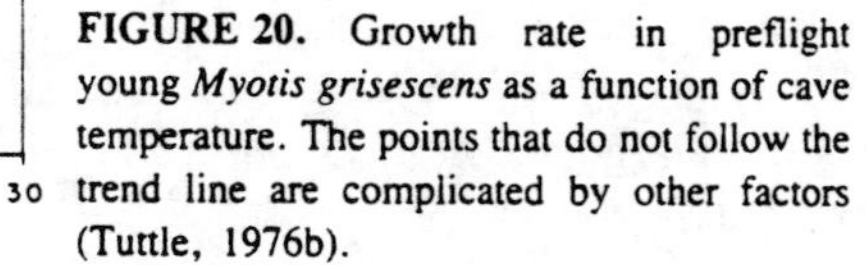
FIGURE 20. Growth rate in preflight young *Myotis grisescens* as a function of cave temperature. The points that do not follow the trend line are complicated by other factors (Tuttle, 1976b).

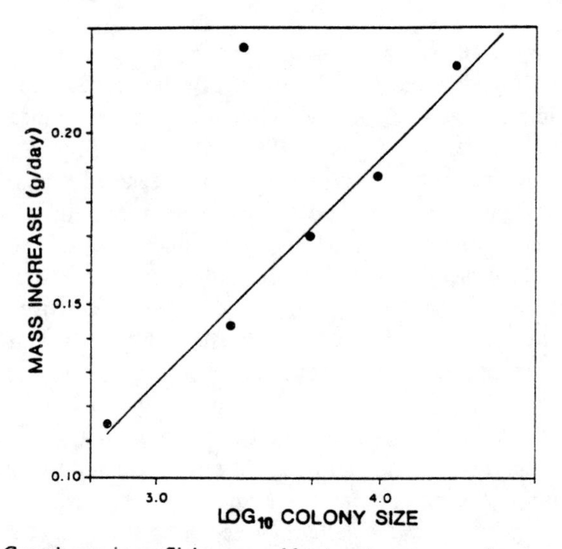

FIGURE 21. Growth rate in preflight young *Myotis grisescens* as a function of colony size as measured by the number of young. The point that does not follow the trend line is complicated by other factors (Tuttle, 1976b).

High roost temperatures can be attained in caves by selecting domes or horizontal passages where heat can accumulate. This behavior is found mainly in species living in warmer climates and tends to be less common in temperate environments, where cave temperatures are generally too cold. Cool caves can be made somewhat more thermally acceptable by bats assembling in large clusters (Dwyer and Harris, 1972; Tuttle, 1975). For example, the presence of some 3000 *Miniopterus schreibersii* in an Australian cave was sufficient to raise the chamber temperature by 8°C when the bats were present during the day; at night the chamber temperature fell by about 4°C while the bats were out of the cave. The level of the effective maternity temperature, as reflected in the growth rate of *Myotis grisescens*, varies with colony size (Fig. 21), as well as with air temperature (Fig. 20). The correlation with colony size appears to reflect an increase in colony temperature. The only way that solitary bats, or bats forming small clusters, such as *Plecotus townsendii*, can maintain their maternity roosts in caves is by finding naturally warm microclimates (Humphrey and Kunz, 1976). In North America few *Myotis* breed in caves (e.g., *M. thysanodes, M. velifer, M. yumanensis, M. austroriparius*, and *M. grisescens*), and they are limited in distribution to the southwest and southeast (Tuttle, 1975). Most of these species cluster. In Europe some bats occupy cold caves in summer (e.g., *Miniopterus schreibersii, Myotis myotis, Rhinolophus euryale*, and *R. ferrumequinum*), but they form large maternity colonies and are much larger bats than North American cave bats.

The most common solution to cool cave temperatures in summer is for cave bats to abandon caves for warmer sites, such as houses, barns, trees, and rock piles. For example, *Myotis thysanodes* and *M. yumanensis* both abandon caves for external maternity colonies in the northern parts of their distribution. Of course, some bats, such as *M. lucifugus* and possibly *M. keenii*, exclusively use buildings as maternity roosts throughout their range. Tuttle and Heaney (1974) showed that *M. leibii*, which normally hibernates in caves in treeless areas, often has its maternity roosts under boulders or in cracks in vertical banks. These roosts have temperatures in summer between 26 and 33°C, which are much warmer than local caves would be. *Myotis sodalis*, which forms large hibernating clusters in caves, disperses widely to wooded areas, where they form small maternity colonies situated under the bark of living and dead trees (Humphrey *et al.*, 1977). These roosts are warmer than the adjacent atmospheric temperature and much warmer than caves would be. In some cases bats may choose maternity roosts that become so warm that the bats must disperse to cooler areas to prevent overheating (Licht and Leitner, 1967). One of the bats studied by Licht and Leitner was *Tadarida brasiliensis;* it, like other molossids, often faces very high temperatures in maternity roosts (Twente, 1956; Henshaw, 1960; Herreid, 1963, 1967).

It is important to realize that the principal reason for the selection of warm microclimates during the maternity period is to accelerate the rate at which newborn bats attain behavioral independence of their parents, thereby permitting both adult and immature bats to store sufficient amounts of fat to ensure survival through winter. In lieu of warm microclimates and sufficiently large clusters to produce high cluster temperatures, pregnant bats would have to regulate their body temperature endothermically, which vespertilionids are capable of doing (Stones and Wiebers, 1967; Studier and O'Farrell, 1972), but only at great expense.

2.3. Energy Budgets

Ideally, an energy budget is a statement of the total amount of energy consumed by an organism over a given period of time and a complete accounting of the energy expended, either in terms of the physical pathways by which energy is lost to the environment or in terms of the biological uses to which the energy is put. It is difficult to give a complete accounting of energy allocation because of the diversity of activities used and because the energy equivalency of many activities is unknown.

There are a limited number of ways to construct an energy budget (for a review see Kunz, 1980). One is to measure the total budget "directly," which can be done through Ransome's method (1973) of total food intake as estimated by the accumulation of feces, through the use of daily or seasonal variations in body mass (Kunz, 1974), or through the use of labeled water (which has not yet

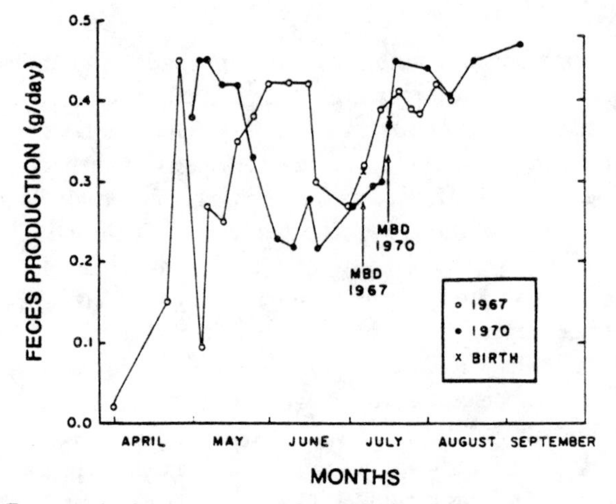

FIGURE 22. Feces production by a colony of *Rhinolophus ferrumequinum* during the period of pregnancy, birth, and postnatal development in 1967 and 1970. Notice that during late pregnancy feces production, and presumably food intake, is reduced, and that this reduction is greater in 1970 than in 1967. (Data from Ransome, 1973.) MBD signifies mean birth date.

been used on bats). For example, Ransome (1973) showed that *Rhinolophus ferrumequinum* in summer consumes enough insects to produce up to 0.45 g of feces per day (Fig. 22); these data can be converted into an energy budget by determining the energy equivalence of feces formation. It is noteworthy in these data, both in 1967 and 1970, that there was a depressed food intake and entrance into torpor during middle to late pregnancy. This reduction may reflect a negative effect of carrying a fetus on the ability of pregnant bats to act as aerial predators (see Kunz, 1974); this interference may increase with the growth of the embryo. The reduction in food intake ended with birth. Ransome noted that the depression of food intake was two to three times greater in 1970 than in 1967 (Fig. 22). which he suggested may have been due to the presence of fewer nonbreeders. Nonbreeders normally maintain high rates of food intake and may contribute to the breeding success of the colony by raising cluster temperature. One consequence of the greater reduction in the rate of food intake in 1970 was that the mean birth date that year was delayed by 10 days as compared to 1967.

Kunz (1974) measured the annual cycle in body mass in male and female *Myotis velifer* (Fig. 23). These bats depend for their energy expenditures mainly upon stored fat in winter and upon feeding during the rest of the year. Kunz showed that the daily variation in body mass indicates two principal feeding periods, one in the early evening and the other in the early morning. Females tend to feed more than males because they are usually pregnant, lactating, or depositing more fat than males. It is of interest that females delay their molt until the end of lactation to reduce the imposition of one costly process on another.

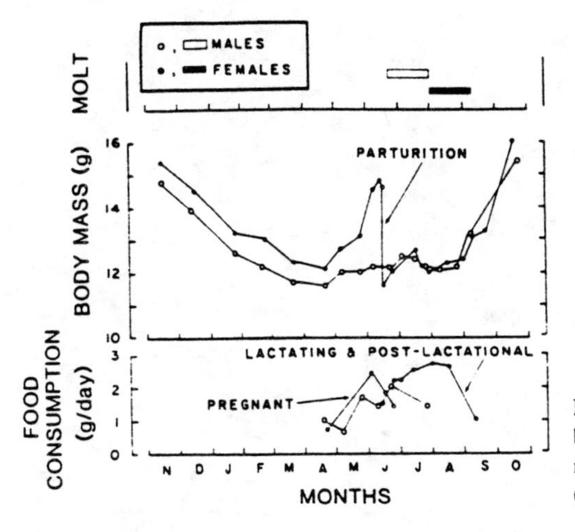

FIGURE 23. Seasonal variations in body mass, food consumption, and molt in male and female *Myotis velifer* (Kunz, 1974).

These data can, with a few modest assumptions, be converted into an annual energy budget.

Another method used to estimate energy budgets is derived from a "time budget," where the time spent on each biologically and energetically important activity is multiplied by the energetic equivalent of the activity, the energy budget being the sum of these products, one for each relevant activity.

One of the earliest attempts to estimate an energy budget for a bat was made by Griffin (1958). Aside from the fact that some assumptions on body mass, the cost of flight, and on flight time would be changed today, this budget has value in demonstrating the extreme to which an annual energy budget can be taken in a small endotherm that enters daily and seasonal torpor. Griffin estimated that a 6-g *Myotis lucifugus* would annually expend about 540 kcal, 78% of which would be used during 2 hr of flight per night for 6 months, that is, during 360 hr per year! Even during winter, when *M. lucifugus* is assumed to spend 98% of its time in torpor, only 54% of the energy expenditure is estimated to have occurred during torpor. Clearly, the short periods of flight in summer and entrance into torpor in winter greatly reduce the energy expenditure of this bat, but it is also clear that the activity that occurs in summer and winter is exceedingly expensive.

More refined energy budgets have recently been calculated from time budgets by other workers, but they too are at best only approximations. Kunz (1980) has summarized these estimates and has shown that the total daily energy expenditure in free-living bats during the nonhibernal period is given by the relation

$$\dot{E}\ (\text{kcal/day}) = 0.92m^{0.767}$$

where m is in g (Fig. 24). The coefficient 0.92 becomes 3.85 if the units for $\dot{E}$ are

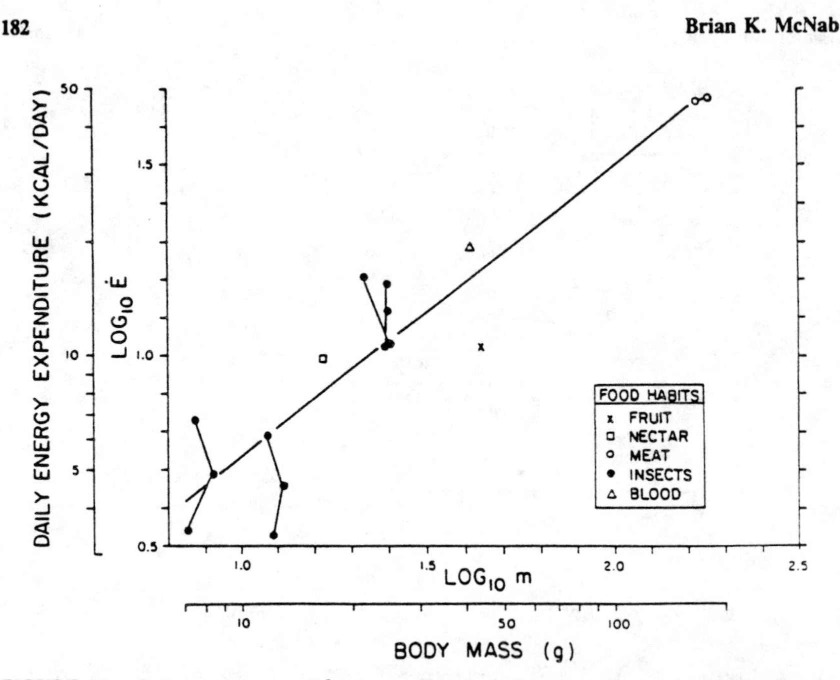

FIGURE 24. Daily energy expenditures as a function of body mass and food habits in bats. Various values for a species are connected with a line; these variations usually reflect differences in reproductive condition. (Data from Kunz, 1980.) Energy expenditures can be converted to kJ/day by multiplying kcal/day by 4.19.

in kJ/day. This mean curve is only about 2.3 times the basal rate expected of mammals, a ratio that is somewhat low for mammals of this size (McNab, 1980b), but which is commensurate with their low basal rates of metabolism and their propensity for torpor. It is of interest, if correct, that temperate insectivorous bats during summer are estimated to have daily energy expenditures similar to those of tropical, thermoregulating, frugivorous bats of the same size (Fig. 24).

2.4. The Evolution of Bat Energetics

As we have seen in bats, food habits act in concert with body size to influence the rate of metabolism and the level and precision of temperature regulation. This scheme can be used to examine the evolution of energetics and temperature regulation in bats, although unfortunately the early evolution of bats is unknown. The earliest bats known already are bats: they may (Van Valen, 1979) or may not (Smith, 1976) have been ancestral to all living bats.

Many opinions have been expressed on the evolution of bat energetics. Hock (1951), Eisentraut (1960), and Kayser (1961) all argued that torpor in bats is a "primitive" character. Twente and Twente (1964) maintained that torpor evolved in tropical bats as a response to cool caves. I suggested (McNab, 1969) that torpor in tropical bats was the consequence of combining a low basal rate with a small mass and that seasonal torpor (hibernation) is a specialized response adaptive to temperate climates. Dwyer (1971) maintained (1) that torpor is adaptive but was an integral part of the original set of chiropteran characteristics, (2) that torpor was maintained through the evolutionary mainstream of bats, and (3) that various evolutionary side branches, which are often associated with a strictly tropical distribution and with the shift to food habits other than insects, abandoned torpor. Given this diversity of views and the recent advances in understanding the energetics of endothermy, the evolution of energetics in bats deserves to be reexamined.

One of the greatest difficulties, of course, is to describe the energetics of the group of mammals that directly gave rise to bats [the "prebat" of Dwyer (1971)]. Van Valen (1979) noted that only dermopterans and tupaiids among living mammals may be related to bats, but such relationships must be distant at best. Van Valen suggested a series of characteristics that might have been present in the ancestor to the earliest bats and possibly present in these bats themselves. They include (1) an insectivorous diet, (2) crepuscular habits, (3) small body size, and (4) a tropical habitat. I would add arboreal habits to this list. It might be possible to estimate the energetics of a prebat by examining a tropical mammal that has all or most of these characteristics. One possible living "model" of the prebat might be tree shrews (Tupaiidae), which by coincidence (or not) may be related to the precursor that gave rise to bats. Living tupaiids differ from Van Valen's primitive ecological characteristics only in terms of diet: they are omnivorous, including both fruits and insects in their diet. This mixed-food habit may be a better estimate of the food habits of the earliest bats or of the prebats themselves [contrary to the view of Gillette (1975), who also argued for insectivorous habits], because it would permit the direct development of both frugivory and insectivory.

Rates of metabolism and temperature regulation have been measured in some tupaiids. For example, the diurnal tree shrews *Tupaia glis* (Bradley and Hudson, 1974) and *T. belangeri* (Weigold, 1979) have basal rates that are approximately 77 and 73%, respectively, of the value expected from the Kleiber relation. *Ptilocercus lowii*, a nocturnal tupaiid, has a basal rate equal to about 60% of the Kleiber value (Whittow and Gould, 1976). Although there are other values on the basal rates of tupaiids, they are higher and appear to be measured at temperatures below thermoneutrality. It can be concluded that a small, active, scansorial omnivore would have had a low basal rate, but as in *Tupaia* it would be high enough, given body mass, to eliminate the necessity of daily torpor (Fig. 5), although smaller tupaiids (such as *Ptilocercus*) may fall to the left of the boundary curve. *Ptilocercus* has not been reported to enter torpor. The arboreal ancestor of bats probably did not enter torpor unless it weighed less than 80 g, and possibly not even then.

The relation of food habits to the generally accepted patterns of chiropteran

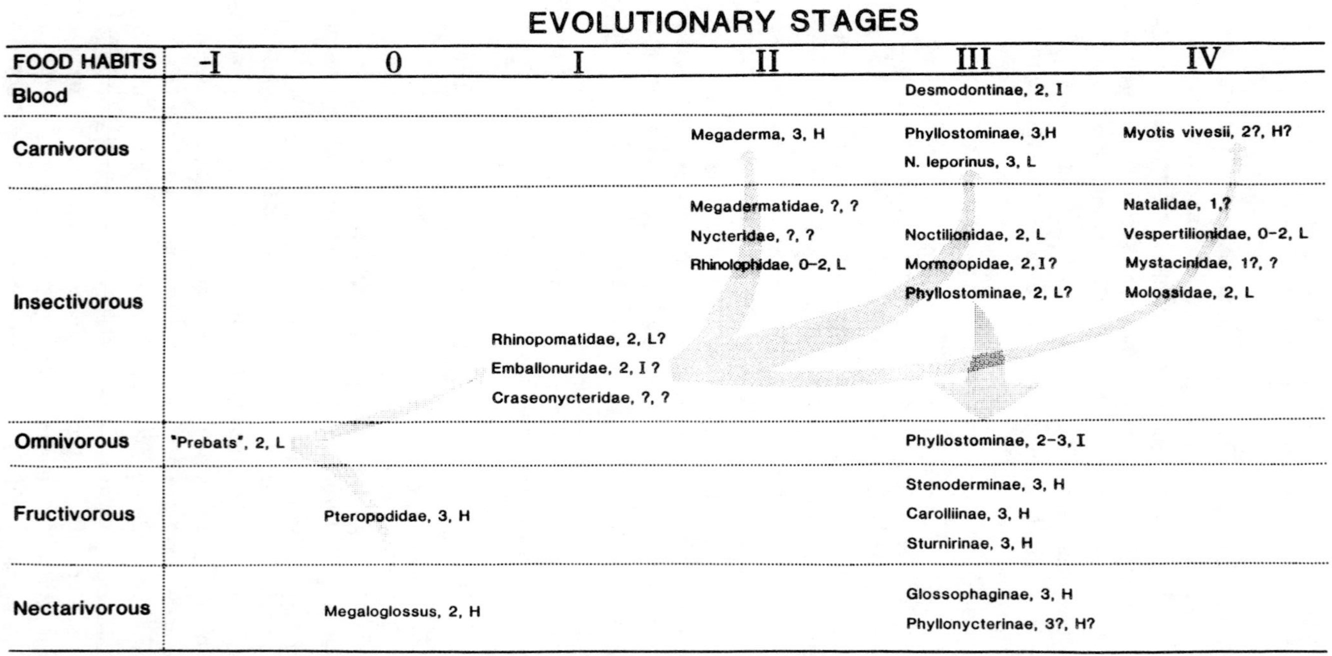

EVOLUTIONARY STAGES

FOOD HABITS	-I	0	I	II	III	IV
Blood					Desmodontinae, 2, I	
Carnivorous				Megaderma, 3, H	Phyllostominae, 3,H N. leporinus, 3, L	Myotis vivesii, 2?, H?
Insectivorous			Rhinopomatidae, 2, L? Emballonuridae, 2, I ? Craseonycteridae, ?, ?	Megadermatidae, ?, ? Nycteridae, ?, ? Rhinolophidae, 0–2, L	Noctilionidae, 2, L Mormoopidae, 2, I ? Phyllostominae, 2, L?	Natalidae, 1,? Vespertilionidae, 0–2, L Mystacinidae, 1?, ? Molossidae, 2, L
Omnivorous	"Prebats", 2, L				Phyllostominae, 2–3, I	
Fructivorous		Pteropodidae, 3, H			Stenoderminae, 3, H Carolliinae, 3, H Sturnirinae, 3, H	
Nectarivorous		Megaloglossus, 2, H			Glossophaginae, 3, H Phyllonycterinae, 3?, H?	

FIGURE 25. Evolution of the basal rate of metabolism and temperature regulation in bats as a function of food habits. The basal rate is indicated by L (<65%), I (65–80%), or H (80–130% of the Kleiber relation), and the type of thermoregulation by 0 ($\Delta T = k$ and $0 < k < 4$°C; hibernation occurs), 1 (as in 0, but without hibernation; if tropical, $m < 10$ g), 2 ($\Delta T \neq k$, $\Delta T \rightarrow 10$°C at $T_a > 10$°C, $T_b < 33$°C), or 3 ($\Delta T \neq k$, $\Delta T \rightarrow 0$°C at $T_a < 10$°C, $T_b > 34$°C). The evolutionary sequence is based on Van Valen (1979).

familial relationships (as modified by Van Valen, 1979) is indicated in Fig. 25, along with the level of the basal rate of metabolism and the thermoregulatory capacities of bats. The basal rates are crudely characterized as low, intermediate, or high, and thermoregulation as being equal to 0, 1, 2, and 3, which, with modification, follows Dwyer (1971): 0 means a bat maintains a very small ΔT and enters seasonal torpor (e.g., many vespertilionids); 1 is similar to 0 but does not enter hibernation (e.g., *Natalus*); 2 represents a limited capacity to regulate body temperature, generally in tropical species (e.g., molossids and *Histiotus*); and 3 represents precise temperature regulation (e.g., *Pteropus, Chrotopterus*, and *Megaderma*).

A pattern can be seen among the level of metabolism, the type of thermoregulation, and the phylogenetic position of bats. The megachiropterans, or flying foxes, are principally frugivorous, with high basal rates and precise thermoregulation, except at smaller body sizes. Thus, the nectarivorous *Megaloglossus* is very small (12 g) and enters daily torpor (Kulzer and Storf, 1980). Group I of the microchiropterans (see Van Valen, 1979) includes three insectivorous families that, unfortunately, are poorly studied. Two of these families apparently have type 2 thermoregulation (Kulzer, 1965) but do not normally appear to enter deep torpor and may have intermediate to low basal rates; the third family, Craseonycteridae, includes only one species, which is the smallest living bat, and that at 2 g undoubtedly ensures type 1 regulation. The family Noctilionidae has usually been placed here, but Van Valen has placed it in group III; in either case it may have been derived from the Emballonuridae. Group II consists of Old World tropical insectivorous families, which in the case of the Rhinolophidae have moved into temperate environments, and in the case of the Megadermatidae have become carnivorous. Again, few bats belonging to these families have been studied; they are tropical species and have low basal rates and type 2 thermoregulation. The carnivore *Megaderma lyra*, however, has a high basal rate and type 3 thermoregulation. Some temperate rhinolophids have type 0 thermoregulation.

Group III bats are exclusively New World and tropical in distribution. Stem noctilionids, mormoopids, and phyllostomids are principally insectivorous in habits, with type 2 thermoregulation and low basal rates. As we have seen, temperature regulation improves with an increase in size and with a switch in food habits to vertebrates, fruit, nectar, and blood. Group IV includes temperate and tropical species from both the Old and the New World. They are highly specialized for feeding on insects, although a very few have become piscivorous. Almost all of these bats probably have low basal rates of metabolism. Most temperate vespertilionids have type 0 thermoregulation, although some tropical species (e.g., *Histiotus velatus*) and most (all?) molossids have type 2 temperature regulation. The remaining bats belong to the extended family Natalidae, or are broken up into a series of smaller families; they all are small, which, coupled with presumably low basal rates, leads to a type 1 form of thermoregulation (where known).

This simplified survey of the energetics of bats leads to several preliminary conclusions. (1) Prebats were tropical, scansorial omnivores with low basal rates and without any marked propensity to enter torpor. (2) The earliest insectivorous bats were tropical in distribution with low basal rates and molossid-type thermoregulation (type 2); they may have facultatively entered torpor, especially if they had a small mass. (3) With the shift of foods from omnivory or insects to meat, fruit, and nectar, either early or late in the phylogeny of bats, basal rates increased, which in turn allowed thermoregulation to become precise and reproduction to be more independent of environmental conditions. (4) The use of hibernation is a derived specialization in response to seasonal variations in the availability of food in temperate environments.

3. THE WATER BALANCE OF BATS

An adequate supply of water is required by organisms, because water is needed as a solvent for all biologically important chemical reactions. Water that is lost from a normally hydrated organism to the environment through evaporation, or through the disposal of wastes, must be replaced. The amount of water income required may therefore be reduced if the magnitude of these pathways of water loss is reduced.

The degree to which evaporation or waste disposal will influence the water budgets of mammals is expected to vary greatly. For example, evaporation should be the greatest in hot, dry environments, so it would be expected that water parsimony would be high in desert bats. Equally, those bats feeding on foods with large electrolyte or protein contents would be expected to make adjustments to conserve water because of their need to dispose of large amounts of soluble wastes. Although some of the adjustments of bats to a water shortage may be behavioral (e.g., the survivorship of water-deprived insectivorous bats increases if they do not feed; Geluso, 1975), they are limited: no bat gives up flight, and high rates of evaporative water loss are directly related to high rates of gas exchange. Consequently, many of the adjustments required for attaining water balance occur in the kidney.

3.1. Kidney Function

Mammals that live in arid environments or that have high electrolyte or protein loads in their diet have kidneys that are capable of producing a concentrated urine (MacMillen, 1972). MacMillen and Lee (1967) showed that some desert mice are capable of producing a urine that is up to 17.9 times as concentrated as their plasma. Unfortunately, there have been relatively few studies of the ability of bats to produce a concentrated urine. Most of the available data are derived from only five papers (Carpenter, 1968, 1969; Geluso, 1975, 1978,

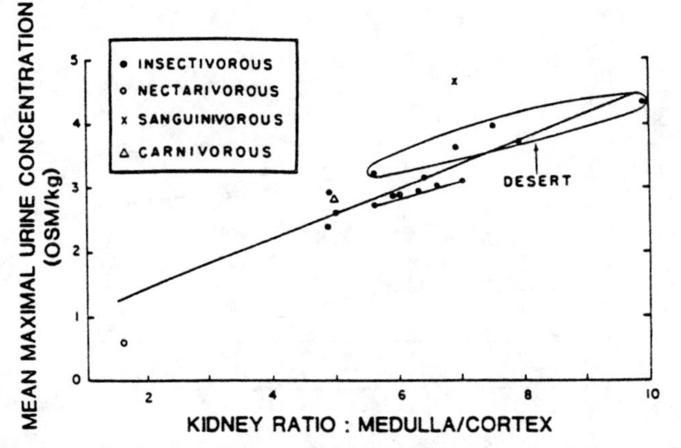

FIGURE 26. Maximal urine concentrations produced by bats in relation to the ratio of kidney medulla thickness to kidney cortex thickness. (Data derived from Carpenter, 1968, 1969; Geluso, 1980.)

1980), and even these studies have included no measurements of plasma concentration during water deprivation and therefore there are no data on the urine/plasma concentration ratios in bats.

As in other mammals, the ability of bats to produce a concentrated urine increases with the length of the loop of Henle and the consequent increase in its attendant osmotic gradient. This dependence in bats is illustrated in Fig. 26, where the mean maximal urine concentration is shown to increase with the ratio of kidney medulla thickness to cortex thickness (M/C). Clearly, desert insectivorous bats can produce more concentrated urine, because they have proportionally longer loops of Henle.

Body size complicates the ability of the M/C ratio to predict the capacity of a kidney to produce a concentrated urine. In general, the M/C ratio decreases with an increase in body mass (Fig. 27), which simply reflects the higher water turnover in small species and the adjustment made to reduce this turnover. The reason why the absolute length of the loops of Henle, or the absolute thickness of the medulla (where the loops are located), does not determine the maximal urine concentration is that the nephron's diameter increases with body size, which permits the filtrate to have a larger volume and to move more rapidly through the nephron than it would in a small animal. Therefore, the urine of large mammals would not be as concentrated as that of small mammals, unless there is a compensatory increase in the length of the loop of Henle.

The meager data presently available suggest that bats have several noteworthy aspects to their kidney form and function (Fig. 27). (1) Bats appear to have higher M/C ratios than do other mammals with the same climatic distribution (i.e., desert bats over other xeric mammals, and mesic bats over other mesic mammals). This difference may be due to the combination of high protein loads

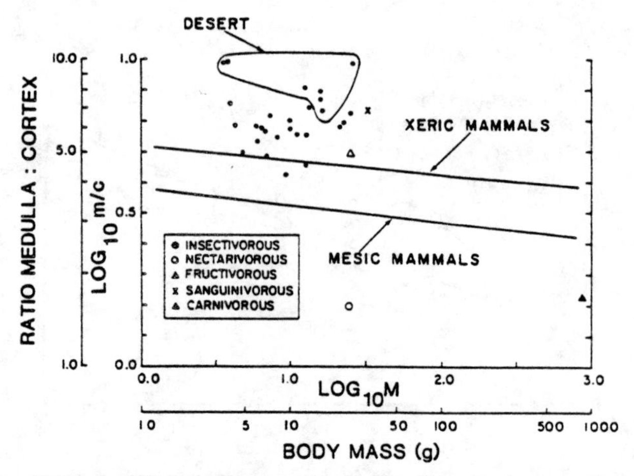

FIGURE 27. Kidney medulla/kidney cortex ratio as a function of body mass. (Data derived from Carpenter, 1969; Geluso, 1980. Curves derived from Greegor, 1975.)

in the diet with high rates of evaporative water loss associated with flight. (2) Preliminary results indicate that if bats feed on foods with a low protein and high water content (such as nectar and fruit), they also have very low M/C ratios and produce urine with very low maximal concentrations, even if they live in arid environments (e.g., *Leptonycteris*). (3) Sanguinivorous (*Desmodus*) and carnivorous (*Myotis vivesi*) bats can produce a concentrated urine, even though their food has a high water content, because it contains a high protein or salt content. Vampires excrete much of the ingested water while feeding, thereby reducing the mass that is transported to the roost (McNab, 1973). As a result, vampires must excrete urea at a time when they have a reduced water supply, a behavior that requires a high renal capacity for producing a concentrated urine. (4) Insectivorous bats, irrespective of whether they are phyllostomids (*Macrotis*), rhinopomatids (*Rhinopoma*), vespertilionids (many), or molossids (*Tadarida*, *Eumops*), have similar M/C ratios that vary primarily with the environment and (presumably) with mass but not with taxonomic affiliation. (5) Such simple renal indices as the M/C ratio may not be adequate to account for all of the variation in the ability of mammals to produce a concentrated urine (see, e.g., *Desmodus*). Some of the complications and indices used have been discussed by Heisinger and Breitenbach (1969) and Geluso (1978).

3.2. Balancing a Water Budget

The significance of all behavioral, anatomical, and physiological adjustments of water flux, of course, is to balance the water budget. At one extreme are the nectarivores and frugivores, which have a large water intake and produce copious and dilute urines; at the other extreme are those species that feed on insects, meat, or blood, and which must excrete large amounts of urea and (in the case of *Myotis vivesi,* which feeds on marine crustaceans) have high salt loads. Some desert insectivorous bats, such as the Namibian *Platymops petrophilus* and *Eptesicus zuluensis,* may be so highly modified for life in the desert that they need not drink water, even though they face dry atmospheres, often at high environmental temperatures (Roer, 1970, 1971). Most other desert bats, however, depend upon surface water and may have recently expanded their distribution into dry areas as a result of the construction of cattle watering holes (Geluso, 1978).

It is clear that much work remains to be done on the water balance and kidney function in bats. One of the greatest gaps in our knowledge is the effect of food habits in conjunction with the influence of body size on water balance. Furthermore, although there is evidence that there may be variations in kidney function at the subspecific level in rodents that reflect geographic variation in environmental conditions (Heisinger *et al.*, 1973), the only evidence of such adaptation in bats was shown by Bassett and Wiebers (1979). The question as to whether bats are locally differentiated in renal function assumes considerable interest when it is realized that many species of *Myotis* are found in xeric and mesic environments in North America. Finally, it should be asked why, if an ability to produce a concentrated urine is of value in balancing a water budget, it is not of greater value to produce an even more concentrated urine? A general answer is that there must be some limitation to the production of a concentrated urine. The nature of this limitation is unclear. Could it be the high cost of producing a concentrated urine? Or is there some other limitation? Whatever the limitation, we can never completely comprehend the nature of adaptation until we understand what conflicting factors contribute to an adaptation and how these conflicts are resolved. At present this analysis is more complete in the study of energetics than it is in the study of water balance.

4. DISTRIBUTIONAL LIMITS TO BATS

Many factors limit the distribution of animals, three of the most important being extreme conditions in the physical environment, food habits, and "competition" with ecologically similar species. It is of interest here to examine the influence of the first two factors on the limits to the geographic distribution of bats. Bats can be roughly classified into temperate and tropical species, although the Vespertilionidae is represented both by temperate and tropical species.

4.1. Temperate Limits of Tropical Bats

In North America members of the insectivorous family Molossidae (Fig. 28) make the farthest incursion into temperate zones of any tropical family

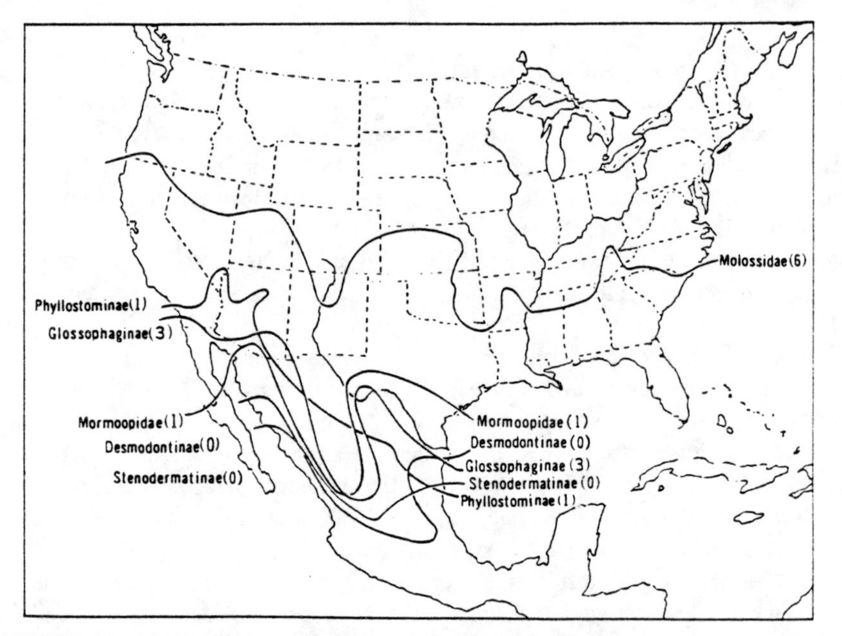

FIGURE 28. Northern limits to distribution of tropical bats in North America. Numbers in parentheses represent the number of species belonging to the family or subfamily that are found in the United States. (Data from Hall and Kelson, 1959; Villa-R., 1966.)

(Henshaw, 1972). This family also has the largest number of species (six) occurring in the United States of any tropical family. Two other tropical families also enter the United States—Mormoopidae (one species, an insectivore) and Phyllostomidae (four species, three nectarivores, and one insectivore)—but they are limited in distribution to regions near the Mexican boundary (Barbour and Davis, 1969). Molossids also penetrate the temperate regions of South America (Fig. 29) and Australia farther than members of other tropical families. In Eurasia the penetration of temperate environments by molossids is exceeded in tropical bats only by the Rhinolophidae, which may be considered of a tropical origin if, as is commonly the case, it is combined with the Hipposideridae. The rhinolophids in temperate environments, unlike the molossids but convergent with the vespertilionids, have adopted hibernation and sperm storage by females.

Bats that do not feed on insects show little penetration into temperate climates (Fig. 30). The nectarivorous species that enter the southwestern United States migrate to Mexico for winter. Their northern limits of distribution are correlated with the northern limits to distribution of the plants that are used as a source of pollen and nectar that constitute the food of these bats. Fruit bats make no appreciable entrance into temperate climates anywhere [e.g., Stenoderminae in North America (Fig. 28), Sturnirinae in Argentina (Fig. 29), and Pteropodidae in Australia], presumably because of an inadequate food supply. Vampire bats

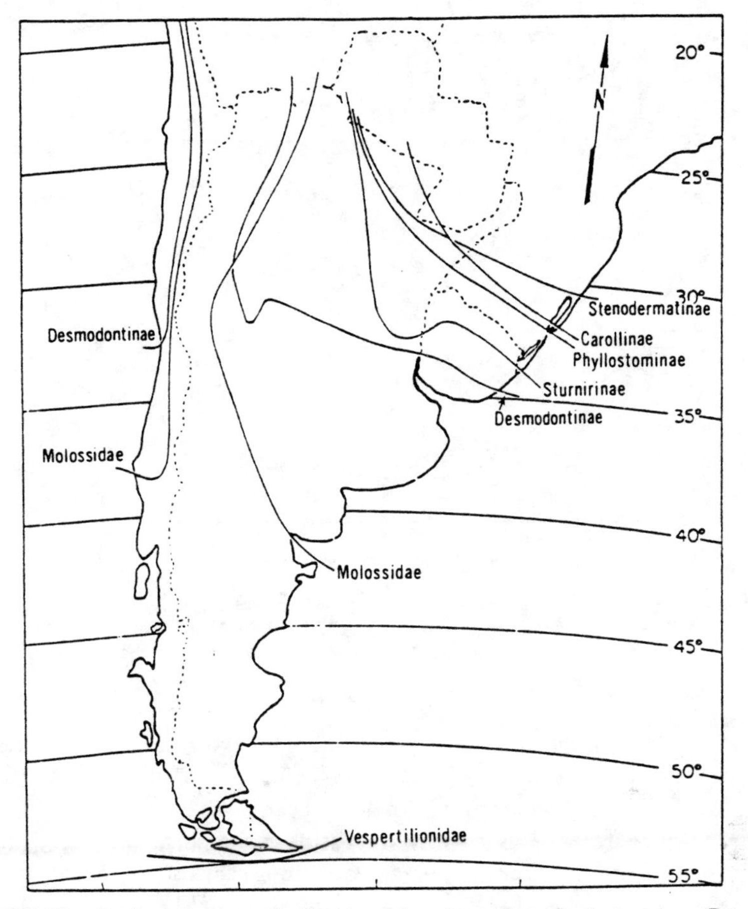

FIGURE 29. Southern limits to distribution of bats in southern South America. (Data from Cabrera, 1957; Mann-F., 1978.)

almost reach the United States border (Fig. 28); in fact, there is one recent record of *Diphylla ecaudata* from the Texas side of the Rio Grande (Reddell, 1968). Vampires also move into subtropical Chile and Argentina (Fig. 29). Bats that feed on blood obviously are not excluded from temperate climates by the absence of prey. It has been suggested (McNab, 1973) that the vampire's temperate limits to distribution result from its inability to transport quantities of food sufficient to maintain a high body temperature in cool to cold caves.

The limits of distribution of tropical bats along an altitudinal transect with respect to food habits are similar to the distribution of food habits found along a latitudinal transect. Tuttle (1970), for example, showed that many species of bats living in the Amazon basin encountered an altitudinal limit on the eastern face of the Andes. Of 101 species of bats recorded from eastern Peru, only 15 have been

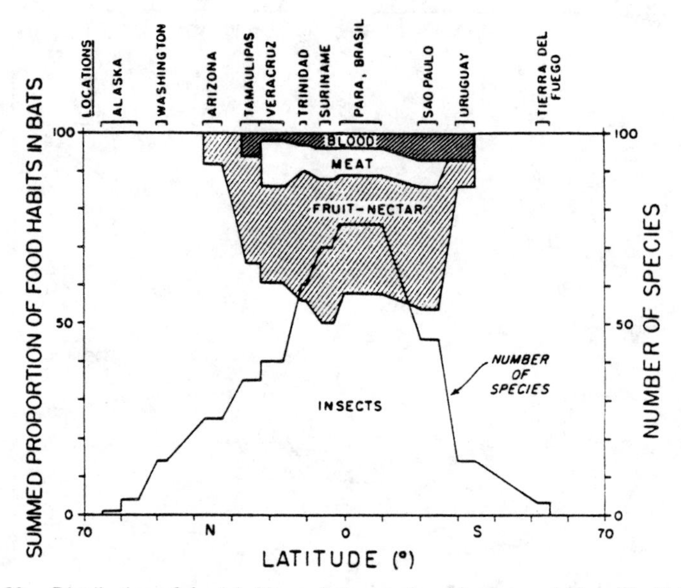

FIGURE 30. Distribution of food habits and the number of species of New World bats as a function of latitude. (Data compiled from Acosta and Lara, 1950; McNab, 1971.)

found above 2000 m, including five vespertilionids, two molossids, two vampires, two nectarivores (*Anoura*), and two *Sturnira;* above 3000 m only three vespertilionids and *Desmodus rotundus* have been found.

The question remains as to why molossids move so far into temperate environments. In part this is due to their insectivorous habits and their propensity to feed at levels high above the ground, where insect abundance may be seasonally highly variable (McNab, 1976). As a result of this variability, some tropical molossids seasonally deposit large amounts of fat (McNab, 1976). The adjustment of molossids to a highly variable food resource in the tropics may have permitted them to tolerate a highly seasonal food supply in temperate environments, at least as long as they do not require extended periods of seasonal torpor.

4.2. Limits to Distribution in Temperate Bats

Temperate bats face two latitudinal limits to distribution: polar and equatorial. Few bats live under arctic conditions. In fact, the polar limits to distribution in vespertilionids appear to be related to the limits to distribution of forests in both Eurasia and North America, probably because of the need for roost sites. The northernmost vespertilionid in North America is *Myotis lucifugus* and in Eurasia it is *Eptesicus nilssoni.* These limits, of course, occur only in summer;

in autumn these species move southward. The northern limit of distribution of *M. lucifugus* in winter is determined by the occurrence of hibernacula that are protected from freezing temperatures (Fenton, 1972).

Most temperate bats do not enter the lowland tropics. In Mexico they are usually limited in distribution to the central plateau, although some species occur south along the Central American cordillera. The only species of bat likely to be of temperate North American origin that have penetrated far into South America belong to the genus *Lasiurus,* including both *L. cinereus* and *L. borealis*. The limits to distribution of temperate bats in central Mexico could in part be due to the competitive exclusion of temperate species by tropical, lowland species, except that a similar decline in temperate-bat diversity occurs along the Florida peninsula (Fig. 31), where there is no tropical influx. In Florida the decline in diversity occurs both in tree and cave bats.

We do not know what is responsible for the tropical limits of temperate bats, but it would be naive to believe that all species were responding to the same factor. In certain cave bats equatorial limits to distribution are related to the requirement for cave temperatures cold enough to permit hibernation (McNab, 1974a). As has been seen, large or clustering bats need cooler cave temperatures for hibernation than do small or solitary bats and are therefore less likely to enter the tropics. Thus such clustering hibernators as *Myotis sodalis* and *M. lucifugus* do not even enter warm temperate environments. The one large, clustering hibernator that does, *M. grisescens,* migrates north in autumn to hibernate in cold caves (Tuttle, 1976a). In contrast, the small hibernator *Pipistrellus subflavus* can hibernate at warm cave temperatures in Florida.

There is, of course, another adjustment that can be made to a warm climate by temperate bats: they can shift copulation to spring so that females need not store spermatozoa. This adjustment was made in *Myotis austroriparius,* at least in central Florida (Rice, 1957; McNab, 1974a). Such a modification may occur in a temperate species only if a subtropical or warm–temperate population has been geographically isolated from temperate populations of the same species. *Myotis austroriparius* may have been a subtropical isolate of *M. lucifugus.*

An interesting, but unstudied, example of the reaction of a temperate species to warm climates is found in *Eptesicus fuscus*. In the northern United States this bat frequently hibernates in clusters in caves and buildings. It is regularly found in caves only as far south as Kentucky (and Tennessee?). Farther south *Eptesicus* is a (solitary?) tree bat. This shift apparently occurs in response to warm cave temperatures, which will not permit a large bat to hibernate. Roosting in trees may permit *Eptesicus* in winter to remain in torpor for longer periods without risking pregnancy, assuming that these southern female *Eptesicus* are also storing spermatozoa. It is fascinating, however, that *E. fuscus* in Puerto Rico is, again, a cave bat, almost as if, as an island isolate, it made an adjustment to a tropical environment, possibly by abandoning sperm storage.

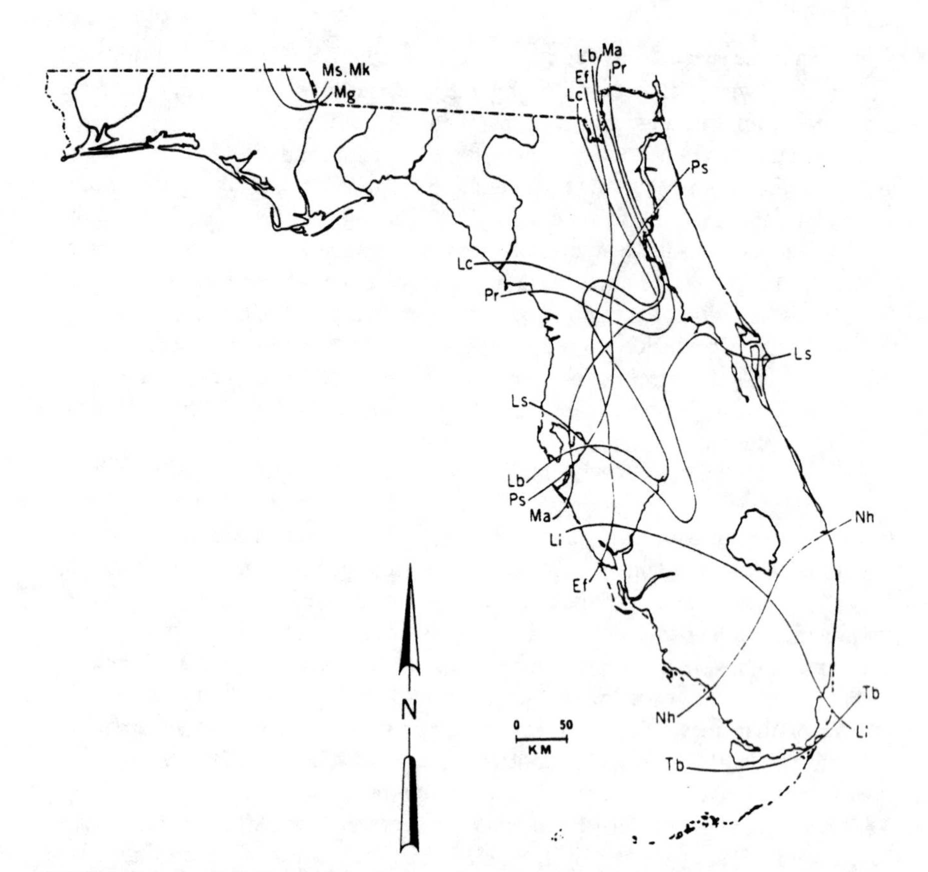

FIGURE 31. Southern limits of distribution of bats in Florida. The following abbreviations are used: Mg, *Myotis grisescens;* Ms, *M. sodalis;* Mk, *M. keenii;* Ma, *M. austroriparius;* Ps, *Pipistrellus subflavus;* Ef, *Eptesicus fuscus;* Lb, *Lasiurus borealis;* Ls, *L. seminolus;* Lc, *L. cinereus;* Pr, *Plecotus rafinesquii;* Nh, *Nycticeius humeralis;* Li, *Lasiurus intermedius;* and Tb, *Tadarida brasiliensis.* In addition, *Eumops glaucinus* reaches its northern limit of distribution near Miami. (Data from Jennings, 1958.)

It is unfortunate that so little is known of the biology of temperate tree bats, but the observation that they tend to drop out along the Florida peninsula suggests that they too often have an obligatory form of seasonal torpor and that most of them appear unable to adjust to a warm winter. Most exceptions appear to be tropical species [*Lasiurus intermedius* (=*floridanus*), *Tadarida brasiliensis, Eumops glaucinus*]. Obviously, much remains to be done on dissecting the influence of physiological factors on the limits to the distribution of bats.

5. SUMMARY

An examination of bat energetics leads to the following conclusions: (1) the rate of energy expenditure in bats is principally determined by body size, food habits, environmental temperature, and activity; (2) endothermic temperature regulation is scaled to body size, that is, small species often show marginal temperature regulation unless the scaling of metabolism with respect to mass is changed from the Kleiber curve at small masses; (3) because rate of metabolism varies with food habits in bats, effective temperature regulation at small masses is also correlated with food habits; (4) the low rates of metabolism in insectivorous bats leads to long life-spans and short reproductive seasons, compared to bats with other food habits; and (5) temperate bats either migrate or enter seasonal torpor to evade the shortage of food during the winter.

Temperate bats select different temperatures during winter than they do during summer. Hibernating cave bats select temperatures that vary with body size and clustering habits to minimize activity, energy use, and the probability of pregnancy during winter. In contrast, during the breeding season bats usually select warm maternity roosts, either inside or outside of caves, as a means of increasing the rate of growth of young before and after birth. Such high growth rates reduce the time that it takes for young bats to gain independence, which permits both young and mother bats to store enough fat for the hibernal period.

Because much of the variation in the level and precision of temperature regulation varies with body mass and food habits, the evolution of energetics and temperature regulation in the phylogeny of bats can be examined. It is suggested that bats evolved from a tropical, arboreal omnivore, which, judging from living examples, was probably characterized by a low basal rate and little or no tendency to enter torpor. Most nectarivorous, frugivorous, and carnivorous bats do not enter torpor, mainly because they generally have high basal rates of metabolism, except for the smallest species, whose high basal rates may be insufficient to compensate for their small masses. All insectivorous bats enter torpor, because they combine a small mass with low basal rates. Seasonal torpor, however, is a specialized response to temperate climates and is not a general feature of bats.

The few studies of water balance in bats indicate that those species that feed on food high in protein or salt or that live in desert environments can produce a concentrated urine; that is, these bats have kidneys with relatively thick medullas. In general, bats have kidneys with medullas thicker than in other mammals living in the same environments, either because they have higher protein loads or because of the high rates of evaporative water loss in association with flight.

Many bats are limited in their distribution to tropical or temperate regions. Tropical bats have a wide range of food habits, but the only food used extensively by temperate bats is insects. Insectivorous bats belonging to the family Molossidae are the most prone of tropical bats to enter temperate environments, in part because they are accustomed to feed on the variable insect populations flying above the forest canopy in the tropics. Some temperate bats appear to be limited to temperate environments because the females store sperm during winter and must remain torpid to prevent pregnancy.

ACKNOWLEDGMENTS. I thank Thomas Kunz of Boston University for the invitation to participate in this book, for a critical review of an early draft of this manuscript, and for his tolerance of my constitutional inability to meet deadlines.

6. REFERENCES

Acosta y Lara, E. 1950. Los quirópteros del Uruguay. Com. Zool. Mus. Hist. Nat. Montevidéo, **3**:1–72.

Baker, R. R. 1978. The evolutionary ecology of animal migration. Holmes and Meier, New York, 1022 pp.

Baker, W. W., S. G. Marshall, and V. B. Baker. 1968. Autumn fat deposition in the evening bat (*Nycticeius humeralis*). J. Mammal., **49**:314–317.

Barbour, R. W., and W. H. Davis. 1969. Bats of America. University of Kentucky Press, Lexington, 286 pp.

Bassett, J. E., and J. E. Wiebers. 1979. Subspecific differences in the urine concentrating ability of the bat, *Myotis lucifugus*. J. Mammal., **60**:395–397.

Beer, J. R., and A. G. Richards. 1956. Hibernation of the big brown bat. J. Mammal., **37**:31–41.

Bourlière, F. 1958. Comparative biology of aging. J. Gerontol. Suppl. I, **13**:16–24.

Bradley, S. R., and J. W. Hudson. 1974. Temperature regulation in the tree shrew *Tupaia glis*. Comp. Biochem. Physiol., **48**:55–60.

Cabrera, L. A. 1957. Catálogo de los mamíferos de América del Sur. Revista del Museu Argentina de Ciencias Naturales "Bernardino Rivadavia," **4**:1–307.

Calder, W. A., III. 1974. Consequences of body size for avian energetics. Pp. 86–151. *in* Avian energetics. (R. A. Paynter, Jr., ed.). Publication 15. Nuttall Ornithological Club, Cambridge, 334 pp.

Carpenter, R. E. 1968. Salt and water metabolism in the marine fish-eating bat, *Pizonyx vivesi*. Comp. Biochem. Physiol., **24**:951–964.

Carpenter, R. E. 1969. Structure and function of the kidney and the water balance of desert bats. Physiol. Zool., **42**:288–302.

Cross, S. P. 1965. Roosting habits of *Pipistrellus hesperus*. J. Mammal., **46**:270–279.

Daan, S. 1973. Activity during natural hibernation in three species of vespertilionid bats. Neth. J. Zool., **23**:1–71.

Davis, W. H. 1970. Hibernation: Ecology and physiological ecology. Pp. 265–300, *in* Biology of bats. Vol. 1. (W. A. Wimsatt, ed.). Academic Press, New York, 406 pp.

Davis, W. H., and H. B. Hitchcock. 1964. Notes on sex ratios of hibernating bats. J. Mammal., **45**:475–476.

Davis, W. H., and W. Z. Lidicker, Jr. 1956. Winter range of the red bat, *Lasiurus borealis*. J. Mammal., **37**:280–281.

Davis, W. H., and O. B. Reite. 1967. Responses of bats from temperate regions to changes in ambient temperature. Biol. Bull., **132**:320–328.

Dwyer, P. D. 1963. The breeding biology of *Miniopterus schreibersi blepotis* (Temminck) (Chiroptera) in north-eastern New South Wales. Aust. J. Zool., **11**:219–240.

Dwyer, P. D. 1964. Seasonal changes in activity and weight of *Miniopterus schreibersi blepotis* (Chiroptera) in north-eastern New South Wales. Aust. J. Zool., **12**:52–69.

Dwyer, P. D. 1971. Temperature regulation and cave-dwelling in bats: An evolutionary perspective. Mammalia, **35**:424–455.

Dwyer, P. D., and J. A. Harris. 1972. Behavioral acclimatization to temperature by pregnant *Miniopterus* (Chiroptera). Physiol. Zool., **45**:14–21.

Eisentraut, M. 1960. Heat regulation in primitive mammals and in tropical species. Bull. Mus. Comp. Zool., **124**:31–43.

Ewing, W. G., E. H. Studier, and M. J. O'Farrell. 1970. Autumn fat deposition and gross body composition in three species of *Myotis*. Comp. Biochem. Physiol., **36**:119–129.

Fenton, M. B. 1970. Population studies of *Myotis lucifugus* (Chiroptera: Vespertilionidae) in Ontario. Life Sci. Contrib. R. Ont. Mus., **77**:1–34.

Fenton, M. B. 1972. Distribution and overwintering of *Myotis leibii* and *Eptesicus fuscus* (Chiroptera: Vespertilionidae) in Ontario. Life Sci. Occas. Pap. R. Ont. Mus., **21**:1–8.

Fleming, T. H., E. T. Hooper, and D. E. Wilson. 1972. Three Central American bat communities: Structure, reproductive cycles, and movement patterns. Ecology, **53**:555–569.

Geluso, K. N. 1975. Urine concentration cycles of insectivorous bats in the laboratory. J. Comp. Physiol., **99**:309–319.

Geluso, K. N. 1978. Urine concentrating ability and renal structure of insectivorous bats. J. Mammal., **59**:312–323.

Geluso, K. N. 1980. Renal form and function in bats: An ecophysiological appraisal. Pp. 403–414, *in* Proceedings of the Fifth International Bat Research Conference. (D. E. Wilson and A. L. Gardner, eds.). Texas Tech Press, Lubbock, 434 pp.

Gillette, D. D. 1975. Evolution of feeding strategies in bats. Tebiwa, **18**:39–48.

Greegor, D. H., Jr. 1975. Renal capabilities of an Argentine desert armadillo. J. Mammal., **56**:626–632.

Griffin, D. R. 1958. Listening in the dark. Yale University Press, New Haven, 413 pp.

Hall, E. R., and K. R. Kelson. 1959. The mammals of North America. Ronald Press, New York, 1083 pp.

Hall, J. S. 1962. A life history and taxonomic study of the Indiana bat, *Myotis sodalis*. Sci. Publ., Reading Pub. Mus. Art Gallery, **12**:1–68.

Heisinger, J. F., and R. P. Breitenbach. 1969. Renal structural characteristics as indexes of renal adaptation for water conservation in the genus *Sylvilagus*. Physiol. Zool., **42**:160–172.

Heisinger, J. F., T. S. King, H. W. Halling, and B. L. Fields. 1973. Renal adaptations to macro- and microhabitats in the family Cricetidae. Comp. Biochem. Physiol., **44A**:767–774.

Heithaus, E. R., T. H. Fleming, and P. A. Opler. 1975. Foraging patterns and resource utilization in seven species of bats in a seasonal tropical forest. Ecology, **56**:841–854.

Henshaw, R. E. 1960. Responses of free-tailed bats to increases in cave temperature. J. Mammal., **41**:396–398.

Henshaw, R. E. 1972. Niche specificity and adaptability in cave bats. Bull. Nat. Speleol. Soc., **34**:61–70.

Henshaw, R. E., and G. E. Folk, Jr. 1966. Relation of thermoregulation to seasonally changing microclimate in two species of bats (*Myotis lucifugus* and *M. sodalis*). Physiol. Zool., **39**:223–236.

Herreid, C. F., II. 1963. Survival of a migratory bat at different temperatures. J. Mammal., **44**:431–433.

Herreid, C. F., II. 1964. Bat longevity and metabolic rate. Exp. Gerontol., **1**:1–9.

Herreid, C. F., II. 1967. Temperature regulation, temperature preferences and tolerance, and metabolism of young and adult free-tailed bats. Physiol. Zool., **40:**1–22.

Hock, R. J. 1951. The metabolic rates and body temperatures of bats. Biol. Bull., **101:**289–299.

Hooper, J. H. D., and W. H. Hooper. 1956. Habits and movements of cave-dwelling bats in Devonshire. Proc. Zool. Soc. London, **127:**1–26.

Humphrey, S. R., and F. J. Bonaccorso. 1979. Population and community ecology. Pp. 409–441, *in* Biology of bats of the New World family Phyllostomatidae. Part III. (R. J. Baker, J. K. Jones, Jr., and D. C. Carter, eds.). Spec. Publ. Mus. Texas Tech Univ., **16:**1–441.

Humphrey, S. R., and T. H. Kunz. 1976. Ecology of a Pleistocene relict, the western big-eared bat (*Plecotus townsendii*), in the southern Great Plains. J. Mammal., **57:**470–494.

Humphrey, S. R., A. R. Richter, and J. B. Cope. 1977. Summer habitat and ecology of the endangered Indiana bat, *Myotis sodalis*. J. Mammal., 58:334–346.

Jennings, W. L. 1958. The ecological distribution of bats in Florida. Unpublished Ph.D. dissertation, University of Florida, Gainesville, 126 pp.

Kayser, C. 1961. The physiology of natural hibernation. Pergamon Press, New York, 325 pp.

Kleiber, M. 1932. Body size and metabolism. Hilgardia, **6:**315–353.

Krulin, G. S., and J. A. Sealander. 1972. Annual lipid cycle of the gray bat, *Myotis grisescens*. Comp. Biochem. Physiol., **42A:**537–549.

Krzanowski, A. 1961. Weight dynamics of bats wintering in the cave at Pulawy (Poland). Acta Theriol., **4:**249–264.

Kulzer, E. 1965. Temperaturregulation bei Fledermäusen (Chiroptera) aus verschieden Klimazonen. Z. Vergl. Physiol., **50:**1–34.

Kulzer, E., and R. Storf. 1980. Schlaf-Lethargie bei dem afrikanischen Langzungenflughund *Megaloglossus woermanni* Pagenstecher, 1885. Z. Säugetierk., **45:**23–29.

Kunz, T. H. 1974. Feeding ecology of a temperate insectivorous bat (*Myotis velifer*). Ecology, **55:**693–711.

Kunz, T. H. 1980. Daily energy budgets of free-living bats. Pp. 369–392, *in* Proceeding of the Fifth International Bat Research Conference. (D. E. Wilson and A. L. Gardner, eds.). Texas Tech Press, Lubbock, 434 pp.

Lasiewski, R. C. 1963. The energetic cost of small size in hummingbirds. Pp. 1095–1103, *in* Proceedings of the 13th International Ornithology Congress. American Ornithology Union, Baton Rouge. 1246 pp.

Lekagul, B., and J. A. McNeely. 1977. Mammals of Thailand. Association for the Conservation of Nature, Bangkok, 758 pp.

Licht, P., and P. Leitner. 1967. Behavioral responses to high temperatures in three species of California bats. J. Mammal., **48:**52–61.

MacMillen, R. E. 1972. Water economy of nocturnal desert rodents. Symp. Zool. Soc. London, **31:**147–174.

MacMillen, R. E., and A. K. Lee. 1967. Australian desert mice: Independence of exogenous water. Science, **158:**383–385.

McNab, B. K. 1969. The economics of temperature regulation in Neotropical bats. Comp. Biochem. Physiol., **31:**227–268.

McNab, B. K. 1970. Body weight and the energetics of temperature regulation. J. Exp. Biol., **53:**329–348.

McNab, B. K. 1971. The structure of tropical bat faunas. Ecology, **52:**352–358.

McNab, B. K. 1973. Energetics and the distribution of vampires. J. Mammal., **54:**131–144.

McNab, B. K. 1974a. The behavior of temperate cave bats in a subtropical environment. Ecology, **55:**943–958.

McNab, B. K. 1974b. The energetics of endotherms. Ohio J. Sci., **74:**370–380.

McNab, B. K. 1976. Seasonal fat reserves of bats in two tropical environments. Ecology, **57:**332–338.

McNab, B. K. 1980a. On estimating thermal conductance in endotherms. Physiol. Zool., **53**:145–156.

McNab, B. K. 1980b. Food habits, energetics, and the population biology of mammals. Am. Nat., **116**:106–124.

McNab, B. K. 1980c. Energetics and the limits to a temperate distribution in armadillos. J. Mammal., **61**:606–627.

McNab, B. K. 1983. Ecological and behavioral consequences of adaptation to various food resources, *in* Recent advances in the study of Mammalian behavior. (J. E. Eisenberg and D. G. Kleiman, eds.). Spec. Publ. Amer. Soc. Mammal. (in press).

McNab, B. K. 1982. Energetics, body size, and the limits to endothermy. J. Zool., **197**.

McNab, B. K., and P. R. Morrison. 1963. Body temperature and metabolism in subspecies of *Peromyscus* from arid and mesic environments. Ecol. Monogr., **33**:63–82.

Mann, T. 1954. The biochemistry of semen. Methuen, London, 240 pp.

Mann-F., G. 1978. Los pequenos mamíferos de Chile. Gayana Univ. Concepcion Zool., **40**:1–342.

Mares, M. A., and D. E. Wilson. 1971. Bat reproduction during the Costa Rican dry season. Bioscience, **21**:471–473.

Morrison, P. R., F. A. Ryser, and A. Dawe. 1959. Studies on the physiology of the masked shrew *Sorex cinereus*. Physiol. Zool., **32**:256–271.

Noll, U. G. 1979. Body temperature, oxygen consumption, noradrenaline response and cardiovascular adaptations in the flying fox, *Rousettus aegyptiacus*. Comp. Biochem. Physiol., **63A**:79–88.

O'Shea, T. J. 1976. Fat content in migratory central Arizona Brazilian free-tailed bats, *Tadarida brasiliensis* (Molossidae). Southwest. Nat., **21**:321–326.

Pagels, J. F. 1975. Temperature regulation, body weight and changes in total body fat of the free-tailed bat, *Tadarida brasiliensis cynocephala* (LeConte). Comp. Biochem. Physiol., **50A**:237–246.

Pearson, O. P. 1948. Metabolism of small mammals, with remarks on the lower limit of mammalian size. Science, **108**:44.

Racey, P. A. 1973. Environmental factors affecting the length of gestation in heterothermic bats. J. Reprod. Fertil. Suppl., **19**:175–189.

Ransome, R. D. 1973. Factors affecting the timing of births of the greater horse-shoe bat (*Rhinolophus ferrumequinum*). Period. Biol., **75**:169–175.

Reddell, J. R. 1968. The hairy-legged vampire, *Diphylla ecaudata*, in Texas. J. Mammal., **49**:769.

Reite, O. B., and W. H. Davis. 1966. Thermoregulation in bats exposed to low ambient temperatures. Proc. Soc. Exp. Biol. Med., **121**:1212–1215.

Rice, D. W. 1957. Life history and ecology of *Myotis austroriparius* in Florida. J. Mammal., **38**:15–32.

Roer, H. 1970. Zur Wasserversorgung der Microchiropteren *Eptesicus zuluensis vansoni* (Vespertilionidae) und *Sauromys petrophilus erongensis* (Molossidae) in der Namibwuste. Bijdr. Dierkd., **40**:71–73.

Roer, H. 1971. Zur Lebensweise einiger Microchiropteren der Namibwuste (Mammalia, Chiroptera). Zool. Abh. (Dresden), **32**:43–55.

Shimoizumi, J. 1959. Studies of the hibernation of bats (2). Seasonal variation in body weight, dry matter, crude fat and moisture in the body. Sci. Rep. Tokyo Kyoiku Daigaku, Sect. B, **9**:1–16.

Shump, K. A., Jr., and A. U. Shump. 1980. Comparative insulation in vespertilionid bats. Comp. Biochem. Physiol., **66A**:351–354.

Smith, J. D. 1976. Chiropteran evolution. Pp. 44–69, *in* Biology of bats of the New World family Phyllostomatidae. Part I. (R. J. Baker, J. K. Jones, Jr., and D. C. Carter, eds.). Spec. Publ. Mus. Texas Tech Univ., **10**:1–218.

Stones, R. C., and J. C. Wiebers. 1967. Temperature regulation in the little brown bat, *Myotis lucifugus*. Pp. 97–109, *in* Mammalian hibernation. (K. C. Fisher, A. R. Dawe, C. P. Lyman, E. Schonbaum, and F. E. Smith, Jr., eds.). American Elsevier, New York, 535 pp.

Strelkov, P. P. 1969. Migratory and stationary bats (Chiroptera) of the European part of the Soviet Union. Acta Zool., Cracov., **14**:393–439.

Studier, E. H., and M. J. O'Farrell. 1972. Biology of *Myotis thysanodes* and *M. lucifugus* (Chiroptera: Vespertilionidae)—I. Thermoregulation. Comp. Biochem. Physiol., **41A**:567–595.

Studier, E. H., and D. E. Wilson. 1970. Thermoregulation in some Neotropical bats. Comp. Biochem. Physiol., **34**:251–262.

Studier, E. H., and D. E. Wilson. 1979. Effects of captivity on thermoregulation and metabolism in *Artibeus jamaicensis* (Chiroptera: Phyllostomatidae). Comp. Biochem. Physiol., **62A**:347–350.

Thomas, D. W., M. B. Fenton, and R. M. R. Barclay. 1979. Social behavior of the little brown bat, *Myotis lucifugus*. I. Mating behavior. Behav. Ecol. Sociobiol., **6**:129–136.

Thomas, S. P. 1975. Metabolism during flight in two species of bats, *Phyllostomus hastatus* and *Pteropus gouldii*. J. Exp. Biol., **63**:273–293.

Thomas, S. P., and R. A. Suthers. 1972. The physiology and energetics of bat flight. J. Exp. Biol., **57**:317–335.

Tracy, C. R. 1977. Minimum size of mammalian homeotherms: Role of the thermal environment. Science, **198**:1034–1035.

Tucker, V. A. 1970. Energetic cost of locomotion in animals. Comp. Biochem. Physiol., **34**:841–846.

Tuttle, M. D. 1970. Distribution and zoogeography of Peruvian bats, with comments on natural history. Univ. Kans. Sci. Bull., **49**:45–86.

Tuttle, M. D. 1975. Population ecology of the gray bat (*Myotis grisescens*): Factors influencing early growth and development. Occas. Pap. Mus. Nat. Hist. Univ. Kans., **36**:1–24.

Tuttle, M. D. 1976a. Population ecology of the gray bat (*Myotis grisescens*): Philopatry, timing and patterns of movement, weight loss during migration, and seasonal adaptive strategies. Occas. Pap. Mus. Nat. Hist. Univ. Kans. **54**:1–38.

Tuttle, M. D. 1976b. Population ecology of the gray bat (*Myotis grisescens*): Factors influencing growth and survival of newly volant young. Ecology, **57**:587–595.

Tuttle, M. D., and L. R. Heaney. 1974. Maternity habits of *Myotis leibii* in South Dakota. Bull. South. Calif. Acad. Sci., **73**:80–83.

Twente, J. W., Jr. 1955. Some aspects of habitat selection and other behavior of cavern-dwelling bats. Ecology, **36**:706–732.

Twente, J. W., Jr. 1956. Ecological observations on a colony of *Tadarida mexicana*. J. Mammal., **37**:42–47.

Twente, J. W., Jr., and J. A. Twente. 1964. An hypothesis concerning the evolution of heterothermy in bats. Ann. Acad. Sci. Fennicae, **71**:435–442.

Van Valen, L. 1979. The evolution of bats. Evol. Theor., **4**:103–121.

Villa-R., B. 1966. Los murciélagos de Mexico. Inst. Biol., Univ. Nac. Auton. Mexico, Mexico City, 491 pp.

Villa-R., B., and E. L. Cockrum. 1962. Migration in the guano bat *Tadarida brasiliensis mexicana* (Saussure). J. Mammal., **43**:43–64.

Weber, N. S., and J. S. Findley. 1970. Warm-season changes in fat content of *Eptesicus fuscus*. J. Mammal., **51**:160–162.

Weigold, H. 1979. Koerpertemperatur, Sauerstoffverbrauch und Herzfrequenz bei *Tupaia belangeri* Wagner, 1841 im Tagesverlauf. Z. Säugetierk., **44**: 343–353.

Western, D. 1979. Size, life history and ecology in mammals. Afr. J. Ecol., **17**:185–204.

Whittow, G. C., and E. Gould. 1976. Body temperature and oxygen consumption of the pentail tree shrew (*Ptilocercus lowii*). J. Mammal., **57**:754–756.

Wilson, D. M. 1977. Environmental monitoring and baseline data from the Isthmus of Panama—1976, Vol. IV. Smithsonian Institution, Washington, D.C., 267 pp.

Chapter 5

Ecological Aspects of Bat Activity Rhythms

Hans G. Erkert

Institut für Biologie III
der Universität Tübingen
D-7400 Tübingen 1
Federal Republic of Germany

1. INTRODUCTION

Terrestrial organisms live in an environment that, by its geophysical nature, is subject to profound rhythmic alteration. The rotation of the earth brings about a 24-hr periodicity in light intensity, temperature, and humidity. As the earth revolves around the sun there are annual changes in day length, temperature, and the duration of twilight, which become more pronounced at increasing latitudes; that is, over the period of a year the form, level, and amplitude of diurnal variables themselves vary to different degrees. Other effects result from the revolution of the moon about the earth. The rhythmic gravitational fluctuations so produced are of little relevance to terrestrial animals, but there are also variations in night-time brightness with lunar periodicity. Organisms have adapted to this temporal structure of their environment in different ways. Very early in evolution they developed endogenous diurnal rhythms with periods of approximately 24 hr. These circadian rhythms were coupled with the external diurnal cycle by special synchronizing mechanisms so that a certain relatively stable phase relation, characteristic of each species, was maintained. In many cases circalunar and circannual endogenous rhythms evolved in addition to the circadian rhythm; these, too, are synchronized with the corresponding environmental variables by special mechanisms.

Because the endogenous rhythms of animals mesh in different ways with the overall periodicity of the environment, the biotic factors of concern to each individual—food supply, competition for food, predator pressure—also vary on a daily, annual, and in some cases, lunar basis. A successful strategy for survival must therefore include optimal adjustment of an animal's activity rhythm to these environmental periodicities. This is true especially of long-lived small species with a low rate of reproduction, such as the Chiroptera. Only recently extensive field and laboratory studies of bats have produced illuminating evidence as to their phylogenetic adaptations. In the course of evolution bats have differentiated greatly not only in feeding habits, foraging strategies, behavior, and social structure (e.g., Fleming *et al.*, 1972; Bradbury and Emmons, 1974; Bradbury and Vehrencamp, 1976; Bradbury, 1977), but also in the regulation of their activity rhythms (Erkert *et al.*, 1980). Both endogenous mechanisms and their control by exogenous factors have been affected; variations have evolved in the circadian timing systems themselves, as well as in the entrainment of endogenous rhythms by particular environmental periodicities and the direct influences that a number of environmental factors impose on endogenous rhythms to inhibit or enhance activity. This chapter summarizes what is currently known about the ways exogenous and endogenous factors interact to regulate the diurnal activity rhythms of bats. The emphasis is on the ecological aspects of this rhythmicity—the optimal adjustment of the activities of individuals and species to the diurnal cycles of the environment—and on the question of the phylogenetic development of the mechanisms involved.

2. METHODS FOR RECORDING THE ACTIVITY OF BATS

The onset, end, and time course of bat activity can be monitored by a wide variety of methods. The simplest, and thus one of the most commonly used, is direct observation. But this approach, as a rule, gives information only about the times when bats leave and return to their roost and/or about the temporal distribution of the frequency of visits to fruit-bearing or blooming trees (by frugivorous and nectarivorous species) or to particular watering spots (Jimbo and Schwassmann, 1967; Start, 1974; Herreid and Davis, 1966). When the size of the colony is known, the number of emergences and returns observed in a particular time sample can provide a basis for making inferences about the flight-activity pattern of the entire colony (Swift, 1980). Observations of this sort can be automated; recording devices have been designed with two parallel electric eyes positioned across the flight path so that emergence and return flights are recorded separately (Kolb, 1959; Böhme and Natuschke, 1967; Engländer and Laufens, 1968; Laufens, 1972; Voûte *et al.*, 1974). Another possibility is to electronically record the movement of individual bats carrying miniaturized resonant circuits, as they enter or leave the roost (Kielmann and Laufens, 1968).

Most of the data on bat-activity patterns have been obtained by captures in mist nets or in special traps (Constantine, 1958; Tuttle, 1974) set up in the immediate vicinity of large colonies (Wimsatt, 1969; Crespo *et al.*, 1972; Kunz, 1974; Start, 1974), or were based on mist-net captures of bats taken while feeding or drinking over bodies of water at various (unknown) distances from roosting places (Cockrum and Cross, 1964; O'Farrell and Bradley, 1970; Jones, 1965; LaVal, 1970; Gaisler, 1973; Kunz, 1973). The advantages of these capture methods are that the activity patterns of several species can be monitored, the bats can be classified according to age, sex, and reproductive state, and one simultaneously gains information about the diversity and abundance of the entire bat fauna in the area concerned. But it can be disadvantageous in that capture and handling may be stressful experiences for the animals. Furthermore, with their good spatial memory (Neuweiler and Möhres, 1967), they may avoid the capture site for the remainder of the night (LaVal, 1970); thus, data so obtained can be misleading. When foraging activity is recorded with ultrasonic detectors (Fenton, 1970; Fenton *et al.*, 1973, 1977; Kunz and Brock, 1975), such interference with normal behavior is avoided. This method, however, monitors only the echolocating activity of the whole population of hunting bats within a narrow sector; species differentiation is possible only to a limited extent, and the activity of species with very penetrating echolocation sounds is weighted more heavily than those of "whispering" bats (see Chapter 7). The most precise information about the activity rhythm of single individuals and species in their habitat is provided by radiotelemetry. Recently, free-living animals bearing miniaturized transmitters have been successfully tracked (Heithaus and Fleming, 1978; Morrison, 1978a). By the modulation of the transmitted signal, it is possible to determine whether or not the animal is moving.

For automated recording of activity under quasinatural or controlled conditions in the laboratory, several possibilities are available, including suspended cages with movement recorders (Griffin and Welsh, 1937; Rawson, 1960; Subbaraj and Chandrashekaran, 1978), electric-eye arrangements (DeCoursey, 1964), cages with hanging sticks mounted on microswitches (Erkert, 1970; Zack *et al.*, 1979), and, the most successful to date, electroacoustic recording devices with microphones sensitive to substrate-conducted sound (Bay, 1976, 1978).

3. ACTIVITY PATTERNS AND TIMING OF FLIGHT ACTIVITY UNDER NATURAL AND CONTROLLED CONDITIONS

3.1. Activity Patterns

Bats are almost all strictly night-active animals. Not until the late evening twilight do most species begin to forage, often returning to their roost long before sunrise or, at the latest, in the twilight of early morning. This behavior offers excellent protection from their chief potential predators, the day-active birds of prey. From the owls that hunt at night bats are protected by other mechanisms. Among these is the tendency to keep the relatively unprotected daily foraging flight as short as possible. This tendency is also favorable from the point of view of energetics, for considerable energy is consumed in flight (see Thomas and Suthers, 1972; Thomas, 1980). Because the duration of flight, for a given energy requirement of the individual, depends (apart from the foraging efficiency of the animal and the abundance of food) primarily on the food's content of energy-containing and ballast substances; one would expect *a priori* that the differences in chiropteran feeding habits would be reflected in their activity patterns. Forms specializing in high-energy food, such as the sanguivore *Desmodus rotundus*, as well as carnivorous and insectivorous species, should require a shorter foraging phase to meet their daily energy requirements than do frugivores, which eat food high in carbohydrates, low in proteins, and containing a high proportion of indigestible material (see Kunz, 1980). The duration of flight also depends on the size of a colony; that is, when more bats are present in a colony, individuals, on the average, must commute farther to their resources in order to minimize competition.

The activity patterns of about 30 of over 850 currently recognized species have been described, and there is remarkable variation among them. However, on closer examination of single species it is evident that activity may be so strongly affected by the prevailing external and physiological conditions that it is problematical to speak of "the activity pattern" of a species (Erkert, 1974; Kunz, 1974; Fenton *et al.*, 1977; Swift, 1980). For this reason caution must be used in comparing patterns of individuals and species if these were not recorded with the same techniques and under the same environmental conditions. For example, Fig. 1 shows a number of activity patterns from the field (a–i) under seminatural conditions (j–l, dotted curves) and in controlled laboratory (m–o) conditions, using various methods with insectivorous (a–f, m–o), frugivorous, and nectarivorous (g–l) species.

Bimodal activity patterns are characteristic of almost all insectivorous species so far studied in the field and of all three species that have been kept in artificial 12:12-hr light-dark cycles (LD 12:12, $10:10^{-4}$ lx). The main peak is at the beginning of the activity phase in each case and the secondary peak is usually toward the end. It can be assumed that this is the basic pattern of insectivorous species, which can be modified in certain cases, depending on the conditions. By contrast, unimodal patterns are dominant among frugivorous and nectarivorous species. For example, all six of the fruit bats studied under artificial conditions (LD 12:12) and two-thirds of the 18 other species studied in the field or under quasinatural conditions had predominantly unimodal activity patterns. Although net captures of Davis and Dixon (1976) implicated three species as frugivores with bimodal patterns (the phyllostomids *Uroderma bilobatum, Vampyressa bidens* and *Artibeus planirostris*), the authors infer that the basic pattern is unimodal. The inference appears justified in that their observations were made between the period of waxing and full moon, and moonlight tends to strongly suppress flight activity and thus alter the activity pattern of certain species. Whether the same reservations apply to the bimodal pattern (Fig. 1g) that Heithaus and Fleming (1978) observed by radio-tracking of the frugivorous *Carollia perspicillata* cannot be decided with the available data. In two separate studies (Wimsatt, 1969; Turner, 1975) the sanguivore *Desmodus rotundus* spent only a few hours each night foraging, returning immediately afterward to the roost. Thus one would expect this bat, under natural conditions, to exhibit a unimodal activity pattern, with a maximum in the first hours of the night. The bimodal distributions in the captured samples of Brown (1968) and Wimsatt (1969) and the observations of Young (1971) appear to contradict this view. Nor are sufficiently well-founded data available on the activity patterns of carnivorous and piscivorous species.

Laufens (1972) and Swift (1980) studied maternity colonies of the insectivorous vespertilionid species *Myotis nattereri* and *Pipistrellus pipistrellus.* Their observations of emergence and return times showed that the flight-activity patterns of such colonies in the relatively far north (50°40′ and 57°13′ north latitude) can follow an annual rhythm associated with the reproductive cycle. During the early stages of pregnancy, in May and June, emergence and return times indicate a weakly bimodal pattern of absence from the roost. However, during lactation in July there was a distinctly prolonged bimodal pattern, and when the young are weaned (August–September) a weakly bi- to unimodal pattern reappears (Fig. 1a–c). Kunz's (1974) trap captures of the North American *M. velifer* indicated no pattern changes of this sort; however, he did find that when young animals first learned to fly, they exhibited a unimodal flight pattern (Fig. 1d, histogram), which gradually shifted to the bimodal form characteristic of adults. On the other hand, a unimodal pattern for the use of separate night roosts by *Eptesicus fuscus* and *Antrozous pallidus,* was reported by Kunz (1973) and O'Shea and Vaughan (1977), with a maximum number of roosting animals in the second half of the night (Fig. 1f, histogram). Therefore it cannot be ruled out that the flight-activity patterns of *M. nattereri* and *P. pipistrellus* are actually also biphasic all the time, and that during lactation the females return to daytime roosts during the night to rest and suckle their young, rather than resting in night roosts as at other times (see Chapter 1).

The activity patterns in Fig. 1, usually constructed on the basis of 1- or 2-hour intervals, are an incomplete representation of the actual time course of activity. Continuous activity recordings in the laboratory (Bay, 1976; Häussler and Erkert, 1978), together with isolated observations (Kunz, 1974; O'Shea and Vaughan, 1977) and certain radio-tracking experiments in the field (Heithaus and

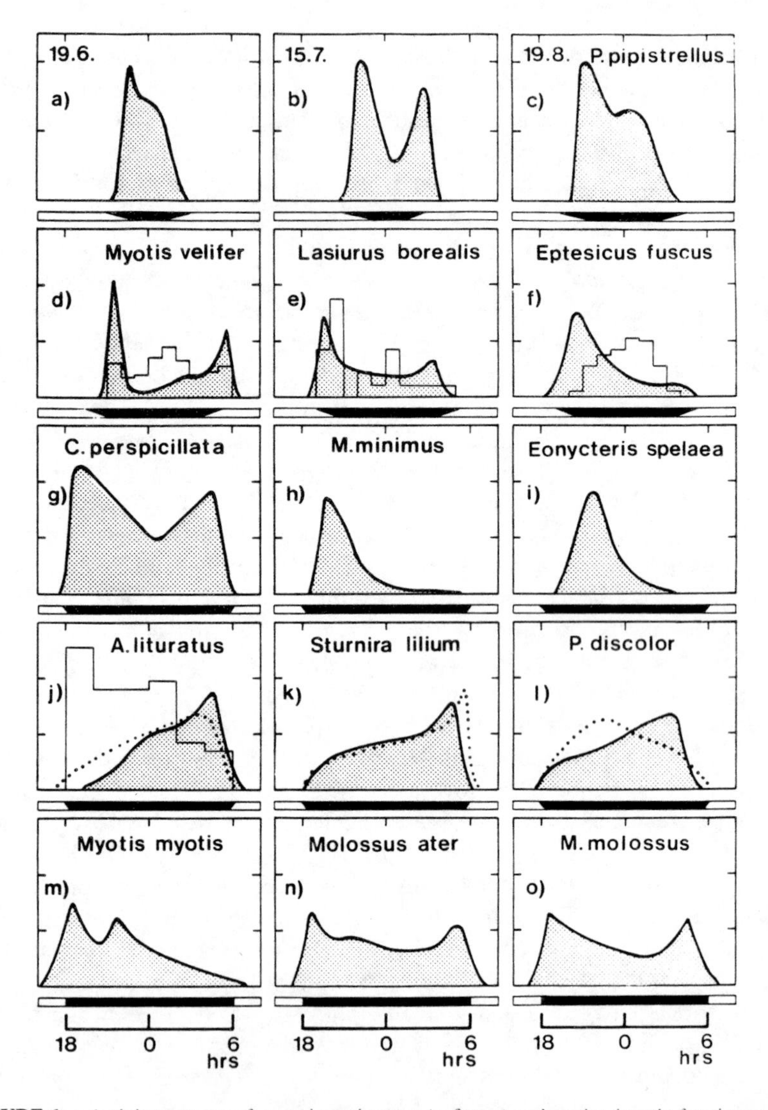

FIGURE 1. Activity patterns of some insectivorous (a–f;m–o) and predominantly fructivorous or nectarivorous (g–l) bat species under natural conditions in the field (a–i), seminatural conditions in the biotope (j–l), or controlled conditions (light–dark cycle 12:12; $10{:}10^{-4}$ lx) in the laboratory (m–o). The histograms in (d) and (e) show the flight-activity patterns of young animals, in (f) the distribution of times of utilization of separate night roosts, and in (j) the temporal distribution of net captures (n = 84) in the biotope. The dotted curves in (j–l) denote the activity pattern in LD 12:12 ($10{:}10^{-4}$ lx). (Diagrams a–c from observations by Swift, 1980; d–f from the results of net and trap captures by Kunz, 1973, 1974; g from radio-tracking results of Heithaus and Fleming, 1978; h and i from observations and the results of net captures by Start, 1974; j–l from unpublished data from activity recordings in Colombia; and m–o from activity recordings by Erkert *et al.*, 1976a.)

Fleming, 1978), reveal that as a rule the daily activity phase is composed of many bouts of movement (foraging flights) and pauses for rest, eating, and digestion, in more or less regular succession. It remains to be learned if, and to what extent, this behavior is the expression of an endogenously programmed ultradian periodicity, such as is found in other species of small mammals (Lehmann, 1976; Daan and Slopsema, 1978).

3.2. Arousal and Timing of Flight Activity

The time when the animals arouse from their daytime rest (or lethargy), when there is no directly disturbing influence, is controlled by the endogenous rhythm synchronized with the external 24-hr oscillation. Before departure from the roost there is, as a rule, an extended phase of grooming and social interaction—as evidenced by the intensive vocalization to be heard for as long as 1–2 hr preceding the emergence of colonial species (see Chapter 1). In the case of cave-dwelling species this phase can be followed by a "light-sampling phase" of varying duration (see below).

The evening emergence phase can last from a few minutes to 2 or 3 hours, depending on species and colony size. In almost all species studied to date, evening emergence proceeds almost in parallel with sunset time (*Pipistrellus pipistrellus:* Venables, 1943; Eisentraut, 1952; Church, 1957; Swift, 1980; *P. mimus:* Prakash, 1962; *Eptesicus serotinus:* Eisentraut, 1952; *Myotis nattereri* and *M. bechsteini:* Laufens, 1972; *M. mystacinus:* Nyholm, 1956, 1965; *M. dasycneme:* Voûte *et al.*, 1974; *Nyctalus lasiopterus:* Maeda, 1974; *Rhinolophus hipposideros:* Gaisler, 1963; *Rhinopoma microphyllum* (=*kinneari*): Gaur, 1980; *Tadarida brasiliensis:* Herreid and Davis, 1966; *Molossus ater, M. molossus* and *Myotis nigricans:* Erkert, 1974; *Rousettus aegyptiacus:* Jacobsen and DuPlessis, 1976; *Pteropus giganteus:* Zack *et al.*, 1979; *Phyllostomus discolor* and *Molossus ater:* Möller, 1979). As a result, the annual variation in emergence time (Fig. 2) becomes greater at increasing latitudes. The timing of return flights to the roost is influenced considerably by foraging success and the weather conditions; it also depends upon whether, and for how long, separate night roosts are occupied. Therefore the return phase of colony members to their daytime roost, in some species, can extend over a considerably longer time span than the emergence phase. For example, Voûte (1972) showed, in his analysis of the effect of the natural light–dark cycle on the activity rhythm of the pond bat *Myotis dasycneme*, that the time between the appearance of the first emerging bat and the return of the first individual to the colony is usually as brief as 1 or 2 hr, whereas the last bats do not return to the roost until 6–10 hours (depending on the season) after the first bat appeared to emerge (Fig. 2, upper right). The return of the last bat to the roost changes in time approximately in parallel with sunrise (Nyholm, 1965; Laufens, 1972; Gaisler, 1963; Voûte *et al.*, 1974; Erkert, 1978; Zack *et al.*, 1979; Möller, 1979).

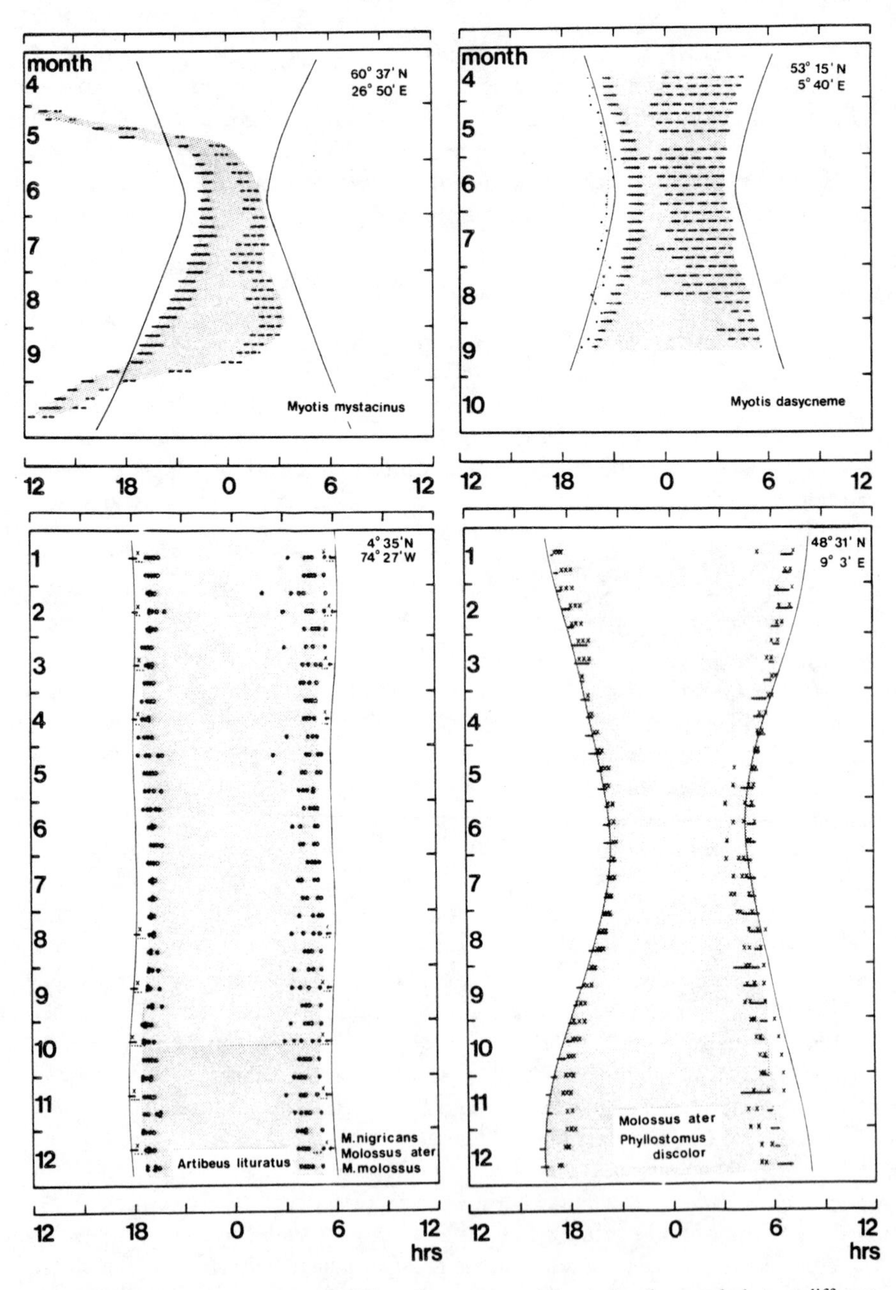

FIGURE 2. Change in timing of flight or locomotor activity during the year in bats at different latitudes. The times of sunset and sunrise are shown by the left and right curves, respectively. The shaded areas indicate the extent of the activity phase of the colony as a whole, from the departure of

There appears to be one exception to the general principle that the beginning and end of flight activity shift roughly in parallel to sunset and sunrise. *Taphozous melanopogon*, to date studied only in India, reportedly emerges at the same time of day throughout the year (Subbaraj and Chandrashekaran, 1977). Transient deviations from the basic timing of foraging activity can occur at very high latitudes. For example, individuals in a colony of *Myotis mystacinus* at 60°37′ north latitude were noted by Nyholm (1965) to be truly night active only from the end of May to the end of September. From the beginning through the middle of May and in October they made foraging flights all day (Fig. 2, upper left). This aberrant behavior can be interpreted as a response to ambient temperature, which at these times of year exceeded the critical temperature (7°C for this species) only during the day, and explained by the fact that the availability of aerial insects is greatest during the day in the cooler months. Some observations suggest that other species in the far north exhibit similar adjustments in the timing of activity to the special biotic and abiotic conditions of their environment that prevail only in spring and fall, immediately preceding and following the hibernation period.

The phase position of the bat-activity cycle, that is, the temporal relation of the onset and end of the flight phase to the arbitrarily chosen reference points of sunset and sunrise in the external cycle not only depends on various environmental factors but also varies among different species. This is clearly evident in the results of a number of comparative studies. Of the central European vespertilionids, Harmata (1960) studied six species and Eisentraut (1952) and Laufens (1972) two each, whereas Jones (1965) studied 18 vespertilionid and one molossid species in the southwestern United States, and Erkert (1978) two molossids and one vespertilionid in central Colombia. The sequence in which the first bats of various species leave their roosts and the last bats return is usually consistent (Fig. 2, bottom; Erkert, 1978; Laufens, 1972). Laufens (1972) marked individuals in a *Myotis nattereri* colony for observation in the field and found that the sequence in which particular members of the colony depart and return is also fixed, as a rule. Thus synchronized activity rhythms differ at both the species and individual level in the details of their phase relation to the external diurnal rhythm under natural conditions. A general rule proposed for birds (As-

---

the first flyer to the return of the last flyer. (Top) The duration of the emergence phase is denoted by solid horizontal lines, and that of the return phase by dashed lines. The dots connected by a light, dotted line in the right diagram indicate the onset of the light-sampling phase of *Myotis dasycneme*. (Left diagram from Nyholm, 1965; right diagram from Voûte *et al.*, 1974.) (Bottom left) Activity times of individual *Artibeus lituratus* kept in captivity under quasinatural conditions (original data). Onset and end of activity are shown by dots. ——, ···, and × mark the emergence and return times of a colony of *Molossus ater*, one of *M. molossus*, and one of *Myotis nigricans*, respectively. (From Erkert, 1978.) (Bottom right) Onset and end of the activity phase of two Neotropical bat species under quasinatural lighting conditions at higher latitudes. (From Möller, 1979.)

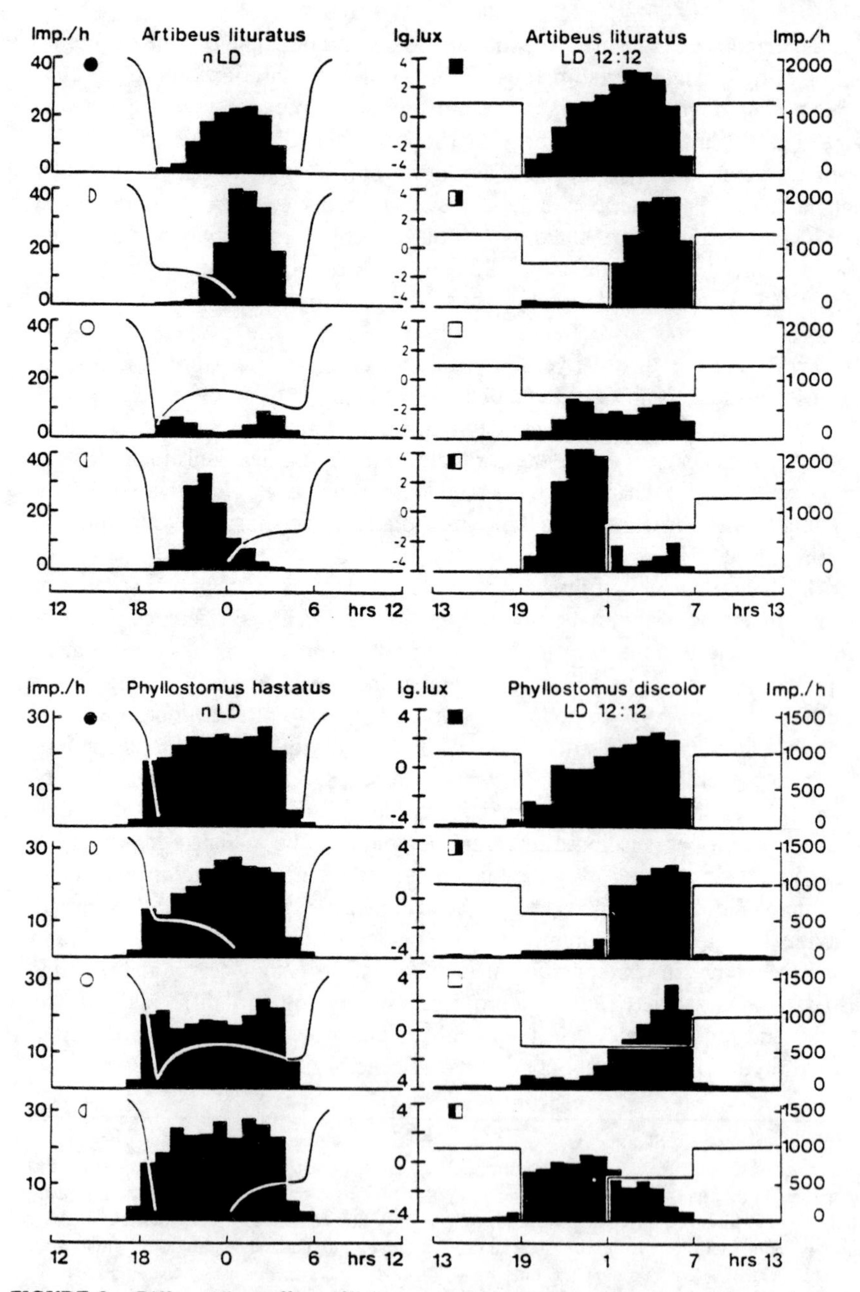

FIGURE 3. Different direct effects of light on entrained activity rhythms of Neotropical leaf-nosed bats. (Left) Activity pattern under natural conditions (nLD) in the normal range (lat. 4°48′ N, long. 74°27′ W), and influence of the nocturnal light profile (curves) in the four lunar phases—new moon,

choff and Wever, 1962), according to which in a given species or individual, the later the resting phase begins, the earlier the animals become active, seems also to hold true for chiropterans (see Laufens, 1972; Erkert, 1978). The sequence of morning return times of individuals and species was exactly the reverse of evening emergence sequences (Fig. 2).

Although the duration of the activity phase of all species studied to date is positively correlated with the night, or dark, phase, it does not always exactly parallel changes in night length over the course of a year. It appears that tropical and subtropical species in particular are incapable of extending and curtailing their activity in such a way as to match the length of night far from the equator under quasinatural conditions. Evidence for this limitation is provided by the activity recordings made by Erkert (1970) and Zack *et al.* (1979) of two Paleotropical fruit bats, *Rousettus aegyptiacus* and *Pteropus giganteus*, and by Möller (1979; Fig. 2, below right) of two Neotropical species, *Phyllostomus discolor* and *Molossus ater*. Even in subarctic species such as *Pipistrellus pipistrellus* the activity phase seems not to be reducible beyond a certain limit by the short summer days in the far north, nor is that of *Myotis nattereri* arbitrarily extensible by the long nights of late autumn (Swift, 1980; Fig. 1b; Laufens, 1972).

In every case changes in the relationship between night length and activity period are linked to annual variations in the temporal relationship (phase-angle difference, PAD) between the onset or end of activity and sunset or sunrise. Depending on the properties of the endogenous timing system, either (1) the PAD of the end of activity can be maintained continuously while that of the onset changes, as in *Pteropus* (Zack *et al.*, 1979), (2) the PAD of the onset (ψ_o) can be kept constant and only that of the end (ψ_e) changed, as in *Molossus ater* at 48°31′ north latitude, or (3) the PADs of both onset and end can change in opposite directions, simultaneously but to different degrees, as in *Phyllostomus discolor* (Möller, 1979) and *Myotis nattereri* (Laufens, 1972). On the basis of such variations in ψ_o and ψ_e, the phase-angle difference for the cycle as a whole (ψ_m, between the middle of the activity period and the middle of the night) can exhibit seasonal changes characteristic of the species. As Daan and Aschoff (1975) showed, using data obtained by Nyholm (1965), Voûte (1972), and Laufens (1972) for four *Myotis* species, the amplitude of these annual variations in whole-cycle PADs increases with increasing latitude. Consistent with this finding, four Neotropical species studied near the equator (Erkert, 1978; Fig. 2) exhibited no annual PAD changes.

---

waxing moon, full moon, and waning moon—on the activity patterns of *Artibeus lituratus* and *Phyllostomus hastatus*. (From Erkert, 1974.) (Right) Light-induced modulation of the activity patterns of *A. lituratus* and *P. discolor*, with the nocturnal light conditions in the four phases of the moon simulated by square-wave light–dark cycles. (From Häussler and Erkert, 1978.)

Correlation analyses show that in many cases the emergence time of the first bat or that of the whole colony is more closely correlated with the occurrence of particular light intensities during the evening twilight (e.g., 10, 1, or 0.1 lx) than with the time of sunset (Fig. 3; Laufens, 1972; Gould, 1961; Voûte *et al.*, 1974; Erkert, 1978). Thus the crucial factor for the timing of emergence is illumination intensity. Interspecific variations in emergence and return times, in fact, provides evidence of species-specific emergence and return light intensities or intensity ranges. For example, under the same environmental conditions *Molossus ater* emerges at relatively high intensities between 30 and 300 lx ($\bar{X}$ = 125 lx), *Molossus molossus* waits until the intensity has fallen to 5–30 lx ($\bar{X}$ = 15 lx), and *Myotis nigricans* hardly leaves its roost at all before the illuminance at the zenith is less than 5 lx (Erkert, 1978). The return to the roost in the morning occurs, as a rule, at lower intensities than does the evening departure. This difference might be a special adaptation of the activity rhythm to avoid the danger presented by day-active birds of prey that begin to hunt early in the dawn twilight. The decisive role of illuminance in eliciting the onset of foraging activity is evident in certain field experiments of DeCoursey (1964), Voûte (1972), and Laufens (1972), in which supplementary illumination at the entrance to the roost, in some cases, induced considerable delay in the evening emergence of three *Myotis* species. The frequent observation that bats leave the roost earlier on dark, overcast evenings than when the sky is clear (e.g., Williams and Williams, 1970) is also consistent with this view.

The range of illuminance characteristic of the emergence and return of different species is not a fixed quantity. This is especially the situation at higher latitudes, where it can change during the year. *Myotis nattereri* (Laufens, 1972), for example, at about 50° north latitude emerges at higher light intensities in summer than in spring and fall. The greater tolerance of ordinarily inhibitory light intensities exhibited by these animals in summer becomes increasingly conspicuous at sites farther north (Nyholm, 1965). It is presumably a compensatory adjustment of the activity rhythm to the shorter and brighter summer nights at these latitudes.

3.3. Light-Sampling Behavior

Despite the high correlation between light intensity and the timing of foraging activity in bats, light in this case does not control the activity rhythm directly; its primary action is synchronization of an endogenous timing system for circadian periodicity. However, this mechanism does not in every case function so precisely as to achieve, by itself, an optimal adjustment of the activity phase of the different individuals and species to the external cycle. There appear to be opportunities for making fine adjustments in the onset of flight activity by reference to actual light conditions, so that bats will not emerge too early and be exposed to diurnal birds of prey. Such an adjustment is relatively simple for free-roosting species, because they receive information about the outside light-intensity cycle while still in their resting place (though the light may be attenuated by a factor of $10–10^3$ or 10^4, depending on the site). Upon a too-early arousal these bats need only to postpone their departure until the twilight intensity has fallen to or below an appropriate level.

For species that live in caves, holes in trees, deep crevices in rocks or (as a secondary modification) in buildings the problem is more difficult. These bats that awake in physiological darkness have evolved a special adaptation, the so-called "light-sampling behavior" (Twente, 1955). Following arousal under circadian control and a subsequent grooming phase, these bats fly or crawl from dark roosting places into illuminated regions near the entrance; in buildings there may be a lighter predeparture roost separate from the actual day-roosting quarters. There they wait, often for a considerable time, testing the outdoor brightness by making short flights or by crawling near the exit holes. Not until the threshold region characteristic of the species has been reached or passed do they finally emerge. Voûte (1972), at his pond-bat colony, recorded light-sampling times between 3 and 132 min ($\bar{X} = 54 \pm 30$), depending on the outdoor light conditions. This variation during the year can be seen in Fig. 2 (upper right). Light sampling or light testing prior to emergence has also been described for *Myotis velifer* (about 15 min) and *Antrozous pallidus* (Twente, 1955), *M. myotis* (DeCoursey and DeCoursey, 1964), *M. mystacinus* (Nyholm, 1965), cave-roosting *Rhinolophus hipposideros* (Gaisler, 1963), *Pteronotus parnellii* (Bateman and Vaughan, 1974), *Rousettus aegyptiacus* (Jacobsen and DuPlessis, 1976), *Desmodus rotundus* (Crespo *et al.*, 1961), and *Artibeus jamaicensis* and *Glossophaga soricina* (Erkert, 1978). This may well be a fundamental feature of all species and colonies that roost in sites where they are unable to adequately keep track of diurnal changes in light intensity.

3.4. Influence of External Factors on Activity Rhythms

3.4.1. Light

The most important environmental factor in the regulation of bat activity is without doubt light. It has a twofold influence on the animals' activity, serving as a so-called Zeitgeber to synchronize their endogenous circadian periodicity, and also, as in the case of light-sampling behavior, exerting a direct intensity-dependent inhibitory effect on the level of activity set by the circadian system (Erkert *et al.*, 1980). In the laboratory experiments can be designed to distinguish between these two effects (Häussler and Erkert, 1978). Under natural conditions they are evident in both the timing of flight and foraging activity and the changes in these in relationship to the lunar cycle.

As noted above, some species emerge earlier under heavy, overcast conditions, when the illuminance is about one-tenth of that under a clear sky. For instance, the evening emergence of *Tadarida brasiliensis*, according to the observations of Herreid and Davis (1966), begins 2.2 ± 28.2 min after sunset in overcast conditions, and on clear days not until 13.6 ± 23.4 min after sunset. *Pipistrellus pipistrellus* also begins its foraging phase 5–10 min earlier than usual on dark evenings with overcast skies (Church, 1957; Stebbings, 1968). The correlation of early emergence with overcast conditions has also been reported for *P. mimus* (Prakash, 1962), *P. javanicus* (=*abramus*) (Funakoshi and Uchida, 1978), *Eptesicus serotinus* (Eisentraut, 1952), *Nyctalus noctula* (van Heerdt and Sluiter, 1965), *Myotis dasycneme* (Voûte *et al.*, 1974), and *Myotis velifer* (Kunz, 1974). For *Molossus ater*, *Molossus molossus*, and *Myotis nigricans* the phase-angle difference of the onset of activity (ψ_o) is positively correlated with the degree of cloudiness, whereas the PAD of its end (ψ_e) is negatively correlated. This means that as the sky is more obscured, these bats not only leave their roost earlier in the evening but also return to it later in the morning (Erkert, 1978). The significance of the inhibitory effect of outdoor light on the timing of emergence is also evident in the fact that different colonies of the same species in a given habitat emerge later if the entrance to the roost is fully exposed to light than when it is hidden and shaded (e.g., *Phyllostomus hastatus:* Williams and Williams, 1970; *Myotis velifer:* Kunz, 1974).

The locomotor activity of bats is not inhibited solely by the relatively high light intensities prevailing before emergence; in some species it can be markedly diminished even by the lower intensities of moonlight (the illuminance of the clear night sky when the moon is full is about $1–5 \cdot 10^{-1}$ lx, but only $2–5 \cdot 10^{-4}$ lx with a new moon). The first indications of this sensitivity were given by the observations of Tamsitt and Valdivieso (1961), Villa-R. (1966), Wimsatt (1969), Schmidt *et al.* (1971), and Crespo *et al.* (1972), who caught considerably fewer foraging vampire bats (*Desmodus rotundus*) and phyllostomids (*Artibeus lituratus*, *Phyllostomus discolor* and *Glossophaga soricina*) in their nets at moonlit times of night than before the moon had risen or after it had set. A direct proof of the inhibitory effect of moonlight was obtained only after months of recording the locomotor activity of several captive bats (*Artibeus lituratus* and *Phyllostomus hastatus*) under natural lighting conditions (Erkert, 1974). The results of that study have since been confirmed with a number of techniques, including the radio-tracking of *A. jamaicensis* (Morrison, 1978a, 1978b), bat-detector recordings of the activity of various African microchiropterans (Fenton *et al.*, 1977), and simulation experiments on *A. lituratus* and *P. discolor* in an artificial light–dark cycle in the laboratory (Häussler and Erkert, 1978). Under natural lighting conditions *A. lituratus* exhibits a pronounced modulation of the activity pattern with lunar periodicity (Fig. 3, upper left diagram). The basic unimodal pattern, peaking in the second half of the night, dominates when the moon is new. As it waxes, the increased brightness in the first half of the night causes a distinct reduction in initial activity, and under a full moon activity is reduced over the entire period, which results in a bimodal pattern. As the moon wanes, the night becomes brighter after midnight, and activity is greatly diminished in the second half of the night. When these lighting conditions are simulated with square-wave light–dark cycles, a similar modulation of the activity pattern of this species is produced (Fig. 3). Moreover, laboratory experiments of Häussler and Erkert (1978) have shown that the activity pattern can be varied at will by a short-term increase in the "dark" light intensity from 10^{-4} to 10^{-1} lx, or a reduction of the "full-moon" intensity from 10^{-1} to 10^{-4} lx. In each case activity is markedly inhibited when the intensity is greater. A striking confirmation of these findings was provided by the field observations of Usman *et al.* (1980) during an eclipse of the moon. There was a sharp increase in the flight activity of various bat species while the moon was in shadow, from about 01.00 to 04.30 h.a.m., and a subsequent fall to the original low full-moon level.

The phase angle of the onset and end of activity in *Artibeus* also varies with the lunar cycle. These light-induced shifts tend to mask the phase angle set by the circadian rhythm. Similar effects are likely to occur in other species, but not all species are as susceptible as *Artibeus* to the inhibitory action of moonlight. *Phyllostomus hastatus*, for example, under natural conditions only exhibits a distinct reduction in activity during the first half of the night, when the moon is waxing, and during the brightest part of the night, when the moon is full (which here, again, converts the basic unimodal pattern to a bimodal one; Fig. 3). *Rousettus aegyptiacus* behaves similarly under the same conditions (Erkert, 1974). The light-induced modulation of the activity pattern of *P. discolor* in artificial light–dark cycles (LD) is also less pronounced than that of *A. lituratus*. Since the observations of *Desmodus, Rousettus, P. discolor*, and *Glossophaga* by Crespo *et al.* (1972), Jacobsen and DuPlessis (1976), and Heithaus *et al.* (1974) have indicated that the timing and temporal distribution of foraging activities in these species are also affected by moonlight, it appears likely that the same holds true of many other species.

We may infer from the variability in behavior of these species under natural and artificial lighting conditions, as from the results of Morrison's (1978a, 1978b, 1980) radio-tracking experiments and from the observations and recordings of Fenton *et al.* (1977) and Usman *et al.* (1980), that during chiropteran phylogeny a broad spectrum of varyingly effective direct inhibitory effects of low light intensities has evolved. It can be assumed that this variability is the manifestation of different ecological adaptations. In this view the strong suppression by moonlight on the activity of *Artibeus lituratus* and *A. jamaicensis* and the general tendency of other species to curtail foraging activity in bright moonlight or to remain in the protective shade of trees and shrubs can be interpreted as an adaptive response to selection pressure from nocturnal predators that orient visually. Both species of *Artibeus* are fairly awkward fliers in contrast to many other bats, as is *Carollia perspicillata*, which forages over much shorter distances (only about 20% as far) when the moon is full than when it is new (Heithaus and Fleming, 1978). These bats may be more vulnerable than others to such predators as owls, which are most active at light intensities near 0.1 lx, equivalent to full moonlight (Erkert, 1969). Thus it would be particularly advantageous for such species to reduce the probability of encountering predators by reserving their own activity for times when the light is so dim that both the predator's ability to orient and its activity level are greatly diminished (Fig. 4). The problem with this strategy is that in parts of the lunar cycle the time available for foraging is severely restricted (see Morrison, 1978a; Fig. 1). Not all species can tolerate curtailment of nightly foraging activity; on energetic grounds alone, one would expect that small frugivorous and nectarivorous species in particular would encounter the most severe difficulties.

For each species there must be a balance reached between the costs of predator avoidance on the one hand and adequate foraging time on the other. This balance may be achieved by the mechanism of species-specific direct inhibition of activity by different light levels. Another possibility is that the differential

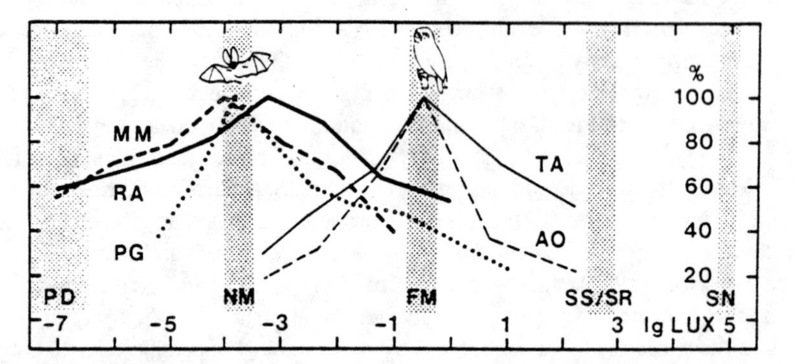

FIGURE 4. Dependence of the locomotor activity of three bat species (heavy lines) on light intensity, as compared with that of two owl species (light lines). The maximum of each curve is set at 100%. The light-intensity scale on the abscissa, from 10^{-7} to 10^{5} lx, includes the entire range of illuminance to which terrestrial animals are exposed under natural conditions, from the region of physiological darkness (PD) at 10^{-6}–10^{-7} lx, to the maximal zenith intensity, about 100,000 lx, in the sunshine at noon under a cloudless sky (SN). Owls become maximally active at intensities corresponding to the light of the full moon (FM), whereas the maximal activity of the bats occurs at intensities equivalent to or less than that of a clear night sky at new moon. Because of the differences in the direct light sensitivity of the activity systems in these two night-active groups, the change in night brightness associated with the moon results in a temporal separation of the activity phases of predator and prey. Bats: MM, *Myotis myotis* (dashed line; from Bay, 1978); RA, *Rousettus aegyptiacus* (solid; from Erkert, 1970); PG, *Pteropus giganteus* (dotted; from Neuweiler, 1962). Owls: TA, *Tyto alba* (solid); AO, *Asio otus* (dashed; both from Erkert, 1969). The area labeled SS/SR denotes the intensity range at times of sunset and sunrise.

sensitivity of bat-activity systems to moonlight could also (or even primarily) serve to reduce the direct interspecific competition among bats specializing on particular food resources by temporal separation of foraging activity based on the lunar cycle. This mechanism has been proposed, for example, by Owings and Lockard (1971) for two *Peromyscus* species with similarly differentiated responses. The available data on bats do not resolve this question, but at present such a function seems doubtful.

3.4.2. Temperature

Ambient temperature has long been recognized as a crucial exogenous factor controlling seasonal timing and the course of hibernation in the heterothermic insectivorous species of temperate zones. During the annual activity period, which depends on species characteristics, the physiological state of the individual, and climatic conditions, the temperature also affects the diurnal rhythm of activity. Many heterothermic species, such as *Myotis nattereri*, *M. bechsteini* (Laufens, 1973), and *Myotis californicus* and *Pipistrellus hesperus* (O'Farrell and Bradley, 1970), hunt for shorter times in cold weather; and when the temperature falls below a certain threshold (especially in spring and fall), they may cease to forage. According to O'Shea and Vaughan (1977), the air-temperature threshold is 10.5°C for *Antrozous pallidus*, 5°C for *P. pipistrellus* (Venables, 1943), and 12–15°C for *P. javanicus* (=*abramus*), depending on location (Funakoshi and Uchida, 1978). For *P. hesperus* in Arizona Jones (1965) and Cross (1965) found inhibition of activity below 14° and 19°C, respectively, although O'Farrell and Bradley (1970) observed this species and *Myotis californicus* in Nevada to remain active even at −8°C. The latter authors also noted that temperature-induced changes in the activity pattern could occur. *Myotis californicus* and *P. hesperus* have a unimodal foraging phase shortened to 4–5 hr on the average at temperatures below 15°C. Only at temperatures above 15°C are they active throughout the night, with the bimodal flight pattern characteristic of insectivorous species. It was mentioned above that certain species (e.g., *M. dasycneme*) not only greatly curtail their activity when it is very cold but also shift it into the warmer parts of the day (Fig. 1; Nyholm, 1965). It remains unclear, however, whether and to what extent this behavior can be regarded as a direct effect of temperature, or whether it is indirectly controlled by temperature-induced changes in the activity times of insect prey.

High ambient temperatures cause the enhanced flight activity of some heterothermic species (O'Farrell *et al.*, 1967), as well as earlier arousal from the daytime lethargy and a slight advancing of the emergence time (Laufens, 1973). Moreover, various field observations indicate that in hot weather bats have a generally higher activity level (or alertness) during the daily resting period. It is still unclear whether and to what extent such effects appear in the intermediate temperature range of 20–30°C. When the heterothermic and homeothermic Neotropical species *Molossus ater* and *Phyllostomus discolor* are kept at LD 12:12 in the laboratory, no clear correlation can be demonstrated between the amount of activity per day and constant T_a's of 20, 25, and 30°C. The activity phase of *M. ater* begins somewhat earlier at 20°C, and in *Phyllostomus* the activity pattern changes, in that at this temperature the peak activity shifts from the first half of the dark period to the second (Erkert and Rothmund, 1981). These findings indicate that the timing and pattern of locomotor activity are slightly affected even by intermediate temperatures, at least in certain species. Not enough data are available as yet to give a satisfactory description of the influence of T_a on bat-activity rhythms.

3.4.3. Precipitation and Wind

Observations made on the effects of precipitation and wind on bat flight activity provide a very heterogeneous picture. For example, Tamsitt and Valdivieso (1961) found no effect on certain phyllostomids, and the same results were obtained by Kunz (1974) for *Myotis velifer* and by Swift (1980) for *Pipistrellus pipistrellus*. However, according to Venables (1943), Church (1957), and Stebbings (1968), heavy rainfall and strong wind cause a distinct reduction in the activity of *P. pipistrellus*. Curtailment of foraging activity by heavy rain, in the form of briefer hunting flights, later emergence, or no flight at all, has been observed in *Rhinolophus hipposideros* (Gaisler, 1963), *M. lucifugus* (Fenton, 1970; O'Farrell and Studier, 1975), *M. nigricans* (Wilson, 1971), *Pipistrellus mimus* (Prakash, 1962), *Desmodus rotundus* (Wimsatt, 1969; Schmidt *et al.*, 1971), *Pipistrellus javanicus* (Funakoshi and Uchida, 1978), *Nyctalus lasiopterus* (Maeda, 1974), *Tadarida femorosacca* (Gould, 1961), *Molossus molossus* and *M. ater* (Erkert, 1978), and *Eptesicus capensis*, *Nycticeius schlieffeni*, and *Scotophilus viridis* (Fenton *et al.*, 1977). Reduced flight activity in very strong, gusty wind has been described for some of the species listed above, as well as for *Pipistrellus pipistrellus*, *Myotis californicus* (O'Farrell *et al.*, 1967), and *Plecotus auritus* (Frylestam, 1970). Together, these observations imply that, in general, light precipitation and wind have little or no effect on the foraging activity of bats; whereas heavy precipitation and strong wind can entirely prevent flight activity in many species. Whether and when this occurs very probably depends not only on the sensitivity of the particular species but to a great extent on the current state of hunger, that is, on the foraging success of previous nights.

3.4.4. Food Supply and Social Factors

No solid data have yet been obtained regarding the influence of food supply on bat-activity rhythms. It is known only that some small phyllostomids, such as

Glossophaga soricina, are capable of surviving periods of starvation lasting one to several days (while the weather prevents foraging) by estivation, in which the metabolic rate is distinctly lowered (Rasweiler, 1973). Apart from this finding, it can be assumed that the availability of food as well as the proximity and dispersal of food sources affect both the duration and time course of the foraging phase. In *Desmodus rotundus*, for instance, Young (1971) found that the searching and detection of prey usually required a longer time in the dry season than in the wet season, and that the vampires in the dry season spent more time at their prey than in the wet season. Insectivorous species, for example, are unlikely to exhibit such a clear-cut bimodal activity pattern as at times of the year when the population density of their insect prey is low. The current state of hunger of the individual could also play a role in the timing of flight activity and, as Nyholm (1965) has shown for *Myotis mystacinus*, in some conditions can overcome the activity-inhibiting action of light and cause earlier emergence from the roost.

The predominantly bimodal activity pattern of the insectivorous species can also be interpreted as a response to the times of food availability. Various studies have shown that the time course of activity and the abundance of flying insects that serve as potential prey are described by a bimodal curve, with maxima in the evening and morning twilight (Swift, 1980; Funakoshi and Uchida, 1978). The insectivorous chiropterans may well have adapted to this by evolving, in turn, a corresponding endogenously programmed bimodal flight-activity pattern. The selective advantage of matching a predator's activity pattern to that of the prey is obvious. Such matching considerably increases the probability that the hunt will give the greatest possible energy yield with the smallest possible expenditure of energy.

The influence, if any, of social factors on the timing and time course of the chiropteran activity rhythm also remains unclear. Laboratory observations of various pteropodids, phyllostomids, and molossids indicate, however, that lower-ranking individuals frequently go to the feeding site in a cage earlier (often well before darkness) than those higher in rank, and that isolated bats develop a much more sharply delimited activity–rest cycle than do individuals living in groups. Groups of bats exhibit repeated intensive social interactions, associated with locomotor activity, even during the resting period. According to observations of several bat colonies, *Myotis myotis* (Mislin, 1942), *Eidolon helvum* (Huggel, 1958), and *Pteropus giganteus* (Neuweiler, 1969), similar socially induced disturbances of the activity rhythm also occur under natural conditions. Since female colonies of *Pipistrellus pipistrellus* and *M. dasycneme* exhibit pronounced changes in the activity pattern during the lactation phase (Swift, 1980; Laufens, 1972), the possibility of socially induced variations in activity during other phases of the reproductive period cannot be ruled out. However, no direct observations of such effects have as yet been made.

4. ACTIVITY RHYTHMS DURING HIBERNATION

A conspicuous feature of the annual rhythm of activity of insectivorous bats in the temperate zones is hibernation, an energetically optimal adaptation to an annually recurring period of extreme food scarcity. Lethargy during this time is not continuous but, rather, is interrupted by several spontaneous arousal and activity phases that follow one another in almost regular succession (for reviews see Nieuwenhoven, 1956; Daan, 1973). Daan (1970, 1973), recording the activity of several *Myotis* species in various hibernation sites in Holland by means of photography and light-beam interruption, found that the relative frequency of flight activity during hibernation at first decreases from October to February and then rises sharply again in March and April.

Arousal and flight frequencies are positively correlated with the ambient temperature. Emergence from the winter roost, often preceded by a more or less long light-sampling phase, occurs only at night and at T_a's above 4°C. Within the hibernaculum, however, bats also fly during the day. The flight activity of the population retains a distinct diurnal rhythm near the beginning of the hibernation period, but as the winter proceeds it becomes increasingly arrhythmic. Not until March–April does a diurnal periodicity again become apparent. Menaker (1959) and Pohl (1961) demonstrated free-running circadian rhythms of body temperature and O_2 consumption, respectively, in *Myotis lucifugus* and *M. myotis* hibernating in the laboratory. It can thus be inferred that the timing of arousal during hibernation is also controlled by the circadian system.

The distinct diurnal rhythm of flying frequency at the onset and toward the end of the hibernation period could be attributed to a synchronization of the circadian rhythm with the external day–night cycle, for light-sampling and flights outside the roost are relatively frequent at this time. The frequency of arousal in December, January, and February is so low that such synchronization is hardly possible. The individual members of the population wake up, at long intervals, at quite different times of day, because of differences in the free-running periods of their endogenous rhythms. Thus the population as a whole presents a picture of aperiodic arousal. This hypothesis is consistent with Ransome's (1968) observation that arousal of the large horseshoe bat (*Rhinolophus ferrumequinum*), which occurs considerably more frequently than in the *Myotis* species studied by Daan (1973), remains to a large extent synchronized with the evening twilight.

Each arousal from dormancy costs the individual an appreciable amount of energy, an amount considerably increased if the animal then proceeds to fly. One indication of this expenditure is the significant positive correlation between loss of body mass during hibernation and the frequency of arousal and "displacement" (Daan, 1973). The biological function of this energy-consuming and survival-endangering behavior remains debatable and is discussed by Daan

(1973). An important aspect, however, is that the length of the periodically occurring (one to several hours) activity phases of hibernating bats is controlled by the endogenous timing system, and the exogenous factor temperature acts chiefly on the frequency of arousal.

5. THE ENDOGENOUS ORIGIN OF BAT ACTIVITY RHYTHMS

5.1. Circadian Activity Rhythms

Even when all information about the daily rhythm of the environment is excluded, bats kept under constant conditions in the laboratory maintain an activity rhythm with a roughly diurnal periodicity. The period of this cycle is almost always systematically different from the 24 hr of the earth's rotation. This fact implies that the activity rhythm in bats, as in other animals, is not induced by exogenous influences but is based on an endogenous basic rhythm that is adjusted to the external daily cycle under natural conditions. The existence of such a "circadian rhythm" of locomotor activity in bats was first demonstrated by Griffin and Welsh (1937) for *Pipistrellus subflavus* and *Eptesicus fuscus*. Circadian rhythms (CRs), autonomously free running under constant conditions, have since been found in *Myotis lucifugus* (Rawson, 1960; Menaker, 1959, 1961), *Rhinolophus ferrumequinum* (DeCoursey and DeCoursey, 1964), the pteropodids *Rousettus aegyptiacus* and *Eidolon helvum* (Erkert, 1970), *M. myotis* (Bay, 1976, 1978), *Taphozous melanopogon* (Subbaraj and Chandrashekaran, 1977, 1978), and *Pipistrellus javanicus* (Funakoshi and Uchida, 1978), as well as in the Neotropical species *Molossus ater, M. molossus, Phyllostomus discolor, Artibeus jamaicensis, Sturnira lilium,* and *Glossophaga soricina* (Erkert and Kracht, 1978; Erkert *et al.*, 1980). As an example, Fig. 5 (top) shows the activity rhythm of single individuals of three frugivorous phyllostomid species during a 20-day period of maintained darkness (DD); at the bottom of the figure is the record of an insectivorous bat, *Molossus ater,* in DD (left) and maintained light (LL) of 10 lx (right). Figure 5 also demonstrates that the circadian rhythms of different species under the same environmental conditions (here DD, 25°C) can differ greatly in period, and that there may be fairly pronounced changes in maintained light.

Menaker (1961) showed that, at least in the case of some hibernating species of the temperate zone, apart from such exogenously induced changes in the course of the year, there can also be pronounced spontaneous variation in the period of the free-running rhythm. Seven of the *Myotis lucifugus* in which he had induced prolonged cold lethargy ("hibernation") during the summer had a free-running body-temperature rhythm with a period of only 22.4 ± 0.9 hr, whereas the periods of 16 individuals of this species studied under identical conditions in

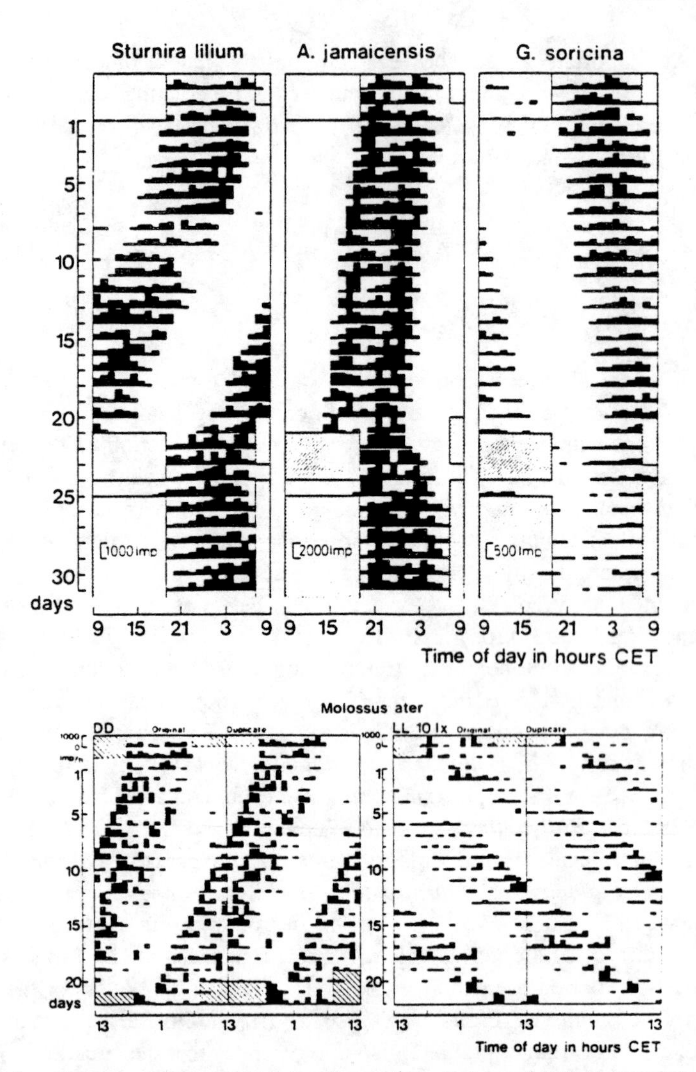

FIGURE 5. Autonomous free-running circadian activity rhythms of individuals of three phyllostomid species maintained in physiological darkness (top), and the light-induced prolongation of the free-running period of a *Molossus ater* maintained in light at 10 lx (bottom right) as compared with the free-running period maintained in darkness (bottom left). The cross-hatched areas denote light times, 10 lx in each case. (After Erkert *et al.*, 1980; Kracht, 1979.)

winter was prolonged to an average of 25.0 ± 1.7 hr. Both the biological significance of such spontaneous seasonal changes in period and the underlying mechanism remain obscure. Even without such knowledge, however, the extent of these changes can be used as an indicator of the plasticity of the endogenous timing system of the species concerned. The same sort of indications are provided by the exogenous light- and temperature-induced variations in period.

5.2. Susceptibility of Period to Exogenous Influences

It has long been known that the CR periods (τ) of both light- and dark-active animals depend on light intensity. According to the "circadian rule" formulated by Aschoff (1960, 1964), the correlation of activity with light intensity is negative in light-active species and positive in dark-active species. The bat circadian activity rhythm conforms to this rule. In almost all the species so far investigated the spontaneous period becomes longer as light intensity increases (Fig. 6, upper diagram). There are marked species-specific differences in the degree to which the CR period depends on intensity. In some species, such as *Myotis myotis*, *Molossus molossus*, and *M. ater*, the variations in CR period are relatively large, whereas in others (e.g., *Glossophaga soricina* and *Rousettus aegyptiacus*) τ is hardly affected by light intensity. The implication is that within the Chiroptera there are considerable differences in the degree of plasticity of the endogenous timing system.

Circadian rhythms are to a great extent temperature compensated (Bünning, 1977); their Q_{10} (the speeding up of the endogenous rhythm by a 10°C increase of T_a), the relevance of which has recently, and justifiably, been questioned (Bünning, 1974), in animals is usually near unity (Pittendrigh, 1960: Q_{10} = 0.9 - 1.2). Rawson (1960) demonstrated that this general rule applies to chiropterans, by recording the activity of *Eptesicus fuscus* and *Myotis lucifugus* over the temperature range 10.4–22°C. The circadian period of the former species varied between 22.47 and 20.19 hr, and that of the latter between 19.54 and 21.48 hr. Although the period of *Myotis lucifugus* appears to bear no clear relation to T_a, that of *Eptesicus* in this temperature range exhibits unequivocal negative correlation with T_a. A different relationship has only recently been found in the two Neotropical species *Phyllostomus discolor* and *Molossus ater*, which have periods positively correlated with T_a in the range 20–30°C (Erkert and Rothmund, 1981). In the heterothermic *M. ater* the change in τ with T_a is considerably greater than in the homeothermic *Phyllostomus;* in the latter case the 30°C value is, on the average, only very slightly above the 20°C value (Fig. 6, below). One must assume that such differences in response will appear in other species as well, and it appears that the circadian timing system of heterothermic species in general is more sensitive to temperature than that of strictly homeothermic species. Differences in sensitivity fluctuations over the course of a year cannot be ruled out, nor can the possibility that the circadian system of some species behaves differently at low ambient temperatures than when it is warmer. However, differences in response like those between *Molossus ater* and *P. discolor* could also be based on species-specific differences in the degree of plasticity of the endogenous timing system. This is suggested, for example, by the fact that in *Molossus* both the period variation induced by temperature and that induced by light are greater than in *Phyllostomus*. For a satisfactory clarification of these relationships, however, wide-ranging comparative studies are required.

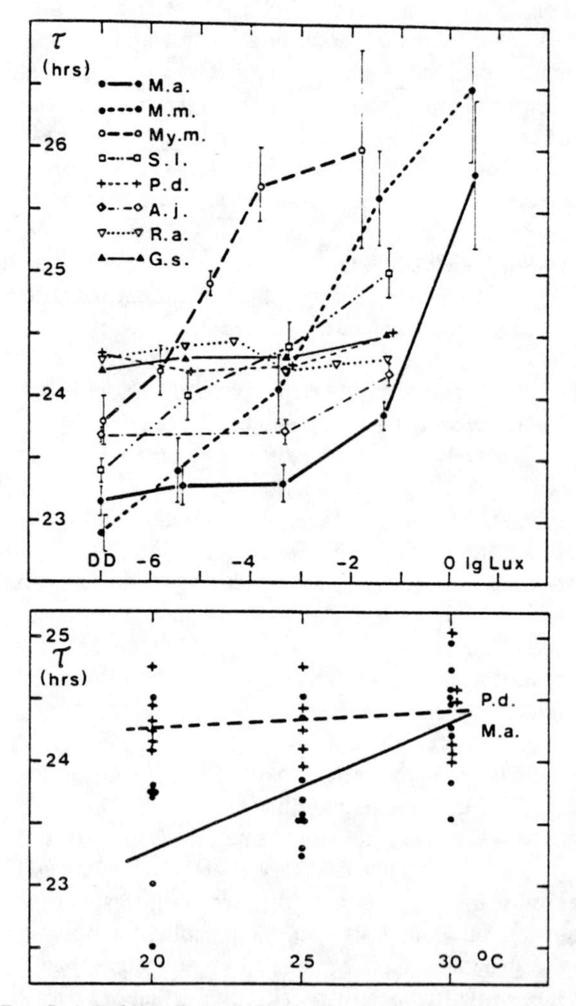

FIGURE 6. (Top) Period of circadian activity rhythm (τ) of eight bat species, free-running under constant conditions, as a function of light intensity. (Means and standard deviations from Erkert *et al.*, 1980.) (Bottom) Influence of ambient temperature on the period of circadian rhythm in one homeothermic and one heterothermic Neotropical bat species, *Phyllostomus discolor* (+, P.d.) and *Molossus ater* (●, M.a.). (From Erkert and Rothmund, 1981.)

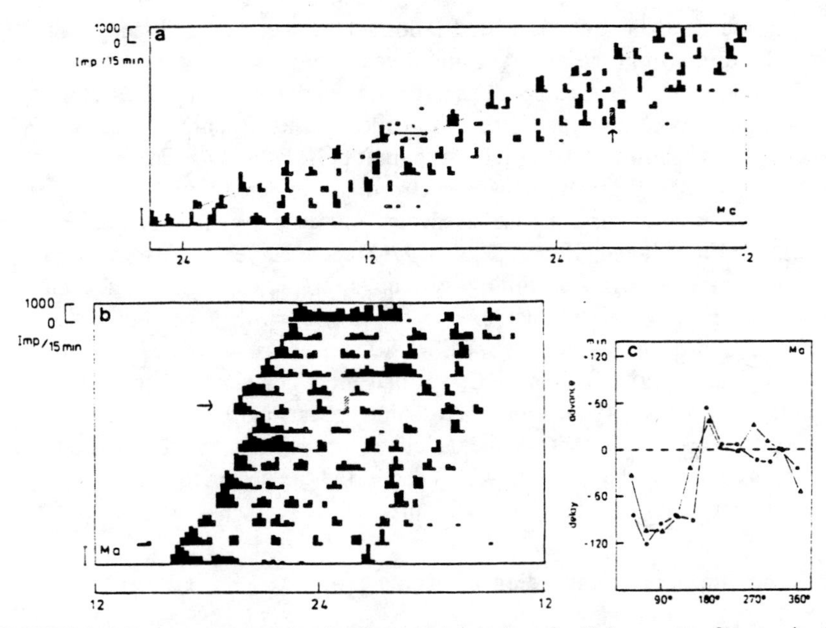

FIGURE 7. Phase response of the circadian activity rhythm of a *Molossus ater*, free running in DD, to 15-min light pulses at 200 lx (shaded areas marked by arrows). (a) An example of an advance shift induced by a light pulse at the beginning of the resting phase. (b) A delay shift induced by light during the activity phase. (c) The phase-response curves of two *M. ater*. Contrary to convention, advance shifts are plotted upward. (From Möller, 1979.)

5.3. The Phase Response of Circadian Activity Rhythms to Light Pulses

Brief increases or decreases in light intensity, during either DD or LL, mostly cause phase shifts in the free-running rhythm of bats kept under constant conditions. The extent and direction (delay or advance) of this "phase response" depend chiefly on the phase of the circadian cycle in which the light pulse is applied. Under DD conditions light pulses presented toward the end of the resting phase and during the phase of activity appear—according to Möller's (1979) studies of two *Molossus* species—to cause varying delays of the subsequent activity phases in chiropterans, as has been found for dark-active rodents (DeCoursey, 1961; Pittendrigh and Daan, 1976). By contrast, light pulses during the resting phase give rise predominantly to advance reactions or only very slight delays (Figs. 7a and 7b). With *Taphozous melanopogon*, Subbaraj and Chandrashekaran (1978) obtained just the opposite responses; however, they did not follow the usual convention of testing the phase response under DD conditions but, instead, used LL of 5 lx. It may be, then, that the difference between the response of their animals and the normal behavior of dark-active small mammals results from the different initial conditions.

By plotting the phase response as a function of the phase of the CR at which the light pulse occurred, one obtains so-called phase-response curves (PRCs, Fig. 7c). These show that sensitivity to a light pulse of a given intensity, duration, and wavelength changes with a circadian periodicity and illustrate the nature of this change for a particular species. Because the shape of the PRC is species specific (Pittendrigh and Daan, 1976), these curves can also be used to characterize the endogenous timing system of a species. This application is problematical, however, in that one is not dealing with a one-dimensional, clearly defined quantity; the interspecific differences are expressed in both the form and the amplitude of the PRC, so that direct comparison becomes more difficult (see Aschoff, 1965). For this reason it appears more useful to characterize the differences in plasticity of the bat circadian system not by the PRC but, in the first instance, by the extent to which the period can be varied, with the range of entrainment and the rate of resynchronization as supplementary parameters.

5.4. Entrainment of Circadian Rhythms

The circadian rhythm, free running under constant conditions with a period different from 24 hr, is synchronized with the external diurnal cycle by certain diurnally varying environmental factors called Zeitgeber, synchronizers, or entraining agents (Aschoff *et al.*, 1965). Their action does not only lengthen or shorten the circadian period so that it matches that of the Zeitgeber, but, in addition, the circadian period is set in a certain stable phase relation to the Zeitgeber cycle, characteristic of the properties of the particular Zeitgeber and circadian system involved. This phase angle between the synchronized CR and the Zeitgeber cycle varies among species in such a way as to optimize the species' ecological situation.

5.4.1. Models of the Mechanism Underlying Synchronization

The mechanism by which circadian activity rhythms are synchronized by Zeitgeber is, at present, as inadequately explained as circadian periodicity itself. It has so far been discussed chiefly in terms of two hypotheses. The most completely elaborated of these is the so-called "threshold-level hypothesis" of Wever (1965). His mathematical oscillation model of biological periodicity is based on the notion that the CR under Zeitgeber conditions behaves like a self-sustained oscillation in the technical sense, which is synchronized by an exciting oscillation. In this view the activity rhythm is equivalent to an oscillatory process divided into two parts by a threshold, the part above this threshold corresponding to the active phase, and the part below it to the resting phase. The synchronizing effect of the Zeitgeber oscillation, in the sense of phase control, would (in highly simplified terms) consist in shifting the level of the biological oscillation with respect to the threshold in a way determined by the oscillatory properties, frequency, amplitude, and average level of the Zeitgeber. A postulated coupling of level and frequency within the organism is then thought to result in an approximation of the two periods, as well as in changes in the form of the biological oscillation. The outcome of these events, in which both proportional and differential effects of the Zeitgeber rhythm play a role, is that the endogenous rhythm settles into a particular phase with respect to the synchronizing Zeitgeber cycle. Depending on whether the endogenous period is appreciably shorter or longer than that of the Zeitgeber, the endogenous rhythm precedes (positive PAD ψ^+) or lags behind (negative PAD ψ^-) the Zeitgeber rhythm by a certain amount. The magnitude of the PAD in each case depends both on the difference in the periods of the two oscillations and on the mean level and oscillation range of the Zeitgeber rhythm. It is thought that the general direction of these effects is that as the oscillation range (intensity) of the Zeitgeber increases, the PAD decreases; whereas an increase in the mean level of the Zeitgeber in dark-active animals results in a reduction of the existing PAD if it is positive, and an increase if it is negative. However, these predictions of Wever's oscillation model have not always been fully confirmed by analyses of control of the activity rhythm in bats (Erkert, 1970; Bay, 1976; Kracht, 1979).

The "phase-response model" of Pittendrigh (1974) for the synchronization of circadian rhythms, which at present enjoys greater popularity, does not allow such detailed predictions about the behavior of circadian systems under Zeitgeber conditions. This model is based on the notion that only differential Zeitgeber phenomena (for example, a rise or fall in light intensity) are effective, and that synchronization is brought about by continual corrective phase shifts. Depending on the phase of the circadian cycle with which the rising and/or falling flank of the Zeitgeber oscillation coincides, the subsequent circadian cycle, in accordance with the shape of the PRC for the stimulus form concerned, will be delayed or advanced. This process is repeated from one cycle to the next and, thus, by a succession of corrective phase shifts of the CR, results in a relatively stable phase relation between the biological and Zeitgeber rhythms.

In certain cases a so-called "splitting" of the free-running CR has been observed in which the activity phase becomes divided into two components, which are for a while independently free running with different periods. In view of this behavior, Pittendrigh has proposed that the CR is controlled not by one oscillator but by two, which he calls the N (night) and M (morning) oscillators. Whereas the N oscillator would respond to light-pulse disturbances only with delays and would give maximal responses at the onset of subjective night, the M oscillator (as Pittendrigh conceives it) would respond only with advance phase shifts to light pulses that occur near the subjective daybreak.

In circadian systems that exhibit both advances and delays, for example, in *Taphozous melanopogon* (Subbaraj and Chandrashekaran, 1978) and *Molossus ater* (Möller, 1979), these two oscillators would, according to Pittendrigh's model, be permanently coupled with one another. The direction and degree of reaction elicited in any particular case would then depend on which of the two oscillators was maximally influenced by the rising and/or falling phase of the Zeitgeber rhythm. In each case the result would be more or less marked phase shifts of the CR, which ultimately would lead to synchronization with the Zeitgeber. The phase angle between the entrained CR and the Zeitgeber would depend on the time course of its sensitivity to the respective Zeitgeber modality, that is, on the form and amplitude of the relevant PRC.

Concerning the basic physiological mechanisms underlying the synchronization of CRs, there have been as few solid ideas generated based on the phase-response model as there have been based on the threshold-level hypothesis. A third model, recently proposed by Enright (1980), suggests that vertebrate-activity rhythms are controlled by an array of coupled pacemaker neurons in the central nervous system. One aspect of this hypothesis is that synchronization of circadian rhythms involves direct or indirect exogenous control by light on the rhythm of synthesis and/or secretion of melatonin.

5.4.2. Zeitgeber for Bat Circadian Activity Rhythms

The chief Zeitgeber that synchronizes the CR with the 24-hr rhythmicity, in bats as in other animals, is the change in light during the course of the day. Two main factors support this conclusion: the free-running activity rhythm under constant conditions gives a phase response to light pulses, and the free-running rhythm can always be entrained by 12:12 hr LD cycles of sufficient amplitude (see Fig. 6; DeCoursey, 1964; Erkert, 1970; Bay, 1976; Erkert and Kracht, 1978; Subbaraj and Chandrashekaran, 1977). The rhythm of the Egyptian fruit bat, *Rousettus aegyptiacus*, can be entrained by intensity differences of only 0.006 lx in a LD 12:12 oscillation of low average intensity (0.041:0.036 lx). By contrast, when the average intensity is relatively high, even LD differences of more than 20 lx (31:3.9 lx) have no entraining effect (Erkert, 1970). Judging from this intensity dependence, it can be inferred that the synchronizing action of light cycles is subject to the same rules that govern the relationship between the difference threshold and the adaptation intensity in the visual system. The remarkably effective Zeitgeber action of 24-hr LDs on bats is also evident by the fact that with some species (*R. aegyptiacus*, *Myotis myotis*) the light time/dark time ratio can be varied between 22:2 and 2:22 hr without loss of entrainment of the activity rhythm (Erkert, 1970; Bay, 1976).

In bats, as in other mammals, the effectiveness of temperature rhythms as a Zeitgeber is slight. The results of a study on the influence of a trapezoidal temperature oscillation 10°C in amplitude (20:30°C) on the free-running circadian activity rhythms of one homeothermic (*Phyllostomus discolor*) and one heterothermic Neotropical species (*Molossus ater*) indicate that such oscillations do not function as a Zeitgeber for homoiothermic species; however, if the oscillation range is large enough (≧10°C; Erkert and Rothmund, 1981) they may completely entrain heterothermic species. Moreover, phase shifts of a *M. ater* activity rhythm, entrained by an LD 12:12, can be induced by temperature. Thus one may conclude that controlling the timing of locomotor activity of heterothermic species under natural conditions involves both Zeitgeber modalities, to degrees corresponding to their effectiveness on the circadian system concerned, and that they jointly determine the phase angle of onset and end of activity.

In some species, at least, the activity rhythm also appears entrainable by social interaction with conspecifics. Marimuthu *et al.* (1978) have drawn this conclusion from the observation that individual *Hipposideros speoris* kept alone in a spring-suspended cage in the DD of a cave inhabited by the "mother" colony exhibited an activity pattern corresponding to the emergence and return phases of the colony on 11 successive days. This observation in itself does not provide unequivocal proof of social entrainment, because the experiment was not preceded by a prolonged session in which the endogenous activity rhythm of the test animal was free running. A further question regarding the prevalence of social entrainment of the circadian rhythm of bats in general is raised by the fact that other species (e.g., *Molossus molossus, M. ater, Phyllostomus discolor, Sturnira lilium* and *Glossophaga soricina*) kept under constant conditions do not synchronize one another, even when they are able to communicate acoustically, but have free-running rhythms with individual periods of different length. These objections, however, fail to rule out the possibility that in individual cases direct disturbance by, and/or social contacts with, already active colony members can play a significant role in the timing of arousal and flight activity under natural conditions.

Bay (1976) has pointed out that when *Myotis myotis* (an insectivorous species) is kept under constant conditions, the time of feeding affects its activity pattern. Accordingly, it seems possible, especially among insectivorous bats, that food supply, which varies with a diurnal periodicity, may take on a Zeitgeber function. Again, however, convincing evidence is lacking.

5.5. Range of Entrainment and Speed of Resynchronization

Whether or not an environmental rhythm that could, in principle, serve as a Zeitgeber (e.g., a LD or temperature oscillation) can entrain the circadian activity rhythm in any specific case depends on several factors. It depends not only on the oscillation range of the potential Zeitgeber but also, and even more importantly, on its period T. Entrainment of the biological oscillation by a

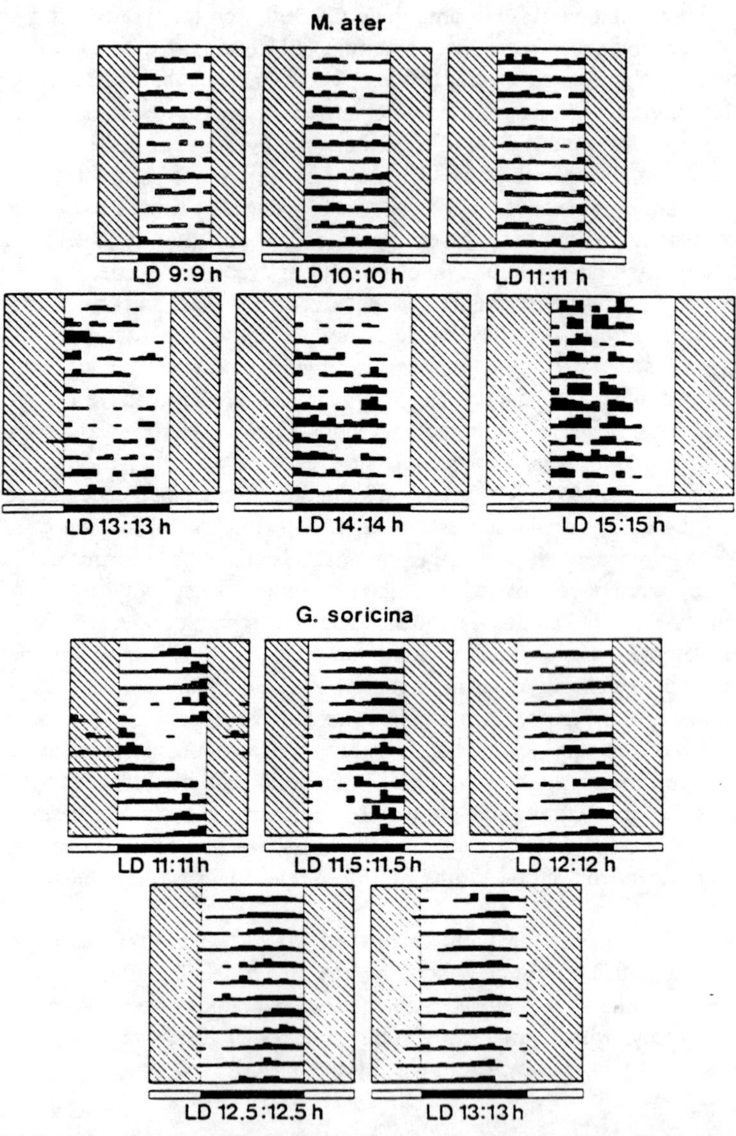

FIGURE 8. Range of entrainment of the circadian activity rhythm of the insectivorous bat *Molossus ater* (top) in comparison with that of the predominantly nectarivorous *Glossophaga soricina* (bottom), with a LD intensity ratio of $10^1:10^{-4}$ lx. In *Molossus* 18- and 30-hr Zeitgeber periods (LD 9:9 and 15:15 hr) have an entraining action; whereas the activity rhythm of *Glossophaga* is free running, with Zeitgeber periods of 23 and 26 hr (LD 11.5:11.5 and 13:13 hr). The shaded areas denote the light phase, and the unshaded areas the dark phase, of the LD cycle. (From Kracht, 1979.)

Zeitgeber is possible only within certain range of periods. This range is more restricted the greater the degree of persistence, or oscillator strength, of the biological rhythm. The size of the effective range can therefore be used as a measure of the plasticity of the circadian system of a species. Experiments by Erkert (1970) on *Rousettus aegyptiacus*, by Bay (1976) on *Myotis myotis*, and by Kracht (1979) on five Neotropical species have shown that the range of entrainment (T_e) can vary widely. Species with free-running CRs whose periods change greatly depending on the constant light level have a considerably larger range of entrainment than those in which τ varies only within very narrow limits. *Myotis myotis*, with τ ranging from 22.5 to 27.5 hr, exposed to a LD ratio of $10{:}10^{-4}$ lx can be fully entrained by LD cycles of 8:8 to 18:18 hr, that is, by Zeitgeber periods ranging from 16 to 36 hr (Bay, 1978). For *Molossus molossus* and *M. ater* the range of entrainment with this LD intensity ratio extends from less than 18 hr to over 30 hr; whereas it is considerably narrower for certain phyllostomids (22.5–27 hr for *Sturnira lilium*, 23.7–27 hr for *Phyllostomus discolor*, and only 23.4–25 hr for *Glossophaga soricina;* Fig. 8; Kracht, 1979; Erkert *et al.*, 1980). *Rousettus*, with a free-running period that varies only between 24.2 and 24.6 hr, also has a very narrow range of entrainment, between about 23 and 25 hr in LD $10{:}10^{-2}$ lx (Erkert, 1970).

Besides the plasticity of a CR, the range (amplitude) of the Zeitgeber oscillation is another factor determining the range of entrainment. The smaller the LD intensity difference, the narrower the range of entrainment. *Molossus ater* and *M. molossus* have a much narrower range in LD $10^{-2}{:}10^{-4}$ lx—2 and 4 hr, respectively (T_e = 23–25 and 22–26 hr). *Glossophaga* in this light regime can be entrained only by 12:12 hr LD rhythms.

The strength of the endogenous component as a controller of the activity rhythm can also be estimated by the speed of resynchronization after phase shifts of the Zeitgeber oscillation. The stronger and more rigid an individual's CR, the longer the organism requires to readjust to a Zeitgeber that has suddenly been shifted by several hours. The resynchronization experiments that have been done on bats, by suddenly shifting a 12:12 hr LD by 8 hr, ahead or back (positive or negative phase shift of the Zeitgeber; $\Delta\Phi + 8$ or $\Delta\Phi - 8$ hr), have shown that in this respect, too, there are marked differences. Species with a wide variance in period and a large range of entrainment resynchronize very quickly; whereas species with small ranges and a τ that changes hardly at all need a considerably greater number of transient cycles of lengthened or shortened period. The first group includes, for example, *Myotis myotis*, *Molossus molossus*, and *M. ater*, which require only 2–3 days to resynchronize with a LD 12:12 ($10{:}10^{-4}$ lx) after 8-hr advance or delay shifts (Bay, 1976; Kracht, 1979). *Rousettus aegyptiacus*, *Eidolon helvum*, and the phyllostomids *Phyllostomus discolor* and *Glossophaga soricina*, by contrast, are not resynchronized with a phase-shifted Zeitgeber until

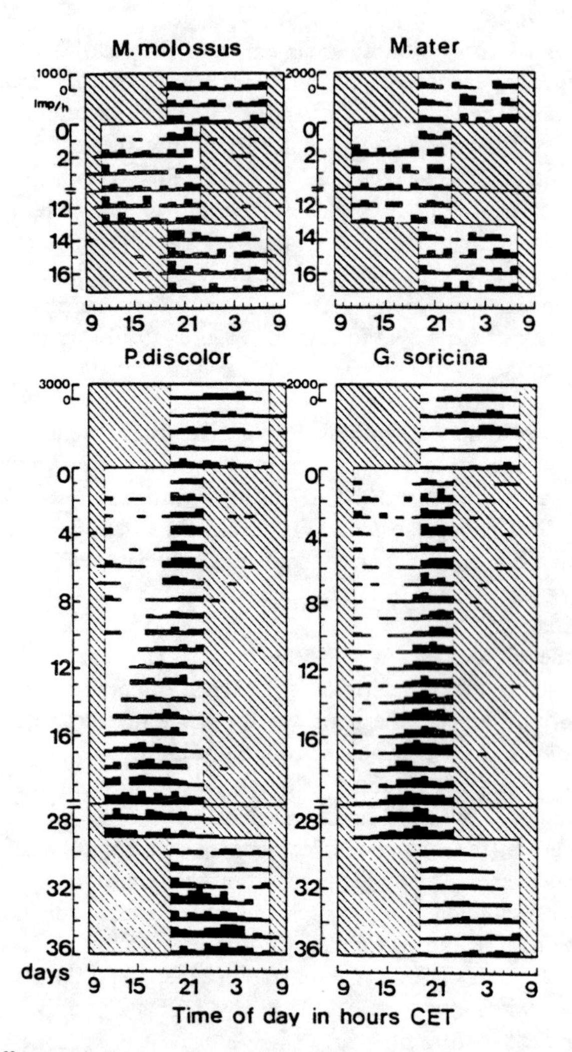

FIGURE 9. Differences in the resynchronization behavior of circadian activity rhythms for two purely insectivorous (top) and two predominantly frugivorous/nectarivorous Neotropical bat species, after phase shifts of the light–dark Zeitgeber (LD 12:12, 10:10^{-4} lx) by $\Delta\Phi = +8$ hr (advance shift) and $\Delta\Phi = -8$ hr (delay shift). (From Kracht, 1979.)

6 to more than 20 days have elapsed, depending on the direction of the phase shift (see Fig. 9; Erkert, 1970; Kracht, 1979).

Bat species with autonomous free-running periods that are predominantly greater than 24 hr as a rule resynchronize more slowly after advance shifts than after delay shifts (e.g., *Rousettus aegyptiacus*, *Phyllostomus discolor*, and *Glossophaga soricina*). This "asymmetry effect" (Aschoff *et al.*, 1975), more appropriately "directional effect" (Erkert, 1981), is just the reverse in species with periods predominantly less than 24 hrs, such as *Eidolon helvum* (Erkert, 1970). The range (amplitude) and the form of the Zeitgeber oscillation are other factors that affect the speed of resynchronization. Resynchronization occurs more slowly the lower the LD intensity ratio, and the farther the ratio of light time to dark time departs from 1:1 (12:12 hr) (Erkert, 1976b, 1982).

6. ECOLOGICAL ADAPTATION OF CIRCADIAN SYSTEMS AND EVOLUTIONARY ASPECTS

The conspicious differences among bat species, in the degree to which the free-running circadian period can be changed, the size of the range of entrainment, and the speed of resynchronization, all indicate that bats have evolved timing mechanisms of varying rigidity. Some circadian systems are very plastic and susceptible to exogenous influences (e.g., *Molossus ater*, *M. molossus*, *Myotis myotis*, and *M. lucifugus*), whereas others are extremely rigid (*Rousettus aegyptiacus*, *Glossophaga soricina*, and *Phyllostomus discolor*). Between these groups are several intermediate forms, represented by *Artibeus jamaicensis* and *Sturnira lilium* (Erkert *et al.*, 1980). It is hardly conceivable that differences of this magnitude can be chance variations that happen to be tolerable in terms of system function and are not eliminated by natural selection. One must ask, then, what adaptive significance such a differentiation might have and how it could have evolved.

In one explanation CR differences are viewed as specific adaptations to different Zeitgeber conditions within an animal's distributional range. Tropical species, which live in fairly constant Zeitgeber conditions throughout the year, could in general have evolved a more rigid circadian system than species living farther from the equator, where the diurnal parameters (day and night length and duration of twilight, or the rate of change of the light intensity between L and D phases) vary greatly in the course of a year. In fact, however, even among purely tropical species some have very plastic internal timing mechanisms, such as *Molossus ater* and *M. molossus*, although others (e.g., *Glossophaga soricina*, *Phyllostomus discolor*, and *Rousettus aegyptiacus*) are relatively rigid, and there are various intermediate forms.

All bat species studied to date and that have particularly plastic CRs are insectivorous; species with very rigid timing systems are either predominantly frugivorous, nectarivorous, or nectarivorous/frugivorous to omnivorous. This distribution at first suggests that flexible circadian periodicity could be a special adaptation to the predatory habit, enabling a predator to better adjust its activity rhythm to the different activity rhythms of its prey species and/or populations as they change during the day and year. But this "predator-adaptation hypothesis" appears problematical in view of the fact that the CRs of rodents can also differ quite widely in their degree of plasticity (Lohmann, 1967; Daan and Pittendrigh, 1976), although the range of variation seems narrower than in bats. Because insectivorous bats, unlike most frugivorous and nectarivorous species, are mostly heterothermic (Chapter 4), a relationship between thermoregulatory behavior and CR plasticity cannot *a priori* be ruled out. But if there were such a relationship, one would expect *Sturnira lilium*, which, according to McNab (1969), is one of the strictly homeothermic species, to exhibit a considerably less variable circadian rhythm than it actually does (Fig. 6; Erkert and Kracht, 1978; Kracht, 1979).

The main advantage of controlling the activity rhythm by an endogenous timing system is generally thought to lie in preadapting the organisms to changes that are to be expected in the environment, and in preparing it for life-preserving responses to stimuli likely to recur periodically (Enright, 1970). Whether or not this is so, the development and phase regulation of very precise circadian rhythms may well be the most economical way, from an energetic point of view, to adapt to diurnal fluctuations in the various biotic and abiotic factors in the environment. From this point of view, one would expect a tendency to evolve an endogenous period close to 24 hr, which requires only very slight phase correction to become synchronized with the 24-hr Zeitgeber period. However, the CR can follow this tendency only when it does not encounter appreciable ecological obstacles, such as wide annual variations in Zeitgeber conditions and/or the necessity of adapting the organism's own activity rhythm to that of changing prey populations. Such obstacles are least likely to confront the frugivorous/nectarivorous to omnivorous tropical species that live in caves, holes in trees, or rock crevices (for example, *Rousettus aegyptiacus*, *Glossophaga soricina*, and *Phyllostomus discolor*). Their range of distribution is near the equator, where environmental and Zeitgeber conditions are quite constant. There they need not adjust their activity rhythm to that of their prey, and as strictly dark-active cave dwellers, they are exposed to relatively low predator pressure. It appears to be of no disadvantage for these bats to have an inflexible circadian system. Indeed, bats that roost in caves during the daytime, where conditions are fairly constant and no information about the progress of the conditions outside can penetrate, would find it particularly advantageous to have a very precise endogenous timing mechanism, with a spontaneous period hardly any different from the external diurnal period. Arousal and the timing of flight and foraging activity would be so exactly prescheduled by such a system that major corrective phase shifts induced by light, when emergence is too early in the evening or return too late in the morning, would become superfluous. Light-sampling flight, during which bats may be endangered by late-flying day-active birds of prey, could thus be minimized.

To date, no solid suggestions have been made on the way particularly rigid circadian timing systems of bats could have evolved. However, if one considers that even protozoa, tissue cultures, and isolated organs of higher metazoan organisms (including mammals) exhibit autonomous, free-running circadian rhythms in metabolic activity, some of which vary considerably in period, it becomes clear that the circadian periodicity of the whole organism, whether in body temperature, O_2 uptake, or activity, can hardly be controlled by one central oscillator, a "central clock" located at a particular place. Rather, the organism itself constitutes a multioscillator system (Bünning, 1978; Wever, 1979). Greater precision and rigidity of the overall endogenous timing system would in this case be brought about not so much by improving the precision of one or more individual oscillators as by improving their reciprocal coupling, thereby improving internal synchronization. Closely coupled elements of a multioscillator system stabilize one another at an intermediate period and thus increase the precision of the overall rhythm. In achieving this situation, both the endocrine system and the central nervous system or certain of their components could play an important role. Given that the endogenous timing mechanisms of bats have evolved different degrees of rigidity on the basis of selective advantages inherent in their environment, the above considerations imply that an underlying phylogenetic mechanism may have led to a progressive improvement in the reciprocal coupling of the many oscillating circadian subsystems. However, neither for the Chiroptera nor for any other group is there as yet experimental evidence bearing on this concept.

7. SUMMARY

In this review of studies on bat-activity rhythms it has become apparent that within the Chiroptera, one of the most successful mammalian orders in terms of the number of species and individuals, this periodicity has undergone many modifications that can be interpreted as phylogenetic adaptations to the ecological requirements of an environment that changes in a 24-hr rhythm. The basic pattern of locomotor activity has been affected, as well as the degree of plasticity the circadian system exhibits in different species, the nature of the direct activity-inhibiting action of light, and the sensitivity of the system to various meteorological factors. All of these together determine the timing and pattern of flight and foraging activity and/or are responsible for characteristic modulations of activity patterns for each species. The selective advantage of such differentiation is twofold: it contributes toward optimization of the energy balance which is particularly important for the chiropterans, and it reduces predator pressure by increasing the avoidance of predators. Another possibility suggested by some authors, that reducing direct interspecific food competition plays a role and that the differences in times of activity onset and peak activity among species in a given habitat are to be interpreted in this light, appears questionable (also see Chapter 7).

It was shown that, as a means of improving overall energy balance, not only does the bimodal activity pattern of insectivorous species become understandable, but also that the development of extremely rigid and precise endogenous timing systems among the predominantly frugivorous and nectarivorous chiropterans can be viewed by such a mechanism as well. Part of the strategy of increasing the chance of avoiding predators by the appropriate timing of activity is to ensure the latest possible emergence and the earliest possible termination of flight and foraging. In the first instance this is achieved by development of a precise circadian rhythm and of its phase setting through synchronization by light as the main Zeitgeber. This mechanism is supplemented by selection for a direct inhibitory action of light on the activity level set by the individual circadian systems, an action graded to meet the "needs" of a given species. One result is that emergence time can be matched as closely as is required to the outdoor light intensity; another is that the activity of bats is inhibited when there is moonlight, times when their chief predators, the visually oriented owls, are maximally active. The degree of inhibition should correspond to the level of selection pressure imposed on a species, and in this way species-specific changes in activity patterns with a lunar periodicity can evolve.

8. REFERENCES

Aschoff, J. 1960. Exogenous and endogenous components in circadian rhythms. Cold Springs Harbor Symp. Quant. Biol., **25**:11–28.

Aschoff, J. 1964. Die Tagesperiodik licht- und dunkelaktiver Tiere. Rev. Suisse Zool., **71**:528–558.

Aschoff, J. 1965. Response curves in circadian periodicity. Pp. 95–111, *in* Circadian clocks. (J. Aschoff, ed.). North-Holland, Amsterdam, 479 pp.

Aschoff, J., and R. Wever. 1962. Beginn und Ende der täglichen Aktivität freilebender Vögel. J. Ornithol., **103**:2–27.

Aschoff, J., K. Klotter, and R. Wever. 1965. Circadian vocabulary. Pp. x–xix, *in* Circadian clocks. (J. Aschoff, ed.). North-Holland, Amsterdam, 479 pp.

Aschoff, J., K. Hoffmann, H. Pohl, and R. Wever, 1975. Reentrainment of circadian rhythms after phase-shifts of the Zeitgeber. Chronobiology, **2**:23–73.

Bateman, G. C., and T. A. Vaughan. 1974. Nightly activities of mormoopid bats. J. Mammal., **55**:45–65.

Bay, F. A. 1976. Untersuchungen zur Lichtsteuerung der Aktivitätsperiodik von Fledermäusen (*Myotis myotis* Borkh. 1797 und *Phyllostomus discolor* A. Wagner 1843). Unpublished dissertation, University of Tübingen, Tübingen, 62 pp.

Bay, F. A. 1978. Light control of the circadian activity rhythm in mouse-eared bats (*Myotis myotis* Borkh. 1797). J. Interdiscipl. Cycle Res., **9**:195–209.

Böhme, W., and G. Natuschke. 1967. Untersuchungen der Jagdflugaktivität freilebender Fledermäuse in Wochenstuben mit Hilfe einer doppelseitigen Lichtschranke und einige Ergebnisse an *Myotis myotis* (Borkhausen, 1797) und *Myotis nattereri* (Kuhl, 1818). Säugetierkd. Mitt., **13:**129–138.

Bradbury, J. W. 1977. Social organization and communication. Pp. 1–72, *in* Biology of bats. Vol. 3. (W. A. Wimsatt, ed.). Academic Press, New York, 651 pp.

Bradbury, J. W., and L. H. Emmons. 1974. Social organization of some Trinidad bats. I. Emballonuridae. Z. Tierpsychol., **36:**137–183.

Bradbury, J. W., and S. L. Vehrencamp. 1976. Social organization and foraging in emballonurid bats. I. Field studies. Behav. Ecol. Sociobiol., **1:**337–381.

Brown, J. H. 1968. Activity patterns of some Neotropical bats. J. Mammal., **49:**754–757.

Bünning, E. 1974. Critical remarks concerning the Q_{10} values in circadian rhythms. Int. J. Chronobiol., **2:**343–346.

Bünning, E. 1977. Die physiologische Uhr—Circadiane Rhythmik und Biochronometrie. Springer-Verlag, Berlin Heidelberg, New York, 176 pp.

Bünning, E. 1978. Evolution der zirkadianen Organisation. Arzneim. Forsch. Drug Res., **28:**1811–1813.

Church, H. F. 1957. The times of emergence of the Pipistrelle. Proc. Zool. Soc. London, **128:**600–602.

Cockrum, E. L., and S. P. Cross. 1964. Time of bat activity over water holes. J. Mammal., **45:**635–636.

Constantine, D. G. 1958. An automatic bat-collecting device. J. Wildl. Manage., **22:**17–22.

Crespo, J. A., J. M. Vanella, B. D. Blood, and J. M. De Carlo. 1961. Observaciones ecologicas del vampiro en el Norte de Cordoba. Cienc. Zool., **6:**131–161.

Crespo, J. A., S. B. Linhart, and G. C. Mitchell. 1972. Foraging behavior of the common vampire bat related to moonlight. J. Mammal., **53:**366–368.

Cross, S. P. 1965. Roosting habits of *Pipistrellus hesperus*. J. Mammal., **46:**270–279.

Daan, S. 1970. Photographic recording of natural activity in hibernating bats. Bijdr. Dierkd., **40:**13–16.

Daan, S. 1973. Activity during natural hibernation in three species of vespertilionid bats. Neth. J. Zool., **23:**1–71.

Daan, S., and J. Aschoff. 1975. Circadian rhythms of locomotor activity in captive birds and mammals: Their variations with season and latitude. Oecologia, **18:**269–316.

Daan, S., and C. S. Pittendrigh. 1976. A functional analysis of circadian pacemakers in nocturnal rodents. II. The variability of phase response curves. J. Comp. Physiol., **106:**253–266.

Daan, S., and S. Slopsema. 1978. Short-term rhythms in foraging behavior of the common vole, *Microtus arvalis*. J. Comp. Physiol., **127:**215–227.

Davis, W. B., and J. R. Dixon. 1976. Activity of bats in a small village clearing near Iquitos, Peru. J. Mammal., **57:**747–749.

DeCoursey, P. J. 1961. Effect of light on the circadian activity rhythm of the flying squirrel, *Glaucomys volans*. Z. Vergl. Physiol., **44:**331–354.

DeCoursey, P. J. 1964. Function of a light response rhythm in hamsters. J. Cell. Comp. Physiol., **63:**189–196.

DeCoursey, P. J., and G. DeCoursey. 1964. Adaptive aspects of activity rhythms in bats. Biol. Bull., **126:**14–27.

Eisentraut, M. 1952. Beobachtungen über Jagdroute und Flugbeginn bei Fledermäusen. Bonn. Zool. Beitr., **3:**211–220.

Engländer, H., and G. Laufens. 1968. Aktivitätsuntersuchungen bei Fransenfledermäusen (*Myotis nattereri*, Kuhl 1818). Experientia, **24:**618–619.

Enright, J. T. 1970. Ecological aspects of endogenous rhythmicity. Ann. Rev. Ecol. Syst., **1**:221–238.

Enright, J. T. 1980. The timing of sleep and wakefulness. Springer-Verlag, Berlin, Heidelberg, New York, 252 pp.

Erkert, H. G. 1969. Die Bedeutung des Lichtsinnes für Aktivität und Raumorientierung der Schleiereule (*Tyto alba guttata* Brehm). Z. Vergl. Physiol., **64**:37–70.

Erkert, H. G. 1974. Der Einfluss des Mondlichtes auf die Aktivitätsperiodik nachtaktiver Saugetiere. Oecologia, **14**:269–287.

Erkert, H. G. 1976a. Lunarperiodic variation of the phase-angle difference in nocturnal animals under natural Zeitgeber-conditions near the equator. Internat. J. Chronobiol., **4**:125–138.

Erkert, H. G. 1976b. Der Einfluss der Schwingungsbreite von Licht–Dunkel-Cyclen auf Phasenlage und Resynchronisation der circadianen Aktivitätsperiodik dunkelaktiver Tiere. J. Interdiscipl. Cycle Res., **7**:71–91.

Erkert, H. G. 1978. Sunset-related timing of flight activity in Neotropical bats. Oecologia, **37**:59–67.

Erkert, H. G. 1982. Effect of the Zeitgeber pattern on the resynchronization behavior of dark-active mammals (*Rousettus aegyptiacus*). Int. J. Chronobiol., (in press).

Erkert, H. G., and S. Kracht. 1978. Evidence for ecological adaptation of circadian systems. Oecologia, **32**:71–78.

Erkert, H. G., and E. Rothmund. 1981. Differences in temperature sensitivity of the circadian systems of homoiothermic and heterothermic Neotropical bats. Comp. Biochem. Physiol., **68A**:383–390.

Erkert, H. G., F. A. Bay, and S. Kracht. 1976. Zeitgeber induced modulation of activity patterns in nocturnal mammals (Chiroptera). Experientia, **32**:560–561.

Erkert, H. G., S. Kracht, and U. Häussler. 1980. Characteristics of circadian activity systems in Neotropical bats. Pp. 95–104, *in* Proceedings of the Fifth International Bat Research Conference. (D. E. Wilson and A. L. Gardner, eds.). Texas Tech Press, Lubbock, 434 pp.

Erkert, S. 1970. Der Einfluss des Lichtes auf die Aktivität von Flughunden (Megachiroptera). Z. Vergl. Physiol., **67**:243–272.

Fenton, M. B. 1970. A technique for monitoring bat activity with results obtained from different environments in southern Ontario. Can. J. Zool., **48**:847–851.

Fenton, M. B., S. L. Jacobson, and R. N. Stone. 1973. An automatic ultrasonic sensing system for monitoring the activity of some bats. Can. J. Zool., **51**:291–299.

Fenton, M. B., N. G. H. Boyle, T. M. Harrison, and D. J. Oxley. 1977. Activity patterns, habitat use, and prey selection by some African insectivorous bats. Biotropica, **9**:73–85.

Fleming, T. H., E. T. Hooper, and D. E. Wilson. 1972. Three Central American bat communities: Structure, reproductive cycles, and movement patterns. Ecology, **53**:555–569.

Frylestam, B. 1970. Studier över langörade fladdermusen (*Plecotus auritus* L.). Fauna Flora, **65**:72–84.

Funakoshi, K., and T. A. Uchida. 1978. Studies on the physiological and ecological adaptation of temperate insectivorous bats. III. Annual activity of the Japanese house-dwelling bat, *Pipistrellus abramus*. J. Fac. Agric. Kyushu Univ., **23**:95–115.

Gaisler, J. 1963. Nocturnal activity in the lesser horseshoe bat, *Rhinolophus hipposideros* (Bechstein, 1800). Zool. Listy., **12**:223–230.

Gaisler, J. 1973. Netting as a possible approach to study bat activity. Period. Biol., **75**:129–134.

Gaur, B. S. 1980. Roosting ecology of the Indian desert rat-tailed bat, *Rhinopoma kinneari* Wroughton. Pp. 125–133, *in* Proceedings of the Fifth International Bat Research Conference. (D. E. Wilson and A. L. Gardner, eds.). Texas Tech Press, Lubbock, 434 pp.

Gould, P. J. 1961. Emergence time of *Tadarida* in relation to light intensity. J. Mammal., **42**:405–407.

Griffin, D. R., and J. H. Welsh. 1937. Activity rhythms in bats under constant external conditions. J. Mammal., **18**:337–342.

Harmata, W. 1960. Ethological and ecological observations made on the bats (Chiroptera) of the Wolski forest near Cracov. Zesz. Nauk. Univ. Jagiellon. Pr. Zool. **33**:163–203.

Häussler, U., and H. G. Erkert. 1978. Different direct effects of light intensity on the entrained activity rhythm in Neotropical bats (Chiroptera, Phyllostomidae). Behav. Proc., **3**:223–239.

Heithaus, E. R., and T. H. Fleming. 1978. Foraging movements of a frugivorous bat, *Carollia perspicillata* (Phyllostomatidae). Ecol. Monogr., **48**:127–143.

Heithaus, E. R., P. A. Opler, and H. G. Baker. 1974. Bat activity and pollination of *Bauhinia pauletia:* Plant–pollinator coevolution. Ecology, **55**:412–419.

Herreid, C. F., II, and R. B. Davis. 1966. Flight patterns of bats. J. Mammal., **47**:78–86.

Huggel, H. 1958. Zum Studium der Biologie von *Eidolon helvum* (Kerr): Aktivität und Lebensrhythmus während eines ganzen Tages. Pp. 141–144, *in* Verh. Schwiez. Naturf. Ges.

Jacobsen, N. H. G., and E. DuPlessis. 1976. Observations on the ecology and biology of the cape fruit bat *Rousettus aegyptiacus leachi* in the Eastern Transvaal. S. Afr. J. Sci., **72**:270–273.

Jimbo, S., and H. O. Schwassmann. 1967. Feeding behavior and the daily emergence pattern of "*Artibeus jamaicensis*" Leach (Chiroptera, Phyllostomidae). Atas. Simp. Biota Amazonica **5**:239–253.

Jones, C. 1965. Ecological distribution and activity periods of bats of the Mogollon mountains area of New Mexico and adjacent Arizona. Tulane Stud. Zool., **12**:93–100.

Kielmann, N., and G. Laufens. 1968. Kennzeichnung mehrerer Individuen durch Kleinstschwingkreise. Experientia, **24**:750–756.

Kolb, A. 1959. Ein Registrierapparat Für Fledermäuse und einige biologische Ergebnisse. Zool. Anz., **163**:134–141.

Kracht, S. 1979. Anpassung circadianer Systeme—Vergleichende Untersuchungen an neotropischen Fledermäusen. Unpublished dissertation, University of Tübingen, Tübingen, 61 pp.

Kunz, T. H. 1973. Resource untilization: Temporal and spatial components of bat activity in central Iowa. J. Mammal., **54**:14–32.

Kunz, T. H. 1974. Feeding ecology of a temperate insectivorous bat (*Myotis velifer*). Ecology, **55**:693–711.

Kunz, T. H. 1980. Energy budgets of free-living bats. Pp. 369–392, *in* Proceedings of the Fifth International Bat Research Conference. (D. E. Wilson and A. L. Gardner, eds.). Texas Tech Press, Lubbock. 434 pp.

Kunz, T. H., and C. E. Brock. 1975. A comparison of mist nets and ultrasonic detectors for monitoring flight activity of bats. J. Mammal., **56**:907–911.

Laufens, G. 1972. Freilanduntersuchungen zur Aktivitätsperiodik dunkelaktiver Säuger. Unpublished dissertation, University of Köln, Köln, 87 pp.

Laufens, G. 1973. Einfluss der Aussentemperaturen auf die Aktivitätsperiodik der Fransen- und Bechsteinfledermäuse (*Myotis nattereri*, Kuhl 1818 und *Myotis bechsteini*, Leisler 1818). Period. Biol., **75**:145–152.

LaVal, R. K. 1970. Banding returns and activity periods of some Costa Rican bats. Southwest. Nat., **15**:1–10.

Lehmann, U. 1976. Short-term and circadian rhythms in the behavior of the vole, *Microtus agrestis* (L.). Oecologia, **23**:185–199.

Lohmann, M. 1967. Ranges of circadian period length. Experientia, **23**:788–790.

McNab, B. K. 1969. The economics of temperature regulation in Neotropical bats. Comp. Biochem. Physiol., **31**:227–268.

Maeda, K. 1974. Eco-éthologie de la grande noctule, *Nyctalus lasiopterus*, a Sapporo, Japan. Mammalia, **38**:461–487.

Marimuthu, G., R. Subbaraj, and M. K. Chandrashekaran. 1978. Social synchronization of the activity rhythm in a cave-dwelling insectivorous bat. Naturwissenschaften, **65**:600.

Menaker, M. 1959. Endogenous rhythms of body temperature in hibernating bats. Nature, **184**:1251–1252.

Menaker, M. 1961. The free-running period of the bat clock; seasonal variations at low body temperature. J. Cell. Comp. Physiol., **57**:81–86.

Mislin, H. 1942. Zur Biologie der Chiroptera. I. Beobachtungen im Sommerquartier der *Myotis myotis* Borkh. Rev. Suisse Zool., **49**:200–206.

Möller, E. 1979. Phasenlage und Phasenresponse der circadianen Aktivitätsperiodik neotropischer Fledermäuse (*Phyllostomus discolor* Wagner 1843, *Molossus ater* Geoffroy 1805, *Molossus molossus* ehem. *Molossus major* Kerr 1792). Unpublished Thesis, University of Tübingen, Tübingen, 68 pp.

Morrison, D. W. 1978a. Foraging ecology and energetics of the frugivorous bat *Artibeus jamaicensis*. Ecology, **59**:716–723.

Morrison, D. W. 1978b. Lunar phobia in a neotropical fruit bat, *Artibeus jamaicensis* (Chiroptera: Phyllostomidae). Anim. Behav., **78**:852–855.

Morrison, D. W. 1980. Foraging and day-roosting dynamics of canopy fruit bats in Panama. J. Mammal., **61**:20–29.

Neuweiler, G. 1962. Bau und Leistung des Flughundauges (*Pteropus g. giganteus* Brünn). Z. Vergl. Physiol., **46**:13–56.

Neuweiler, G. 1969. Verhaltensbeobachtungen an einer indischen Flughundkolonie (*Pteropus g. giganteus* Brünn). Z. Tierpsychol., **26**:166–199.

Neuweiler, G., and F. P. Möhres. 1967. Die Rolle des Ortsgedächtnisses bei der Orientierung der Grossblatt-Fledermaus *Megaderma lyra*. Z. Vergl. Physiol., **57**:147–171.

Nyholm, E. S. 1956. Über den Tagesrhythmus der Nahrungsjagd bei der Bartfledermaus, *Myotis mystacinus* Kuhl, während des Sommers. Arch. Soc. Zool. Bot. Fen. Vanamo, **12**:53–58.

Nyholm, E. S. 1965. Zur Ökologie von *Myotis mystacinus* (Leisl.) und *M. daubentoni* (Leisl.) (Chiroptera). Ann. Zool. Fenn., **2**:77–123.

O'Farrell, M. J., and W. G. Bradley. 1970. Activity patterns of bats over a desert spring. J. Mammal., **51**:18–26.

O'Farrell, M. J., and E. H. Studier. 1975. Population structure and emergence activity patterns in *Myotis thysanodes* and *M. lucifugus* (Chiroptera: Vespertilionidae) in northeastern New Mexico. Am. Midl. Nat., **93**:368–376.

O'Farrell, M. J., W. G. Bradley, and G. W. Jones. 1967. Fall and winter bat activity at a desert spring in southern Nevada. Southwest. Nat., **12**:163–171.

O'Shea, T. J., and T. A. Vaughan. 1977. Nocturnal and seasonal activities of the pallid bat, *Antrozous pallidus*. J. Mammal., **58**:269–284.

Owings, D. H., and R. B. Lockard. 1971. Different nocturnal activity patterns of *Peromyscus californicus* and *Peromyscus eremicus* in lunar lighting. Psychonom. Sci., **22**:63–64.

Pittendrigh, C. S. 1960. Circadian rhythms and the circadian organization of living systems. Cold Spring Harbor Symp. Quant. Biol., **25**:159–184.

Pittendrigh, C. S. 1974. Circadian oscillations in cells and the circadian organization of multicellular systems. Pp. 437–458, *in* The neurosciences. (F. O. Schmitt, and F. G. Worden, eds.). MIT Press, Cambridge, 1107 pp.

Pittendrigh, C. S., and S. Daan. 1976. A functional analysis of circadian pace-makers in nocturnal rodents. IV. Entrainment: Pacemaker as clock. J. Comp. Physiol., **106**:291–331.

Pohl, H. 1961. Temperaturregulation und Tagesperiodik des Stoffwechsels bei Winterschläfern. (Untersuchungen an *Myotis myotis* Borkh., *Glis glis* L und *Mesocricetus auratus* Waterh.). Z. Vergl. Physiol., **45**:109–153.

Prakash, I. 1962. Times of emergence of the Pipistrelle. Mammalia, **26**:133–135.

Ransome, R. D. 1968. The distribution of the greater horse-shoe bat, *Rhinolophus ferrum-equinum*, during hibernation, in relation to environmental factors. J. Zool., **154**:77–112.

Rasweiler, J. J. 1973. Care and management of the long-tongued bat, *Glossophaga soricina* (Chiroptera: Phyllostomatidae), in the laboratory with observations on estivation induced by food deprivation. J. Mammal., **54**:391–404.

Rawson, K. S. 1960. Effects of tissue temperature on mammalian activity rhythms. Cold Spring Harbor Symp. Quant. Biol., **25**:105–113.

Schmidt, U., A. M. Greenhall, and W. Lopez-Forment. 1971. Ökologische Untersuchungen der Vampirfledermäuse (*Desmodus rotundus*) im Staate Puebla, Mexiko. Z. Säugetierk., **36**:360–370.

Start, A. N. 1974. The feeding biology in relation to food sources of nectarivorous bats (Chiroptera: Megaloglossinae) in Malaysia. Unpublished Ph.D. thesis, University of Aberdeen, Aberdeen, 259 pp.

Stebbings, R. E. 1968. Measurements, composition and behaviour of a large colony of the bat *Pipistrellus pipistrellus*. J. Zool., **156**:15–33.

Subbaraj, R., and M. K. Chandrashekaran. 1977. 'Rigid' internal timing in the circadian rhythm of flight activity in a tropical bat. Oecologia, **29**:341–348.

Subbaraj, R., and M. K. Chandrashekaran. 1978. Pulses of darkness shift the phase of a circadian rhythm in an insectivorous bat. J. Comp. Physiol., **127**:239–243.

Swift, S. M. 1980. Activity pattern of Pipistrelle bats (*Pipistrellus pipistrellus*) in north-east Scotland. J. Zool., **190**:285–295.

Tamsitt, J. R., and D. Valdivieso. 1961. Notas sobre actividades nocturnas y estados de reproducción de algunos Quirópteros de Costa Rica. Rev. Biol. Trop., **9**:219–225.

Thomas, S. P. 1980. The physiology and energetics of bat flight. Pp. 393–402, *in* Proceedings of the Fifth International Bat Research Conference. (D. E. Wilson and A. L. Gardner, eds.). Texas Tech Press, Lubbock, 434 pp.

Thomas, S. P., and R. A. Suthers. 1972. The physiology and energetics of bat flight. J. Exp. Biol., **57**:317–335.

Turner, D. C. 1975. The vampire bat. Johns Hopkins University Press, Baltimore and London, 145 pp.

Tuttle, M. D. 1974. An improved trap for bats. J. Mammal., **55**:475–477.

Twente, J. W. 1955. Some aspects of habitat selection and other behavior of cavern-dwelling bats. Ecology, **36**:706–732.

Usman, K., J. Habersetzer, R. Subbaraj, G. Gopalkrishnaswamy, and K. Paramanandam. 1980. Behavior of bats during a lunar eclipse. Behav. Ecol. Sociobiol., **7**:79–81.

van Heerdt, P. E., and J. W. Sluiter. 1965. Notes on the distribution and behaviour of the noctule bat (*Nyctalus noctula*) in the Netherlands. Mammalia, **29**:463–477.

van Nieuwenhoven, P. J. 1956. Ecological observations in a hibernation-quarter of cave-dwelling bats in South-Limburg. Publ. Natuurhist. Gen. Limburg, **9**:1–56.

Venables, L. S. V. 1943. Observations at a pipistrelle bat roost. J. Anim. Ecol., **12**:19–26.

Villa-R., B. 1966. Los murcielagos de Mexico. Univ. Nac. Autó. México City, 491 pp.

Voûte, A. M. 1972. Bijdrage tot de Oecologie van de Meervleermuis (*Myotis dasycneme*, Boie, 1825). Unpublished doctoral dissertation, Universiteit Utrecht, Utrecht, 159 pp.

Voûte, A. M., J. W. Sluiter, and M. P. Grimm. 1974. The influence of the natural light–dark-cycle on the activity rhythm of pond bats (*Myotis dasycneme* Boie, 1825) during summer. Oecologia, **17**:221–243.

Wever, R. 1965. A mathematical model for circadian rhythms. Pp. 112–124, *in* Circadian clocks. (J. Aschoff, ed.). North-Holland, Amsterdam, 479 pp.

Wever, R. 1979. The circadian system of man. Results of experiments under temporal isolation. Springer-Verlag, New York, Heidelberg, Berlin, 276 pp.

Williams, T. C., and J. M. Williams. 1970. Radio tracking of homing and feeding flights of a Neotropical bat, *Phyllostomus hastatus*. Anim. Behav., **18**:302–309.

Wilson, D. E. 1971. Ecology of *Myotis nigricans* (Mammalia: Chiroptera) on Barro Colorado Island, Panama Canal Zone. J. Zool., **163**:1–13.

Wimsatt, W. A. 1969. Transient behavior, nocturnal activity patterns and feeding efficiency of vampire bats (*Desmodus rotundus*) under natural conditions. J. Mammal., **50**:233–244.

Young, A. M. 1971. Foraging of vampire bats (*Desmodus rotundus*) in Atlantic wet lowland Costa Rica. Rev. Biol. Trop., **18**:73–88.

Zack, S., D. Czeschlik, and G. Helmaier. 1979. Daily activity pattern of fruit bats under natural light conditions. Naturwissenschaften, **66**:322–323.

Chapter 6

Ecological Significance of Chiropteran Morphology

James S. Findley

Department of Biology
University of New Mexico
Albuquerque, New Mexico 87131

and

Don E. Wilson

U.S. Fish and Wildlife Service
National Museum of Natural History
Washington, D.C. 20560

1. INTRODUCTION

That form is related to function in animal life is a concept that goes back at least to Aristotle. Relating structure to function has been a central occupation of anatomists, and the subject of functional morphology has a dedicated following. Yet the suggestion that an organism's ecological role can be predicted from its morphology often elicits opposition. Because behavior is so flexible, individual animals may do many things that a study of their anatomy would not predict. Ducks with bills highly specialized for filtering small particles from water and mud may, nevertheless consume small vertebrates, centipedes, lettuce, corn, and dog food with great skill. The objection to making ecological predictions from morphology seems to be that although we can assert that a bat eats fruit because its teeth are constructed in a certain way, because of the observed variability in dietary proclivities of individual fruit bats we cannot make more refined predictions about the size and kind of fruit consumed, the extent to which animals are eaten, the nature of seasonal changes in diet, and so on. Our contention is that because the selective regimen imposed on a species through evolutionary time includes these minor variations, they are probably reflected in a bat's phenotype. By comparing the morphology of members of a feeding guild of bats, it should be possible to predict ways in which the several species differ in food habits, foraging styles, shelter seeking, and other ecological traits.

2. THE TROPHIC NICHE

Certain aspects of morphology, especially overall size and the size of trophic structures, have been used to make predictions about the size of food items captured or ingested. Generally a positive relationship exists, but it is complicated by differences in foraging methods and energetics (Hespenheide, 1973). The idea that ecological divergence of close competitors in areas of sympatry leads to morphological divergence has been formalized in the concept of character displacement (Brown and Wilson, 1956), reviewed by Grant (1972) and Hespenheide (1973). That morphological variation reflects ecological constraints, and hence may be a measure of niche width, was expressed as the niche-variation hypothesis by Van Valen (1965). Although not all studies support this relationship (e.g., Soule and Stewart, 1970), enough have so that a belief in its reality persists. If morphological variability matches ecological variability, and if competition must be minimized to permit coexistence, then morphological relationships between species in stable communities or guilds should be predictable, or at least should follow some pattern, as has been suggested by Ricklefs and O'Rourke (1975), McNab (1971a), and Findley (1976). Recently, very precise relationships between intraspecific morphological variation and foraging and feeding ecology have been shown for *Peromyscus* by Smartt and Lemen (1980).

Ecomorphological studies of bats have focused on a number of attributes, including wing morphology, jaw structure, brain size, general external dimensions, and geographic variation. Here we review some published and unpublished work on ecomorphological correspondences in bats and conclude that, at least in this group of mammals, these correspondences are strong enough to make refined ecological predictions from morphology quite feasible.

2.1. Flight and Wing Morphology

One of the first studies relating wing morphology to the flight characteristics of bats was that of Revilliod (1916). Poole (1936) studied differences in wing loadings and wing areas among various bats. His comparative comments included noting that fast-flying bats such as lasiurines were the only kinds that approached birds in wing loading. Vaughan (1959) made a detailed comparison of the locomotor myology and osteology of *Eumops perotis*, *Myotis velifer*, and *Macrotus californicus* integrated with his extensive personal knowledge of the bats' natural history. His is perhaps the pioneering functional morphological study for bats. Vaughan noted the relationship of the long, flat, narrow, high-aspect-ratio wings of *Eumops* to the rapid, enduring flight of that species, which he believed foraged continuously for about 6½ hr per night over a range of 5–15 miles. He related the slow, maneuverable flight of *Myotis* and *Macrotus* to the short, broad, highly cambered wings of those species, both of which forage for short periods of time per night in a fairly limited area. The flattened, elongate, keelless sternum of *Eumops* provides adequate area for the origin of pectoral muscles in this crevice-roosting species, whereas the shorter, keeled sternum of *Macrotus* provides muscle origin area in a species that roosts in a pendent position. Differences in the structure of the pelvic limbs were clearly related to the sophisticated crawling ability of *Eumops,* and the inability of *Macrotus* to crawl. The functional–morphological trends Vaughan described have greatly assisted subsequent investigators. His studies of wing morphology and adaptations for flight have continued, and more recent reviews can be found in Vaughan (1966, 1970a, 1970b) and Vaughan and Bateman (1970, 1980).

Struhsaker (1961) measured aspect ratios and flight-muscle volumes in 12 species of bats, noting some consistencies in morphological relationships as well as some differences that he related to mode of flight. For example, he proposed that the relatively large *serratus magnus* (*superior*) of *Choeronycteris mexicana* may permit that species to hover, inasmuch as the muscle had been shown by Vaughan (1959) to be related to upstroke initiation. Indeed, our own observations and those of J. S. Altenbach (personal communication) suggest that the terminal portion of the glossophagine upstroke (the "flick phase") is a power-generating phase used in hovering.

Hayward and Davis (1964) estimated flight speeds in 17 species of bats of western North America and found that speed was positively correlated with forearm length, thus establishing a basis for prediction of the speed of unstudied species.

Findley *et al.* (1972) assembled measurements of wing morphology for 136 species of bats representing 15 families. They noted that wing area and wing loading were positively correlated with overall size but that wing length was negatively related to these variables. The aspect ratio index (forearm and wing tip/digit V) and the relative length of the wing tip (comprising the third metacarpal and phalanges; referred to as tip index by these authors) varied independently and were related to flight mode, speed, and habitat. The aspect ratio index was the most useful variable in predicting speed according to the relationship

$$\text{Speed} = 17.1(\text{aspect ratio index}) - 23.2$$

Comparison of wing morphology with known habitat predilections suggested that species with low aspect ratios were forest dwellers, whereas those with high aspect ratios foraged in open areas or were migratory.

Lawlor (1973) analyzed relationships between the weight, wing surface area, linear dimensions, aerodynamic properties, and flight and feeding behavior of 25 species of Neotropical bats. Smith and Starrett (1979) provided extensive descriptions of the morphology of 433 species. This wealth of descriptive material remains to be interpreted functionally.

Mortensen (1977) conducted a detailed ecomorphologic study of the phyllostomid subfamily Phyllostominae. He selected traits that he believed had ecological significance and grouped them into four categories: flight, food comminution, jaw mechanics, and sound production and reception. Using multivariate techniques, he grouped the 23 species of bats on the basis of these traits, made ecological predictions from these groupings, and then attempted to test the predictions using original and published observations. In a factor analysis of the 20 characters chosen to estimate aerodynamic properties, the first three factors accounted for 77.0% of the variance. The first factor (33.6%) represented the relative area of flight membranes, the second (25.7%) indexed low-speed maneuverability and lift capacity, and the third (17.7%) indicated wing loading and uropatagial size and control. The placement of these bats on the three factors produced a model that was used to predict flight behavior. Observations on flight and foraging were available for 14 of the 23 species. For each species the data available fit the predictions. Moreover, bats that were similar in wing morphology were also similar in observed flight behavior, whereas morphologically dissimilar species were dissimilar in flight. Mortensen's analysis, although highly suggestive, does lack precise predictive utility because of the nonquantitative nature of the data on flight characteristics.

More recently, M. Schmidt, J. S. Altenbach, and J. S. Findley (unpublished data) attempted to quantify some aspects of flight behavior and to then compare two closely related bats, *Myotis thysanodes* and *M. yumanensis,* that differ in certain aspects of wing structure. Both bats have about the same body size (52–53 mm head and body length), but *M. thysanodes* has longer wings (wing span about 285 mm as opposed to 225 mm) and a greater flight-membrane area (about 139 cm^2 as opposed to 88 cm^2). They predicted that *M. thysanodes* would have more maneuverable flight. The bats were flown in an enclosure in which the walls were marked off in grids. By studying flim strips of the flying bats, they identified the precise point that the bat occupied in each frame, the bat's direction, and the length of each wing-beat cycle. The predictions from morphology were verified. *Myotis yumanensis* flew continuously back and forth in a narrow ellipse at about the same vertical elevation, whereas *M. thysanodes* flew adroitly throughout the chamber. A Shannon–Wiener measure (H') of the diversity of cube use and flight direction in the chamber was 0.99 for *M. thysanodes* and 0.45 for *M. yumanensis*. Wing-beat cycles were longer in *M. thysanodes,* averaging 18 frames for completion (169 wing-beat cycles measured), as opposed to 14 in *M. yumanensis* (218 cycles measured). Clearly the opportunity exists for a precise quantitative study of the relationship between flight performance and morphology in bats.

2.2. Jaw Morphology and Diet

The premise that the sizes of prey and predator are positively related is widely assumed by ecologists, and many studies support it. Hutchinson (1959) presented evidence suggesting that species of birds with similar feeding habits had to differ in bill length by a factor of 1.2–1.4 in order to coexist. Tamsitt (1967) noted that sympatric congeners among Neotropical phyllostomids differed by ratios ranging from 1.07 to 1.32 in the mean length of the mandible. Tamsitt studied frugivorous, nectarivorous, and carnivorous bats, and hence implicitly assumed that prey size, both plant and animal, is directly related to the size of the trophic structures of the consumer.

Freeman (1979) conducted a detailed study of tooth and jaw structure in 80 species of bats of the family Molossidae. She claimed that bats specializing on "hard-shelled" prey such as beetles can be distinguished from bats consuming chiefly soft-bodied insects such as moths. Presumed beetle eaters are characterized by thick dentaries, well-developed cranial crests, and few but large teeth, whereas moth eaters have thin jaws, small cranial crests, and more but smaller teeth. The available information on feeding for the presumed beetle feeders (the *Molossus*-like group) and moth feeders (the *Nyctinomops*-like group) supported her predictions on the diets of these bats. Similar trends in jaw and tooth morphology are seen in other families of bats, notably, the Vespertilionidae, and suggest that dietary predictions can be made for insectivorous bats in general. Indeed, a multivariate comparison of seven bat families (Freeman, 1981) confirms the relationships obtained earlier for the molossids.

Mortensen (1977) analyzed 14 dental and 16 jaw structure traits. Three factors based on the dental traits accounted for 54% of the morphological variance in dental traits. These three factors arrayed the species of phyllostomines along axes that Mortensen interpreted as indicating the following: (1) the texture (brittle or soft) of the food being pierced, cut, and comminuted; (2) the slicing of tough food versus the grinding of softer items; and (3) incisive picking, tearing, and scooping of softer tissue versus reduced incisor use coupled with the ability to handle more brittle tissue. Three factors based on jaw mechanics accounted for 48% of the morphological variance in jaw structure. These factors seemed to express jaw strength, gape, and relative strength at various positions of jaw closure, and lateral versus vertical cutting or grinding. To test the relationship of morphological placement to predicted diet Mortensen assembled some original and published data for all 23 species studied. Of the 14 species with sufficient data to test, nine agreed strongly and five moderately well with predictions. Mortensen's statement regarding *Trachops cirrhosus* seems especially prophetic in view of M. D. Tuttle's (personal communication) findings concerning the extent to which this species preys upon frogs: "Very strong jaws, wide gape, and greatest jaw strength at wide gape suggest that this bat does kill large or strong prey and carries it some distance before eating it."

2.3. Brain Size

Descriptive morphological studies of bat brains revealed substantial variation both in the size of the entire brain and in the sizes of the component parts (see Henson, 1970a, and McDaniel, 1976, for reviews). Impetus for the study of the brain and comparison of its parts was provided by the discovery that most bats had highly developed echolocation systems that presumably developed at the expense of other sensory modalities (Henson, 1970b; Suthers, 1970; Novick, 1977). As the nature of the variation became evident, interest was focused on ellucidating functional or ecological correlates by comparing species from diverse phylogenetic and ecological groups.

Mann (1963) observed that relative brain size may decrease with increasing flight speed but concluded that brain size *per se* was not obviously correlated with the habits of bats. Findley (1969) measured brain volume in bats representing 30 genera, six families, and both suborders, and found that brain size in bats increases more rapidly with a measure of fat-free body size (foramen magnum area; see Radinsky, 1967) than it does in monkeys, prosimians, rodents, or insectivores.

Pirlot and Stephan (1970) documented variation in relative brain size in nine families of bats. They presented an average encephalization index for each family, based on a sample of 51 species. Pteropodidae, Desmodontinae, and Noctilionidae had high indices of encephalization. These results were extended by Stephan and Pirlot (1970) to a comparison of 11 brain structures in 18 species from eight families. They compared the bats to a group of primitive insectivores including *Suncus*, *Sorex*, *Tenrec*, and *Hemicentetes*. In spite of a great deal of variation, they were able to demonstrate a correlation between brain size and feeding habits. The degree of encephalization was correlated with neocortical development, which in turn is related to dietary specialization. Again, noctilionids, desmodontines, and pteropodids had highly developed neocortices, but so did several phyllostomids. Pirlot and Pottier (1977) provided additional information on encephalization and the comparison of brain components in relation to natural history.

Eisenberg and Wilson (1978) studied 225 species of bats and expressed relative brain size by regressing $\log_{10}$ cranial volume against $\log_{10}$ body weight. They assigned bats to the trophic categories of Wilson (1973) and found that frugivores have a higher y intercept ($b = -0.05$) than aerial insectivores ($b = -0.31$), and consequently relatively larger brains. The largely insectivorous Vespertilionidae show considerable variation in relative brain size, but much of this scatter is attributed to the genus *Myotis*. For example, *Myotis auriculus* and *M. thysanodes*, two large-eared, maneuverable, probably gleaning species, have relatively larger brains than their congeners. Findley (1972) has shown that species of *Myotis* scoring high on a factor expressing ear size, membrane area, and rostral length are likely to be hovering gleaners. Eisenberg and Wilson regressed relative brain size in *Myotis* against Findley's factor I and obtained a significant correlation of 0.87. Eisenberg and Wilson concluded that large brains are related to "a foraging strategy based on locating relatively large pockets of energy-rich food that are unpredictable in temporal and spatial distribution."

Baron and Jolicoeur (1980) applied multiple-discriminant analysis to volumetric data on five areas of the brain. They showed all areas, except for the olfactory region, to be larger in bats than in primitive insectivores. Megachiropterans were shown to be mainly reliant on vision and olfaction, but microchiropterans had the expected largest development of the auditory region. The phyllostomids had well-developed olfactory and cerebellar regions. General vestibular development was explained as an adaptation to flight. Plant feeders rely on vision and olfaction, but animalivorous species specialize in audition and echolocation. They suggested that the well-developed cerebellum of phyllostomids may be an adaptation for climbing.

Lemen (1980) has shown that in mice of the genus *Peromyscus* there is a positive correlation between relative brain size and tail length. Tail length in *Peromyscus* is an index of climbing ability and indirectly a measure of habitat complexity.

The hypothesis that brain size is related to habitat complexity is intriguing. The study of Schmidt, Altenbach, and Findley (unpublished) mentioned earlier in the discussion of wing structure tested the hypothesis that closely related bats with different brain sizes differed also in flight complexity. *Myotis yumanensis*, which has a much less diverse flight behavior in the laboratory than *M. thysanodes*, also has a relatively smaller brain. Brain volume $\times 10^3$ /(head and body length)^3 = 0.0080 for the former, versus 0.0117 for the latter. Qualitative observations on the flight of *M. evotis* (relative brain size = 0.0119) and *M. volans* (relative brain size = 0.068) indicate that the flight of these two differs as predicted, with *M. evotis* exhibiting a more complex flight behavior. The possibility of indexing habitat complexity and trophic behavior by relative brain size is a valid one.

2.4. General Morphology and Feeding

Findley and Black (1979) studied general morphology and diet in a Zambian insectivorous bat community. Black collected samples throughout the year for nine species of insectivorous bats near Lusaka, Zambia. Species involved were *Cloeotis percivali, Hipposideros caffer, Rhinolophus simulator, R. swinnyi, R. blasii, Nycteris thebaica, N. woodi, N. macrotis*, and *Miniopteris schreibersii*.

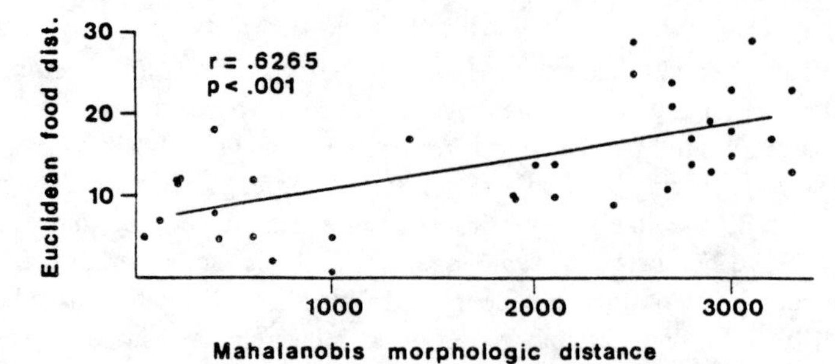

FIGURE 1. The relationship between morphological differences and differences in diet in a Zambian insectivorous bat community. Each dot represents a comparison between two of the nine species in the community. The total number of comparisons is 36. The more different the pair in morphology, the more different they are in diet.

The bats were measured for forearm length, length of third digit and third metacarpal combined (wing tip), length of fourth digit and fourth metacarpal, length of fifth digit and fifth metacarpal, tibial length, hind foot length, tail length, calcar length, uropatagial length, head and body length, and ear length. Each species of bat was scored for 19 food attributes, based on the stomach contents of 332 specimens analyzed by Whitaker and Black (1976). Findley and Black (1979) found that (1) the degree of morphological resemblance between species is highly correlated with the degree of resemblance of their diets (Fig. 1), (2) the degree of morphological variability is predictive of the degree of dietary variability (Fig. 2), (3) the variability of diet increases with the mean distance of a given species of bat from others in the community based on the multivariate analysis of shape, (4) the variability of diet increases with the size of the species

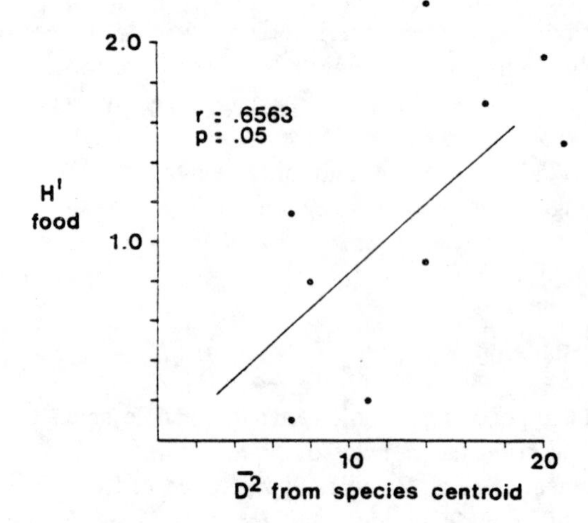

FIGURE 2. The relationship between morphological and deitary variability in a nine-species community of Zambian insectivorous bats. Mean Mahalanobis distance of individuals from their species-morphologic centroids (D^2) are plotted against the diversity of food intake for that species (H' food). Morphologically variable species (high D^2) have more variable diets (high H' food).

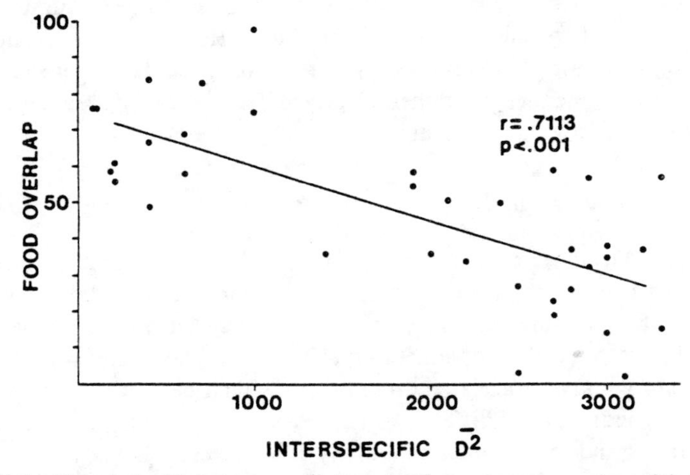

FIGURE 3. The relationship between morphological difference between species and the degree of dietary overlap among species. The interspecific morphologic Mahalanobis distance (D^2) is plotted against the dietary overlap for all comparisons in a nine-species Zambian insectivorous bat community. Bats more dissimilar in morphology overlap less in diet.

of bat involved, (5) the variability of diet increases with distance from other species in the community based on the food intake of the species, (6) the morphology of the bat is predictive of the major functional and taxonomic categories of its prey, and (7) dietary overlap decreases as morphologic resemblance decreases (Fig. 3). So striking are these relationships that the temptation is strong to attempt analyses of the ecological structure of bat communities by using morphological traits as indices of ecological relationships.

3. MORPHOLOGY AND COMMUNITY STRUCTURE

Competition theory asserts that sympatric species differ enough in ecological requirements so that fitness depression caused by competitive interactions cannot drive any of the coexistent kinds to local extinction. This requirement for ecological isolation (Lack, 1971) is dogma to many naturalists. The limiting extent of similarity was explored on a theoretical basis by MacArthur and Levins (1967), and the notion that for coexisting guild members the size ratio of trophic parts must be in the range 1.2–1.4 was proposed by Hutchinson (1959) on empirical grounds and was further explored by Schoener (1965) for birds and by Tamsitt (1967) for bats. If there is a limiting similarity in size and other features, then those communities that have endured long enough for competitive interactions to affect ecological relationships should show regularities in ecological structure. But is this indeed true? If so, do temperate and tropical communities differ in ways predicted by community theory? Answers to these questions are difficult to obtain if we must rely on ecological data. However, if the assumption that morphology closely reflects ecology is a safe one, and the evidence for bats suggests that it is, then we can rather quickly and easily learn a great deal about the ecological structure of communities.

3.1. Species Packing in Temperate versus Tropical-Bat Communities

In this conceptual context Fenton (1972) arrayed 242 species of aerial insectivorous bats from Canada, Guyana, Cameroon, and the Philippines in a bivariate scattergram, the coordinates of which were the ratios ear length/forearm length and metacarpal III length/metacarpal V length. The former was an index of echolocatory characteristics, and the latter of flight behavior; the two ratios together were indicative of differences in prey selection. Fenton concluded that emballonurines and mormoopids essentially replace vespertilionids in the Neotropics, that some phyllostomines are Neotropical equivalents of rhinolophids, nycterids, and megadermatids, and that diclidurines parallel *Taphozous*.

Findley (1973) computed taxonomic distances among 114 kinds of *Myotis*, based on 48 morphological traits, and used the intertaxon distances to assess morphological and ecological diversity in myotine faunas from Europe, Africa, China, New Mexico, Panama, and the eastern United States. He used mean nearest-neighbor distance for each fauna to show that New World faunas were less diverse (nearest-neighbor distances 0.83–0.93) than those from the Old World (distances ranging from 1.12 to 1.28). This study generated several interesting questions: Is the ecological diversification of New World guilds circumscribed by competing species? Are the interstices between Old World species occupied by other kinds of bats? Are all New World bat faunas less diverse than Old World ones? Or is the lack of divergence of New World *Myotis* faunas a function of recency of arrival? The answer to the last question is tentatively positive. New World *Myotis* may bear the same relationship to European *Myotis* as exists between American and European tits of the genus *Parus*. Lack (1969) noted that New World species of these birds were more similar in morphology and ecology than were European ones and in general displayed less geographical overlap. A tentative explanation was recency of arrival of the genus in North America.

Pursuing an understanding of bat community structure on a more general level, Findley (1976) conducted distance analyses of 220 species of insectivorous/carnivorous bats from New Mexico, Europe, West Africa, west Malaysia, Venezuela, and two sites in Costa Rica, based on 20 morphological traits.

Mean nearest-neighbor distances for these faunas shows that there is geographical variation in diversity, but that New World faunas may be just as diverse as Old World ones. The New Mexican fauna, however, is much less diverse than the similarly sized European one (0.362 versus 0.566). The interstices in the European *Myotis* fauna are not filled with other kinds of bats. If they were, the total European fauna would be more packed. There is no indication that tropical faunas are more packed than temperate ones, but they do occupy more ecomorphological space.

All of the faunas studied are alike in consisting of a majority of species that resemble each other strongly and a few which are distant. Those species occupying the central, densely packed part of the community are assumed to have circumscribed niche dimensions, whereas peripheral, more remote kinds should have broader niches. If the niche-variation model of Van Valen (1965) is operable here, then central taxa should be least variable and peripheral ones most variable. Indeed, Findley (1976) showed that nearest-neighbor distance, mean distance to faunal associates, and distance from the faunal centroid are all highly correlated with the summed coefficients of variation of the 20 variables. In a multiple-regression analysis nearest-neighbor distance alone accounts for 49% of the variance in the summed coefficients of variation. Findley concluded that in the communities he studied, most species were closely packed, occupied narrow niches, and were rather invariable, whereas a few experienced competitive release in broader niches and were more variable. Morphological, and presumably ecological, overlap (α) was strongly correlated with nearest-neighbor distance: packed species overlapped their neighbors greatly (modal α = 0.4–0.5, where maximum possible is 1.0), whereas distant ones overlapped little or none.

3.2. Results from Principal-Components Analyses

Michael Schum (manuscript) used 18 of the variables measured by Findley (1976) to study 103 species of North American bats arrayed in 31 communities ranging in size from three to 50 species, and extending geographically from Canada to Panama. He tested the null hypothesis that tropical and temperate communities do not differ from one another in species packing. He computed intertaxon distances as had Findley (1973, 1976) and used nearest-neighbor distances to compare communities. Recognizing, however, the disadvantages of using many highly correlated characters, Schum conducted a principal-component analysis of the data and computed distances based on three and on six of the principal components as well as on the original 18 variables. Schum dealt also with two other shortcomings in the methods of Findley. Because the distribution of nearest-neighbor values in most communities is distinctly nonnormal, being leptokurtic and right skewed, the arithmetic mean is an inappropriate measure of central tendency. Schum therefore made comparisons using the median also. Moreover, since the configuration of many tropical communities is strongly biased by the presence of outliers, that is, species that are aberrant morphologically, Schum used the methods of Sarhan and Greenberg (1962) to define and remove outliers and then recomputed the community distance matrices. Finding that the removal of outliers removed all departures from normality, he then compared communities, also using mean nearest-neighbor distance in 6-principal-component space (PCS).

The results were a striking demonstration of the effect of methodology. In an analysis using 18 variables or 6-PCS and mean nearest-neighbor values there was no obvious relationship between community size and packing, and the null hypothesis is supported. In a similar analysis using 3-PCS with mean-distance values, larger communities appeared to be more densely packed, an effect that was intensified when 3-PCS and median values were used. 6-PCS representations based upon communities from which outliers had been removed showed an increase in packing with community size in the temperate samples, but not in the tropical ones. Schum tentatively concluded that packing increases in temperate communities as species are added but believed that the evidence did not support this relationship in the tropics. Tropical communities did appear to be more packed than temperate ones.

4. SEXUAL DIMORPHISM

Many bats show sexual dimorphism, usually in size. In many epomophorine pteropodids and in some phyllostomids males are larger than females. An extreme of sexual dimorphism among bats is seen in *Hypsignathus monstrosus*, in which males are nearly twice as large as females and have a swollen muzzle with flaring lip flaps and a hypertrophied larynx that fills more than half the body cavity. Male *Hypsignathus* form leks from which they call. Females are attracted to the leks and seemingly select a male with which to mate. In one lek studied by Bradbury (1977) 6% of the calling males accounted for 79% of the copulations. Thus sexual selection seems to be strongly implicated in causing sexual dimorphism in this species.

For many kinds of bats, especially vespertilionids and emballonurids, females are larger than males. This is true in the western pipistrelle, *Pipistrellus hesperus*, studied by Findley and Traut (1970), in which males are up to 6% smaller than females, depending upon the feature measured. These authors noted that spring-taken female pipistrelles failed to go into hypothermia at 4°C in the laboratory, whereas males did. Reasoning that pregnant females might resist hypothermia to avoid interruptions of embryonic development, and that such resistance would be easier for larger bats, Findley and Traut concluded that size dimorphism in this species was a function of differing metabolic demands upon males and females.

Ralls (1976) reviewed data on mammals in which females are larger than males (as in many bats). She noted that the phenomenon seems not to be related to a large degree of male parental investment, polyandry, greater aggressiveness in females than males, greater development of female weapons, female dominance, or matriarchy and concluded that size dimorphism is not the result of sexual selection acting on females. She advanced the "big mother hypothesis," that big mothers are better mothers, and also believed that there may be unusually strong competition for some resource among females of those species in which females are larger than males.

Myers (1978) reviewed sexual dimorphism for 28 taxa of vespertilionid bats. He found a significant positive correlation between size dimorphism as expressed by the ratio of female to male forearm length and the number of young in typical litters. Those taxa producing two or three young displayed greater sexual size dimorphism. Myers concluded that this resulted from the aerodynamic problems resulting from flying with an increased fetal load. He also noted that wing area was greater in females of several species, even after adjustments were made for size differences.

Williams and Findley (1979) examined correlates of sexual size dimorphism in 18 species of vespertilionids. They noted longer forearms in females of six species, after absolute size differences had been eliminated, and longer forearms in males of one species. Failing to substantiate the correlation between dimorphism and the number of young that Myers (1978) had shown, these authors reemphasized the probable importance of the differential thermoregulation hypothesis advanced by Findley and Traut (1970), while noting that female vespertilionids may yet prove to have relatively greater wing areas than those of males when more data are analyzed.

5. GEOGRAPHIC VARIATION

That much geographic variation in mammals, as well as in other organisms, is adaptive has not been seriously questioned. The view that thermoregulatory problems account for the familiar relationship between size and latitude, the Bergmann effect, has been called into question by McNab (1971b), as indeed has the reality of the relationship. A few studies of bats have revealed ecomorphologic trends that have been interpreted in various ways.

Findley and Traut (1970) studied geographic variation in over 2500 specimens of the western pipistrelle, *Pipistrellus hesperus*, from throughout the range of the species. They found a strong positive correlation between color intensity and the precipitation/evaporation ratio of the site from which their specimens came. Thus regions with greater precipitation and/or less evaporation produced darker bats. This was attributed to the selective advantage of protective coloration in regions of pale or dark soils. Bogan (1975) similarly showed a close correspondence between the color of *Myotis californicus* and precipitation. Why a flying nocturnal mammal would benefit from protective coloration is not clear unless the protective advantage is in force chiefly in the roost. Since western pipistrelles and California *Myotis* shelter in well-concealed situations, the question of the advantage of protective coloration seems not to have been satisfactorily dealt with.

Western pipistrelles do exhibit considerable geographic variation in size (Findley and Traut, 1970). Populations west of the continental divide are composed of bats smaller than those east of the divide. These authors attributed the large size of the eastern bats to adaptation to the rigorous continental conditions of a postulated Chihauhuan refugium during Wisconsin time. Western populations show a close correspondence in size to a climatic severity index compounded of latitude and altitude, larger bats occurring in colder climates. This was attributed to the Bergmann effect. That male size showed a closer correspondence to climatic severity than that of females seemed related to the fact that males were more active in colder months than were females. Within the eastern area little relationship existed between size and climatic severity; Findley and Traut postulated failure to adjust to the climatic gradient in the east as an explanation, reasoning that much of the region had been occupied only in postglacial times by pipistrelles. In view of the fact that house sparrows (*Passer domesticus*) have developed ecomorphological clines in North America since their introduction in the 19th century (Johnston and Selander, 1964), it would seem that the 10,000 years since the retreat of Wisconsin glaciation should have been ample for ecomorphological changes in the bats.

Stebbings (1973) detected size clines in the bat *Pipistrellus pipistrellus* in Great Britain that he showed to be related to climatic trends, and invoked the Bergmann effect to explain his observations. Bogan (1975) found no size–climate relationship in *Myotis californicus*. It appears that the evidence for a size–climate relationship in bats is not especially convincing.

An enigmatic relationship between skull size and species diversity was revealed by Findley and Jones (1967) in their taxonomic review of the species *M. fortidens* and *M. lucifugus*. In these bats a very strong positive correlation (0.97–0.99) exists between skull size and the number of other species of *Myotis* existing at the collection site, over a range of 0–5 coexisting species. This effect was attributed rather vaguely to competition. That some biotic effect is at work in this situation seems a reasonable possibility.

6. SUMMARY

The critical assumption in reasoning from morphology to ecology is not simply that form confers ability on its possessor, but that slight differences in form connote slight differences in function such that predictive models of ecological behavior may be formulated. To the extent that behavior is flexible and adaptable, this assumption would seem to be a dangerous one. Nonetheless, evidence from some of the studies cited in this review suggests that, at least for bats, fairly precise prediction may be possible. Certainly individual bats of the same species must vary in behavior as a result of learning, genetic predispositions, and the like. Perhaps, however, since bats develop rapidly and must become independent at an early age, the role of experience and learning in structuring behavior is minimized, and the dominance of inherited factors, likely to be detected and measured by the morphologist, is overriding. To test this hypothesis it should be possible to study the morphological–ecological correspondence in a series of mammals along a spectrum ranging from those in which play and parental involvement are important (e.g., carnivores) to those, such as bats, that quickly become independent of parental influence.

In attempting to conduct ecomorphological studies, it is clear that the greatest possible array of characters should be utilized. Although one certainly should select characters suspected of having ecological valence, other characters should be used as well, since one cannot necessarily understand ecomorphological relationships *a priori*. Albeit a variety of mathematical techniques have been used in handling morphological and ecological data, the concept of the multidimensional niche (Hutchinson, 1957) suggests that multivariate techniques may be most useful in ecomorphological studies. Because many traits likely to be measured by the biologist are correlated at levels that are unknown *a priori*, the technique of principal-components analysis, which extracts independent axes of variation from suites of variables, is a useful one for summarizing and simplifying large amounts of data. Distances based upon placement in principal-component morphological space indicate differences in the magnitude of variables and seem clearly indicative of ecological differences. Correlations within and between taxa are expressive of shape resemblances and differences but have not yet been used extensively in ecomorphological studies.

If, as Findley (1976) suggested, the organism is a cast of the environmental mold, then most morphological traits used by systematists are ecological indices, and taxonomic classifications are basically classifications of the ecological behavior of organisms. The question should then be asked, are descriptions of phylogenetic relationships, to the extent that we can really obtain them, anything more or different?

The structures of the wings, ears, brains, jaws, and teeth of bats is indicative of some unknown, but presumably high, level of precision of their ecological functioning. The possibility exists, through the use of multivariate morphological studies, of quantifying the Hutchinsonian niches of bats and of analyzing and comparing species and their communities.

ACKNOWLEDGMENTS. We thank J. Scott Altenbach, Hal Black, Ken Mortensen, Mary Schmidt, and Michael Schum for their contributions to the previously unpublished studies discussed in this chapter, as well as Michael Bogan and Alfred Gardner who critically reviewed the manuscript.

7. REFERENCES

Baron, G., and P. Jolicoeur. 1980. Brain structure in Chiroptera: Some multivariate trends. Evolution, **34**:386–393.

Bogan, M. A. 1975. Geographic variation in *Myotis californicus* in the southwestern United States and Mexico. U.S. Fish and Wildl. Serv. Res. Rep., **3**:1–31.

Bradbury, J. W. 1977. Lek mating behavior in the hammer-headed bat. Z. Tierpsychol., **45**:225–255.

Brown, W. L., Jr., and E. O. Wilson. 1956. Character displacement. Syst. Zool., **5**:49–64.

Eisenberg, J. F., and D. E. Wilson. 1978. Relative brain size and feeding strategies in the Chiroptera. Evolution, **32**:740–751.

Fenton, M. B. 1972. The structure of aerial-feeding bat faunas as indicated by ears and wing elements. Can. J. Zool., **50**:287–296.

Findley, J. S. 1969. Brain size in bats. J. Mammal., **50**:340–344.

Findley, J. S. 1972. Phenetic relationships among bats of the genus *Myotis*. Syst. Zool., **21**:31–52.

Findley, J. S. 1973. Phenetic packing as a measure of faunal diversity. Am. Nat., **107**:580–584.

Findley, J. S. 1976. The structure of bat communities. Am. Nat., **110**:129–139.

Findley, J. S., and H. L. Black. 1979. Ecological and morphological aspects of community structure in bats. Am. Zool., **19**:989.

Findley, J. S., and C. J. Jones. 1967. Taxonomic relationships of bats of the species *Myotis fortidens*, *M. lucifugus*, and *M. occultus*. J. Mammal., **48**:429–444.

Findley, J. S., and G. L. Traut. 1970. Geographic variation in *Pipistrellus hesperus*. J. Mammal., **51**:741–765.

Findley, J. S., E. H. Studier, and D. E. Wilson. 1972. Morphologic properties of bat wings. J. Mammal., **53**:429–444.

Freeman, P. W. 1979. Specialized insectivory: Beetle-eating and moth-eating molossid bats. J. Mammal., **60**:467–479.

Freeman, P. W. 1981. Correspondence of food habits and morphology in insectivorous bats. J. Mammal., **62**:166–173.

Grant, P. R. 1972. Convergent and divergent character displacement. Biol. J. Linn. Soc., **4**:39–68.

Hayward, B. J., and R. P. Davis. 1964. Flight speed in western bats. J. Mammal., **45**:236–242.

Henson, O. W., Jr. 1970a. The central nervous system. Pp. 58–152, *in* Biology of bats. Vol. 1. (W. A. Wimsatt, ed.). Academic Press, New York, 477 pp.

Henson, O. W., Jr. 1970b. The ear and audition. Pp. 181–264, *in* Biology of bats. Vol. 1. (W. A. Wimsatt, ed.). Academic Press, New York, 477 pp.

Hespenheide, H. A. 1973. Ecological inferences from morphological data. Ann. Rev. Ecol. Syst., **4**:213–229.

Hutchinson, G. E. 1957. Concluding remarks. Cold Spring Harbor Symp. Quant. Biol., **22**:415–427.

Hutchinson, G. E. 1959. Homage to Santa Rosalia or why are there so many kinds of animals? Am. Nat., **93**:145–159.

Johnston, R. F., and R. K. Selander. 1964. House sparrows: Rapid evolution of races in North America. Science, **144**:548–550.

Lack, D. 1969. Tit niches in two worlds, or homage to G. E. Hutchinson. Am. Nat., **103**:43–49.

Lack, D. 1971. Ecological isolation in birds. Harvard University Press, Cambridge, 404 pp.

Lawlor, T. E. 1973. Aerodynamic characteristics of some Neotropical bats. J. Mammal., **54**:71–78.

Lemen, C. 1980. Relationship between relative brain size and climbing ability in *Peromyscus*. J. Mammal., **61**:360–364.

MacArthur, R. H., and R. Levins. 1967. The limiting similarity, convergence, and divergence of coexisting species. Am. Nat., **101**:377–385.

McDaniel, U. R. 1976. Brain anatomy. Pp. 147–200, *in* Biology of bats of the New World family Phyllostomatidae. Part I, (R. J. Baker, J. K. Jones, Jr., and D. C. Carter, eds.). Spec. Publ. Mus. Texas Tech Univ., **10**:1–218.

McNab, B. K. 1971a. The structure of tropical bat faunas. Ecology, **52**:353–358.

McNab, B. K. 1971b. On the ecological significance of Bergmann's rule. Ecology, **52**:845–854.

Mann, G. 1963. Phylogeny and cortical evolution in the Chiroptera. Evolution, **17**:589–591.

Mortensen, B. K. 1977. Multivariate analyses of morphology and foraging strategies of phyllostomatine bats. Unpublished Ph.D. dissertation, University of New Mexico, Albuquerque, 188 pp.

Myers, P. 1978. Sexual dimorphism in size of vespertilionid bats. Am. Nat., **112**:701–711.

Novick, A. 1977. Acoustic orientation. Pp. 73–289, *in* Biology of bats, Vol. 3. (W. A. Wimsatt, ed.). Academic Press, New York, 651 pp.

Pirlot, P., and J. Pottier. 1977. Encephalization and quantitative brain composition in bats in relation to their life-habitats. Rev. Can. Biol., **36**:321–336.

Pirlot, P., and H. Stephan. 1970. Encephalization in Chiroptera. Can. J. Zool., **48**:433–444.

Poole, E. L. 1936. Relative wing ratios of bats and birds. J. Mammal., **17**:412–413.

Radinsky, L. 1967. Relative brain size: A new measure. Science, **155**:836–838.

Ralls, K. 1976. Mammals in which females are larger than males. Q. Rev. Biol., **51**:245–276.

Revilliod, P. 1916. A propos de l'adaptation au vol chez les microchiroptères. Verh. Naturforsch. Ges. Basel, **27**:156–183.

Ricklefs, R. E., and K. O'Rourke. 1975. Aspect diversity in moths: A temperate tropical comparison. Evolution, **29**:313–324.

Sarhan, A., and B. Greenberg. 1962. Contributions to order statistics. John Wiley and Sons, New York, 482 pp.

Schoener, T. W. 1965. The evolution of billsize differences among sympatric congeneric species of birds. Evolution, **19**:189–213.

Smartt, R. A., and C. Lemen. 1980. Intrapopulational morphological variation as a predictor of feeding behavior in deermice. Am. Nat., **116**:891–894.

Smith, J. D., and A. Starrett. 1979. Morphometric analysis of chiropteran wings. Pp. 229–316, *in* Biology of bats of the New World family Phyllostomatidae. Part III. (R. J. Baker, J. K. Jones, Jr., and D. C. Carter, eds.). Spec. Publ. Mus. Texas Tech Univ., Lubbock, **16**:1–441.

Soulé, M., and B. Stewart. 1970. The "niche-variation" hypothesis: A test and alternatives. Am. Nat., **104**:85–97.

Stebbings, R. E. 1973. Size clines in the bat *Pipistrellus pipistrellus* related to climatic factors. Period. Biol., **75**:189–194.

Stephan, H., and P. Pirlot. 1970. Volumetric comparisons of brain structures in bats. Z. Zool. Syst. Evolutionsforsch., **8**:200–236.

Struhsaker, T. T. 1961. Morphological factors regulating flight in bats. J. Mammal., **42**:152–159.

Suthers, R. A. 1970. Vision, olfaction, taste. Pp. 265–310, *in* Biology of bats. Vol. 1. (W. A. Wimsatt, ed.). Academic Press, New York, 477 pp.

Tamsitt, J. R. 1967. Niche and species diversity in Neotropical bats. Nature, **213**:784–786.

Van Valen, L. 1965. Morphological variations and width of ecological niche. Am. Nat., **99**:377–390.

Vaughan, T. A. 1959. Functional morphology of three bats: Eumops, Myotis, and Macrotus. Univ. Kans. Publ. Mus. Nat. Hist., **12**:1–153.

Vaughan, T. A. 1966. Morphology and flight characteristics of Molossid bats. J. Mammal., **47**:249–260.

Vaughan, T. A. 1970a. Flight patterns and aerodynamics. Pp. 195–216, *in* Biology of bats. Vol. 1. (W. A. Wimsatt, ed.). Academic Press, New York, 406 pp.

Vaughan, T. A. 1970b. Adaptations for flight in bats. Pp. 127–143, *in* About bats. (B. H. Slaughter and D. W. Walton, eds.). Southern Methodist University Press, Dallas, 339 pp.

Vaughan, T. A., and G. C. Bateman. 1970. Functional morphology of the forelimb of mormoopid bats. J. Mammal., **51**:217–235.

Vaughan, T. A., and M. M. Bateman. 1980. The molossid wing: Some adaptations for rapid flight. Pp. 69–78, *in* Proceedings of the Fifth International Bat Research Conference. (D. E. Wilson and A. L. Gardner, eds.). Texas Tech Press, Lubbock, 434 pp.

Whitaker, J. O., Jr., and H. L. Black. 1976. Food habits of cave bats of Zambia, Africa. J. Mammal., **57**:198–204.

Williams, D. F., and J. S. Findley. 1979. Sexual size dimorphism in vespertilionid bats. Am. Midl. Nat., **102**:113–126.

Wilson, D. E. 1973. Bat faunas: A trophic comparison. Syst. Zool., **22**:14–29.

Chapter 7

Echolocation, Insect Hearing, and Feeding Ecology of Insectivorous Bats

M. Brock Fenton

Department of Biology
Carleton University
Ottawa K1S 5B6, Canada

I. INTRODUCTION

To obtain a complete picture of the feeding ecology of insectivorous bats requires, among other things, a detailed knowledge of the actions of individuals, where they hunt, what they catch relative to what is available there, and how these patterns vary over time. The purpose of this paper is to consider the echolocation of bats and the hearing-based defenses of insects related to echolocation in the context of the question, are insectivorous bats specialists?

The study of feeding by insectivorous bats has been approached in different ways, from examination of what they have caught and eaten (by analysis of stomach contents, feces, or insect remains deposited under roosts) to the monitoring of feeding behavior, using bat detectors, or by the observation of tagged individuals (light tags or reflective bands). Other studies have attempted to sample the insects available to bats and to compare these samples with those that are captured. Studies on the morphology of bats have led to various predictions about feeding and resource partitioning, drawing on studies of diet to support inferences from morphology (see Chapter 6).

It is difficult to draw valid conclusions about prey selection and the partitioning of food resources from many of these studies; however, what they have demonstrated is that insectivorous bats include a wide range of insects in their diets (e.g., Ross, 1967; Belwood and Fenton, 1976; Anthony and Kunz, 1977) and consume large amounts of food each day (for adults 20–50% of prefeeding mass; e.g., Kunz, 1974; Anthony and Kunz, 1977; Kunz, 1980). Studies where feces or stomach samples from bats netted or trapped at one or more sites usually lack information about where trapped individuals have fed, and comparisons of "selected" prey with samples of insects, however obtained, are relatively meaningless unless insect samples are representative of the prey population available to hunting bats.

The sampling of insect prey has been a perpetual problem, even when feeding sites are known. The entomological literature abounds with insect traps designed to sample a wide range of insects, yet usually each trap is most effective for a particular kind of insect. Flying insects may be sampled by visual counts, sweep nets, tow nets, sticky traps, Malaise traps, suction traps, light traps, whirligig traps, and so on. Bat biologists commonly have used light, Malaise, suction, or sticky traps (e.g., Black, 1974; Buchler, 1976a; Bradbury and Vehrencamp, 1976a; Belwood and Fenton, 1976; Anthony and Kunz, 1977), and currently suction traps appear to be most popular. However, suction traps were originally designed to sample flying aphids and they are less effective for large insects (Taylor, 1962); whirligig traps (Nicholls, 1960) are probably superior. The choice of insect-sampling method is also influenced by convenience and power requirements, and none seems ideal. The sampling of nonflying insects is a different problem, with at least as many pitfalls as in sampling flying insects. Although data on prey availability are crucial to understanding feeding ecology, the problem will continue to be an important hurdle in studies of insectivorous bats.

Therefore, in spite of considerable effort, many basic questions about the feeding of insectivorous bats remain unanswered, but we now have a better idea of what to do next. This background information, coupled with various technological aids (see below), should permit bat biologists to assemble accurate pictures of feeding behavior and to better assess possible partitionings of food resources where and when they may occur.

Because the insectivorous bats studied to date (albeit few relative to the number of species described) use echolocation to locate their prey, apparently even when hunting in broad daylight (e.g., Fenton and Thomas, 1980), biologists can monitor these calls to determine when many bats are actually attempting to catch food. For high-intensity echolocators (see below), a broad-band microphone coupled to a zero-crossing period meter and portable oscilloscope (Simmons *et al.*, 1979a) permits viewing of the frequency–time structures of the calls. This information can be used to identify many bats to species (Bell, 1980; Fenton and Bell, 1981), and to unambiguously ascertain when and where a bat is attempting to feed (Simmons *et al.*, 1979b).

Monitoring bat echolocation calls requires a "bat detector" (a microphone sensitive to the frequencies in the calls) and depends upon the intensity of the calls (see below). Bat detectors may be broad band or narrow band, depending upon the range of frequencies to which they respond (Simmons *et al.*, 1979a).

Broad-band detectors include models whose outputs are the incoming frequencies divided by ten, and others whose outputs are the entire wave form of the sounds; the latter can be used in conjunction with the period meter (Simmons *et al.*, 1979a). Narrow-band detectors may be tuned or tunable to specific frequencies (e.g., "leak detectors" or the QMC Mini bat detector, respectively) and cannot be used with the period meter. The ears of some insects are "bat detectors" sensitive to the frequencies in echolocation calls and warning insects of approaching bats (Roeder, 1967).

Individuals can be marked with light tags (Buchler, 1976b) or reflective bands (Bradbury and Emmons, 1974), permitting the simultaneous monitoring of echolocation calls, feeding behavior, and habitat use, and these data used to identify areas that should be sampled for insect abundance. In some situations individuals may be captured to obtain a picture of what known bats had caught and eaten in areas from which a sample of insects had been obtained. The use of radio tags on larger insectivorous bats (Bradbury *et al.*, 1979) and various night-viewing devices can permit a more detailed analysis of feeding behavior, including associated echolocation calls and prey selection. These viewing techniques may also provide a much needed evaluation of capture efficiency in a comparison of attempted versus successful captures. Vaughan (1977) reported that over 30% of the captures attempted by *Hipposideros commersoni* were successful, while the capture rate of *Myotis auriculus* was over 67% (Fenton and Bell, 1979). Further data are obviously required to provide a more representative picture of this important topic.

The technology associated with low light–level observation, the monitoring of echolocation calls, and the marking of individual bats for studies of behavior combine to make further detailed studies of the feeding behavior and associated patterns of resource partitioning by insectivorous bats more feasible than ever before, and it is useful to consider in more detail the information that has been collected to date.

2. ECHOLOCATION CALLS

Insectivorous bats use a variety of signals to obtain information about their environment, including the presence, position, course, speed, and identity of potential prey (Simmons *et al.*, 1979b; Simmons and Stein, 1980). Although insectivorous bats have been classified by the types of calls they produce (e.g., Simmons *et al.*, 1975), and while the terms *constant frequency* (CF), *frequency modulated* (FM), and some combination of these (e.g., CF–FM) describe the signals, they do not necessarily describe the echolocation strategies (Simmons *et al.*, 1978), nor connote the plasticity of the signals.

2.1. Call Structure

FM echolocation calls were the first described (Griffin, 1958). These are short (1–5 msec) pulses of sound sweeping from high to low frequencies (Fig. 1), and were considered typical of vespertilionids and some other bats. FM pulses can be steep or shallow (Fig. 1), and each provides different information about the target (Simmons and Stein, 1980). CF signals recorded from rhinolophids and some other rhinolophoid bats are typically longer (to 100 msec), with most of their energy in a narrow frequency band (Fig. 1). However, the CF bats include FM components in their calls (Fig. 1), and some FM bats incorporate CF components in their calls (Fig. 1; Simmons *et al.*, 1978). Laboratory research has indicated that the CF components may serve to assist in the detection of targets and, in some species, to encode Doppler-shift information, while FM components encode data about the position and surface detail of the target (Simmons *et al.*, 1979b; Simmons and Stein, 1980).

The distinction between CF and FM bats is artificial, and field studies of bats hunting insects will probably provide more evidence of departures from "normal," for example, the use of CF calls by the supposed FM species, such as *Plecotus phyllotis* (Simmons and O'Farrell, 1977) or *Taphozous mauritianus* (Fenton *et al.*, 1980). However, it is appropriate to distinguish between bats that are capable of Doppler-shift compensation and those that are not with respect to

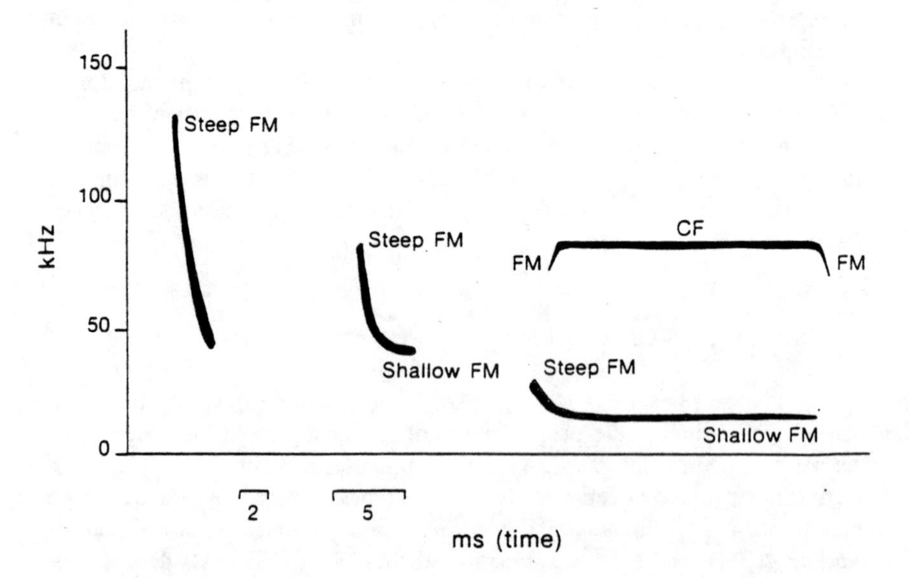

FIGURE 1. Patterns of frequency change over time in typical echolocation calls observable on the display of a zero-crossing period meter (Simmons *et al.*, 1979a). These patterns do not include harmonics.

the function of the CF portion of their calls (Gustafson and Schnitzler, 1979). If *Rhinolophus ferrumequinum* is representative of rhinolophids, an acoustic fovea is central to Doppler-shift compensation (Schuller and Pollak, 1979). This feature, which is structural and neurological, permits these animals to exploit Doppler-shift information about insect wing beats, using the CF portion of the call as a carrier frequency (Schnitzler, 1978; Vater, 1980). The echolocation performance and call structure of *Asellia tridens* is not fundamentally different from that of rhinolophids, for which there are data (Gustafson and Schnitzler, 1979), and if it is typical of hipposiderids, then these species are also capable of Doppler-shift compensation. Observations of *R. ferrumequinum* feeding in the field in Italy (Griffin and Simmons, 1974) indicate that they only attack insects whose wings are beating, and there are similar laboratory data for the Neotropical bat, *Pteronotus parnellii* (Goldman and Henson, 1977).

2.2. Intensity

Another characteristic by which echolocating bats are grouped is the intensity of their calls, initially a distinction between "whispering" and loud bats (Griffin, 1958). Many vespertilionids, molossids, emballonurids, rhinolophids,

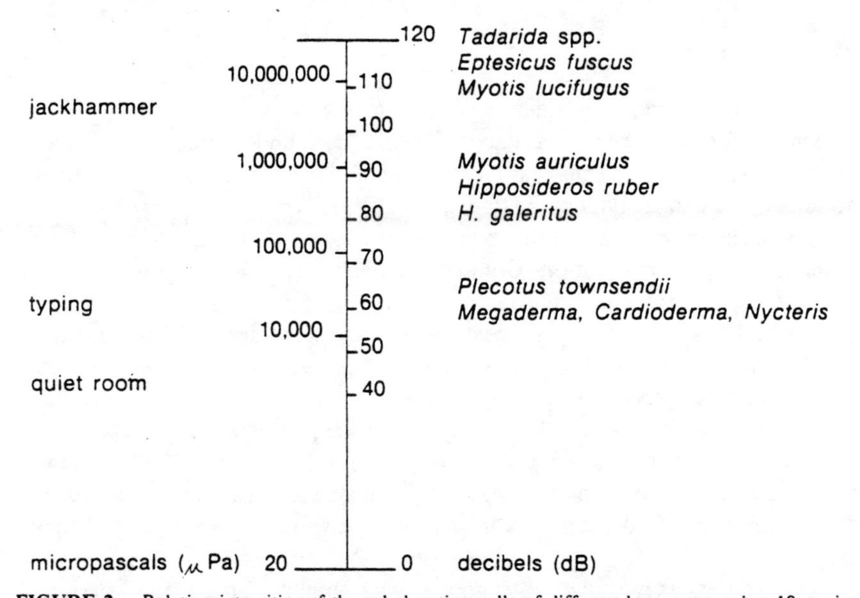

FIGURE 2. Relative intensities of the echolocation calls of different bats measured at 10 cm in front of the bat and compared to common sounds in decibels (dB) or micropascals (μPa). The values for different bats are not absolute, and variations between individuals are not detailed. For further comparison, a "typical" smoke detector produces a signal of 105 dB at 10 cm. (Sound intensity scale adapted from Hassall and Zaveri, 1978.)

and mormoopids use loud calls (Fig. 2), while nycterids and megadermatids, and some vespertilionids and hipposiderids, produce much less intense calls (Fig. 2). However, other vespertilionids, such as *Myotis auriculus* (Fenton and Bell, 1979), and some hipposiderids (Griffin, 1971; Fenton and Fullard, 1979) produce calls of intermediate intensity (Fig. 2). There is an intensity gradient in bat echolocation calls and a number of species, notably hipposiderids, that fall between the two extremes.

The categorization of echolocation strategies by call intensity applies to some bats, but lack of knowledge about variability and the presence of "intermediate" intensities make an accurate classification of most species dubious. Intensity is difficult to measure in the field, as it is influenced by the position of the bats relative to the microphone (e.g., Griffin, 1958; Schnitzler and Grinnell, 1977), as well as by atmospheric attenuation of high-frequency sounds (Griffin, 1971). A useful but tentative classification of the intensity of bat calls for field work using broad-band microphones (Simmons *et al.*, 1979a) when the bats are oriented directly towards the microphone is the following: I, detectable at about 10 m, often more; II, detectable at around 1.5–2 m; and III, detectable only at less than 0.5 m. Bats in these different categories are not equally sampled by broad-band microphones, and studies of feeding or habitat use based on these microphones are biased accordingly (Fenton and Bell, 1981).

2.3. Frequency

Rhinolophids and hipposiderids with acoustic foveas use a narrow range of frequencies for echolocation, although Pye (1972) found a bimodal frequency pattern in some hipposiderids and Gustafson and Schnitzler (1979) documented variations between individuals in the calls of *Asellia tridens*. *Pteronotus parnellii*, another species equipped with an acoustic fovea, also uses a narrow range of frequencies for echolocation (Goldman and Henson, 1977). Laboratory studies show that *Eptesicus fuscus* use similar calls during discrimination tasks (Fullard *et al.*, 1979), which may suggest that there is some degree of "specialization" for a particular call within species of echolocating bats. Furthermore, observation of the echolocation calls of bats in the field via period meter and oscilloscope (Simmons *et al.*, 1979a) reinforces the impression that species may be identified by their calls (Fenton and Bell, 1981). However, it is obvious that each species of bat does not have its own frequency and that both the Doppler-shift-compensating and other bats include a FM sweep in some phase of their orientation sounds.

Higher-frequency sounds have shorter wavelengths than do lower-frequency sounds and provide better resolution of target detail, but they are also more directional (e.g., Shimozawa *et al.*, 1974) and more subject to environmental attenuation (Griffin, 1971). Attenuation, as well as initial call intensity, can strongly influence the effective range of echolocation. There is no clear correlation between the size of a bat and the frequency of its echolocation calls (Novick, 1977), although larger bats tend to use calls of lower frequency than those of smaller bats, a trend that is clearest in the rhinolophids and hipposiderids (Novick, 1977). In the vespertilionids the calls of some sympatric pairs of similar-sized *Pipistrellus* spp. show an interesting pattern of bimodality, with one species producing high-frequency calls, and the other lower-frequency calls (e.g., *P. coromandra* and *P. ceylonicus*, Novick, 1958; *P. nanus* and *P. kuhlii*,

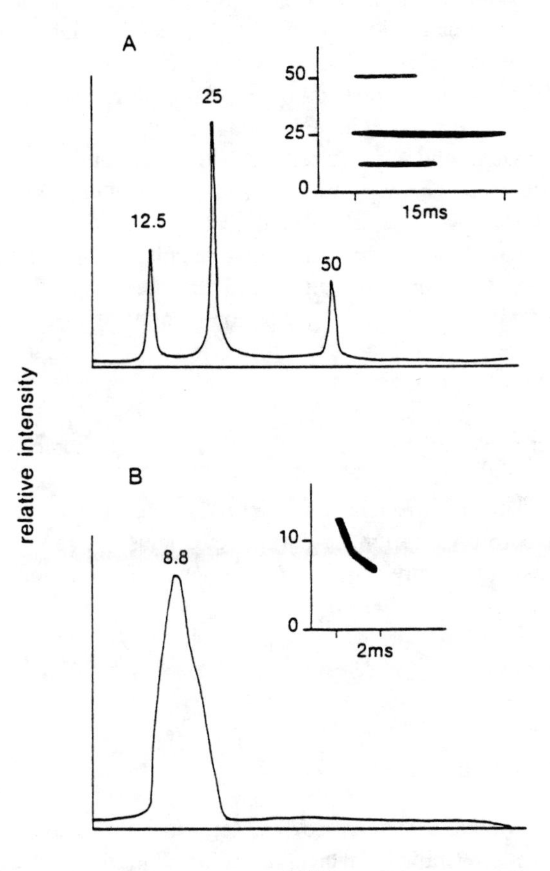

FIGURE 3. Power spectra and sonograms of search-phase echolocation calls of (A) *Taphozous mauritianus* and (B) *Euderma maculatum*. Numbers on the power spectra represent the frequencies (kHz) with the most energy (linear scale); time is in milliseconds and the vertical scale is in kilohertz. Note that in *T. mauritianus* the fundamental frequency is below 15 kHz and in the human audible range, but the bat puts most of its energy in the second harmonic 25 kHz (Fenton *et al.*, 1980). In *E. maculatum* the entire call is audible (Fenton and Bell, 1981).

Fenton, 1975; high and low, respectively). Theoretically, larger bats should be able to use lower-frequency calls, as they would not require the same degree of target resolution as do small bats hunting smaller targets.

Biologists commonly associate echolocation by bats with ultrasonic sound, however, not all insectivorous bats use ultrasonic echolocation calls when hunting prey. There is often an audible click associated with the production of ultrasonic pulses, the "ticklaut," which is described in some detail by Griffin (1958) and which was used by the Dutch biologist Sven Dijkgraaf (1946, 1949) in his studies of the ability of bats to avoid obstacles in total darkness. However, there are bats in many parts of the world that use echolocation sounds that have most of their frequency below 20 kHz and which are therefore entirely audible to many humans (Fig. 3). In Arizona and other parts of its range *Tadarida macrotis* uses low-frequency calls for echolocation; in Zimbabwe the same is true of *Otomops martiensseni;* and in south central British Columbia this is also the case for *Euderma maculatum* (Fenton and Bell, 1981). The emballonurids *Peropteryx macrotis* from Colombia (Pye, 1973) and *Taphozous peli* from Central Africa (Bradbury, 1977) also use audible echolocation calls, and in the Ivory Coast in West Africa I could readily distinguish between three species of bats by their audible echolocation calls. For any of the aforementioned species it was always possible to hear frequency-modulated pulses, including feeding buzzes associated with attempts to catch insects. The emballonurid *Taphozous mauritianus* was slightly different in this regard, as the fundamental frequencies of the calls were audible but most of the energy was in the second harmonic, which was not audible (Fenton *et al.*, 1980).

As in every aspect of the behavior of insectivorous (and other) bats, we need more field data before we can make definitive conclusions about the specificity of echolocation calls for any species. At the level of resolution of the period meter some species are consistent in the calls they produce, both with respect to frequency components, frequency–time patterns, and duration, a situation that makes it possible to identify these bats by their orientation calls (Fenton and Bell, 1981).

2.4. Pulse Repetition Rates

The rate at which bats produce their echolocation calls varies with the situation. When approaching obstacles, attacking flying insects, or approaching a roost site, the pulse repetition rate increases, often dramatically (Fig. 4). Increased pulse repetition rates in these situations appear to be typical of rhinolophids, noctilionids, mormoopids, vespertilionids, and molossids. However, *Myotis auriculus,* when gleaning insects from a surface, did not increase its pulse repetition rate, although it did when chasing flying insects (Fenton and Bell,

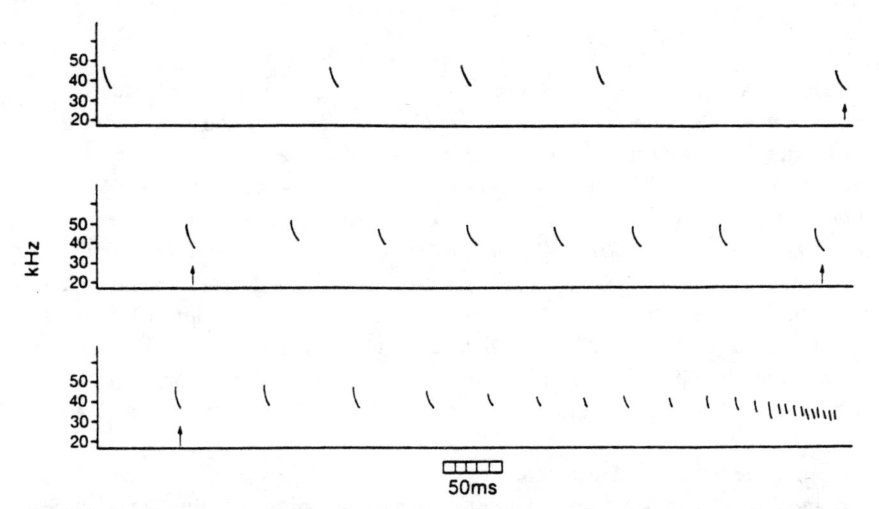

FIGURE 4. The sequence of echolocation calls produced by an *Eptesicus fuscus* in the field as it searched (to the third row), approached (calls 4–10 in the third row), and attacked a June beetle. Pulses marked with arrows are repeated to preserve the time sequence. This display is derived from a period-meter display and does not show harmonics (*E. fuscus* usually includes the third harmonic and suppresses the second).

1979), suggesting that the high pulse repetition rate (feeding buzz) associated with an insect-catching maneuver served to apprise the bat of last-instant changes in the position of prey.

2.5. Harmonics

Simmons *et al.* (1978) showed that *Tadarida brasiliensis* added harmonics to their echolocation calls when hunting in cluttered surroundings but only used fundamental frequencies in uncluttered situations. Harmonics provide the bats with more detail about their surroundings (Simmons and Stein, 1980), and species foraging in cluttered situations and using low-intensity calls often include many harmonics in their cries (e.g., Bradbury, 1970). *Myotis lucifugus*, which uses the fundamental and two harmonics during the approach phase to a target, includes only the fundamental and one harmonic in the terminal phase or feeding buzz (Fenton and Bell, 1979) and does not show the same flexibility in call design as *T. brasiliensis*. The flexibility of the echolocation system of *T. brasiliensis*, especially with respect to harmonic structure, has important implications for echolocation (Simmons *et al.*, 1978; Simmons and Stein, 1980).

2.6. Effective Range

Definitive data on the maximum range of echolocation are conspicuous by their absence. Responses of bats to a target or obstacle provide data about the minimum range, and estimates obtained in this fashion vary from 15 (*Taphozous peli;* Brosset, 1966) to 10 m (*Hipposideros commersoni;* Vaughan, 1977; *Myotis volans;* Fenton and Bell, 1979), to around 2 m (*Myotis auriculus;* Fenton and Bell, 1979), or 1 m (*Myotis lucifugus;* Griffin *et al.*, 1960; *Myotis californicus;* Fenton and Bell, 1979). Novick (1977) used Suthers' (1965) data to calculate a detection distance of 1.8 m for *Noctilio leporinus,* while Schnitzler (1968) used approaches to landing sites to calculate a range of 5–5.5 m for *Rhinolophus ferrumequinum* and Bradbury (1970) obtained a value of 2.2 m for *Vampyrum spectrum.* Griffin (1971) estimated an effective range of 2 m for *Hipposideros galeritus.*

Since these estimates use behavioral responses of bats as criteria for detection, they fail to indicate the maximum range at which detection may have occurred. When compared with vision, echolocation in air appears to occur at relatively short ranges and is influenced by the attenuation of sound, which may account for the preponderance of small sizes (≤40 g) among the insectivorous Microchiroptera (Fenton and Fleming, 1976).

3. HEARING AND INSECT DEFENSE

Roeder and Treat (1960) demonstrated that moths with functional ears had 40% less of a chance of being captured by echolocating bats than did moths lacking functional ears, and Miller and Olesen (1979) have similar data for green lacewings. Pye (1968) immortalized the evolution of bats, echolocation, and insect hearing, and it is obviously only a matter of time before we discover what other insects have "bat detectors." It is likely that ears sensitive to ultrasonic sound are widespread in the Insecta, since most families of Lepidoptera examined to date have tympanate representatives, and in any area well over 50% (to over 90%) of the macrolepidopteran species have them (Fenton and Fullard, 1979). Moiseff *et al.* (1978) found that crickets showed a negative phonotaxis to 30–70 kHz sounds, a possible bat-avoidance response, and any insect that produces ultrasonic communication calls (e.g., Suga, 1966) should be able to detect bats under some conditions. In lacewings the ears are located on the wings (Miller, 1971), whereas moths have them located in different places, either on mouth parts, the thorax, or the abdomen (Roeder, 1974); the ability to detect the echolocation calls of bats is polyphyletic in the Insecta.

In the Lepidoptera, for which data are available, the response of a moth to a bat, or simulated bat, is determined by the intensity of the bat call (Roeder,

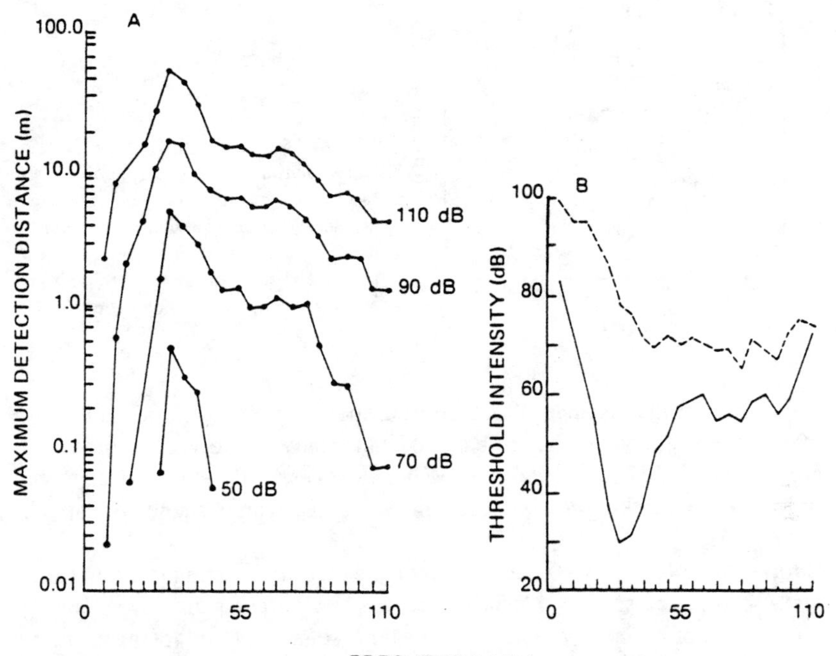

FIGURE 5. (A) The distances at which the African notodontid *Scalmicauda bisecta* would detect sounds of different frequency and intensity and (B) audiograms for a very sensitive African geometrid (*Gorua apicata;* ——) and a relatively insensitive thyretid (*Balacra inflammata;* - - -). Calculations of the distance of detection reflect atmospheric attenuation and the initial intensities (dB) typical of bat echolocation calls. (Modified from Fenton and Fullard, 1979.)

1967). Responses to very intense calls include evasive maneuvers culminating in a dive to the ground, while the response to less-intense sounds is negative phonotaxis (Roeder, 1967). Green lacewings adjust their responses to the intensity of the bat calls and to the rate at which pulses are produced, allowing a "last-chance response" to a feeding buzz (Miller and Olesen, 1979).

At least two factors determine the intensity of a bat call as perceived by an insect: the intensity of the call produced by the bat and the audiogram of the insect. Intensity of calls, hearing sensitivity, along with atmospheric attenuation and the orientation of the bat to the moth will determine the distance of detection (Fig. 5). In most of North America insects with ears tuned as in Fig. 6 will detect most of the bats with which they are sympatric at distances of over 20 m, as the bats generally include some sound between 60 and 30 kHz in their high-intensity echolocation calls. However, data are only available for the Lepidoptera. When the effects of frequency and intensity are combined and corrected for attenuation, the influence of an insect's audiogram and the initial intensity of bat calls is clear

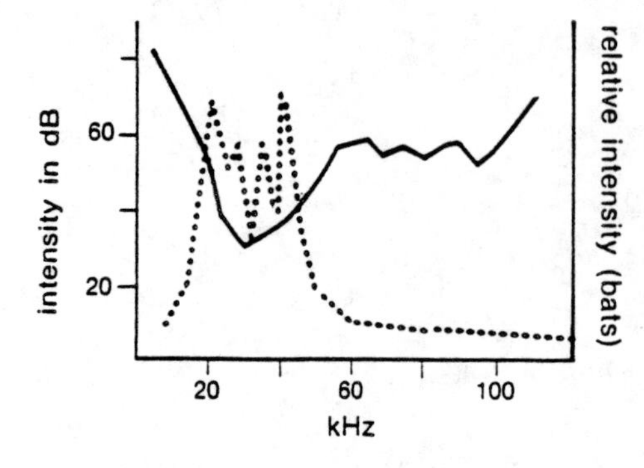

FIGURE 6. Audiogram of the Ivorian *Gorua apicata* (——) superimposed on a fast Fourier-transform analysis of 400 bat echolocation calls recorded in the field in the Ivory Coast (...). The moth is most sensitive in the frequency range with the greatest energy of echolocation calls. (Modified from Fenton, 1980.)

(Fig. 5). Because the distance of detection influences the time to possible capture, the acoustic characteristics of bat calls and moth ears are vital to predators and prey. Roeder and Treat (1960) and Miller and Olesen (1979) showed that the ability to hear an attacking bat reduces a moth's or lacewing's chance of capture by up to 40%.

Moths in the family Arctiidae not only have ears for detecting bats but some species also produce sounds that influence the bats' attack behavior (Dunning and Roeder, 1965). These moths produce their sounds by collapsing a metathoracic episternite (Blest *et al.*, 1963; Fullard and Fenton, 1977). The sounds represent part of a graded series of defensive behavior ranging from negative phonotaxis at low-intensity stimulation, to flight cessation and sound production at stimulations by higher-intensity sounds (Fullard, 1979). The sounds produced by arctiids have been considered examples of aposematic signals by some workers (e.g., Dunning, 1968), but more recently Fullard *et al.* (1979) suggested that these sounds jammed the echolocation of some insectivorous bats by interfering with their information processing. Sound production by arctiids may be analagous to the last-chance response of green lacewings. It is possible that the clicking sounds of some butterflies serve to protect these insects against bats (Møhl and Miller, 1976).

4. RESPONSES OF BATS TO INSECT HEARING

The influence of intensity and hearing sensitivity in the range at which hunting bats are detected by tympanate insects suggests two immediate responses open to echolocating bats: reduction of the intensity of their outgoing pulses and adjustment of the frequencies in their orientation sounds to outside the most sensitive area of the insects' auditory systems (Fenton and Fullard, 1979; Fig. 5). Bats using higher-frequency signals (such as rhinolophids and many hipposiderids) or sounds of low or intermediate intensity (many hipposiderids and nycterids) are less detectable by moths than are high-intensity echolocating bats (Fenton and Fullard, 1979), but it remains to be determined just how many of these species exploit this apparent advantage and feed heavily on tympanate insects. Two other alternatives could be used by bats to overcome any apparent handicap, namely, either increasing their flight speeds and accordingly reducing the time available to insects for evasive maneuvers, or the cessation of echolocation calls (Fenton and Fullard, 1979). Few data are available on the flight speeds of hunting bats, but some appear to fly faster than others, for example, many molossids (Vaughan, 1970).

The use of cues other than echolocation for the detection of prey by insectivorous bats has not been thoroughly explored. Several studies have indicated that bats, especially megadermatids (Vaughan, 1976; Fiedler, 1979) use sounds made by prey moving across a substrate to locate their prey, and this also appears to be true of *Myotis myotis* (Fiedler, 1979; Bauerova, 1978). *Trachops cirrhosus* appears to use echolocation and mating calls to locate the frogs on which it feeds (Barclay *et al.*, 1981). Furthermore, a variety of workers has suggested that insectivorous bats use the sounds of insects to locate prey (e.g., Nyholm, 1965), whether flight sounds or calling sounds (e.g., Rentz, 1975), but these hypotheses are not supported by quantitative data. The role of olfaction in the location of prey by insectivorous bats is unclear, although Kolb (1976) suggested that it is used by *M. myotis*, and Bhatnagar (1975) has shown that olfactory ability is variously developed in the Chiroptera. Vehrencamp *et al.* (1977) suggested that *Vampyrum spectrum* could have used olfaction to locate avian prey. The role of vision in location of prey by insectivorous bats appears to be unexplored.

Some bats are much less conspicuous to insects than others, suggesting that insect hearing may have exerted and continues to exert an important pressure on the echolocation systems of bats (Fenton and Fullard, 1979), but other counter-strategies that bats might adopt to avoid the defensive tactics of tympanate insects remain largely unexplored. One possible approach by bats would depend upon the defensive responses of insects only involving flight. Fenton and Bell (1979) observed *Myotis auriculus* gleaning stationary moths from walls and tree branches in Arizona. Although some of their victims belonged to families with tympanate representatives, none showed signs of defensive behavior, leading to the suggestion that the selection of sitting moths might allow bats to circumvent bat-oriented defensive behavior (Fenton and Bell, 1979). Moths in several families, when walking on a substrate, freeze when approached by an echolocating bat (or facsimile thereof), and some sitting moths grip the substrate more closely when so approached (Werner, 1981). Thus a sitting moth that is often unable to fly immediately for thermal reasons (Bartholomew and Heinrich, 1973) may not, in fact, be vulnerable to a foraging bat.

5. BATS AS SPECIALISTS

The preceding information could lead to predictions about what some bats might eat based on their echolocation calls or on the defensive behavior of insects. Lamentably, the data base currently available does not provide sufficient challenge to the predictions.

5.1. By Time

Schoener (1974) reported that predators more often partition time than do other animals, but there are conflicting data for insectivorous bats. Kunz (1973) and Reith (1980) used captures of bats in mist nets as the source of data for suggesting the partitioning of time by some vespertilionids. However, sampling by the monitoring of echolocation calls yielded no evidence of significant time partitioning (Bell, 1980). In each case the study sites were different (Iowa, Oregon, and Arizona, respectively), but some species were common to the three studies.

Since feeding bats are often difficult to catch in mist nets or traps, data collected by monitoring echolocation calls may be more representative of real activity than captures. Although in some cases levels of activity determined by captures are similar to those obtained by monitoring echolocation calls (Kunz and Brock, 1975), I suspect that these are exceptions. In many habitats in North America, Africa, and Australia I have detected the presence of feeding bats by listening to their echolocation calls when intensive and extensive netting and trapping produced few or no captures.

The accurate identification of bats by their calls necessitates the capture and study of known individuals (Fenton and Bell, 1981), and any study should include both approaches. However, the calls of high-intensity echolocating bats can always be used to gather data on feeding activity and habitat use, but only in some cases will capture by mist nets, traps, or shooting be possible.

5.2. By Diet

Although the data from a number of studies have been used to support the suggestion that some insectivorous bats specialize to some level on particular types of insects, the evidence does not justify these suggestions. When data on the feeding behavior, including feeding sites and prey available there, do not accompany analyses of prey selected, whether from stomach contents or feces, then the conclusion of specialization by diet is unsupportable.

The idea of specialization by diet, with its implications for resource partitioning and competitive exclusion, is attractive, and it is easy to emphasize differences between species or individuals to make the results more compatible with predictions about competitive exclusion (e.g., Wiens, 1977). Data that some colleagues and I had gathered on prey selection and patterns of habitat use by sympatric African insectivorous bats provide an excellent example. Data from the wet season (Fenton *et al.*, 1977), when insects were superabundant, seemed to support a beetle-moth pattern of partitioning, as suggested by Black (1972, 1974); but subsequent work in the cold dry season, when food was scarce, overturned the earlier impression (Fenton and Thomas, 1980; see also Anthony and Kunz, 1977). When food was scarce, the insectivorous bats ate whatever insects were available. The change in diet may reflect a change in strategy by the bats or different prey abundances; the point is that one could not accurately label the bats as specialists by diet.

However, there are indications that some insectivorous bats may feed heavily on particular groups of insects. The best example is provided by *Cloeotis percivali* (Black, 1979), but since the analyses of stomach contents were not accompanied by field observations on where or when the bats were feeding, let alone the availability of insects there, this example provides only a suggestion. Other suggestions of at least seasonal specialization by diet are provided by data on *Myotis myotis* (Bauerova, 1978) and *Hipposideros commersoni* (Vaughan, 1977). There are many examples of short-term specializations by diet (e.g., Buchler, 1976a).

5.3. By Foraging Strategy

Several foraging strategies are evident from work done on insectivorous bats. Brosset (1966) distinguished between long- and short-range bats, a distinction supported by subsequent research (e.g., Fenton and Bell, 1979). Some species react to prey at greater distances than do others (e.g., *Taphozous peli, T. mauritianus, Myotis volans*) and usually only make one attempt to catch a prey item per pass through a feeding area. Other species operate at short ranges, for example, *Myotis lucifugus* or *M. californicus*, less than 1 m, and make several attempts to capture prey on any pass through a feeding site. Bats with large ears and broad wings have traditionally been considered gleaners, taking their food from surfaces, whether water (Dwyer, 1970; Novick and Dale, 1971), the ground (Vaughan, 1976; O'Shea and Vaughan, 1977), or foliage (Fenton and Bell, 1979). Vaughan (1977) described a flycatcher strategy adopted by feeding *Hipposideros commersoni*, and this is also used by young *Myotis lucifugus* during their early feeding activity (Buchler, 1980). Bauerova's (1978) data on *Myotis myotis* and Orr's (1954) observations on *Antrozous pallidus* suggest that these bats may actually forage on the ground. Within any fauna of bats several foraging strategies may be represented, providing an indication of how food resources may be subdivided (O'Shea and Vaughan, 1980).

Different foraging strategies probably result in different prey being selected, but two important caveats must be appended to information on foraging strategies. First of all, there are relatively few data on variability in the foraging strategies of individuals, but that which is available (e.g., Vaughan, 1976; Fenton and Bell, 1979) unequivocally indicates that it is probably unwise to categorize bats by foraging strategy. Secondly, the data base for most bats is so meager as to make any classification of them by foraging strategy an easy but futile and meaningless exercise. Further observation of known individuals in the field could provide a useful indication of flexibility in foraging behavior, such as that recently published for tyrant flycatchers (Fitzpatrick, 1980).

5.4. By Space

A fundamental ploy used by many animals in partitioning is the division of space (e.g., Hespenheide, 1971). Most studies of habitat use (partitioning by space) by insectivorous bats have involved capture by mist nets, Tuttle traps, or shooting, and have, by definition, only involved animals that were captured. Not all species are equally susceptible to capture; for example, in Arizona both *Myotis volans* and *M. auriculus* hunted around a building, but only the former was taken in a Tuttle trap set in the immediate area (Fenton and Bell, 1979).

Casual observations (e.g., Wallin, 1969) or monitoring the behavior of known individuals (e.g., Nyholm, 1965) may provide an impression of habitat partitioning by insectivorous bats. However, systematic sampling is required to accurately assess the validity of these impressions. Reports of habitat partitioning based on captures (e.g., Jones, 1965; Black, 1974) are not necessarily as accurate as data obtained by monitoring echolocation calls when the bats are high-intensity echolocators, for reasons discussed above.

One study of habitat use by sympatric insectivorous bats by monitoring echolocation calls showed no evidence of habitat specialization in a community of 10 species in Arizona (Bell, 1980). The study area included three distinct habitat types, but the bats appeared to treat the area as one continuous community, and associations between species were strong enough to suggest mutual cuing on food resources (Bell, 1980). These data were collected by monitoring echolocation calls with the period meter, permitting an assessment of the levels of activity and feeding behavior of the species involved. His data also clearly demonstrated that all species of insectivorous bats in the area at least occasionally responded to localized concentrations of food. In total, Bell's (1980) data support less-detailed studies conducted elsewhere (e.g., Fenton and Thomas, 1980). None of these results mean that the partitioning of resources by habitat does not occur in other communities of insectivorous bats, but they do not foster or support interpretations of resource partitioning by habitat.

On an intraspecific basis, four studies indicate some spatial partitioning of conspecific insectivorous bats, either defined as a defended area (*Saccopteryx leptura;* Bradbury and Emmons, 1974; *Myotis grisescens;* M. D. Tuttle, personal communication) or as an exclusive feeding area (*Cardioderma cor;* Vaughan, 1976; *Myotis mystacinus* and *M. daubentoni;* Nyholm, 1965). Other studies have produced no evidence of the maintenance of exclusive feeding areas by defense or mutual avoidance (e.g., Bell, 1980; Fenton and Bell, 1979). Until there are more field data on the hunting patterns and home ranges of insectivorous bats, our picture of intraspecific partitioning of feeding space will remain incomplete.

5.5. By Morphology

A superficial examination of different insectivorous bats can produce a strong impression of variety, especially when one considers the relative sizes and shapes of eyes, ears, and wings. More detailed considerations of skull and mandible structures, teeth, and associated muscles (e.g., Freeman, 1979; Belwood, 1979) may reinforce the impression of variety. However, a more detailed examination of the species that occur in an area (e.g., Findley, 1976) conveys an impression of general similarity. Findley (1976) has pointed out that his examination of tropical and temperate bat faunas shows that they resemble one another in faunal statistics, with each including many similar, closely packed, overlapping, relatively invariable species, and a few distinctive, distant, isolated kinds. By the selection of particular species for a quick glance or more detailed examination (e.g., Freeman, 1979) it is possible to emphasize differences rather than similarities. However, the fact that *Tadarida brasiliensis* commuting to a feeding site have narrow wings and rapid flight, while those at feeding sites have much lower aspect ratios (Edgerton *et al.*, 1966), provides a warning that things may not be what they seem.

Studies that have emphasized morphological differences have inferred faunal structure, supporting these inferences with the patchy data on the diets of insectivorous bats. It remains to be determined whether or not the differences are more important than the similarities.

5.6. As Rapid Feeders

Few if any of the preceeding studies unequivocally support the conclusion that insectivorous bats specialize by time, diet, foraging strategy, or habitat use. Nonetheless, there are behavioral (Griffin *et al.*, 1960; Fenton and Morris, 1976; Fenton *et al.*, 1977; Gould, 1978; Vaughan, 1980; Bell, 1980), ecological (Kunz, 1974; Belwood and Fenton, 1976; Anthony and Kunz, 1977), anatomical (Kallen and Gans, 1972), and digestive (Buchler, 1975) data that strongly indicate that at least some insectivorous bats are specialists in rapid feeding and food processing.The propensity of some species to feed on concentrations of insects (e.g., Fenton and Morris, 1976), to rapidly chew food (Kallen and Gans, 1972), which may be stored in cheek pouches (Murray and Strickler, 1975), and to rapidly pass food through their digestive tracts (Buchler, 1975) all suggest fast feeding, a property compatible with the predictions of optimal foraging strategy (Pyke *et al.*, 1977). This pattern of feeding is appropriate for animals that are small and have a large surface area relative to their volume and a limited energy budget (see Kunz, 1980). Rapid, opportunistic feeding may produce specialized or generalized (by taxon) prey selection, along with selection of prey in a particular size range (Anthony and Kunz, 1977), and would serve to explain the mosaic of variable and specialized diets typical of many studies of prey selected by insectivorous bats (reviewed in Fenton and Fullard, 1979). Furthermore, rapid feeding permits individuals to exploit ephemeral food resources and can result in a lack of partitioning of food and/or space, which may also reflect seasonally superabundant food resources.

6. OTHER CONSIDERATIONS

The gregarious nature of many bats has important implications for their feeding ecology. Some insectivorous species are "typical" refuging forms (Hamilton and Watt, 1970), and this pattern of behavior affects their population density, patterns of dispersal from roosts (e.g., Kunz, 1974; Anthony and Kunz, 1977; Morrison, 1978), and may have other effects on feeding behavior. The social structure of colonies of bats can influence patterns of habitat use and hunting behavior (e.g., Bradbury and Vehrencamp, 1976a, 1976b, 1977a, 1977b). If individuals that roost together also feed together and represent a cohesive social unit, the capture success of and prey selected by some members of the colony may influence capture success and prey selection of other individuals. Alternatively, if the members of the colony are merely roost mates by aggregation, the influence of the feeding behavior of one on another may be minimal. A further complication of either situation is provided by data on the responses of *Myotis lucifugus* to the echolocation calls of other bats. In this species individuals respond to the echolocation calls of conspecifics and other species when searching for food or roosts (R. M. R. Barclay, personal communication).

Also germane to the questions about specializations in feeding by different species of insectivorous bats and about their feeding ecology in general are several other points. Increasing knowledge about the echolocation systems of bats and the information processing of which they are capable (see Busnel and Fish, 1980) leads one to predict that many bats should be able to tell a great deal about their targets through echolocation. Do species capable of Doppler-shift compensation use wing-beat information to identify specific insects by their signatures (e.g., Schnitzler, 1978)? Many insectivorous bats appear to have the capability to be very specific in their selection of prey.

I know of two examples of insectivorous bats learning about their prey. Miller (personal communication) found that one of his experimental *Pipistrellus pipistrellus* learned to foil the evasive behavior of green lacewings by modifying its approach behavior. The bat was clearly using information gathered by echolocation to recognize and thus exploit particular characteristics of its prey. Dunning (1968) found that one of her experimental *Myotis lucifugus* learned to distinguish between two species of arctiids by the sounds they produced, but in her experimental setting it is difficult to determine how much information the bat had gathered by echolocation.

There is no reason to presume that either of these observations is exceptional with respect to the capacity of insectivorous bats to gather information about prey and to adjust their pattern of behavior. However, it is crucial to appreciate that Miller's lacewings and Dunning's arctiids were in the worst possible situations from the insects' point of view; a one- or two-species prey population at high density over relatively long periods of time exposed to a single predator. Under natural conditions insectivorous bats may rarely encounter large numbers of one species of insect or even one order of insects from one night to the next. Consistency in prey populations, especially by species composition, may be rare (e.g., Hebert, 1980), making it difficult if not impossible for insectivorous bats to exploit the full information-gathering capacity of their echolocation (or their ability to learn) in the context of identifying and selecting particular prey types.

Demonstration of the ability of an insectivorous bat to identify an insect by its echo does not mean that this happens under natural conditions. Furthermore, the variability in responses of tympanate moths to stimuli resembling hunting bats (Roeder, 1975) could make it practically impossible for a bat to associate a defensive maneuver with particular (by order or by species) insects. Variability in defensive behavior of insects is adaptive for them, given the acuity of bat echolocation (Fenton and Fullard, 1979).

For their size, insectivorous bats use expensive modes of transportation during hunting, a situation which, coupled with factors relating to surface area and volume, may seriously impose on a balanced energy budget. One means of avoiding a deficit is to minimize flight time during feeding, and this could be accomplished by adopting a flycatcher strategy (e.g., *Hipposideros commersoni;* Vaughan, 1977), by feeding on the ground (e.g., *Myotis myotis;* Bauerova, 1978), or by minimizing intercapture intervals by feeding in areas where prey density is high. These considerations, coupled with variability in insect populations and defensive behavior make it relatively easy to accept the lack of specialization observed in the feeding behavior of bats or the duality of diet common in some chiropterans (Gillette, 1975) as representative. It remains for further field work to resolve the still unanswered question of how sympatric species of insectivorous bats coexist and whether or not some of the species have adopted a system that permits specialization in terms of time of activity, habitat use, diet, or foraging strategy.

7. SUMMARY

The echolocation calls of different species of insectivorous bats vary in call structure, intensity, frequency, pulse repetition rates, and harmonic structure, depending upon the situation in which the bats are hunting. Echolocation appears to work best at short ranges, and the maximum range may not exceed 20 m. A variety of insects, notably, moths and lacewings, have ears that are sensitive to the echolocation calls of bats, providing these potential prey with a means of detecting and thus avoiding hunting bats. Differences in the frequency and intensity characteristics of the echolocation calls influence the distance at which bats are detected by insects, supporting the suggestion that insect hearing may influence the design of echolocation calls. In some situations hunting bats do not use echolocation to locate prey; in others they use a combination of echolocation and sounds produced by prey. A consideration of the available literature provides no convincing evidence that bats specialize by the timing of activity, diet, use of habitat, foraging strategy or morphology. In many communities of insectivorous bats there is extensive overlap among all these parameters. However, many insectivorous bats are specialized to feed and process food rapidly, a situation that could explain the overlap observed in other studies. Social factors can strongly influence hunting behavior and the prey selected. It is possible that bats can use echolocation to distinguish between different species of insects, but whether or not they achieve this in the field remains to be determined. There is still no clear picture of how sympatric insectivorous bats partition food resources, or if, indeed, they do.

ACKNOWLEDGMENTS. I am particularly grateful to Eleanor Fenton for her understanding during the times when I have been in the field trying to make sense of the feeding ecology of insectivorous bats. Robert Barclay, Gary Bell, Jackie Belwood, David Cumming, James Fullard, Tom Kunz, Donald Thomas, Christine Thomson, and Tracey Werner kindly read this manuscript and made helpful suggestions. My work on bats has been supported by National Science and Engineering Research Council operating and equipment grants.

8. REFERENCES

Anthony, E. L. P., and T. H. Kunz. 1977. Feeding strategies of the little brown bat, *Myotis lucifugus*, in southern New Hampshire. Ecology, **58**:775–786.

Barclay, R. M. R., M. B. Fenton, M. D. Tuttle, and M. J. Ryan. 1981. Echolocation calls produced by *Trachops cirrhosus* (Chiroptera: Phyllostomatidae) while hunting for frogs. Can. J. Zool., **59:**750–753.

Bartholomew, G. A., and B. Heinrich. 1973. A field study of flight temperatures in moths in relation to body weight and wing loading. J. Exp. Biol., **58:**123–135.

Bauerova, Z. 1978. Contribution to the trophic ecology of *Myotis myotis*. Folia Zool., **27:**305–316.

Bell, G. P. 1980. Habitat use and response to patches of prey by desert insectivorous bats. Can. J. Zool., **58:**1876–1883.

Belwood, J. J. 1979. Feeding ecology of an Indiana bat community with emphasis on the endangered Indiana bat, *Myotis sodalis*. Unpublished M.S. thesis, University of Florida, Gainesville.

Belwood, J. J., and M. B. Fenton. 1976. Variation in the diet of *Myotis lucifugus* (Chiroptera: Vespertilionidae). Can. J. Zool., **54:**1674–1678.

Bhatnagar, K. P. 1975. Olfaction in *Artibeus jamaicensis* and *Myotis lucifugus* in the context of vision and echolocation. Experientia, **31:**856.

Black, H. L. 1972. Differential exploitation of moths by the bats *Eptesicus fuscus* and *Lasiurus cinereus*. J. Mammal., **53:**598–601.

Black, H. L. 1974. A north temperate bat community: Structure and prey populations. J. Mammal., **55:**138–157.

Black, H. L. 1979. Precision in prey selection by the trident-nosed bat (*Cloeotis percivali*). Mammalia, **43:**53–57.

Blest, A. D., T. S. Collett, and J. D. Pye. 1963. The generation of ultrasonic signals by a New World arctiid moth. Proc. R. Soc. London B, **158:**196–207.

Bradbury, J. W. 1970. Target discrimination by the echolocating bat, *Vampyrum spectrum*. J. Exp. Zool., **173:**23–46.

Bradbury, J. W. 1977. Social organization and communication. Pp. 1–72, *in* Biology of bats. Vol. 3. (W. A. Wimsatt, ed.). Academic Press, New York, 651 pp.

Bradbury, J. W., and L. H. Emmons. 1974. Social organization of some Trinidad bats. I. Emballonuridae. Z. Tierpsychol., **36:**137–183.

Bradbury, J. W., and S. L. Vehrencamp. 1976a. Social organization and foraging in emballonurid bats. I. Field studies. Behav. Ecol. Sociobiol., **1:**337–381.

Bradbury, J. W., and S. L. Vehrencamp. 1976b. Social organization and foraging in emballonurid bats. II. A model for the determination of group size. Behav. Ecol. Sociobiol., **1:**383–404.

Bradbury, J. W., and S. L. Vehrencamp. 1977a. Social organization and foraging in emballonurid bats. III. Mating systems. Behav. Ecol. Sociobiol., **2:**1–17.

Bradbury, J. W., and S. L. Vehrencamp. 1977b. Social organization and foraging in emballonurid bats. IV. Parental investment patterns. Behav. Ecol., Sociobiol., **2:**19–29.

Bradbury, J. W., D. W. Morrison, E. Stashko, and E. R. Heithaus. 1979. Radio-tracking methods for bats. Bat Res. News, **20:**9–17.

Brosset, A. 1966. La biologie des chiroptères. Masson, Paris, 237 pp.

Buchler, E. R. 1975. Food transit time in *Myotis lucifugus* (Chiroptera: Vespertilionidae). J. Mammal., **56:**252–255.

Buchler, E. R. 1976a. Prey selection by *Myotis lucifugus* (Chiroptera: Vespertilionidae). Am. Nat., **110:**619–628.

Buchler, E. R. 1976b. A chemiluminescent tag for tracking bats and other small nocturnal animals. J. Mammal., **57:**173–176.

Buchler, E. R. 1980. The development of flight, foraging, and echolocation in the little brown bat (*Myotis lucifugus*). Behav. Ecol. Sociobiol., **6:**211–218.

Busnel, R. -G., and J. F. Fish (eds.). 1980. Animal sonar systems. NATO Advanced Study Institutes Series A, Vol. 28. Plenum Press, New York, 1135 pp.

Dijkgraaf, S. 1946. Die Sinneswelt der Fledermause. Experientia, **3:**206–208.

Dijkgraaf, S. 1949. Spallanzani und die Fledermause. Experientia, **5:**90.

Dunning, D. C. 1968. Warning sounds of moths. Z. Tierpsychol., **25:**129–138.

Dunning, D. C., and K. D. Roeder. 1965. Moth sounds and the insect-catching behavior of bats. Science, **147**:173–174.

Dwyer, P. D. 1970. Foraging behaviour of the Australian large-footed *Myotis* (Chiroptera). Mammalia, **34**:76–80.

Edgerton, H. E., P. F. Spangle, and J. K. Baker. 1966. Mexican freetail bats: Photography. Science, **153**:201–203.

Fenton, M. B. 1975. Observations on the biology of some Rhodesian bats, including a key to the Chiroptera of Rhodesia. Life Sci. Contr. R. Ont. Mus., **104**:1–27.

Fenton, M. B. 1980. Adaptiveness and ecology of echolocation in terrestrial (aerial) systems. Pp. 427–446, *in* Animal sonar systems. NATO Advanced Study Institutes Series A, Vol. 28. (R. -G. Busnel and J. F. Fish, eds.). Plenum Press, New York, 1135 pp.

Fenton, M. B., and G. P. Bell. 1979. Echolocation and feeding behaviour in four species of *Myotis* (Chiroptera). Can. J. Zool., **57**:1271–1277.

Fenton, M. B., and G. P. Bell. 1981. Recognition of species of insectivorous bats by their echolocation calls. J. Mammal., **62**:233–243.

Fenton, M. B., and T. H. Fleming. 1976. Ecological interactions between bats and nocturnal birds. Biotropica, **8**:104–110.

Fenton, M. B., and J. H. Fullard. 1979. The influence of moth hearing on bat echolocation strategies. J. Comp. Physiol., **132**:77–86.

Fenton, M. B., and G. K. Morris. 1976. Opportunistic feeding by desert bats (*Myotis* spp.). Can. J. Zool., **54**:526–530.

Fenton, M. B., and D. W. Thomas. 1980. Dry season overlap in activity patterns, habitat use, and prey selection by sympatric African insectivorous bats. Biotropica, **12**:81–90.

Fenton, M. B., N. G. H. Boyle, T. M. Harrison, and D. J. Oxley. 1977. Activity patterns, habitat use, and prey selection by some African insectivorous bats. Biotropica, **9**:73–85.

Fenton, M. B., G. P. Bell, and D. W. Thomas. 1980. Echolocation and feeding behaviour of *Taphozous mauritianus* (Chiroptera: Emballonuridae). Can. J. Zool., **58**:1774–1777.

Fiedler, J. 1979. Prey catching with and without echolocation in the Indian false vampire bat (*Megaderma lyra*). Behav. Ecol. Sociobiol., **6**:155–160.

Findley, J. S. 1976. The structure of bat communities. Am. Nat., **110**:129–139.

Fitzpatrick, J. W. 1980. Foraging behavior of Neotropical tyrant flycatchers. Condor, **82**:43–57.

Freeman, P. W. 1979. Specialized insectivory: Beetle-eating and moth-eating molossid bats. J. Mammal., **60**:467–479.

Fullard, J. H. 1979. Behavioral analyses of auditory sensitivity in *Cycnia tenera* Hübner (Lepidoptera: Arctiidae). J. Comp. Physiol., **129**:79–83.

Fullard, J. H., and M. B. Fenton. 1977. Acoustic and behavioural analyses of the sounds produced by some species of Nearctic Arctiidae (Lepidoptera). Can. J. Zool. **55**:1223–1224.

Fullard, J. H., M. B. Fenton, and J. A. Simmons. 1979. Jamming bat echolocation: the clicks of arctiid moths. Can. J. Zool., **57**:647–649.

Gillette, D. D. 1975. Evolution of feeding strategies in bats. Tebiwa, **18**:39–48.

Goldman, L. J., and O. W. Henson, Jr. 1977. Prey recognition and selection by the constant frequency bat, *Pteronotus parnellii parnellii*. Behav. Ecol. Sociobiol., **2**:411–420.

Gould, E. 1978. Opportunistic feeding by tropical bats. Biotropica, **10**:75–76.

Griffin, D. R. 1958. Listening in the dark. Yale University Press, New Haven, 413 pp.

Griffin, D. R. 1971. The importance of atmospheric attenuation for the echolocation of bats (Chiroptera). Anim. Behav., **19**:55–61.

Griffin, D. R., and J. A. Simmons. 1974. Echolocation of insects by horseshoe bats. Nature, **250**:731–732.

Griffin, D. R., F. A. Webster, and C. R. Michael. 1960. The echolocation of flying insects by bats. Anim. Behav., **8**:141–154.

Gustafson, Y., and H. -U. Schnitzler. 1979. Echolocation and obstacle avoidance in the hipposiderid bat, *Asellia tridens*. J. Comp. Physiol., **131**:161–167.

Hamilton, W. J., III, and K. E. F. Watt. 1970. Refuging. Ann. Rev. Ecol. Syst., **1**:263–286.

Hassall, J. R., and K. Zaveri. 1978. Application of B & K equipment to acoustic noise measurement. Bruel and Kjaer DK-2850 Naerum, 280 pp.

Hebert, P. D. N. 1980. Moth communities in montane Papua New Guinea. J. Anim. Ecol., **49**:593–602.

Hespenheide, H. A. 1971. Food preference and the extent of overlap in some insectivorous birds, with special reference to the Tyrannidae. Ibis, **113**:59–72.

Jones, C. 1965. Ecological distribution and activity periods of bats of the Mogollon Mountains area of New Mexico and adjacent Arizona. Tulane Stud. Zool., **12**:93–100.

Kallen, F. C., and C. Gans. 1972. Mastication in the little brown bat, *Myotis lucifugus*. J. Morphol., **136**:385–420.

Kolb, A. 1976. Funktion und Wirkungsweise der Reichlaute der Mausohrfledermaus, *Myotis myotis*. Z. Säugetierk., **41**:226–236.

Kunz, T. H. 1973. Resource utilization: temporal and spatial components of bat activity in central Iowa. J. Mammal., **54**:14–32.

Kunz, T. H. 1974. Feeding ecology of a temperate insectivorous bat (*Myotis velifer*). Ecology, **55**:693–711.

Kunz, T. H. 1980. Daily energy budgets of free-living bats. Pp. 369–392, *in* Proceedings of the Fifth International Bat Research Conference. (D. E. Wilson and A. L. Gardner, eds.). Texas Tech Press, Lubbock, 434 pp.

Kunz, T. H., and C. E. Brock. 1975. A comparison of mist nets and ultrasonic detectors for monitoring flight activity of bats. J. Mammal., **56**:907–911.

Miller, L. A. 1971. Physiological responses of green lacewings (*Chrysopa* Neuroptera) to ultrasound. J. Insect Physiol., **17**:491–506.

Miller, L. A., and J. Olesen. 1979. Avoidance behavior in green lacewings. I. Behavior of free flying green lacewings to hunting bats and ultrasound. J. Comp. Physiol., **131**:113–120.

Møhl, B., and L. A. Miller. 1976. Ultrasonic clicks produced by the peacock butterfly: A possible bat repellant. J. Exp. Biol., **64**:639–644.

Moiseff, A., G. S. Pollack, and D. R. Hoy. 1978. Steering responses of flying crickets to sound and ultrasound: Mate attraction and predator avoidance. Proc. Nat. Acad. Sci., U.S.A., **75**:4052–4056.

Morrison, D. W. 1978. On the optimal searching strategy for refuging predators. Am. Nat., **112**:925–934.

Murray, P. F., and T. Strickler. 1975. Notes on the structure and function of cheek pouches within the Chiroptera. J. Mammal., **56**:673–676.

Nicholls, C. F. 1960. A portable, mechanical insect trap. Can. Entomol., **92**:48–51.

Novick, A. 1958. Orientation in paleotropical bats. I. Microchiroptera. J. Exp. Zool., **138**:81–154.

Novick, A. 1977. Acoustic orientation. Pp. 74–289, *in* Biology of bats. Vol. 3. (W. A. Wimsatt, ed.). Academic Press, New York, 651 pp.

Novick, A., and B. A. Dale. 1971. Foraging behavior in fishing bats and their insectivorous relatives. J. Mammal., **52**:817–818.

Nyholm, E. S. 1965. Zur Ökologie von *Myotis mystacinus* (Leisl.) und *M. daubentoni* (Leisl.) (Chiroptera). Ann. Zool. Fenn., **2**:77–123.

Orr, R. T. 1954. Natural history of the pallid bat, *Antrozous pallidus* (LeConte). Proc. Calif. Acad. Sci., **28**:165–246.

O'Shea, T. J., and T. A. Vaughan. 1977. Nocturnal and seasonal activities of the pallid bat, *Antrozous pallidus*. J. Mammal., **58**:269–284.

O'Shea, T. J., and T. A. Vaughan. 1980. Ecological observations on an east African bat community. Mammalia, **44**:485–496.

Pye, J. D. 1968. How insects hear. Nature, **218**:797.

Pye, J. D. 1972. Bimodal distribution of constant frequencies in some hipposiderid bats (Mammalia, Hipposideridae). J. Zool., **166**:323–335.

Pye, J. D. 1973. Echolocation by constant frequency in bats. Period. Biol., **75**:21–26.

Pyke, G. H., H. R. Pulliam, and E. L. Charnov. 1977. Optimal foraging: A selective review of theory and tests. Q. Rev. Biol., **52**:137–154.

Reith, C. C. 1980. Shifts in times of activity by *Lasionycteris noctivagans*. J. Mammal., **61**:104–108.

Rentz, D. C. 1975. Two new katydids of the genus *Melanonotus* from Costa Rica with comments on their life history strategies (Tettigoniidae: Pseudophyllinae). Entomol. News, **86**:129–140.

Roeder, K. D. 1967. Nerve cells and insect behavior. Rev. ed., Harvard University Press, Cambridge, 238 pp.

Roeder, K. D. 1974. Acoustic sensory responses and possible bat-evasion tactics of certain moths. Pp. 71–78, *in* Proceedings of the Canadian Society of Zoology, Annual Meeting. (M. D. B. Burt, ed.). University of New Brunswick, Fredericton, 135 – 64 pp.

Roeder, K. D. 1975. Neural factors and evitability in insect behavior. J. Exp. Zool., **194**:75–88.

Roeder, K. D., and A. E. Treat. 1960. The acoustic detection of bats by moths. Paper presented at the XI International Entomological Congress in Vienna.

Ross, A. 1967. Ecological aspects of the food habits of insectivorous bats. Proc. West. Found. Vert. Zool., **1**:205–263.

Schnitzler, H. -U. 1968. Die Ultraschall-Ortungslaute der Hufeisen-Fledermause (Chiroptera, Rhinolophidae) in ve schiedenen Orientierungssituationen. Z. Vergl. Physiol., **57**:376–408.

Schnitzler, H. -U. 1978. Die detektion von Vewegungen durch Echoortung bei Fledermausen. Verh. Dtsch. Zool. Ges., **1978**:16–33.

Schnitzler, H. -U., and A. D. Grinnell. 1977. Directional sensitivity of echolocation in the horseshoe bat, *Rhinolophus ferrum-equinum*. I. Directionality of sound emission. J. Comp. Physiol., **116**:51–61.

Schoener, T. W. 1974. Resource partitioning in ecological communities. Science, **185**:27–39.

Schuller, G., and G. Pollak. 1979. Disproportionate frequency representation in the inferior colliculus of Doppler-compensating greater horseshoe bats: Evidence for an acoustic fovea. J. Comp. Physiol., **132**:47–54.

Shimozawa, T., N. Suga, P. Hendler, and S. Schuetze. 1974. Directional sensitivity of echolocation system in bats producing frequency modulated signals. J. Exp. Biol., **60**:53–69.

Simmons, J. A., and M. J. O'Farrell. 1977. Echolocation by the long-eared bat, *Plecotus phyllotis*. J. Comp. Physiol., **122**:201–214.

Simmons, J. A., and R. A. Stein. 1980. Acoustic imaging in bat sonar: Echolocation signals and the evolution of echolocation. J. Comp. Physiol., **135**:61–84.

Simmons, J. A., D. J. Howell, and N. Suga. 1975. Information content of bat sonar echoes. Am. Sci., **63**:204–215.

Simmons, J. A., W. A. Lavender, B. A. Lavender, J. E. Childs, K. Hulebak, M. R. Rigden, J. Sherman, B. Woolman, and M. J. O'Farrell. 1978. Echolocation by free-tailed bats (*Tadarida*). J. Comp. Physiol., **125**:291–299.

Simmons, J. A., M. B. Fenton, and M. J. O'Farrell. 1979a. Echolocation and pursuit of prey by bats. Science, **203**:16–21.

Simmons, J. A., M. B. Fenton, W. R. Ferguson, M. Juttin, and J. Palin. 1979b. Apparatus for research on animal ultrasonic signals. Life Sci. Misc. Publ. R. Ont. Mus., 31 pp.

Suga, N. 1966. Ultrasonic production and its reception in some Neotropical Tettigoniidae. J. Insect Physiol., **12**:1039–1050.

Suthers, R. A. 1965. Acoustic orientation by fish-catching bats. J. Exp. Zool., **158**:319–348.

Taylor, L. R. 1962. The absolute efficiency of insect suction traps. Ann. Appl. Biol., **50**:405–421.

Vater, M. 1980. Coding of sinusoidally frequency-modulated signals by single cochlear nucleus neurons of *Rhinolophus ferrum-equinum*. Pp. 999–1001, *in* Animal sonar systems. NATO Advanced Study Institutes Series A. Vol. 28. (R. -G. Busnel and J. F. Fish, eds.). Plenum Press, New York, 1135 pp.

Vaughan, T. A. 1970. Flight patterns and aerodynamics. Pp. 195–216, *in* Biology of bats. Vol. 1. (W. A. Wimsatt, ed.). Academic Press, New York, 477 pp.

Vaughan, T. A. 1976. Nocturnal behavior of the African false vampire bat (*Cardioderma cor*). J. Mammal., **57**:227–248.

Vaughan, T. A. 1977. Foraging behaviour of the giant leaf-nosed bat (*Hipposideros commersoni*). East Afr. Wildl. J., **15**:237–249.

Vaughan, T. A. 1980. Opportunistic feeding by two species of *Myotis*. J. Mammal., **61**:118–119.

Vehrencamp, S. L., F. G. Stiles, and J. W. Bradbury. 1977. Observations on the foraging behavior and avian prey of the Neotropical carnivorous bat, *Vampyrum spectrum*. J. Mammal., **58**:469–478.

Wallin, L. 1969. The Japanese bat fauna. Zool. Bidrag, Uppsala, **37**:223–440.

Werner, T. K. 1981. Responses of non-flying moths to ultrasound: The threat of gleaning bats. Can. J. Zool. **59**:525–529.

Wiens, J. A. 1977. On competition and variable environments. Am. Sci., **65**:590–597.

Chapter 8

Foraging Strategies of Plant-Visiting Bats

Theodore H. Fleming

Department of Biology
University of Miami
Coral Gables, Florida 33124

I. INTRODUCTION

Approximately 250 of the 850 known species of bats (about 29%) are partially or wholly dependent upon plants as a source of food. These bats are involved in a mutualistic exploitation system with plants in which the bats obtain food in the form of nectar, pollen, or fruit while providing mobility for the plant's pollen grains or seeds. Coevolutionary aspects of these bat–plant interactions are discussed in detail (Chapter 9). In this chapter I will discuss our present knowledge of the foraging strategies of plant-visiting bats and will be minimally concerned with the potential consequences of chiropteran foraging behavior on plant populations.

With few exceptions, bat–plant interactions are restricted to tropical parts of the world. Plant-visiting bats are found in two families, the Phyllostomidae of the New World tropics and the Pteropodidae of the Old World tropics. Bats of these families pollinate or disperse the seeds of hundreds of species of plants, including many economically important species such as *Ceiba pentandra* (kapok), *Durio zibethinus* (durian), *Eucalyptus* spp., *Ficus* spp. (figs), *Mangifera indica* (mango), *Manilkara zapota* (chicle), *Musa paradisiaca* (wild banana), and *Ochroma lagopus* (balsa). Plant-visiting bats have been interacting evolutionarily with angiosperms for somewhat less than 50 million years, during which they have had an appreciable influence on angiosperm diversity at the generic and specific (but not familial) level (Fleming, 1979). Increased plant diversity, in turn, has undoubtedly had a diversifying effect on bats.

TABLE I
General Comparisons between Pteropodid and Phyllostomid Bats[a]

Characteristic	Pteropodids	Phyllostomids
1. Evolutionary age	Late Eocene	Late Oligocene/early Miocene
2. Geographic range	Old World tropics	New World tropics
3. Present taxonomic diversity	38 genera, 150 species	49 genera, 137 species
4. General size (see Fig. 1)	Relatively large (up to 900 g)	Small to medium (up to 190 g)
5. Major dietary items	Fruit, flowers, nectar, leaves	Diverse: insects, fruit, nectar, pollen, vertebrates
6. Echolocation ability	Absent except in *Rousettus* (which produces low-frequency clicks)	Present in all species, but many produce low-intensity ultrasounds
7. Brain features	Relatively large brain, enlarged visual and olfactory regions	Relatively large brain, enlarged visual, olfactory, and cerebellar regions
8. Wing and flight features	Considerable uniformity within family: wings short with average aspect ratio, high wing loading, and short wing tips for fast, direct flight with little hovering ability	Considerable diversity within family: generally wings have average-to-low aspect ratio, average-to-high wing loading, and long wing tips for slow, maneuverable flight; many can hover

[a]Sources of information for this table include Baron and Jolicoeur (1980), Eisenberg and Wilson (1978), Findley *et al.* (1972), Jones and Carter (1976), Koopman and Cockrum (1967), Smith (1976), and Sussman and Raven (1978).

The two families of plant-visiting bats offer an interesting series of contrasts regarding their evolutionary histories and, consequently, their anatomy, physiology, and behavior. These differences, which are summarized in Table I, might lead us to expect major differences in the foraging strategies of phyllostomid and pteropodid bats. Pteropodid bats, classified in the suborder Megachiroptera (hence the name *megabats*), probably arose in the late Eocene (or possibly late Oligocene/early Miocene—see Van Valen, 1979) and have always been confined to the Old World. In contrast, phyllostomid bats, classified in the suborder Microchiroptera (hence the name *microbats*), arose somewhat later (in late Oligocene/early Miocene) from an insectivorous ancestor and have always been confined to the New World. Both families are represented by similar numbers of species, but the generic diversity of the phyllostomids is greater (Table I). As their name implies, megabats, on the average, are larger than microbats (Fig. 1). The largest pteropodid, *Pteropus vampyrus*, weighs up to 900 g and has a wingspan of about 1.7 m, whereas the largest phyllostomid, *Vampyrum spectrum* (which is a carnivore, not a frugivore) weighs up to 190 g and has a wingspan of about 1 m.

Despite their greater uniformity of size, phyllostomid bats are more diverse morphologically, behaviorally, and ecologically than are pteropodid bats. Because they can echolocate, phyllostomids have a wider spectrum of potential food items available to them. Although many species are basically frugivorous or

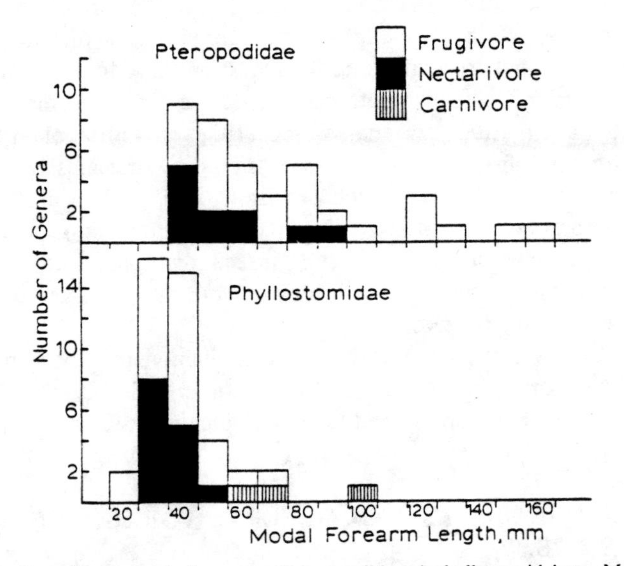

FIGURE 1. Size distributions of genera of pteropodid and phyllostomid bats. Modal forearm lengths were taken from Walker (1964); food habits of the bats follow Wilson (1973). Insectivorous phyllostomid genera were eliminated from consideration.

nectarivorous, their diets often include insects (Arata *et al.*, 1967; Ayala and D'Alessandro, 1973; Fleming *et al.*, 1972; Gardner, 1977). Certain phyllostomids, including the 11-g *Glossophaga soricina* and the 90-g *Phyllostomus hastatus*, are omnivorous, feeding on nectar, fruit, insects, and, in the case of *P. hastatus*, vertebrates. Pteropodids, in contrast, are plant feeders and do not include animals in their diets. Because they do not produce ultrasounds, aerial insectivory is not available as a foraging mode to pteropodids.

In addition to major dietary differences, these two bat families differ in their flight characteristics (Table I). Based on an examination of nine species (of 9 genera) of pteropodids and 43 species (25 genera) of phyllostomids, Findley *et al.* (1972) concluded that megabats are more uniform regarding wing morphology and flight characteristics than are microbats. Pteropodid wings are relatively short and heavily loaded and have short wing tips. Their wings seem to be designed for speed and direct flight, and most species have limited hovering ability. Phyllostomid wings, in contrast, are morphologically more diverse but seem designed for a slower, more maneuverable flight. Many species are good-to-excellent hoverers.

A final morphological feature, relative brain size, sets pteropodid and phyllostomid bats apart from other kinds of bats. Members of both families are notable for their relatively large brains, resulting in large part from their expanded olfactory, visual, neocortical, and, in phyllostomids, cerebellar regions (Eisenberg and Wilson, 1978; Baron and Jolicoeur, 1980). Compared with frugivorous microbats, megabats have relatively larger visual and olfactory regions, which they use for flight orientation and food location. In addition to using visual and olfactory information, microbats can, of course, orient themselves using auditory information. Eisenberg and Wilson (1978: p. 750) explain the large brains of chiropteran frugivores as follows: "Foraging strategies involving the location of rich food resources which are isolated in small pockets seem to require a large brain weight relative to body mass." Providing support for this conclusion is the fact that foliage-gleaning insectivorous bats, which probably hunt for food in much the same way as frugivores, have larger brains than similarly sized aerial insectivores.

Before discussing the foraging behavior of plant-visiting bats in detail, I will briefly summarize two topics that significantly influence this behavior: patterns of plant food availability and the basic components of optimal foraging strategies.

2. FOOD AVAILABILITY AND GENERAL FORAGING STRATEGIES

2.1. Food Availability

Plant-visiting bats have four major concerns regarding the nature of their food supply: what kinds of plant foods are available, how much is there of each kind, where is it located, and for how long is it available? The answers to these questions reflect in part the outcome of the coevolution between bats and plants. On the botanical side this coevolution has affected the morphology of flowers and fruits and numerous phenological traits, including the timing of flower and fruit production, the degree of intraspecific synchrony in floral presentation or fruit maturation, the size of the flower or fruit crop, the amount of pollen and nectar or ripe fruit available per plant per night, and the nutritional characteristics of pollen, nectar, and fruits. Numerous authors (e.g., Baker, 1961; Heinrich, 1975; Howe and Estabrook, 1977; Janzen, 1978; McKey, 1975; and Stiles, 1978a) have discussed the evolution and coevolution of these botanical traits.

Regarding overall phenological patterns, Gentry (1974), working with plants of the Bignoniaceae, recognizes four basic flowering patterns, ranging from the production of a large number of flowers over a short period ("big bang") to the production of a small number of flowers over an extended time period ("steady state"). The remaining two strategies include "cornucopia" (a rather large number of flowers over a month's time) and "multiple bang" (several widely spaced, large flower crops per year). Fruit strategy analogues of each of these flowering patterns also exist.

The two extreme patterns, big bang (BB) and steady state (SS), represent significantly different phenological strategies in a number of ways. BB plants, which usually attract large numbers of opportunistic plant visitors, are probably less predictable in space and time than are SS plants. Owing to a rather high diversity of visitors, many of whom will not be effective pollen or seed vectors, many pollen grains or seeds will not be removed from the parent plant. Thus, the probability that each "propagule" will be effectively dispersed will generally be very low. BB plants sacrifice the quality of animal service to maximize the quantity of visitations.

SS plants, in contrast, appear to maximize the quality of the services their visitors provide while sacrificing the quantity of propagules dispersed. SS plants attract a much smaller number of more specialized visitors, which are thought to provide more reliable dispersal services (McKey, 1975). This reliability means that the probability of each "propagule" being effectively dispersed will be much higher than in BB plants. In general, the nutritional rewards provided by SS plants should be more predictable in time and space.

With this background in mind, we can examine the reproductive strategies of bat-visited plants, beginning with flowering strategies. Like most tropical plants (Frankie *et al.*, 1974; Croat, 1975), plants producing bat flowers tend to reproduce seasonally. In the dry tropical forest of western Costa Rica, for example, more species of bat flowers are available in the dry season (mid-November to mid-May) than in the wet season (Fig. 2). The mean length of the flowering season in 25 Costa Rican species is 3.8 months, and there is a strong trend towards temporally displaced flowering periods, perhaps to reduce interspecific competition for high-quality pollination services (Heithaus *et al.*, 1975). A dry-

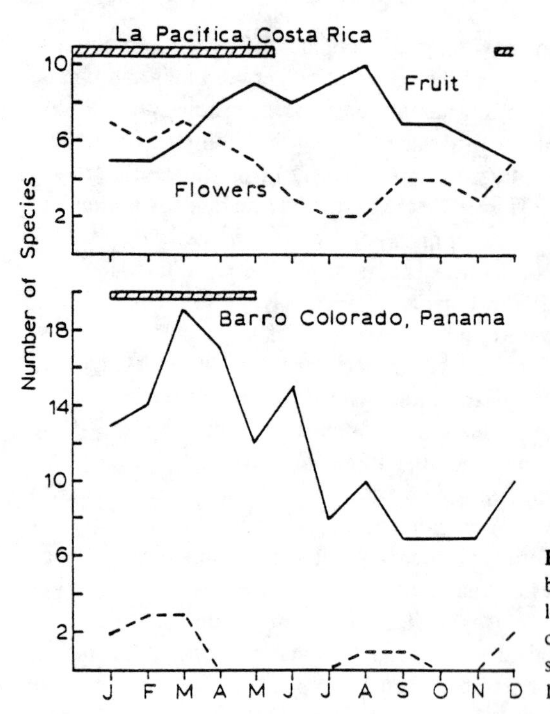

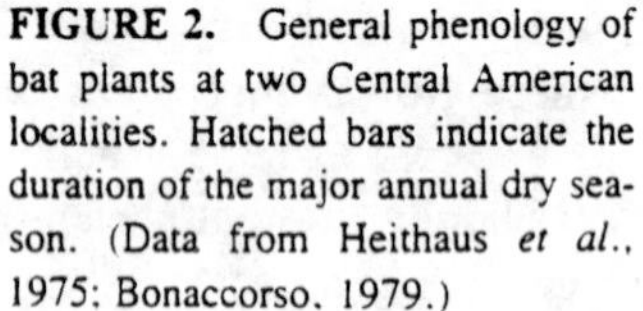
FIGURE 2. General phenology of bat plants at two Central American localities. Hatched bars indicate the duration of the major annual dry season. (Data from Heithaus *et al.*, 1975; Bonaccorso, 1979.)

season flowering peak also occurs in central Panama (Fig. 2). Seasonal flowering patterns are probably also the rule in Old World bat plants (e.g., Start and Marshall, 1976; Lack, 1978). This generalization can also be extended to flowers visited by nectarivorous birds. In the lowland wet forest at La Selva, Costa Rica, for example, flowering species visited by a guild of hummingbirds have relatively short blooming periods ($\bar{x}$ = 3.3 months) that are temporally displaced (Stiles, 1975, 1977, 1978b).

Can BB and SS flowering strategies be recognized among bat-visited plants? Although there is considerable diversity in the blooming patterns of chiropterophilous plants, a majority of such species appear to be either BB or SS plants. Examples of BB species in the Neotropics include *Ceiba pentandra*, *C. acuminata*, and *Lafoensia glyptocarpa* (Baker *et al.*, 1971; Sazima and Sazima, 1977). In the Paleotropics *Ceiba pentandra*, *Parkia* spp., *Durio zibethinus*, and *Sonneratia* spp., to name a few, produce large flushes of flowers over short time periods (Start and Marshall, 1976). Neotropical SS flower species include *Bauhinia pauletia*, *B. ungulata*, *Crescentia cujete*, and *Passiflora mucronata* (Gentry, 1974; Heithaus *et al.*, 1974; Sazima and Sazima, 1978). Species of *Bauhinia* in western Costa Rica produce two to three flowers per inflorescence every 2–3 days, and *Passiflora mucronata* produces one flower per branch per night. In the Old World *Musa* spp., *Cocos nucifera*, *Arenga* spp., *Durabanga grandiflora*, and *Oroxylum indicum* are SS bloomers (Start and Marshall, 1976). *Oroxylum* trees bear up to 40 active inflorescences ($\bar{x}$ = 7.5), and each inflorescence contains one to four open flowers each night (Gould, 1978).

In addition to seasonal patterns of flower availability, nectarivorous bats face nightly interspecific variation in the timing and amounts of nectar production. Some flowers (e.g., *Durio zibethinus*, *Musa acuminata*, and *Bauhinia pauletia*) begin nectar secretion at or before sunset, whereas others (e.g., *Oroxylum indicum*, *Passiflora mucronata*, and *Agave palmeri*) do not begin secreting until well after sunset (Table II). Howell (1977) suggested that temporal secretion differences between sympatric species have evolved to reduce interspecific competition for the same pollinating agent. As summarized in Table II, nightly nectar production for a variety of flowers ranges from 0.4 to 2.5 ml or more. *Parkia clappertoniana*, whose inflorescences secrete up to 5 ml per night (Baker and Harris, 1957), is a particularly productive species. The energetic value of these nectars is somewhat less than 4.2 kJ/ml, so that a flower or inflorescence secreting 5 ml per night yields about 21 kJ.

Recent work by Herbert and Irene Baker (e.g., Baker, 1978) indicates that the chemical composition of Neotropical bat plant nectar differs significantly from nectar in other pollination syndromes (e.g., bee or butterfly nectar). It tends to be low in sugar ($\bar{x}$ concentration = 17%, N = 5 species), total amino acids ($\bar{x}$ = 275 μmole/l, N = 15 species), and nonprotein amino acids (which might possibly serve as deterrents against nectar thieves). The former two characteristics seem to reflect the foraging modes of nectarivorous microbats. Since they make brief visits to flowers, these bats cannot easily handle concentrated nectar; moreover, they often supplement their plant diets with insects and hence have an alternate source of amino acids. The significance of the third trait is not currently understood. Similar chemical features also characterize the nectar of hummingbird flowers (Baker, 1978).

As in the case of bat flowers, plants producing bat fruits usually reproduce seasonally. In many species, fruit production and maturation tends to be highly synchronous. Janzen (1978) and Smythe (1970) attribute this to seed-predator satiation. Synchronous seasonal fruit production reduces the chances that either predispersal seed predators, such as bruchid beetles (Coleoptera, Bruchidae), or postdispersal seed predators, such as rodents, can build up populations large enough to destroy entire seed crops. Predator avoidance has also apparently selected for interannual flowering and fruiting (Janzen, 1978). Examples of bat resources not available every year in the Neotropics include flowers of *Hymenaea courbaril* and fruits of *Andira inermis*. Intraspecific asynchrony in fruit production is characteristic of many species of figs (*Ficus*) throughout the world. This asynchrony, which makes fig crops rather patchy in time and space, is the product of coevolution between figs and their pollinators, tiny wasps of the family Agaonidae (Janzen, 1979; Milton *et al.*, 1982).

TABLE II
Rates and Amounts of Nectar Production by Bat-Visited Flowers

Species and family	Location	Period of nectar flow	Total volume of nectar per flower per night (range) (ml)	kJ/ml	References
Oroxylum indicum (Bignoniaceae)	Malaya	~2045–0130 h	1.8 (0.8–2.3)	4.0	Gould (1978)
Durio zibethinus (Bombacaceae)	Malaya	before sunset–2400 h	0.4 (0.1–1.4)	3.6	Gould (1978)
Musa acuminata (Musaceae)	Malaya	1830–0200 h	0.6 (0.5–0.8)	2.7	Gould (1978)
Maranthes polyandra (Chrysobalanaceae)	West Africa	1730–0300 h	0.4		Lack (1978)
Bauhinia pauletia (Leguminosae)	Costa Rica	1815–2300+ h	~2.5		Heithaus *et al.* (1974)
Bauhinia ungulata (Leguminosae)	Costa Rica	2200–0500 h	~0.4		Howell (1977)
Inga vera (Leguminosae)	Costa Rica	1830–0200 h	~0.4		Howell (1977)
Passiflora mucronata (Passifloraceae)	Southeast Brazil	0100–0700 h	0.5+[a]		Sazima and Sazima (1978)
Agave palmeri (Agavaceae)	Southern Arizona	2000–2200 h	0.6[b]	2.2	Howell (1979)

[a]Maximum capacity of the floral chamber.
[b]Dehiscent flowers only.

Although individual species have restricted fruiting seasons, several kinds of bat fruits are generally available at all times of the year in tropical habitats. In Costa Rican dry tropical forests, for example, at least five and as many as 10 species of bat fruits are available each month, and on Barro Colorado Island, Panama, monthly totals range from seven to 19 species (Fig. 2). The overall biomass of bat fruits in these habitats, and possibly in other tropical habitats (e.g., La Selva in Costa Rica; Frankie *et al.*, 1974), tends to be greatest in the early-to-mid wet season and lowest late in the wet season.

At least three fruiting strategies can be recognized among bat-dispersed plants: BB, cornucopia, and SS. The BB (or multiple-bang) strategy is perhaps best characterized by various *Ficus* species. On Barro Colorado Island *F. yoponensis* and *F. insipida* produce enormous crops of 5000–50,000 fruits, which are depleted in about 1 week by a wide variety of vertebrate frugivores. Individual fig trees fruit once or twice a year but rarely do so at the same time every year (Milton *et al.*, 1982). An example of a "cornucopia" fruit species is *Chlorophora tinctoria* (Moraceae). In western Costa Rica individuals of this species annually produce large crops (thousands of fruits) that ripen and are eaten over a period of about 1 month; individual trees bear ripe fruits for several weeks (Heithaus and Fleming, 1978). An Old World example of this fruiting strategy is *Celtis duranda* (Ulmaceae), which in Uganda produces large crops that last for several months (Strushaker, 1978).

SS bat-dispersed species are probably common in all tropical forests. Neotropical dry forest examples include *Muntingia calabura, Cecropia peltata*, and *Piper* spp. (Heithaus and Fleming, 1978, unpublished data). *Muntingia* is unusual among dry forest species in that it produces ripe fruit year-round, although its dry season production is considerably lower than its daily wet season production of 10–50 fruits per tree. Similarly, certain populations of *C. peltata* produce a few ripe fruits per plant nearly year-round, whereas other populations located on drier sites restrict their fruiting to the wet season. A third example of Neotropical SS fruiting plants are members of the genus *Piper* (Piperaceae). Most *Piper* tend to reproduce seasonally, but certain moist or wet forest forms flower and fruit throughout the year (Croat, 1978; Opler *et al.*, 1980). Temporal replacement series are known to exist in the Costa Rican dry forest and the moist forest in central Panama (Fleming *et al.*, 1977; Bonaccorso, 1979). At Santa Rosa National Park, Costa Rica, for example, *P. amalago* fruits in June and July, *P. pseudofuligineum* in July and August, *P. marginatum* in December and January, and *P. tuberculatum* in March and April. Each night individuals of *P. amalago* and *P. pseudofuligineum* bear only one to four ripe fruits (about 5% of a plant's total fruit crop), so that nightly fruit density is low (but spatially and temporally predictable).

Although the general phenological patterns of bat-visited plants have been described for at least two tropical locations (Heithaus *et al.*, 1975; Bonaccorso,

1979), we still lack knowledge of spatiotemporal changes in the total bat resource density for any locality. Data needed for these calculations include the density and dispersion patterns of bat-visited plants and their daily levels of resource production. A partial picture of the bat plant resource landscape during two wet season months (July and August) at Santa Rosa National Park, Costa Rica, has been painted by Heithaus and Fleming (1978). With the exception of *Ficus* species, each of the eight species they mapped had clumped dispersion patterns, and some species (e.g., *Muntingia calabura, Cecropia peltata,* and *Chlorophora tinctoria*) occur in mixed-species clumps (Fleming and Heithaus, 1981). Plant densities ranged from less than 0.5 per hectare in the small tree *Solanum hayesii* to 200 individuals per hectare in the shrub *Piper amalago*. Daily numbers of ripe fruits produced per hectare varied from about 1 in *P. jacquemontianum* to nearly 2000 in *Ficus* spp. and *Chlorophora tinctoria*, and the daily production of utilizable kilocalories ranged from about 2 per hectare in *Solanum hayesii* to 3150 per hectare in *Chlorophora tinctoria* (Lockwood *et al.*, 1982). To my knowledge, similar estimates have not yet been made for bat flower resources.

Bats cannot live on energy alone. In addition to knowing the energetic values of nectar, pollen, and fruit and flower parts eaten by bats, we need to know more about their other nutritional characteristics, including amino acid, fat, and carbohydrate compositions and levels of critical elements such as calcium and phosphorus. On theoretical grounds, we might expect plants to produce fruit that lack certain dietary items so that animals have to feed on more than one species (in perhaps different locations) to obtain a balanced diet (Smith, 1977). A corollary of this hypothesis is that species producing fruit at the same time of the year and utilizing the same dispersal agents should have complementary nutritional deficiencies, whereas species fruiting at different times might converge in their nutritional rewards. Plants gain a twofold advantage from this strategy. First, they gain greater mobility for their seeds (unless plants occur in mixed-species clumps) and, secondly, they experience less interspecific competition for seed-dispersal services. Unfortunately, data needed to test this hypothesis for guilds of frugivorous bats do not exist (but see Hladik *et al.* (1971) for the composition of monkey fruits on Barro Colorado Island, Panama). Future studies on the food habits of frugivorous bats should direct more attention to the nutritional characteristics of different fruit species.

2.2. General Foraging Strategies

During the past 15 years a considerable literature dealing with the evolution of foraging behavior has appeared (see major reviews by Schoener, 1971; Pyke *et al.*, 1977; Krebs, 1978, 1979). Under the assumption that natural selection favors individuals that forage efficiently, numerous theoretical and empirical studies have addressed the question, what behavioral tactics maximize foraging efficiency (see Krebs, 1979, for a discussion of the link between foraging efficiency and increased individual fitness)? Basic tactics or components of foraging strategies include diet breadth, search strategy, foraging group size, and allocation of time between different feeding areas. Most authors use maximization of net energy gained per unit of foraging time as a reasonable measure of foraging efficiency, but clearly other measures, such as maximization of the intake rate of a crucial nutrient or minimization of the risk of not finding any food, might be appropriate evolutionary goals under certain conditions. The time frame of most foraging models is brief, usually one or a few days, but long-term (e.g., seasonal) considerations are undoubtedly also important (Katz, 1974). It is likely that actual foraging strategies involve trade-offs between maximization of short-term versus long-term energetic efficiency. Therefore, depending on the time frame utilized, foraging strategies may not appear to be "optimal."

Optimal diet breadth has been a central issue in many discussions of foraging behavior and deals with the question, what conditions promote the evolution of specialized rather than generalized diets? Variables that influence diet breadth are numerous and include the following: (1) the abundance, diversity, and seasonality of different kinds of prey; (2) the relative costs of searching for, pursuing, handling, and assimilating different prey species; (3) the nutritional characteristics of alternate prey; and (4) the predation risks involved in finding and eating different prey species. Because they involve rather restricted conditions, such as constant availability, balanced nutritional composition, and high density of preferred "prey," completely specialized (monospecific) diets should be uncommon among plant-visiting bats. Owing to the seasonal occurrence and diversity of sizes, shapes, biochemistry, and abundance of bat fruits and flowers, generalized (or at least polyspecific) diets should be the rule among these bats. Dietary generalization should be expected, particularly if different fruit or flower species are nutritionally complementary. Resource seasonality might also promote sequential specialization in which a bat species concentrates its foraging efforts on a limited number of food species at different times of the year. This foraging strategy would offer the benefits that accrue to specialists (e.g., more efficient searching and food-handling behavior) while retaining some of the resource flexibility that characterizes generalists.

In Schoener's (1969) terminology, plant-visiting bats are pure searchers rather than pursuers. Their "prey" items expect to be found but often occur in patchy distributions and are temporally ephemeral. Given these conditions, how should a bat search for food so as to maximize its net rate of energy intake? Refuging roosting strategies can potentially complicate this matter, because resource availability rates are likely to vary with distance from the central roost, especially if hundreds of conspecific bats roost together. In general, we expect resources located close to the central roost to disappear faster during the night than those located farther from the roost, because bat density should decrease with increasing distances from the roost (Hamilton and Watt, 1970). Thus a plant-visiting bats' food search strategy involves decisions regarding (1) how far to travel from the roost before beginning to look for food (this is usually called the "commute" distance), (2) whether or not to be constantly alert for food both while commuting and during the actual search, and (3) whether or not to search along a regular path (a "trap line") or an irregular one that might include searching currently unproductive areas for upcoming "blooms" of flowers or fruits.

The optimal solutions to these problems involve a number of variables, including the spatiotemporal predictability of resources and predator risk. Regarding commute distances, refuging theory suggests that when resources are homogeneously distributed, commuting different distances from the refuge involves a trade-off between travel costs and expected resource availability. Short distances involve low travel costs but also low food density (because of high intra-roost competition), whereas longer distances involve higher travel costs but relatively higher food density. Under these conditions we might predict that short-commuting bats should visit a greater number of (less rich) feeding areas than longer-commuting bats to fulfill their daily energy requirements. When resources are heterogeneously distributed, clear-cut predictions regarding trade-offs between travel costs and energetic gains are difficult to make but could be modeled by computer simulation.

Morrison (1978c) has presented a geometrical model of the optimal search strategy of refuging predators that was inspired by his observations of the foraging behavior of the frugivorous bat *Artibeus jamaicensis*. Under the rather unrealistic assumptions that prey (fruiting trees) are randomly distributed and that predators move a constant distance between random turns in their search for food, his model predicts that nonrefuging predators should travel rather long distances in relation to the size of their perceptual field before turning so as to minimize the overlap of different search fields. Refuging predators, in contrast, should travel short distances relative to their perceptual field, at the cost of higher overlap between search fields, to minimize the commute distance between roost and feeding areas. A numerical simulation indicates that turning rate (and hence overlap of search fields) should increase with an increase in the number of expected commutes to a feeding area. An unobvious prediction of the model is that refuging species need not always search for the food patch closest to the roost to forage optimally. Commute distances several times greater than the distance between the roost and the nearest patch can result in higher rates of energy gain than would occur if a predator always carefully searched for the closest patch. This would be especially true if the location of food patches changed considerably on a seasonal basis.

A second question related to searching strategies is, when should food-searching behavior take place—whenever the bat leaves its roost or only in the vicinity of known or suspected food locations? The spatiotemporal distributions of different food species will obviously influence this decision. For example, we might predict that bats should be constantly alert for food that is uniformly distributed and/or is available for long time periods, whereas they should rely more on locational memory and not be constantly alert for food when it occurs in low density, ephemeral patches (Fleming *et al.*, 1977). The former strategy requires that commute and search behaviors occur simultaneously, whereas the latter involves the separation of these two behaviors.

A final aspect of search behavior is the travel path taken between roost and different feeding areas. The two end points of what is probably a behavioral continuum are a regular path with stops at the same places each night over long time periods ("trap-lining" behavior) and an irregular path with frequent visits to new areas in search of upcoming food resources. Constancy of resource availability will importantly influence this aspect of foraging behavior. Whenever food locations are highly predictable in time and space, trap-lining behavior, which minimizes commute and search distances and costs, is likely to evolve. On the other hand, whenever food locations are less predictable, selection should favor those foraging paths in which some time is spent looking for future food patches. Because of their greater length, irregular foraging paths will be energetically more expensive than trap lines and, to the extent that longer flight times increase a bat's exposure to predators, they are riskier.

Since most plant-visiting bats live gregariously in day roosts, they have a potentially wide range of group sizes in which to forage. The two basic choices, solitary versus group foraging, provide contrasting advantages and disadvantages to individuals. At first glance, solitary foraging might appear to be the less advantageous mode because it does not provide three of the advantages often ascribed to group foraging (Wilson, 1975): increased protection from predators while commuting and searching for food, an increased sensory area with which to locate patchily distributed food, and increased knowledge of which food patches have already been visited by conspecifics each night. Loss of these advantages must be weighed against the advantages of potentially reduced feeding interference and/or reduced competition for food within a feeding area that accrues to solitary foragers. Resource densities and distributions will again influence this aspect of foraging behavior. Group foraging should maximize individual foraging efficiency when resources occur in rich but ephemeral, widely spaced patches. A predator-rich environment will reinforce the advantage of group foraging when resources are patchily distributed. In contrast, solitary foraging should be optimal when resources are more predictable in space and time; predator-free environments should also favor solitary foraging. As a final comment, resource predictability might possibly influence another aspect of chiropteran sociality, namely, the strong tendency to live in gregarious rather than solitary day roosts. Ward and Zahavi (1973) have postulated that the colonial roosts of a wide array of birds serve as "information centres" in which individuals parasitize each other's knowledge of the location of good feeding sites. Rich but ephemeral food patches—the same conditions that might favor the evolution of group foraging—appears to be the key factor that promotes the evolution of colonial roosts as information centers by individual (rather than group) selection. Plant-visiting bats may be gregarious for similar reasons.

Once a plant-visiting bat arrives at a feeding site, it is faced with two important behavioral decisions: whether or not to defend that site (either individually or as part of a group) against other bats and how long to remain at the site before moving to another food patch, back to its roost, or elsewhere. Decisions concerning site defense, which increases overall foraging costs, center around the area's "economic defendability" (Brown, 1964). Theoretical discussions (e.g., Case and Gilpin, 1974; Carpenter and MacMillen, 1976) and empirical work (e.g., Gill and Wolf, 1975; Kodrich-Brown and Brown, 1978) indicate that resource density is a key variable in determining whether or not a consumer should defend a food site. In two situations, low and high resource density, defense costs outweigh the advantages of defending a food site, and consumers should tolerate the intrusions of other individuals. At intermediate resource densities, on the other hand, the advantages of exclusive access to known amounts of resource outweigh the costs of resource defense, and territorial behavior should be expressed.

Allocation of time among different feeding areas has received considerable theoretical attention (Hassell and May, 1974; Charnov, 1976). Intuitively, we expect the residence time of a consumer in a feeding area to be directly proportional to the food density in that area, because searching effort should always be concentrated in those areas in which expectation of reward is greatest. In addition to the food value of a consumer's current food patch, however, residence time is also influenced by a relative value of alternate potential food patches (Charnov, 1976). Assuming that resource renewal rates are low, as is likely to be the case for fruit or flower resources each night, residence time per patch should be lower when alternate patches are food-rich than when they are food-poor. This is true because after a consumer has begun to deplete its present feeding area, it is more likely to increase its net rate of food intake by moving to a new area in a food-rich environment than in a food-poor one. Additional factors that influence feeding-area residence times include distances between suitable food patches, specific dietary needs, and predation pressure. Other things being equal, the residence time per patch should increase as the distance between patches increases, provided that search costs within a patch are less than travel costs between patches (whose location the consumer is assumed to already know).

Acquisition of a nutritionally balanced diet could cause consumers to move to other patches before the current one has been depleted if different foods occur in different locations. Fear of predator attack could also terminate a visit to a food patch before it has been depleted (Gentry, 1978; Howe, 1979). This should be especially true for food-rich patches, which might serve as predator "flags" and thereby be a more likely site for a predator attack. In a predator-rich environment individuals might actually increase their fitness (or at least their survival) by avoiding rich food patches and feeding in less obvious places. Finally, the relative importance of the location of a likely predator attack—in a feeding area or in flight while traveling between areas—will influence rates of site changes. The former situation will select for reduced food-patch residence time, whereas the latter situation will select for increased residence time.

To summarize this section, plant-visiting bats utilize food resources that are "bound to be found" but that are often patchily distributed. Significant seasonal variation occurs in the diversity and abundance of bat flowers and fruits. Some plants produce copious but short-lived bursts of flowers or fruits (the BB strategy), whereas others produce few resources per unit time over extended time periods (the SS strategy). Given this variation in resource availability both within and between seasons, how should the foraging behavior of plant-visiting bats evolve so as to maximize individual foraging efficiency? Although it is probably impossible to formulate general foraging rules that will cover all situations, I have attempted to make predictions concerning the foraging tactics that will maximize net energy yield to individuals under four idealized environmental conditions (Table III).

Optimal foraging tactics depend on levels of environmental variability, which have at least two important components, spatial and temporal. For simplicity's sake, I recognize two variability states for each component: low and high variability. Along the temporal dimension low and high variability represent aseasonal and seasonal resource environments, respectively. Along the spatial dimension low and high variability represent uniform and patchy distributions, respectively. As discussed previously, environments that exhibit low temporal variability will favor the evolution of specialized diets and trap-lining behavior, whereas generalized diets and irregular foraging paths will be favored in temporally variable environments (Table III). Low spatial variability will select for solitary foraging and foraging flights in which commute and search behaviors are combined. High spatial variability will select for group foraging and the separation of commuting and searching behaviors (Table III). These predictions explicitly ignore the effects of food density and predation—additional factors that can probably override the importance of spatiotemporal variability in certain situations. Nevertheless, it will be useful to examine the actual foraging behav-

TABLE III
Predicted Behavioral Tactics That Maximize Individual Foraging Efficiency in Four Idealized Environments[a]

	Temporal dimension	
Spatial dimension	Low variability	High variability
Low variability	Specialized diet Solitary forager Combined commute and food search Regular foraging path	Generalized diet Solitary forager Combined commute and food search Irregular foraging path
High variability	Specialized diet Group forager Separate commute and food search Regular foraging path	Generalized diet Group forager Separate commute and food search Irregular foraging path

[a]See text for further explanation.

iors of plant-visiting bats to see whether or not these general predictions actually hold in nature.

3. THE FORAGING BEHAVIOR OF PLANT-VISITING BATS

Despite their high species diversity and often large population sizes, plant-visiting bats are poorly known ecologically and behaviorally. At present most of our knowledge about these bats concerns the foods they eat, but we know comparatively little about how this food is obtained (Fenton and Kunz, 1977). The bulk of our knowledge concerning the foraging behavior of phytophagous bats comes from detailed radio-tracking studies of only four species, three phyllostomids (*Phyllostomus hastatus, Carollia perspicillata,* and *Artibeus jamaicensis*) and one pteropodid (*Hypsignathus monstrosus*). Less-detailed information is available for two other phyllostomids (*A. lituratus* and *Vampyrodes caraccioli*). Two pteropodids, *Epomops buettikoferi* and *Micropteropus pusillus*, are currently being intensively studied in the Ivory Coast, Africa (D. W. Thomas, personal communication). Questions about the optimality of different foraging patterns have been asked of only three phyllostomid species, *Leptonycteris sanborni, C. perspicillata* and *A. jamaicensis*. Despite our current dearth of knowledge about these bats, they appear to be excellent subjects for studies of plant-animal coevolution and the evolution of social systems and foraging in relation to resource availability.

3.1. Food Habits and Diet Breadth

Available information on food habits, which is more extensive for phyllostomid bats (Gardner, 1977) than for pteropodids, indicates that most species are food generalists rather than food specialists. In species that have been extensively studied (i.e., collected) food habits are known to change seasonally and geographically. The diet of *Artibeus jamaicensis*, an abundant and widely distributed 50-g Neotropical frugivore, illustrates these points well. An overall summary of its known diet (Gardner, 1977) indicates that it consumes parts of at least 92 plant taxa (84 kinds of fruit, nectar, and pollen of eight flowers, and the leaves of one species, *Ficus religiosa*), but throughout its range figs (*Ficus* spp.) form the bulk of its diet. In the dry tropical forest of northwestern Costa Rica, *A. jamaicensis* visits at least 21 species of flowers during the dry season and eats at least 12 kinds of fruits (Heithaus *et al.*, 1975). *Ceiba pentandra*, a very "fecund" BB species, is its major flower species, but individuals often visit two or more different floral species in one night. In the moist tropical forest on Barro Colorado Island, Panama, 78% of *A. jamaicensis*' diet is composed of two large-fruited fig species, *F. insipida* (with 9-g fruits) and *F. yoponensis* (3-g fruits) (Morrison, 1978a; Bonaccorso, 1979). It switches to other fruit species such as *Quararibea asterolepis* and *Spondias* spp. in the late wet and early dry seasons, when the absolute abundance of figs is low, and visits several species of flowers in the dry season. Finally, in Trinidad, *A. jamaicensis* is known to eat at least 51 species of fruit (Goodwin and Greenhall, 1961).

Other well-studied phyllostomid species show similar trends. *Glossophaga soricina*, an 11-g glossophagine bat (members of the Glossophaginae are the most specialized nectar and pollen feeders among phyllostomids), consumes nectar, pollen, fruit, and insects. In Mexico its diet includes 34 flower or pollen species (Alvarez and Gonzalez, 1970); in northwestern Costa Rica it is basically nectarivorous year-round and visits 20 flower species, but it also consumes the fruits of seven species in the wet season (Heithaus *et al.*, 1975). Another nectarivore, *Leptonycteris sanborni* (17 g), consumes at least 28 species of pollen in Mexico, with seasonal dietary shifts being evident in the state of Guerrero (Alvarez and Gonzalez, 1970). When it migrates to southern Arizona and Texas in the temperate-zone summer, however, its diet consists of only two flower species, a paniculate agave (*Agave palmeri*) and the saguaro cactus (*Carnegeia gigantea*) (Howell, 1974). As a final example, the 20-g *Carollia perspicillata* is basically frugivorous year-round and consumes the fruit of at least 37 plant species (Gardner, 1977). When they are available, it tends to concentrate on the fruits of various *Piper* species, but, like *Artibeus jamaicensis*, it also visits several species of flowers in the dry season (Heithaus *et al.*, 1975; Heithaus and Fleming, 1978). In Trinidad its diet includes at least 23 fruit species, many of which are also eaten by *A. jamaicensis* and *A. lituratus* (Goodwin and Greenhall, 1961).

Broad diets and seasonal dietary shifts also characterize pteropodid bats. Two abundant African species, *Rousettus aegyptiacus* and *Eidolon helvum*, eat a wide variety of soft fruits (of which figs are especially important), buds and young leaves, and the nectar and pollen of several BB species, including *Ceiba pentandra* (also a phyllostomid favorite), *Bombax malabrachium*, *Adansonia digitata*, *Kigelia africana*, and *Eucalyptus* spp. (Kingdon, 1974). In coastal Kenya the January diet of *Epomophorus wahlbergi* includes fruits of *Ficus sycomorous*, *Diospyros squamosa*, *Anacardium occidentale* (whose New World congener *A. excelsum* is a phyllostomid favorite), *Psidium guajava*, and *Terminalia catappa*, and leaves of *Balanites wilsoniaria* (possibly for the steroidal sapogenins, diosgenin and yamogenin, they contain) (Wickler and Seibt, 1976). Of the three species of Malayan flower-visiting bats studied by Start and Marshall (1976), *Eonycteris spelaea* visits 31 flower species, with 11 BB species comprising the bulk of its diet. In contrast, *Macroglossus minimus* and *M. sobrinus* have more restricted diets. The former bat visits three species of *Sonneratia*, two of which are BB plants, and the latter visits various SS species of *Musa*. Finally, the diet of Africa's largest pteropodid, *Hypsignathus monstrosus*, is apparently more restricted than many other species. Between May and October in Gabon, the bulk of its diet comes from fruits of two *Ficus* species and *Anthocleista;* lesser amounts of *Musanga cecropiodes* and *Solanum* spp. fruits are also eaten (Bradbury, 1977).

The energetic, nutritional, and ecological bases of food choice in plant-visiting bats are currently poorly known. Why, for example, do *Artibeus* species and other related stenodermine phyllostomids tend to specialize on *Ficus* fruits, whereas *Carollia* species tend to specialize on *Piper* fruits? Do these tendencies have a nutritional basis or do they merely represent opportunistic responses to particularly abundant and/or predictable food sources? Certain aspects of food choices in phyllostomids and pteropodids probably represent opportunistic feeding. This appears to be especially true in the case of the many basically frugivorous bats that visit BB flowers. In Costa Rica, for example, each of five frugivorous species visit flowers of the BB species *Ceiba pentandra* more heavily than other flowering species (Heithaus *et al.*, 1975). Likewise, African and Malayan pteropodids such as *Epomophorus gambianus*, *Nannonycteris veldkampi*, *Eidolon helvum*, and *Eonycteris spelaea* tend to respond opportunistically with short but intense visitation periods to BB plants such as *Parkia* spp. and *Ceiba pentandra* (Baker, 1973; Start and Marshall, 1976).

Because they are known to eat a wide variety of fruits throughout their geographical range, the tendency for species of *Artibeus* to concentrate on *Ficus* fruits in places such as Barro Colorado Island, where fig density is high (2.0–2.8 trees per hectare for *F. insipida* and *F. yoponensis*, respectively), may represent an opportunistic response to an abundant food supply rather than a nutritional dependence. Support for this hypothesis comes from the fact that dietary specialization on *Ficus* is higher (65–78% of the total diet) in the two larger species of

Artibeus (*A. jamaicensis* and *A. lituratus*) than in the 13-g *A. phaeotis*, whose diet consists of only 30% *Ficus* fruits (Bonaccorso, 1979). Because cost of transport increases exponentially with decreasing size (Thomas, 1975), the two larger species can more easily afford to search out patchy but rich food sources than can the small species.

Food choice has been particularly well studied in Costa Rican *Carollia perspicillata* (Lockwood *et al.*, 1982). Four fruit species, *Piper amalago* (which comprises 32% of its diet), *P. pseudofuligineum* (28%), *Cecropia peltata* (14%), and *Chlorophora tinctoria* (11%), dominate its diet in July and August. When captive bats in a flight cage were given a choice of up to seven different fruits to eat (*P. pseudofuligineum* was not offered), *P. amalago* and *Cecropia peltata* had the highest preference rankings. Food preference was neither correlated with the energy content of different fruits nor with the net rate of energy gain per unit of handling time, but instead correlated most strongly with the spatiotemporal predictability of the different fruit species. These results suggest that ecological, rather than nutritional, factors play the major role in this bat's food choice. In support of this hypothesis, L. Herbst (personal communication) has found that short-term survival (for up to 14 days) and maintenance of mass and a high body temperature in *Carollia perspicillata* is as high on a low-ranking fruit species (*Muntingia calabura*) as on a high-ranking one (*Piper amalago*).

Leaf eating appears to be more common in pteropodids than in phyllostomids, and this dietary difference may have a nutritional basis. D.W. Thomas (personal communication) has analyzed the nutritional content of fruits (particularly *Ficus* spp.) eaten by two West African bats (*Epomops buettikoferi* and *Micropteropus pusillus*) and finds that they are rich in carbohydrates but notably deficient in total protein (about 0.5% by weight). By way of comparison, the total protein content of fruits of *Cecropia peltata*, which is eaten by many Neotropical frugivores, is 8.1% (Glander, 1980). Thomas suggests that since pteropodids do not obtain significant amounts of protein through insectivory, they turn to leaf eating to increase their protein intake. In contrast, frugivorous phyllostomids occasionally eat insects, which are a much richer source of proteins and essential amino acids than are leaves, and hence do not need to eat leaves to meet their protein requirements. Interestingly, the Neotropical howler monkey *Alouatta palliata*, whose diet on Barro Colorado Island, Panama, contains 48% leaves by fresh weight, is dependent on fruits to supplement the proteins it obtains from leaves (Milton, 1979)—a situation opposite to that in pteropodid bats.

3.2. Foraging Behavior

In this section I will discuss general aspects of the ways in which plant-visiting bats search for and obtain food, including roosting strategies (see Chapter 1 for a detailed account of this topic), foraging group sizes, search paths and distances moved between roosts and feeding areas, and longer-distance movements. Following this, I present four case histories. General predictions made in Table III will be compared with actual observations to determine whether or not the behavior of well-studied species appears to be "optimal."

3.2.1. Roosting Behavior

Most bats roost gregariously in a wide variety of natural or man-made shelters. Colony size varies from a few to millions of individuals. Different species of plant-visiting bats span most of this range of colony sizes. For example, small groups of up to 20 individuals that roost in the foliage of trees or shrubs are often characteristic of species of stenodermine phyllostomids (e.g., *Artibeus, Uroderma,* and *Vampyrodes*) and epomorphorine and macroglossine pteropodids (e.g., *Hypsignathus, Epomophorus, Epomops,* and *Macroglossus*) (Start and Marshall, 1976; Wickler and Seibt, 1976; Bradbury, 1977; Morrison, 1980a). Intermediate-sized colonies of tens or hundreds of individuals living in caves, culverts, hollow trees, and so forth are characteristic of certain glossophagines (e.g., *Glossophaga* and *Leptonycteris*) and *Carollia* species. Very large colonies containing up to 500×10^3 individuals are found in certain pteropodids, including the cave-dwelling *Rousettus aegyptiacus* and *Eonycteris spelea* and the tree-roosting *Eidolon helvum* and various *Pteropus* species (Rosevear, 1965; Kingdon, 1974; Start and Marshall, 1976). Gregarious roosting undoubtedly serves numerous biological functions, including protection from predators and from stressful environmental conditions. It also affects foraging success in at least two different ways: (1) it facilitates the transfer of information about the location of good or new feeding areas and (2) it promotes increased competition for food in the vicinity of large roosts.

Roost-site fidelity varies among different phytophagous bats. Small groups of foliage-roosting bats (e.g., bachelor groups of *Artibeus jamaicensis,* harems of *A. lituratus* and *Vampyrodes caraccioli,* and mixed-sex groups of *Hypsignathus monstrosus*) seem to change their roosting sites every few days (Bradbury, 1977; Morrison, 1980a). There appears to be two main reasons for these changes: (1) they reduce the chances of the bats being located by predators during the day and (2) they allow bats to roost closer to their food supplies. Bats that roost in caves or other traditional sites, in contrast, often return to the same roost for several seasons or years. This occurs in phyllostomids such as *Phyllostomus hastatus* and *Carollia perspicillata* and pteropodids such as *Eidolon helvum* and *Eonycteris spelaea* (Start and Marshall, 1976; Heithaus and Fleming, 1978; McCracken and Bradbury, 1980).

3.2.2. Foraging Group Sizes

Gregarious living provides bats with a choice of foraging singly or in groups. In the case of plant-visiting bats, resource abundance and distribution patterns appear to significantly influence foraging group size. This point is well illustrated by the behavior of *Phyllostomus discolor*, which eats a variety of fruits and also visits flowers in Costa Rica and Brazil (Heithaus *et al.*, 1974; Sazima and Sazima, 1977). When feeding at flowers of the patchily distributed shrub *Bauhinia pauletia*, *Phyllostomus discolor* travels in groups of 2–12 bats. It also travels in groups to visit the flowers of *Parkia gigantea*, *P. auriculata*, *Hymenaea courbaril*, and *Lafoensia glyptocarpa*. At the latter species, feeding group size declines as the number of flowers per plant declines; groups contain 10–15 bats when trees bear more than 60 flowers, but decline to 1–2 when trees bear less than 10 flowers. Additional species that are known to forage in groups include the following flower visitors: *Leptonycteris sanborni* (at *Agave* and *Carnegeia* flowers), *Epomophorus gambianus* (at *Adansonia* flowers), *Eonycteris spelaea* (at various BB flowers), and probably Australian *Pteropus* (at *Eucalyptus* flowers) (Nelson, 1965; Ayensu, 1974; Start and Marshall, 1976; Howell, 1979).

In the case of *Leptonycteris sanborni*, Howell (1979) argues that the main advantage of flock foraging is that it increases individual foraging efficiency on a patchily distributed flower resource. In southern Arizona its main food, *Agave palmeri*, occurs in clumps that are often separated by tens of kilometers and that differ in flowering stage and hence in their nectar and pollen richness. Under these conditions flock foraging probably increases the rate of patch discovery (because a group should possess a wider sensory field than a single individual) and decreases the chance that individuals will waste time visiting already depleted plants. Flock foraging also increases the potential for feeding interference among individuals, but individuals of *L. sanborni* avoid this by "taking turns" dipping into flowers while circling the plant in an orderly fashion. Howell (1979) states that this orderly pattern of flower visitation represents a form of cooperative or altruistic behavior, but I believe it also can be interpreted as selfish behavior. Failure to visit flowers in an orderly fashion would perhaps be analogous to driving up a down ramp on a Los Angeles expressway at rush hour! In addition to increased search efficiency, which has yet to be rigorously demonstrated in this species, group foraging also serves a physiological function. After a 20-min foraging bout all members of a group roost together for about 0.5 hr. Group roosting decreases thermoregulatory costs, reduces evaporative water loss, and increases the rate of digestion (Howell, 1979).

Group foraging for fruit is probably common in pteropodids that live in large colonies (e.g., *Eidolon helvum*, *Rousettus aegyptiacus*, and *Pteropus* spp.), but it has not been well documented (but see Thomas and Fenton, 1978). Among phyllostomid frugivores *Artibeus jamaicensis* and *Phyllostomus hastatus* have been reported flying in groups (Dalquest, 1953; Goodwin and Greenhall, 1961), but, as described below, both of these species are basically solitary foragers.

3.2.3. Foraging Distances

The distances that plant-visiting bats fly between dry roosts and feeding areas undoubtedly reflect the interaction between several variables, including body size, the distribution and abundance of food sources, and the number of potential consumers of that food. Detailed knowledge of foraging distances requires that individual bats be followed, typically those capable of carrying miniature radio transmitters. As mentioned previously, only a handful of species has been so studied at present. Less-detailed observations on a number of other species provide us with a general idea of how far these bats fly to obtain food. On energetic grounds we expect small species of bats to commute shorter distances to their feeding areas than larger species. Rosevear (1965) speculates that this is the case among West African pteropodids, and Heithaus *et al.* (1975) provide indirect evidence for this in the form of a positive correlation between the mean recapture distances of banded bats and their body sizes in several Costa Rican phyllostomids. In the Costa Rican wet forest, however, LaVal and Fitch (1977) failed to find a similar correlation in some of the same or closely related species.

A second general expectation is that species living in large groups should commute longer distances than those living either solitarily or in small groups. Evidence supporting this expectation comes from several Paleotropical species. Individuals of two species, the 250–300 g *Eidolon helvum* and the 40–70 g *Eonycteris spelaea*, travel 20–40 km to feeding areas; in the latter species adult males feed at greater distances from the roost than do adult females or subadults (Kingdon, 1974; Start and Marshall, 1976). Large, gregarious Australasian *Pteropus* species also commute tens of kilometers to their feeding areas (Nelson, 1965). In contrast, individuals of the 100–120 g *Epomophorus wahlbergi* in coastal Kenya forage within a few kilometers of their roosts, which typically contain 3–100 individuals (Wickler and Seibt, 1976). Finally, groups of *Rousettus aegyptiacus* in Rhodesia apparently travel at least 10 km from their caves to visit fruiting *Diospyros senesis* shrubs in the dry season (Thomas and Fenton, 1978).

3.2.4. Behavior within Feeding Areas

Once in a feeding area, one to hundreds of bats can feed simultaneously in the same flowering or fruiting tree. Megabats typically differ from microbats in their food-gathering behavior in that they often cling to flowers and fruits while lapping nectar or ingesting soft pulp and fruit juices and clamber about the branches in their food trees. Considerable feeding interference can occur as bats jostle and knock fruits away from one another (Ayensu, 1974; Kingdon, 1974). Megabats may or may not remove fruits from a tree to a nearby night roost to feed. Microbats, in contrast, usually make fleeting visits lasting only 1–2 sec to flowers and pluck fruits from trees for consumption at either a solitary or, less often, a gregarious night roost (Jimbo and Schwassmann, 1967; Janzen *et al.*, 1976; Sazima and Sazima, 1978; Morrison, 1978a; Heithaus and Fleming, 1978). Night-roosting behavior, which can significantly increase total nightly flight distances, probably has evolved in these bats to reduce feeding interference and to reduce predator risk at "conspicuous" locations (Humphrey and Bonaccorso, 1978; Heithaus and Fleming, 1978; Howe, 1979; Morrison, 1980a). Different feeding times in the same trees, as occurs in *Epomophorus gambianus* and *Nannonycteris veldkampi* at flowering *Parkia clappertoniana* trees (Baker and Harris, 1957) and in *Artibeus jamaicensis, A. lituratus, A. phaeotis* and *Vampyrodes caraccioli* at fruiting fig trees (Bonaccorso, 1979), is an additional way whereby bats reduce the probability of feeding interference and predation at their food sources.

Despite potential advantages to be gained from reduced feeding interference and access to greater amounts of food, aggressive defense of flowering or fruiting plants does not appear to be common among plant-visiting bats. Other than Gould's (1978) description of Malayan *Pteropus vampyrus* defending flowering *Durio zibethinus* (a BB species) against conspecific intruders by audible growls and spread-wing displays, territorial behavior has not been reported in these bats. The general absence of territorial behavior in phytophagous bats suggests that their food resources are not ecologically defendable, either because of their rich but ephemeral nature or their low nightly densities. Similarly, frugivorous birds typically do not defend fruiting plants (Snow, 1976). It should be noted that although the defense of feeding areas is apparently uncommon in phyllostomid and pteropodid bats, males of many species actively defend roost sites or, in the case of the lek-forming *Hypsignathus monstrosus*, calling sites, and thereby gain reproductive access to females. Territorial behavior is focused more strongly on mate acquisition than on food defense in these bats.

Since plant-visiting bats often visit more than one feeding area in a night, the question arises, do changes in feeding sites represent attempts to maximize rates of food intake, to obtain a nutritionally balanced diet, to avoid competitors or predators, or to attain some combination of these goals? Except for *Leptonycteris sanborni*, reasons for these changes have not been elucidated. Field observations in southern Arizona (Howell, 1979) indicate that *L. sanborni* forages in cohesive flocks that abandon one *Agave* panicle and fly to another before its inflorescences are completely drained. The decision to switch panicles is made by one bat, and the others follow it without confirming that nectar reward levels are low. To see whether these bats use a panicle "switching rule" that maximizes individual energetic rewards, Howell and Hartl (1980) conducted laboratory experiments with a flock of six bats that could feed at two identical 20-"flower" artificial panicles and computer simulation experiments. Their results show that the flock switches panicles after each bat makes about 35 dips

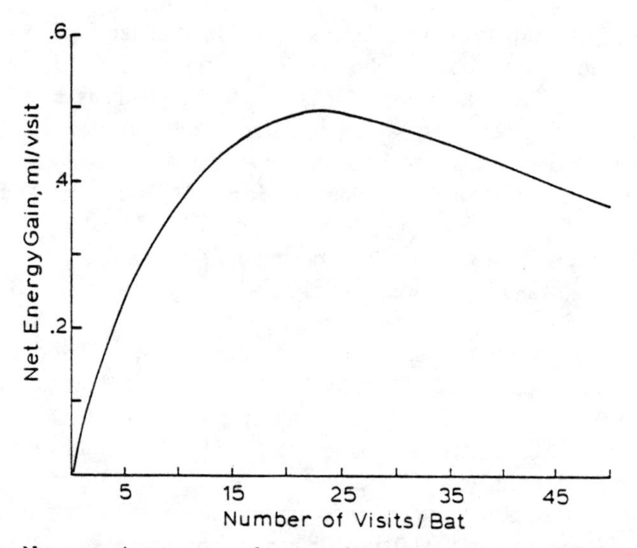

FIGURE 3. Net energetic return as a function of the number of successful visits to artificial flowers by *Leptonycteris sanborni* in the laboratory. (Modified from Howell and Hartl, 1980.)

into randomly selected "flowers" (nectar-filled funnels) of one panicle. On the average, 23 of the dips are "wet" (yield a nectar reward) and 12 are "dry." The bats abandon a panicle after removing approximately 75% of its available nectar. A computer simulation indicates that the switching rule that best describes this behavior is "leave after eight dips yield no reward," and an energetic cost-benefit analysis suggests that this rule maximizes individual energetic gains. Individuals will gain less net energy if they leave the panicle before or after making 21–26 successful dips (Fig. 3). The decision to change feeding sites in these bats thus conforms to predictions of optimal foraging theory.

3.2.5. Longer Foraging Movements

In many tropical areas seasonal changes occur in the kinds and locations of flower and fruit resources. At the local level these changes will cause foragers to spend time and energy searching for new feeding areas. When they occur over broader geographical areas, these changes can select for migratory behavior. Some bats (e.g., *Eonycteris spelaea* and *Macroglossus* spp. in Malaya) are known to return to the same foraging areas repeatedly for many months (Start and Marshall, 1976), but others undoubtedly change their feeding areas regularly. Individuals of *Eidolon helvum*, for example, can quickly locate newly planted fruit crops outside their normal geographical range (Kingdon, 1974). This ability suggests that some individuals regularly scout out new feeding areas and probably transmit their knowledge to fellow roost mates. This kind of feeding cooperation would seemingly be selectively advantageous to individuals only if the information were transmitted, initially at least, to close relatives and/or if the feeding areas were rich enough to preclude food competition among roost mates. Reciprocity among individuals of stable, long-lived social groups could also favor feeding cooperation (McCracken and Bradbury, 1980). Behavioral mechanisms that underly the discovery of new feeding areas in plant-visiting bats are currently poorly understood and warrant increased attention.

Local and sometimes extensive seasonal migrations are known to occur in several plant-visiting bats. In the New World two nectarivorous glossophagines, *Leptonycteris sanborni* and *Choeronycteris mexicana*, regularly migrate from northern Central America and Mexico into subtropical and warm temperate regions of southern Arizona and Texas to feed at flowers of paniculate agaves and certain species of cactus (Barbour and Davis, 1969). Elsewhere in the Neotropics *Phyllostomus discolor* apparently makes local migrations from the Costa Rican dry tropical forest to more flower-rich areas in the wet season (Heithaus *et al.*, 1975), and on Barro Colorado Island, Panama, several frugivores migrate to the mainland in the late wet and early dry seasons, when fruit abundance is low (Bonaccorso, 1979). Paleotropical plant visitors known or suspected to migrate seasonally include *Rousettus aegyptiacus*, *Eidolon helvum*, *Epomophorus wahlbergi*, and three species of eastern Australian *Pteropus* (Allen, 1939; Nelson, 1965; Kingdon, 1974). At Kampala, Uganda, *Eidolon helvum* abandons a large roost in July and August. Since this abandonment does not coincide with local shortages of fruit, Kingdon (1974) suggests that it occurs when females begin to teach their young how to search for food. The Australian *Pteropus* species move between summer and winter "camps," sometimes located more than 100 km apart, in response to changes in the location of concentrations of flowers (especially *Eucalyptus* spp.) and fruits (especially figs) (Nelson, 1965).

3.2.6. Trap-Lining

In the absence of marked seasonal changes in the kinds and locations of plant resources, selection should favor the evolution of trap-lining behavior. Several species of flower-visiting bats are known or suspected to be trap-liners, including groups of *Phyllostomus discolor* when they feed on flowers of *Bauhinia pauletia* and *Glossophaga* spp. visiting flowers of *Mucuna andreana* and *Trianea* sp.; each of these flower species is an SS plant (Baker, 1973; Heithaus *et al.*, 1974). Because they are visited each night in a pulsed fashion, Gould (1978) has suggested that Malayan flowers of *Musa* spp. and *Oroxylum indicum* are trap-lined by *Macroglossus sobrinus* and *Eonycteris spelaea*, respectively. Pulsed visitation patterns are suggestive of trap-lining behavior, but, because bats often rest in night roosts between foraging bouts, these patterns do not provide rigorous proof that trap-lining exists. Direct observation (e.g., by chemiluminescent tags or by radio tagging) is the only convincing way to document this behavior.

To summarize this section, plant-visiting bats utilize a variety of roosting and foraging tactics to exploit their patchily distributed food resources. Considerable variation (both within as well as between species) can occur regarding parameters such as roost type and colony size, foraging group sizes, foraging distances, and degree of fidelity to roost sites and feeding areas. Reproductive cycles, social organization, and changes in food distributions are factors that importantly influence foraging behavior. The interaction of these factors will become especially clear in the four case histories presented below.

3.3. Case Histories

3.3.1. *Hypsignathus monstrosus*

Hypsignathus monstrosus, a large and sexually dimorphic epomophorine pteropodid, has been studied in the lowland rain forest of Gabon by Bradbury (1977). Adult males weigh nearly twice as much as adult females (420 versus 234 g). Size dimorphism is the result of sexual selection in this lek-forming bat (Fig. 4A). Males assemble twice annually for 1–3 months at traditional calling sites (leks) and attract potential mates by issuing loud, harmonically rich calls for several hours each night. As is true in other lekking bird and mammal species, relatively few males account for the vast majority of matings. This social system strongly influences foraging behavior, because both adult males and females spend time each night away from foraging areas at the calling sites. During the day individuals roost cryptically in the foliage of widely scattered canopy trees, either singly or in small groups of mixed ages and sexes. The locations and compositions of these roosts change every 5–9 days. New roosts are located up to 4 or more km apart, and males roost up to 8 km from their lek area.

At night during the calling season adult males spend up to 4 hr at the lek before beginning to forage. They fly an average of 6.7 km from lek to feeding area, forage on figs and *Anthocleista* fruits until near dawn, and then either return directly to their day roost or to the lek site for additional calling. Noncalling males and females fly about 5 km between day roost and their feeding areas. Individuals occasionally make long, nonforaging flights in the middle of the night.

In summary, *Hypsignathus monstrosus* is a nongregarious, relatively long-distance forager. Social concerns strongly influence the foraging distances of females and especially adult males, which fly over twice as far each night as

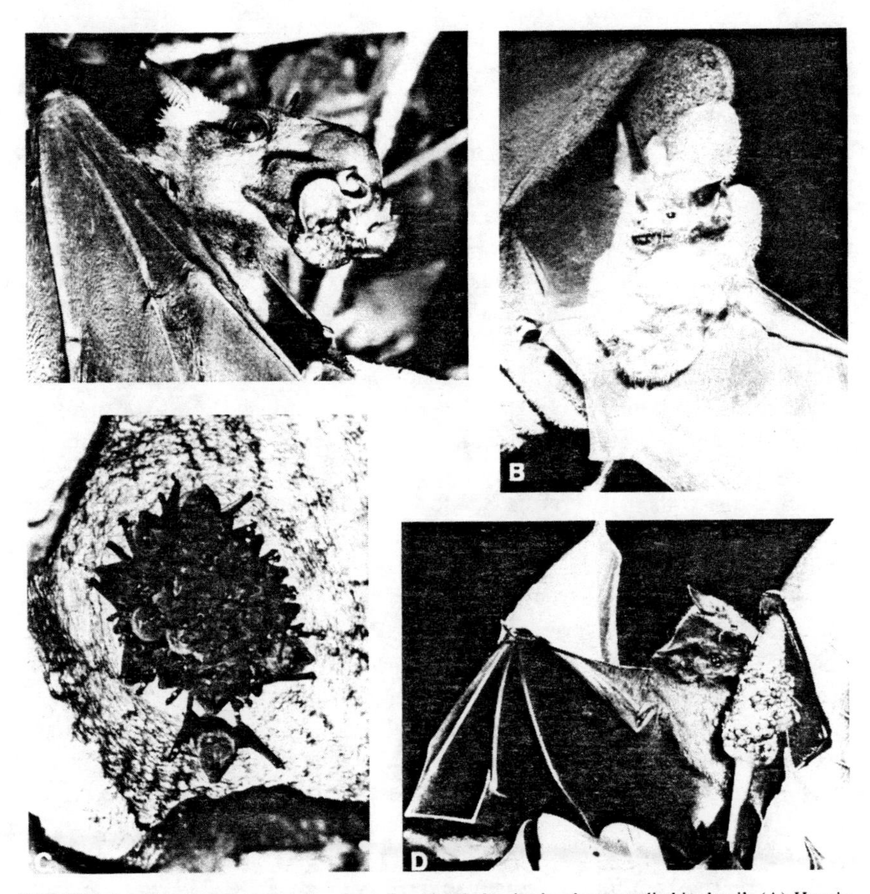

FIGURE 4. Four frugivorous bats whose foraging behavior has been studied in detail. (A) *Hypsignathus monstrosus* male; (B) *Carollia perspicillata;* (C) *Phyllostomus hastatus* harem (note male at the periphery); (D) *Artibeus jamaicensis*. [Photo credits: (A) A. R. Devez, CNRS (France); (B) T. Fleming; (C) G. McCracken; (D) M. Tuttle.]

other sex/age classes. Individuals appear to have rather narrow diets, forage singly, and probably spend considerable time and energy scouting out the locations of new fruiting trees (the prolonged, nonforaging flights?). To judge from the well-known phenology of its major food source, figs (Janzen, 1979), *H. monstrosus* in lowland Gabon probably lives in an environment of low temporal but high spatial variability. Its behavior differs from that predicted in Table III in that it does not forage in groups. Its unique (among bats) social system probably precludes it from being a group forager.

3.3.2. *Artibeus jamaicensis*

Artibeus jamaicensis, a 50-g stenodermine phyllostomid (Fig. 4D), has been studied in the moist tropical forest on Barro Colorado Island (BCI), Panama, and in the dry tropical forest in Mexico by Morrison (1978a, 1978b, 1978c, 1978d, 1979, 1980b). Although they differ considerably in size and social organization, *A. jamaicensis* and *Hypsignathus monstrosus* are similar in at least two respects: figs are a major item in their diets and they both roost in small groups in trees (although *A. jamaicensis* also roosts in caves) (Tuttle, 1976; Vazquez-Yanes *et al.*, 1975). *A. jamaicensis* is a harem–polygynous bat, and on BCI adult males defend tree holes containing 4–11 adult females and their dependent young against intrusions by other males. Bachelor males and immature females roost in small groups in canopy foliage. The location of foliage roosts changes frequently, and nonharem bats typically roost closer to their food trees than do harem bats.

Bats leave their day roosts 30–40 min after sunset and fly solitarily to a fruiting tree that is usually located less than 1 km from the roost. During the dark half of the month individuals make a total of 10–15 feeding passes from one or more night roosts to as many as five fruiting trees; the distance between night roosts and feeding trees averages to 175 m (range, 25–400 m). Individuals eat about their own weight in fruit each night and spend up to 80% of their time away from the day roost, resting quietly in the night roost. Unusually long flights of 20–45 min, in which newly ripening fruit crops are apparently located, are occasionally made on dark nights. During the light half of the month bats visit only one to two trees, return to the day roost for 1–7 hr when the moon is at its zenith, and usually forage again for 1–2 hr before sunrise. In the Mexican state of Jalisco, where the fig density is much lower, the distance between feeding trees and the day roosts of four radio-tagged females averaged 8 ± 2 km, compared to a distance of only 0.6 ± 0.4 km on BCI. Foraging distances and roosting habits are clearly responsive to resource distances in this and presumably other species of bats. Except for being less lunar phobic, two other fig-eating BCI stenodermines, *Artibeus lituratus* and *Vampyrodes caraccioli*, show foraging patterns similar to those of *A. jamaicensis* (Morrison, 1980a).

In summary, *Artibeus jamaicensis* on BCI lives in a fig-rich environment, and, as a consequence, it has a relatively specialized diet and forages rather close to its day roosts. Because the location of fruiting trees changes frequently, it must spend some time searching for new feeding areas, which, however, are seldom located more than about 1 km apart. Fruit relocation experiments conducted in Costa Rica suggest that fig eaters, including *A. jamaicensis*, which is common at the study site, rely on locational memory to find fruiting trees and hence separate the commute and food-search components of foraging behavior (Fleming *et al.*, 1977). Except for foraging solitarily—probably because overall food density is relatively high on BCI—*A. jamaicensis* appears to "view" its environment as though it were temporally invariable but spatially variable (Table III). It would be interesting to know whether this species forages in groups in Jalisco, Mexico, where the food density is much lower and the climatic seasonality much greater.

3.3.3. *Phyllostomus hastatus*

Phyllostomus hastatus, a 90-g phyllostomid, has been intensively studied in the wet tropical forest of Trinidad, the West Indies, by Williams and Williams (1970) and McCracken and Bradbury (1977, 1980). *Phyllostomus hastatus* is a gregarious cave dweller (Fig. 4C). Although it is a generalized feeder and eats fruits, vertebrates, and nectar, the bulk of its diet consists of several species of fruit, including *Cecropia peltata*, *Piper* spp., *Gurania spinulosa*, and *Psiguria umbrosa*. Like *Artibeus jamaicensis*, it is harem–polygynous, and two basic social configurations exist in caves: (1) harems containing 15–25 females and their dependent young, which are defended by a single male; and (2) groups of bachelor males. Once they disperse from their natal groups, individuals reside in one roost for most of their lifetimes. Females, which are more likely to leave their birth caves than males, form long-term, stable associations with their female harem mates. Genetic relatedness among these females, however, is no greater than would be expected by chance alone. Males can defend the same harem for as long as 3 years.

The foraging behavior of individuals reflects their social status. Harem males forage closest (sometimes within 1 km) to the roost and are gone from the cave for only about 1 hr each night. Bachelor males, in contrast, forage up to 9 km away from the roost and are usually out of the cave all night. Harem females fly individually to a feeding area that they use throughout the year and are gone from the roost for about 2 hr. The feeding areas of female harem mates are close together and often overlap extensively. There is some evidence suggesting that females in the same harem occasionally exchange information about the location of fruiting trees: a pair was once observed to arrive simultaneously at a new feeding site. Feeding areas of different harems, which are located 1–5 km from the cave, do not overlap, but there is no evidence that females defend their feeding areas against nonharem members. Harem males clearly do not patrol or defend the feeding areas of their females.

In summary, *Phyllostomus hastatus* shows considerable stability in its social relationships and foraging behavior. It responds to its environment in Trinidad as though it were temporally variable but spatially stable. Its generalized food habits appear to reflect significant seasonal variations in food availability. However, the apparent spatial predictability of their food supplies allows harem members to spend a minimum of time in searching for food. Long life-spans (some individuals are known to live 10 years) and constant use of the same patch of forest probably means that these bats can rely extensively on locational memory to increase their food-searching efficiency.

3.3.4. *Carollia perspicillata*

Carollia perspicillata, a 20-g phyllostomid (Fig. 4B), has been intensively studied in the dry tropical forest of Santa Rosa National Park, Costa Rica, by Fleming and associates (Fleming *et al.*, 1977; Jamieson, 1977; Heithaus and Fleming, 1978, unpublished data). Like *Phyllostomus hastatus*, *C. perspicillata* is a harem–polygynous cave dweller. At Santa Rosa and elsewhere, however, some individuals also roost in hollow trees, old wells, culverts, and so forth (Tuttle, 1976). Two major cave roosts located 3.5 km apart each contain over 200 individuals with bats being distributed in two major social configurations in the wet season: (1) harems containing one male and up to 14 females and their babies and (2) groups of bachelor males. Harem males defend their roost territories for months, even in the dry season when most females have left the cave. Unlike *P. hastatus*, females do not appear to form long-term associations with each other or with a particular male; they frequently change harems within one roost and they appear to change roosts on a regular seasonal basis.

At Santa Rosa *Carollia perspicillata* has a broad diet that reflects major changes in the seasonal availability of different fruits. When available, it feeds extensively on fruits of five species of *Piper*, but its diet also includes at least 12 additional species. It probably also visits flowers in the dry season.

As in *Phyllostomus hastatus*, social status influences the foraging behavior of *Carollia perspicillata*. Throughout the year harem males remain outside the day roost each night for short periods of time and forage within 1 km of the roost. In contrast, bachelor males and harem females generally stay away from the roost all night, even on nights of bright moon. In the wet season bachelor males and females individually visit two to six different feeding areas located no more than 3 km away from the roost each night, and most individuals feed less than 2 km from the roost. The total nightly flight distances of these bats average 1.6 km between roost and first and last feeding areas, 1.5 km between different feeding areas, and 1.6 km for gathering fruits within feeding areas. In the dry season individuals are concentrated along moist creek beds, and feeding areas are about twice as far from the roost as they are in the wet season.

Throughout the year bats eat fruits and rest in night roosts within 100 m of fruiting plants. They consume about their own weight in fruit pulp each night and forage every night, regardless of the weather or moonlight conditions. Individuals return to the same feeding areas for at least 2–3 weeks and occasionally scout out new feeding areas in prolonged flights. They gradually phase out old feeding areas over a several-day period. Harem females do not appear to forage near each other or near their harem males.

At Santa Rosa *Carollia perspicillata* experiences an environment of high spatiotemporal variability. Its foraging behavior, however, conforms to only two of the four predictions (food generalist and irregular foraging paths) listed under that kind of environment in Table III. It is a solitary forager, probably because its preferred food plants produce only small amounts of fruit nightly, and fruit relocation experiments indicate that it is constantly alert for *Piper* fruits and thus probably combines commute and food-search behaviors when feeding on these fruits. Since the preferred fruit species of *Carollia* are SS plants, its food choice helps to reduce the spatiotemporal variation of what is otherwise a very seasonal environment.

4. SUMMARY AND GENERAL CONCLUSIONS

What generalizations regarding foraging behavior emerge from this four-species comparison? What predictions can we make about the foraging behavior of currently unstudied species if we know something about the spatiotemporal variability of their environments? Major features of the social and foraging behavior of the four species, summarized in Table IV, lead to the following generalizations:

1. Polygynous, but not promiscuous, mating systems are common in plant-visiting bats. Males defend territories (either with or without continuous female presence) in day roosts or at nocturnal assembly sites.
2. Foraging budgets and locations are influenced by a bat's social status. Reproductively dominant males have different foraging patterns than do other age–sex combinations.
3. Diet breadth can be broad but tends to be specialized whenever seasonably possible. Such specialization will increase food search efficiency in long-lived, relatively sedentary species.
4. Despite differences in the degree of gregariousness, frugivores tend to forage solitarily.
5. Foraging distances (from day and night roosts) are probably positively related to body size. They also reflect the distribution and abundance of preferred food.
6. Individuals show high fidelity to a limited number of feeding sites. This fidelity increases as seasonal changes in food distribution decrease.
7. Food-search behavior is responsive to the spatiotemporal distribution of preferred foods. When foods are widely distributed, commuting and

TABLE IV
Summary of the Major Aspects of the Social and Foraging Behavior of Four Species of Frugivorous Bats

Parameter	*Hypsignathus monstrosus* (200–400 g)	*Artibeus jamaicensis* (45–50 g)	*Phyllostomus hastatus* (80–100 g)	*Carollia perspicillata* (18–22 g)
Social system	Lek–polygynous	Harem–polygynous	Harem–polygynous	Harem–polygynous
Diet breadth	Specialized (figs)	Specialized (figs)	Generalized	Generalized
Roost size and location	Small, in foliage	Small, in foliage or tree holes, caves	Large, in caves	Large, in caves
Foraging group size	Solitary	Solitary	Solitary	Solitary
Foraging distances	≤ 10 km from roost	≤ 2 km from roost	≤ 9 km from roost	≤ 3 km from roost
Frequency of feeding site changes	Not reported	Frequent	Infrequent	Relatively frequent, especially on a seasonal basis
Night roost used?	Yes?	Yes	Yes	Yes
Food-search strategy	Not reported	Commute and food search separate	Not reported	Commute and food search combined
Trap-line behavior	Probably not	No	Probably not?	No
Prolonged food searches?	Probably	Yes	Probably not?	Yes
Lunar phobic?	No	Yes	No	No

searching behaviors will be combined, whereas they will be separated when food occurs in widely spaced patches.

8. Frugivores probably do not regularly follow "trap lines" in visiting their food plants. They must occasionally increase their flight time looking for new food sources, and the frequency of such scouting trips probably increases with increased climatic (and food) seasonality.
9. Species probably minimize the time spent in actual flight and spend considerable time resting in night roosts. In Schoener's (1971) terminology, frugivorous bats are probably time minimizers rather than energy maximizers in their search for food.

I would like to stress that the above generalizations are based on a very small sample size of plant-visiting bats and hence must be regarded as tentative hypotheses. Information already available suggests that some of these generalizations are not universally true. For example, the foraging behavior of many pteropodids is likely to differ in numerous ways, including foraging group sizes and fidelity to a limited number of feeding areas, from the behavior of the above species. Finally, we particularly need detailed foraging data on flower-visiting bats. With the exception of *Leptonycteris sanborni* and perhaps *Eonycteris spelaea*, we know very little about how these graceful bats make their living.

In concluding this section, I will address the following question: are the foraging behaviors of the four well-studied species "optimal" in the sense of maximizing net rates of energy acquisition? As Cody (1974) and many others (e.g., Lewontin, 1978) have pointed out, this question is really tautological, because the answer must be yes if foraging behavior is sensitive to natural selection, which tends to act as an "efficiency expert." Since the efficient acquisition of food must have a positive effect on fitness, measured as net reproductive success, there is no reason to believe that foraging behavior is not sensitive to natural selection. This does not mean, however, that foraging behavior is always being optimized, especially when only short time spans (one or a few nights) are considered. Some aspects of the behavior of the four species (and others) clearly do not lead to short-term foraging optimization. An example of this is when *Carollia perspicillata* shuttles back and forth between different feeding areas several times in one night. Since the nightly fruit density in these areas will only decline through time because depletion rates far exceed renewal rates, one wonders why a bat spends extra time and energy returning to a site that is becoming less food rich through time. Why does an individual not remain in one area and deplete it to a certain level before moving elsewhere for the rest of the night? Since not every area contains the same potential food sources, one reason for these shuttles might be to increase dietary diversity, but this fails to explain why the patches are not visited in a more "economical" fashion. If predation risk increased linearly with time spent near a fruit tree, then frequent departures might reduce this risk. But we currently know very little about the predation rates on plant-visiting bats, and this hypothesis is currently untested.

Another example of "uneconomical" foraging is the long flight distances of the territory-holding male *Hypsignathus monstrosus*. Males (and adult females visiting leks) certainly fly far greater distances than is necessary to simply acquire sufficient food. The potential reproductive payoffs of advertising at and visiting lek sites clearly make this less-than-optimal foraging pattern easily understandable.

We clearly need more information about bat predators, because this ecological factor appears to have influenced two aspects of bat foraging behavior in ways that potentially reduce foraging efficiency (but increase pollen or seed mobility). These aspects include the use of night roosts, which significantly increases flight distances (and energy costs), and various degrees of lunar phobia, which either temporarily eliminates foraging behavior or reduces its intensity. There are two plausible reasons why bats prefer to eat fruits away from fruiting plants in a secluded location: (1) to reduce feeding interference by other bats and (2) to reduce predation risk. To my knowledge, neither of these hypotheses, which are not mutually exclusive, has been critically examined. Lunar phobia is strongly expressed in *Artibeus jamaicensis* (Morrison, 1978b) and less strongly expressed in *Carollia perspicillata* (Heithaus and Fleming, 1978). This behavior results in restricted or reduced flight times during moon-bright nights of the month and extra flights between roosts and feeding areas. Avoidance of visually hunting predators such as owls and the carnivorous phyllostomid *Vampyrum spectrum* has been the usual explanation for this behavior. A critical test of this hypothesis is needed.

Social concerns also constrain foraging behavior, especially in reproductively dominant males. In each of the four intensively studied species, the behavior of territory-holding males differs significantly from that of other individuals. Generally speaking, dominant males must be more efficient foragers than others because they usually need to spend time each night defending the territories that give them nearly exclusive reproductive access to females. Except in *Hypsignathus monstrosus,* these males forage closer to their day roosts and spend much less time away from the day roost than do females and bachelor males. Harem males tend to be the older males in populations of these species, and their intimate knowledge of good feeding areas, built up over seasons or years of "apprenticeship," probably contributes significantly to their ability to acquire and successfully defend their reproductive status.

It is likely that plant-visiting bats are efficient foragers, particularly when the effects of resource seasonality, social status, and predator pressure are taken into consideration. Although it will always be profitable to study foraging behavior with predictions of optimal foraging behavior in mind, it should be clear that foraging behavior cannot be isolated from other aspects of animal's lives. Links between social behavior and foraging appear to be especially strong in the bats reviewed in this chapter, and it would therefore seem unwise to study one of these aspects of chiropteran biology without giving considerable attention to the other.

ACKNOWLEDGMENTS. I wish to thank the Costa Rican National Park Service for permission to study bat–plant interactions at Santa Rosa National Park and for the many courtesies that the park personnel has extended to my field parties and myself over the years. I also thank Doug Morrison for his useful comments on a previous draft of the manuscript. My work with tropical bats has been generously supported by the U.S. National Science Foundation. This chapter is dedicated to my wife, Marcia, and my children, Michael and Cara, for cheerfully accepting my long absences in the field.

5. REFERENCES

Allen, G. M. 1939. Bats. Harvard University Press, Cambridge, 368 pp.

Alvarez, T., and L. Gonzalez Q. 1970. Analisis polinico del contenido gastrico de murcielagos Glossophaginae de Mexico. An. Esc. Nac. Cienc. Biol., Mexico City, **18:**137–165.

Arata, A. A., J. B. Vaughn, and M. E. Thomas. 1967. Food habits of certain Colombian bats. J. Mammal., **48:**653–655.

Ayala, S. C., and A. D'Alessandro. 1973. Insect feeding behavior of some Colombian fruit-eating bats. J. Mammal., **54:**266–267.

Ayensu, E. S. 1974. Plant and bat interactions in West Africa. Ann. Mo. Bot. Gard., **61:**702–727.

Baker, H. G. 1961. The adaptation of flowering plants to nocturnal and crepuscular pollinators. Q. Rev. Biol., **36:**64–73.

Baker, H. G. 1973. Evolutionary relationships between flowering plants and animals in American and African tropical forests. Pp. 145–159, *in* Tropical forest ecosystems in Africa and South America: A comparative review. (B. J. Meggers, E. S. Ayensu, and W. D. Duckworth, eds.). Smithsonian Institute Press, Washington, 350 pp.

Baker, H. G. 1978. Chemical aspects of the pollination biology of woody plants in the tropics. Pp. 57–82, *in* Tropical trees as living systems. (P. B. Tomlinson and M. H. Zimmerman, eds.). Cambridge University Press, Cambridge, 675 pp.

Baker, H. G., and B. J. Harris. 1957. The pollination of *Parkia* by bats and its attendent evolutionary problems. Evolution, **11:**449–460.

Baker, H. G., R. W. Cruden, and I.Baker. 1971. Minor parasitism in pollination biology and its community function: The case of *Ceiba acuminata*. Bioscience, **21:**1127–1129.

Barbour, R. W., and W. H. Davis. 1969. Bats of America. University Press of Kentucky, Lexington, 286 pp.

Baron, G., and P. Jolicoeur. 1980. Brain structure in Chiroptera: Some multivariate trends. Evolution, **34:**386–393.

Bonaccorso, F. J. 1979. Foraging and reproductive ecology in a Panamanian bat community. Bull. Fla. State Mus. Biol. Sci., **24:**359–408.

Bradbury, J. W. 1977. Lek mating behavior in the hammer-headed bat. Z. Tierpsychol., **45:**225–255.

Brown, J. L. 1964. The evolution of diversity in avian territorial systems. Wilson Bull., **76:**160–169.

Carpenter, F. L., and R. E. MacMillen. 1976. Threshold model of feeding territoriality and test with a Hawaiian honeycreeper. Science, **194**:639–642.

Case, T. J., and M. E. Gilpin. 1974. Interference competition and niche theory. Proc. Nat. Acad. Sci., U.S.A., **71**:3073–3077.

Charnov, E. L. 1976. Optimal foraging: The marginal value theorem. Theor. Pop. Biol., **9**:129–136.

Cody, M. L. 1974. Optimization in ecology. Science, **183**:1156–1164.

Croat, T. B. 1975. Phenological behavior of habit and habitat classes on Barro Colorado Island (Panama Canal Zone). Biotropica, **7**:270–277.

Croat, T. B. 1978. The flora of Barro Colorado Island. Stanford University Press, Stanford, 943 pp.

Dalquest, W. W. 1953. Mexican bats of the genus *Artibeus*. Proc. Biol. Soc. Wash., **66**:61–66.

Eisenberg, J. F., and D. E. Wilson. 1978. Relative brain size and feeding strategies in the Chiroptera. Evolution, **32**:740–757.

Fenton, M. B., and T. H. Kunz. 1977. Movements and behavior. Pp. 351–364. *in* Biology of bats of the New World family Phyllostomatidae. Part II. (R. J. Baker, J. K. Jones, Jr., and D. C. Carter, eds.). Spec. Publ. Mus. Texas Tech Univ., Lubbock, **13**:1–364.

Findley, J. S., E. H. Studier, and D. E. Wilson. 1972. Morphologic properties of bat wings. J. Mammal., **53**:429–444.

Fleming, T. H. 1979. Do tropical frugivores compete for food? Am. Zool., **19**:1157–1172.

Fleming, T. H., and E. R. Heithaus. 1980. Frugivorous bats, seed shadows, and the structure of tropical forests. Biotropica, **13**:45–53.

Fleming, T. H., E. T. Hooper, and D. E. Wilson. 1972. Three Central American bat communities: Structure, reproductive cycles, and movement patterns. Ecology, **53**:653–670.

Fleming, T. H., E. R. Heithaus, and W. B. Sawyer. 1977. An experimental analysis of the food location behavior of frugivorous bats. Ecology, **58**:619–627.

Frankie, G. W., H. G. Baker, and P. A. Opler. 1974. Comparative phenological studies of trees in tropical wet and dry forests in the lowlands of Costa Rica. J. Ecol., **62**:881–919.

Gardner, A. L. 1977. Feeding habits. Pp. 293–350. *in* Biology of bats of the New World family Phyllostomatidae. Part II. (R. J. Baker, J. K. Jones, Jr., and D. C. Carter, eds.). Spec. Publ. Mus. Texas Tech Univ., Lubbock, **13**:1–364.

Gentry, A. H. 1974. Flowering phenology and diversity in tropical Bignoniaceae. Biotropica, **6**:64–68.

Gentry, A. H. 1978. Anti-pollinators for mass-flowering plants? Biotropica, **10**:68–69.

Gill, F. B., and L. L. Wolf. 1975. Economics of feeding territoriality in the Golden-winged Sunbird. Ecology, **56**:333–345.

Glander, K. E. 1980. Feeding patterns in mantled howling monkeys. Pp. 231–257. *in* Foraging behavior: Ecological, ethological, and psychological approaches. (A. Kamil, ed.). Garland Press, New York, 534 pp.

Goodwin, G. G., and A. M. Greenhall. 1961. A review of the bats of Trinidad and Tobago. Bull. Am. Mus. Nat. Hist., **122**:187–302.

Gould, E. H. 1978. Foraging behavior of Malaysian nectar-feeding bats. Biotropica, **10**:184–193.

Hamilton III, W. J., and K. E. F. Watt. 1970. Refuging. Ann. Rev. Ecol. Syst., **1**:263–286.

Hassell, M. P., and R. M. May. 1974. Aggregation of predators and insect parasites and its effect on stability. J. Anim. Ecol., **43**:567–594.

Heinrich, B. 1975. Energetics of pollination. Ann. Rev. Ecol. Syst., **6**:139–170.

Heithaus, E. R., and T. H. Fleming. 1978. Foraging movements of a frugivorous bat, *Carollia perspicillata* (Phyllostomatidae). Ecol. Monogr., **48**:127–143.

Heithaus, E. R., P. A. Opler, and H. G. Baker. 1974. Bat activity and pollination of *Bauhinia pauletia:* Plant-pollinator coevolution. Ecology, **55**:412–419.

Heithaus, E. R., T. H. Fleming, and P. A. Opler. 1975. Foraging patterns and resource utilization in seven species of bats in a seasonal tropical forest. Ecology, **56**:841–854.

Hladik, C. M., A. Hladik, J. Bousett, P. Valdebouze, G. Viroben, and J. Delort-Laval. 1971. Le régime alimentaire de primate de l'ile de Barro Colorado (Panama). Folia Primatol., **16**:95–122.

Howe, H. F. 1979. Fear and frugivory. Am. Nat., **114**:925–931.

Howe, H. F., and G. F. Estabrook. 1977. On intraspecific competition for avian dispersers in tropical trees. Am. Nat., **111**:817–832.

Howell, D. J. 1974. Bats and pollen: physiological aspects of the syndrome of chiropterophily. Comp. Biochem. Physiol., **48**:263–276.

Howell, D. J. 1977. Time sharing and body partitioning in bat-plant pollination systems. Nature, **270**:509–510.

Howell, D. J. 1979. Flock foraging in nectar-feeding bats: advantages to the bats and to the host plants. Am. Nat., **114**:23–49.

Howell, D. J., and D. L. Hartl. 1980. Optimal foraging in Glossophagine bats:When to give up. Am. Nat., **115**:696–704.

Humphrey, S. R., and F. J. Bonaccorso. 1978. Population and community ecology. Pp. 409–441, *in* Biology of bats of the New World family Phyllostomatidae. Part III. (R. J. Baker, J. K. Jones, Jr., and D. C. Carter, eds.). Spec. Publ. Mus. Texas Tech Univ., Lubbock, **16**:1–441.

Jamieson, R. W. 1977. Foraging behavior in *Carollia perspicillata*, a neotropical frugivorous bat. Unpublished M. S. thesis, University of Missouri, St. Louis, 39 pp.

Janzen, D. H. 1978. Seeding patterns of tropical trees. Pp. 83–128, *in* Tropical trees as living systems. (P. B. Tomlinson and M. H. Zimmerman, eds.). Cambridge University Press, Cambridge, 675 pp.

Janzen, D. H. 1979. How to be a fig. Ann. Rev. Ecol. Syst., **10**:13–51.

Janzen, D. H., G. A. Miller, J. Hackforth-Jones, C. M. Pond, K. Hooper, and D. P. Janos. 1976. Two Costa Rican bat-generated seed shadows of *Andira inermis* (Leguminosae). Ecology, **57**:1068–1075.

Jimbo, S., and H. O. Schwassmann. 1967. Feeding behavior and the daily emergence pattern of "*Artibeus jamaicensis*" Leach (Chiroptera, Phyllostomatidae). Atas Simp. Biota Amazonica, **5**:239–253.

Jones, J. K., Jr., and D. C. Carter. 1976. Annotated checklist, with keys to subfamilies and genera. Pp. 7–38, *in* Biology of bats of the New World family Phyllostomatidae. Part I. (R. J. Baker, J. K. Jones, Jr., and D. C. Carter, eds.). Spec. Publ. Mus. Texas Tech Univ., Lubbock, **10**:1–218.

Katz, P. L. 1974. A long-term approach to foraging optimization. Am. Nat., **108**:758–782.

Kingdon, J. 1974. East African mammals. Vol. 2. Part A. Academic Press, London, 341 pp.

Kodrich-Brown, A., and J. H. Brown. 1978. Influence of economics, interspecific competition, and sexual dimorphism on territoriality of migrant Rufous Hummingbirds. Ecology, **59**:285–296.

Koopman, K. F., and E. L. Cockrum. 1967. Pp. 109–150, *in* Recent mammals of the world. (S. Anderson and J. K. Jones, Jr., eds.). Ronald Press, New York, 453 pp.

Krebs, J. R. 1978. Optimal foraging: decision rules for predators. Pp. 34–63, *in* Behavioural ecology: An evolutionary approach. (J. R. Krebs and N. B. Davies, eds.). Blackwell Science, Oxford, 694 pp.

Krebs, J. R. 1979. Foraging strategies and their social significance. Pp. 225–270, *in* Handbook of behavioral neurobiology. Vol. 3. (P. Marler and J. G. Vandenbergh, eds.). Plenum Press, New York, 411 pp.

Lack, A. 1978. The ecology of the flowers of the savanna tree *Maranthes polyandra* and their visitors, with particular reference to bats. J. Ecol., **66**:287–295.

LaVal, R. K., and H. S. Fitch. 1977. Structure, movements and reproduction in three Costa Rican bat communities. Occas. Pap. Mus. Nat. Hist. Univ. Kans., **69**:1–27.

Lewontin, R. C. 1978. Fitness, survival, and optimality. Pp. 3–21, *in* Analysis of ecological systems. (C. H. Horn, R. Mitchell, and G. R. Stairs, eds.). Ohio State University Press, Columbus, 312 pp.

Lockwood, R., E. R. Heithaus, and T. H. Fleming. 1982. Fruit choices among free-flying captive *Carollia perspicillata*. J. Mammal., (in press).

McCracken, G. F., and J. W. Bradbury. 1977. Paternity and genetic heterogeneity in the polygynous bat, *Phyllostomus hastatus*. Science, **198**:303–306.

McCracken, G. F., and J. W. Bradbury. 1980. Social organization and kinship in the polygynous bat *Phyllostomus hastatus*. Behav. Ecol. Sociobiol., **8**:11–34.

McKey, D. 1975. The ecology of coevolved seed dispersal systems. Pp. 159–191, *in* Coevolution of animals and plants. (L. E. Gilbert and P. H. Raven, eds.). University of Texas Press, Austin, 246 pp.

Milton, K. 1979. Factors influencing leaf choice by howler monkeys: a test of some hypotheses of food selection by generalist herbivores. Am. Nat., **114**:362–378.

Milton, K., D. M. Windsor, D. W. Morrison, and M. A. Estribi. 1982. Fruiting phenologies of two neotropical *Ficus* species. Ecology, **63**:752–762.

Morrison, D. W. 1978a. Foraging ecology and energetics of the frugivorous bat *Artibeus jamaicensis*. Ecology, **59**:716–723.

Morrison, D. W. 1978b. Lunar phobia in a Neotropical fruit bat, *Artibeus jamaicensis* (Chiroptera: Phyllostomidae). Anim. Behav., **26**:852–855.

Morrison, D. W. 1978c. On the optimal searching strategy for refuging predators. Am. Nat., **112**:925–934.

Morrison, D. W. 1978d. Influence of habitat on the foraging distances of the fruit bat, *Artibeus jamaicensis*. J. Mammal., **59**:622–624.

Morrison, D. W. 1979. Apparent male defense of tree hollows in the fruit bat, *Artibeus jamaicensis*. J. Mammal., **60**:11–15.

Morrison, D. W. 1980a. Foraging and day roosting dynamics of canopy fruit bats in Panama. J. Mammal., **61**:20–29.

Morrison, D. W. 1980b. Efficiency of food utilization by fruit bats. Oecologia, **45**:270–273.

Nelson, J. E. 1965. Movements of Australian flying foxes (Pteropodidae: Megachiroptera). Aust. J. Zool., **13**:53–73.

Opler, P. A., G. W. Frankie, and H. G. Baker. 1980. Comparative phenological studies of treelets and shrubs in the tropical wet and dry forests in the lowlands of Costa Rica. J. Ecol., **68**:167–188.

Pyke, G. H., H. R. Pulliam, and E. L. Charnov. 1977. Optimal foraging: A selective review of theory and tests. Q. Rev. Biol., **52**:137–154.

Rosevear, D. R. 1965. The bats of West Africa. Br. Mus. Nat. Hist. London, 418 pp.

Sazima, I., and M. Sazima. 1977. Solitary and group foraging: Two flower-visiting patterns of the lesser spear-nosed bat *Phyllostomus discolor*. Biotropica, **9**:213–215.

Sazima, M., and I. Sazima. 1978. Bat pollination of the passion flower, *Passiflora mucronata*, in southeastern Brazil. Biotropica, **10**:100–109.

Schoener, T. W. 1969. Models of optimal size for solitary predators. Am. Nat., **103**:277–313.

Schoener, T. W. 1971. Theory of feeding strategies. Ann. Rev. Ecol. Syst., **2**:369–404.

Smith, C. C. 1977. Feeding behavior and social organization in howling monkeys. Pp. 97–126, *in* Primate ecology: Studies of feeding and ranging behaviour in lemurs, monkeys, and apes. (T. H. Clutton-Brock, ed.). Academic Press, London, 631 pp.

Smith, J. D. 1976. Chiropteran evolution. Pp. 49–70, *in* Bats of the New World family Phyllostomatidae. Part I. (R. J. Baker, J. K. Jones, Jr., and D. C. Carter, eds.). Spec. Publ. Mus. Texas Tech Univ., Lubbock, **10**:1–218.

Smythe, N. 1970. Relationships between fruiting seasons and seed dispersal methods in a neotropical forest. Am. Nat., **104**:25–35.

Snow, D. W. 1976. The web of adaptations. Collins, London, 176 pp.

Start, A. N., and A. G. Marshall. 1976. Nectarivorous bats as pollinators of trees in West Malaysia.

Pp. 141–150, *in* Tropical trees: Variation, breeding, and conservation. Linnean Society Symposium Series No. 2. (J. Burley and B. T. Styles, eds.). Academic Press, London, 243 pp.

Stiles, F. G. 1975. Ecology, flowering phenology, and hummingbird pollination of some Costa Rican *Heliconia* species. Ecology, **56:**285–301.

Stiles, F. G. 1977. Coadapted competitors: Flowering seasons of hummingbird food plants in a tropical forest. Science, **198:**1177–1178.

Stiles, F. G. 1978a. Ecological and evolutionary implications of bird pollination. Am. Zool., **18:**715–727.

Stiles, F. G. 1978b. Temporal organization of flowering among the hummingbird foodplants of a tropical wet forest. Biotropica, **10:**194–210.

Struhsaker, T. T. 1978. Food habits of five monkey species in Kibale Forest, Uganda. Pp. 225–248, *in* Recent advances in primatology. Vol. 1. (D. J. Chivers and J. Herbert, eds.). Academic Press, London, 980 pp.

Sussman, R. W., and P. H. Raven. 1978. Pollination by lemurs and marsupials: An archaic coevolutionary system. Science, **200:**731–736.

Thomas, D. W., and M. B. Fenton. 1978. Notes on the dry season roosting and foraging behaviour of *Epomophorus gambianus* and *Rousettus aegyptiacus* (Chiroptera: Pteropodidae). J. Zool., **186:**403–406.

Thomas, S. P. 1975. Metabolism during flight in two species of bats, *Phyllostomus hastatus* and *Pteropus gouldii*. J. Exp. Biol., **63:**273–293.

Tuttle, M. D. 1976. Collecting techniques. Pp. 71–88, *in* Biology of bats of the New World family Phyllostomatidae. Part I. (R. J. Baker, J. K. Jones, Jr., and D. C. Carter, eds.). Spec. Publ. Mus. Texas Tech Univ., Lubbock, **10:**1–218.

Van Valen, L. 1979. The evolution of bats. Evol. Theory, **4:**103–121.

Vazquez-Yanes, C., A. Orozco, G. Francois, and L. Trejo. 1975. Observations on seed dispersal by bats in a tropical humid region in Veracruz, Mexico. Biotropica, **7:**73–76.

Walker, E. P. 1964. Mammals of the world. Vol. 1. Johns Hopkins Press, Baltimore, 644 pp.

Ward, P., and A. Zahavi. 1973. The importance of certain assemblages of birds as "information centres" for food-finding. Ibis, **115:**517–534.

Wickler, W., and U. Seibt. 1976. Field studies of the African fruit bat, *Epomophorus wahlbergi*, with special reference to male calling. Z. Tierpsychol., **40:**345–376.

Williams, T. C., and J. M. Williams. 1970. Radio tracking of homing and feeding flights of a Neotropical bat, *Phyllostomus hastatus*. Anim. Behav., **18:**302–309.

Wilson, D. E. 1973. Bat faunas: A trophic comparison. Syst. Zool., **22:**14–29.

Wilson, E. O. 1975. Sociobiology. Belknap Press, Cambridge, 697 pp.

Chapter 9

Coevolution between Bats and Plants

E. Raymond Heithaus

Biology Department
Kenyon College
Gambier, Ohio 43022

I. INTRODUCTION

More than 250 species of bats eat some fruit, nectar, or pollen. In doing so, bats often provide dispersal and pollination services to at least 130 plant genera (Howell and Hodgkin, 1976). These interactions have profoundly influenced the evolution of some bats and plants. For example, the leaf-nosed bat, *Leptonycteris sanborni*, has morphological and physiological traits that allow it to subsist entirely on flowers and fruits, whereas the same traits apparently preclude the extensive use of insects (Howell and Hodgkin, 1976). On the other side of the interaction, plants such as *Crescentia alata* (the calabash tree) are so efficiently pollinated by bats that they are unable to reproduce without bat assistance. The structure of its flowers and the timing of nectar presentation preclude pollination by other animals.

These species illustrate the potential changes produced by evolutionary responses to interactions between bats and plants. Either group may become very specialized, but the cost of specialization is decreased flexibility. Given the trade-off between the effects of specialization on efficiency and flexibility, we can expect many degrees of dependence between bats and plants.

Broad generalizations about bat–plant interactions are difficult to formulate because the participants and selective pressures on them are diverse. The dependence of bats on plants for nutrition varies, especially among the Phyllostomidae. For example, among the Glossophaginae, *Leptonycteris* rarely eats animals, while *Glossophaga soricina* is an omnivore. Other bats, such as most *Micronycteris* spp. (Phyllostominae) only occasionally divert their feeding to fruits, away from their usual foliage gleaning for insects (Wilson, 1971; Bonaccorso, 1979). Dependence on bats for pollination or dispersal also varies among plants. Additional diversity in interactions is caused by differences between New World and Old World systems, which have different evolutionary histories (Baker, 1973). Descriptions of "typical" interactions and morphological coadaptations (Faegri and van der Pijl, 1966; Vogel, 1969; and van der Pijl, 1972) tend to focus on the more specialized species; to balance this and to reduce repetition I will emphasize variability.

"Coevolution" describes many evolutionary responses to interactions between species. The concept of coadaptation based on species interactions can be traced to Darwin (1872), who first studied pollination systems with an evolutionary approach. The term *coevolution* was used by theoreticians as early as 1958 (Mode, 1958). Popular use of the term was stimulated by Ehrlich and Raven (1964), who used the concept to interpret patterns of coadaptation between herbivorous butterflies and their food plants.

Coevolution implies more than a simple response to another species; it involves response and counter-response. It is the process by which one population makes an evolutionary change in response to a second population; this change is followed by an evolutionary response by the second population (Ehrlich and Raven, 1964). An excellent example of coevolution is described by Gilbert (1975). Plants in the genus *Passiflora* contain toxic chemicals in their leaves, and the storage of these probably evolved in response to herbivory. One group of herbivores, butterflies in the genus *Heliconus*, counter-responded by incorporating these chemicals for their own defense against predators. Further coadaptations have evolved among these plants and insects (Gilbert, 1971, 1977), but this example illustrates the process. For bats and plants the sequence of response and counter-response is rarely seen between just two species, or even two genera; rather, bats and plants show "diffuse coevolution" (Janzen, 1980), because each group may include many species.

Studies of coevolution are nearly as diverse as the interactions between bats and plants. Various approaches emphasize coupled speciation, complex coadaptations, or the consequences of coevolution on community attributes such as stability, diversity, and production efficiency. Some theoretical models have attempted to integrate coevolutionary effects on selection and population dynamics (Roughgarden, 1979; Slatkin and Maynard-Smith, 1979; Levin and Udovic, 1977), providing still another way to look at coevolution. I will try to relate each of these approaches to coevolution between bats and plants.

2. COUPLED SPECIATION

2.1. Evolutionary Origins of Frugivory and Nectarivory

The evolutionary history of bats and plants is the best evidence for their coevolution. Responses and counter-responses that are recorded in the fossil record are far less ambiguous than simple correlations seen between extant species. Even though few fossils of bats have been found, so that there are no clear transitional forms (J. D. Smith, 1976, 1977), several patterns have emerged.

Bats became significant pollination or dispersal agents only after flowering plants had evolved for at least 75 million years (Sussman and Raven, 1978). The first bats were insectivorous, appearing in the early Eocene or perhaps the Paleocene (Jepson, 1970; Van Valen, 1979). In the Old World the megachiropterans had appeared by the Oligocene (Jones and Genoways, 1979) and were already specialized for a diet of fruits. The Neotropical frugivores (Phyllostomidae) had evolved by the late Miocece (J. D. Smith, 1977) and were probably derived from insect-eating microchiropterans that moved through Europe and North America before entering the New World tropics (Baker, 1973; Van Valen, 1979).

The later evolution of frugivory and nectarivory among the New World bats means that they were exposed to plants preadapted for visitation by bats. Plants there were pollinated by megachiropterans and were incorporated into the Neotropical flora as African and South American continental plates drifted apart (Baker, 1973). Other South American plants were probably pollinated or dispersed by nocturnal vertebrates before the evolution of phyllostomids, and the structures associated with interactions with nonflying mammals overlap with characteristics now thought to typify "bat" plants (Sussman and Raven, 1978). Bats must have made the first response in the coevolutionary scheme in the New World.

Variable feeding patterns, including diets of animals, fruits, and nectar, are a major factor influencing the adaptive radiation among bats (Jepson, 1970; Gillette, 1975; Pirlot, 1977; Van Valen, 1979). Among the megachiropterans, diets range from mixtures of fruits, flowers, leaves, and insects to specialized flower feeding (Start and Marshall, 1976). Several small members of the Pteropodidae are highly specialized flower visitors, with many characteristics convergent with hummingbirds. Three subfamilies of the Neotropical Phyllostomidae showed adaptive radiation following independent evolution of frugivory: the Stenodermatinae, Phyllonycterinae, and Glossophaginae (van der Pijl, 1972). Feeding at flowers influenced the diversification of the Glossophaginae and Phyllonycterinae as well. A point to be emphasized is that both specialized and varied feeding mechanisms were successful (Gillette, 1975; Van Valen, 1979).

The transition from insectivory to eating nectar or fruit probably involved bats catching insects near or on plants (Jepson, 1970; Gillette, 1975; Yalden and Morris, 1975). According to one hypothesis, bats sought insects that were attracted to ripe fruits (or flowers). Supplemental ingestion of fruit or nectar led to a gradual incorporation of plant parts into the diets of bats. The behavior of two insectivorous bats supports this intuitively attractive hypothesis. An African insectivore, *Hipposideros commersoni*, has been observed attacking figs to extract weevil larvae (Yalden and Morris, 1975), and in the New World *Antrozous pallidus* (Vespertilionidae) sometimes shows behavior that matches actions expected of an incipient frugivore (Howell, 1980). The insectivorous *Antrozous* usually do not eat fruit, but at one site in Arizona Howell (1980) found fruits and seeds in one quarter of their fecal remains. *Antrozous pallidus* is generally a gleaner and prefers moths to other foods. The frugivory at the site described by Howell occurs because these bats are gleaning noctuid moths from inside ripe fruits of the organ-pipe cactus. *Antrozous* apparently has extended its diet as it captured moths associated with the fruit, which lends credence to the dual-feeding hypothesis of developing frugivory.

Slightly different starting points in the "feeding duality" pathway may explain major differences between megachiropterans and fruit-eating phyllostomids. Gillette (1975) hypothesized that ancestors of frugivorous megachiropterans were foliage-gleaning insectivores. This may have preadapted them for feeding in trees, rather than aerially, because they would already have developed flight patterns that deemphasize aerial maneuverability. Instead, their morphologies promoted long-distance flight capacity. Other characteristics, including size and highly developed vision, would follow from a foliage-gleaning origin. In contrast, the early microchiropterans were adapted for aerial insectivory. Modern frugivores and nectarivores in this suborder retain sonar orientation and take fruits and nectar while in flight (Novick, 1977).

2.2. Effects of Bats on Plant Diversification

Animal vectors of pollen and seeds were critical elements in the adaptive radiation of angiosperms (Raven, 1977; Regal, 1977; Mulcahy, 1979). Although bats were not present to contribute to the displacement of tree ferns, gymnosperms, and horsetails by flowering plants in the Cretaceous, later diversifications was partly correlated with exploiting bats for pollination or dispersal. For example, the bat-pollinated family Sonneratiaceae has an excellent fossil record, but no fossils predate flower-visiting bats (Sussman and Raven, 1978). The family Sonneratiaceae is exceptional in that seven species in it are pollinated by bats, and several are dispersed by bats, too. Usually bat pollination is one of several pollination mechanisms in a plant family, indicating independent evolution of this trait. For example, *Cobaea scandens* and *C. trianaei* are the only two species pollinated by bats among the 327 species in the phlox family; they are probably derived from a moth-pollinated form (Grant and Grant, 1965). Bat pollination has arisen independently in at least 27 New World families (Vogel, 1969), affecting more than 500 species. Some families with several bat-pollinated species include the Bignoniaceae, Cactaceae, Leguminosae, Bombacaceae, Gesneriaceae, Solanaceae, Lobelioidae, Bromeliaceae, Agavaceae, Sapotaceae, and Musaceae (Vogel, 1969; van der Pijl, 1956; Start and Marshall, 1976).

Fruits that encourage dispersal by bats also arose independently many times. Several plant families include some species that are dispersed by bats, other species specialized for bird dispersal, and still other species with fruits that are dispersed by a wide variety of animals (van der Pijl, 1972). The Moraceae, including the genera *Cecropia, Chlorophora,* and *Ficus,* is a pantropical family that shows such diversity.

2.3. Coupled Speciation through Coevolution?

The diffuse coevolution between some bats and plants has led to few species-specific systems. No bat relies on just one species of plant for all of its nutrition, in spite of the large numbers of bats that eat fruits or flowers. Many plants depend on bats for pollination or seed dispersal, but dependence on a particular kind of bat is rare. There is just one published report, to my knowledge, of a plant that has flowers that exclude all but one species of bat (Gould, 1978). Diversification has not depended on the special characteristics of just one species from the interacting group.

In spite of the fact that diffuse coevolution has promoted plant diversification, not all mutually dependent species are coevolved. Preadaptation is an alternate route to mutual dependence. This would short-circuit the response-counter–response that defines coevolution. For example, the pantropical trees *Ceiba pentandra* and *Parkia* spp. depend on bats for pollination on both sides of the Atlantic. Their floral characteristics have been used to define "bat" flowers (Faegri and van der Pijl, 1966), and New and Old World bats use the flowers as a major source of food (Baker and Harris, 1959; Heithaus *et al.*, 1975). Baker (1963, 1973) points out, however, that continental drift would have effectively separated the Neotropical and Paleotropical populations by the Eocene epoch, well before plant-specialized megachiropterans or microchiropterans appeared. The phenotypes of these plants were established before nectar-feeding bats had evolved in either hemisphere. Although bats are now the principle pollinators in both Africa and South America, they must have displaced some other pollinating agent. These well-adapted, bat-dependent plants cannot represent examples of coevolved diversification if "the plants appear to have been ready for the bats before the bats were available" (Baker, 1963; p. 881). Several other species with "bat-pollinated" flowers were isolated in South America before phyllostomid bats arrived (Vogel, 1969b), so the likelihood of preadaptation is not restricted to *Ceiba* and *Parkia.*

Independent evidence supports Baker's claim that correlations in morphology and dependence do not justify a conclusion that all diversification has resulted from coevolution. For plants, floral traits similar to those that attract bats attracted some nonflying mammals that were specialized pollinators before flower-feeding bats appeared (Sussman and Raven, 1978). Marsupials have probably eaten nectar since the upper Cretaceous. Several mammals from the early Tertiary, such as the extinct Picrodontidae, were specialized flower feeders, and these small primates showed many characteristics that converged with phyllostomid bats (Sussman and Raven, 1978). Pollination by nonflying mammals appears to promote floral characteristics similar to those seen in bat-pollinated plants; flowers are large with dull colors and produce copious amounts of nectar. Sussman and Raven (1978) suggest that the late-appearing bats competitively displaced many of these groups. Where bats dominate today, the nonflying mammals may well have left a legacy of plant species preadapted for pollination or dispersal by bats.

3. COMPLEX COADAPTATIONS BETWEEN BATS AND PLANTS

Correlated traits between bats and plants include coadaptations in morphology, physiology, and behavior. For example, in Malaysia *Oroxylum iridicum* (Bignoniaceae) has flowers that precisely match the morphology and behavior of *Eonycteris spelaea* (Pteropodidae) (Gould, 1978). *Eonycteris* migrates from caves to feeding areas, so its feeding begins well after sunset; *Oroxylum* flowers open 2½ hr after sunset and drop before sunrise, so early-feeding bats and diurnal nectarivores are excluded. Just after opening, the width of the cup-shaped flower matches the width of *Eonycteris'* shoulders; bats larger than *Eonycteris* are discouraged, if not excluded, from the nectar at first. The location of nectar-soaked hairs in the flower closely matches the spot that an adult *Eonycteris* can reach with its fully extended tongue as its body is pushed into the flower. These nectar-soaked hairs are charged with nectar only after the flower is tipped, but the flower is on a rigid stalk. Even an immature *Eonycteris* lacks sufficient weight to depress the flower, so bats smaller than adult *Eonycteris* cannot easily obtain nectar. *Oroxylum* parcels nectar in small doses and in just a few flowers per night: *Eonycteris* can efficiently forage by visiting scattered flowers repeatedly, unlike many other larger pteropodids. Gould (1978) describes this system as a "lock and key" specific pollination system.

The remarkable congruence seen in the *Oroxylum–Eonycteris* example is unusually specific for bat–plant systems. Flowers and fruits are normally accessible to several species of bats, and often to organisms other than bats, but the *Oroxylum–Eonycteris* example illustrates some general points. Mutual congruence is often cited as evidence for coevolution (Janzen, 1980), but even in this system a specific response and counter-response is not evident. Although *Oroxylum* appears to have responded very specifically to *Eonycteris*, this bat visits other plants and shows no specific evolutionary response to *Oroxylum*. The example also indicates that efficient use of plant resources may require special morphology, physiology, and behavior. The following sections examine the specific traits of bats that are responses to selection for nectar, pollen, or fruit eating.

3.1. Coadaptations

3.1.1. Morphological Responses

The teeth and palate of many plant-visiting bats are modified, compared to ancestral, insectivorous species. The extent of divergence from ancestral morphologies is correlated with the degree of dependence on plant resources. Usually frugivores crush fruits in their mouths, so they have muscular tongues that can force food against a series of transverse ridges on the palate (Yalden and Morris, 1975). Bats that primarily consume fruits have molars that are flatter and broader than those of insectivorous species. In very specialized frugivores sharp cusps are lost, but omnivores retain portions of sharp cusps on their molars. These features evolved independently in the fruit-eating Pteropodidae and Phyllostomidae (Yalden and Morris, 1975). Molars from nectar-feeding bats may be greatly reduced or lost, with reduction correlated with dependence on floral resources. Specialist flower visitors may also lose incisors, especially if they have long tongues for lapping nectar. The length of the muzzle also increases with specialization for using nectar and pollen. These patterns suggest that extreme adaptation to a plant-eating life involves sacrificing features that permit other diets.

The tongues of bats also show variation in modifications for collecting nectar. Bats that are dependent upon nectar may have very long tongues. The distal portion of the tongue is covered with filiform papillae in bats that use nectar extensively. Fewer papillae are on the tongues of species that usually eat fruit but visit flowers seasonally; no filiform papillae are on the tongues of insectivorous bats (Howell and Hodgkin, 1976). Again, this pattern evolved independently in the New and Old World bats (Yalden and Morris, 1975).

Hair has also been modified from ancestral forms and may serve as pollen collectors. The hairs of flower-visiting bats, both mega- and microchiropterans, have projecting scales (Benedict, 1957; Vogel, 1969; Howell and Hodgkin, 1976). In contrast, insectivorous bats have smooth hair (the scales are flat against the shaft), and omnivorous bats have hairs that are intermediate in form (Howell and Hodgkin, 1976). The projecting scales increase the pollen-carrying capacity of a bat's pelage, much as plumose hairs on bees increase the ability to collect pollen. The selective pressures for holding pollen may be similar in these different groups; like bees, bats groom pollen from their bodies for food (Howell, 1974a, 1974b; Howell and Burch, 1974; Gould, 1977; Daniel, 1979).

A frugivorous diet also is correlated with modifications of the intestines. Fruit-eating members of the Phyllostomidae have more highly differentiated lymphoid tissue (Peyer's patches) than insectivorous phyllostomids. Frugivores also have more Peyer's patches in their intestines, relative to bats with other primary diets (Forman *et al.*, 1979). These modifications are likely to aid the assimilation of nutrients. Assimilation may be a limiting factor in foraging efficiency, because fruits are usually in one end and out the other in 20–40 min. (Klite, 1965; Morrison, 1978a; Lockwood *et al.*, 1981). Fruits retain much of their characteristic odor, texture, and mass after passing through a bat, so there is little outward sign of a bat receiving substantial benefit. Lymphoid tissue may facilitate the assimilation of fats, which are more likely to be in short supply than the carbohydrates that form most of a fruit (van der Pijl, 1972).

Structures that are associated with finding and harvesting fruits can be expected to reflect plant characteristics. Fruit-eating bats, in general, have short, broad wings that aid in carrying loads (fruits are often carried to roosts) and allow slow flight in cluttered environments (Findley *et al.*, 1972; Yalden and Morris, 1975). Sonar characteristics are also related to environmental clutter (Simmons *et al.*, 1979), but habitats filled with foliage are not unique to plant-visiting bats. Consequently, some insectivores and frugivores share echolocation patterns (Gould, 1977; Novick, 1977). The specialized flower-visiting bats tend to have better hovering ability, which is facilitated by wing tips that are long relative to the rest of the wing (Findley *et al.*, 1972). But there is much variation within these general patterns. Smith and Starrett (1979) show that specific wing conformations cannot be predicted by difference in diet for phyllostomids; instead, differentiation is most closely related to taxonomic differences at the subfamilial level. In their study only specialized flower feeders were a uniquely differentiated group.

3.1.2. Physiological Responses

Food habits influence the basal metabolic rates and thermoregulation of bats (Chapter 4). To summarize, for a given body weight, nectarivores have higher metabolic rates than frugivores, and insectivores have the lowest rates (McNab, 1969, 1980). The metabolic rates of omnivorous bats are intermediate between those of frugivores and insectivores. Frugivory also facilitates strict homeothermy compared to insectivory (Bartholomew *et al.*, 1964; McNab, 1969). These differences may be related to the seasonality of resources rather than the nutritional qualities of fruit or nectar *per se*. McNab (1980) suggests that frugivores can switch foods, so some resources are consistently available; insectivores must contend with fluctuating food supplies, even in the tropics. But not all patterns are consistent with the hypothesis that the seasonality of resources and basal metabolic rates are negatively correlated. By this hypothesis, omnivores should be least sensitive to environmental fluctuations and would be expected to have basal metabolic rates equal to, not lower than, frugivores.

Some flower-feeding bats have developed the physiological mechanisms to subsist entirely on nectar and pollen. Bat-pollinated flowers on saguaro and some agaves from North America produce pollen that is much richer in proteins than pollen from insect-pollinated congeners (Howell, 1974a). *Leptonycteris sanbor-*

ni obtains a sufficient protein supply by ingesting pollen (Howell, 1974a). Pollen is protected by a durable covering that cannot be digested, but an acidic, high-sugar environment promotes pollen germination, which makes protein available. Nectar provides the correct carbohydrate conditions in the gut of *Leptonycteris*, while special acid-secreting glands in the gastrointestinal submucosa complete the germination requirements (Howell, 1974a). This is another example of very specific mutual congruence between plants and pollinating bats. The specificity appears to extend to plants in providing specific amino acids (proline and tyrosine) required by bats (Howell, 1974a). Whether this pattern is typical is still unknown, but it clearly is not universal. For example, the epiphyte *Markea neurantha* (Solanaceae) does not provide a full complement of amino acids in its pollen, although it is pollinated by relatively specialized flower-visiting bats (Voss *et al.*, 1980).

3.1.3. Behavioral Responses

While morphological and physiological responses to plant characteristics have increased the potential of bats to exploit fruits, pollen, and nectar, the behavior of bats determines the benefits received by both groups. What bats eat, how they treat their food and where they go with it are the critical components to the interactions between bats and plants. The variation in bat behavior is considerable and it is not trivial with respect to potential coevolution. For example, the effectiveness of pollination by different species of bats varies, depending on the chance that a particular bat will contact stamens and then pistils on conspecific plants (Heithaus *et al.*, 1975). Fruit eaters also vary in effectiveness; *Pteropus* may spend an entire night feeding in a *Durio* tree, failing to move seeds beyond the tree's canopy, while smaller bats move extensively (Gould, 1978).

The feeding behavior of plant-visiting bats is responsive to many aspects of fruiting and flowering, as reviewed by others (Baker, 1973; Start and Marshal, 1976; and Fleming, Chapter 8). Many specific feeding behaviors are not unique to plant-visiting bats. Insectivorous and frugivorous bats share many searching, harvesting, and social patterns when resource distributions are similar (Bradbury and Vehrencamp, 1976a, 1976b; Fleming, Chapter 8). A major difference between fruits or nectar and other food types is that there is little selection for fruits and nectar to escape their bat consumers. Evolutionary responses in behavior that are specific to eating fruits and nectar are usually related to this difference.

Major behaviors that can be specifically related to plants include the choice of trap-lining or opportunistic foraging (Baker, 1973) and food-handling behavior. These patterns are discussed in some detail elsewhere (Chapter 8), but I will summarize them briefly here. When plants produce flowers or fruits for prolonged periods, sometimes for weeks or months, bats will learn the locations of plants and return to them regularly (e.g., Vogel, 1968, 1969; Baker *et al.*, 1971;

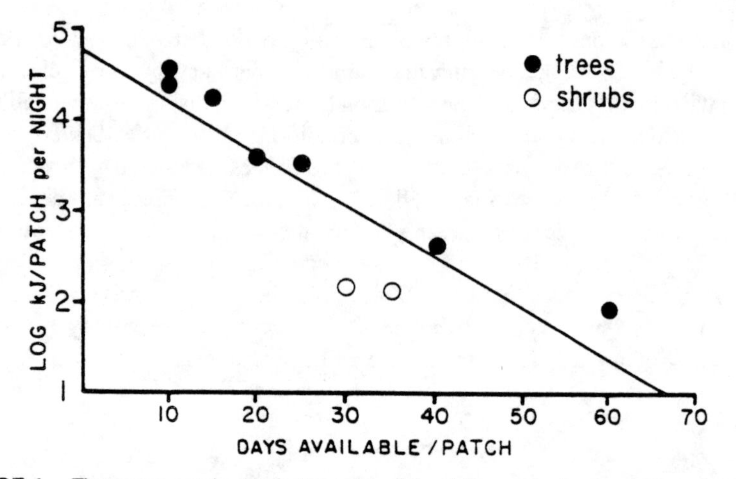

FIGURE 1. The energy contained in fruits produced per night compared to the number of days that individual plants produce fruit for plants visited by *Carollia perspicillata* in Santa Rosa National Park, Costa Rica. (Data from Heithaus and Fleming, 1978; Lockwood *et al.*, 1981.)

Baker, 1973; Start and Marshall, 1976; Heithaus and Fleming, 1978; Gould, 1978). Usually these plants will present only a few fruits or flowers per night (Fig. 1), so trap-lining bats usually travel alone or in very small groups (e.g., Heithaus *et al.*, 1974; Howell, 1979).

When plants produce many fruits or flowers, and these are present only for a few nights (Fig. 1), food becomes relatively unpredictable. Under these conditions bats forage opportunistically, where changes in foraging sites occur relatively frequently (2–10 days) to match new, abundant resources. In seasonal, African habitats, where fruiting and flowering often favors opportunistic foraging, huge flocks of fruit bats may converge on one or a few trees (Baker and Harris, 1959; Harris and Baker, 1958, 1959; Ayensu, 1974).

Although the foraging behavior of bats may appear coevolved to be optimally adapted for harvesting from a plant (e.g., Howell, 1974a, 1979, 1980), most bat behavior is not fixed. Flexible behavior allows bats to efficiently exploit a variety of plants. In the Neotropics the number of flowers produced by *Lafoensia glypotocarpa* and two *Bauhinia* spp. influenced changes in the feeding behavior of *Phyllostomus discolor* (Sazima and Sazima, 1975, 1977). These bats forage singly or in groups, depending on resource abundance. In Malaysia *Eonycteris spelaea* may trap-line or forage opportunistically, depending on which plants are producing food at the time (Gould, 1978). Behaviors for collecting and consuming plant resources are also flexible. For example, *Carollia perspicillata* may land on a plant to eat a fruit in place, land to pick the fruit before transporting it elsewhere, pluck a fruit while airborne, or take fallen fruit from the ground. It can eat fruits whole, bite off parts and waste the rest, or nibble sections off a hard central stalk as if the fruit were corn on the cob (Lockwood *et al.*, 1981). A complete catalog of fruit- and flower-handling behaviors is well beyond the scope of this chapter, but it should be clear from these examples that variation in handling behavior may have a major influence on where bats deposit seeds.

Coevolutionary responses by plants to bats may be complicated, especially when a single resource distribution elicits different foraging behaviors. This may occur when food "distribution," from a bat's perspective, is dependent on bat behavior. For example, *Phyllostomus discolor* and *Glossophaga soricina* show different foraging patterns when feeding at the same patch of *Bauhinia pauletia* (a shrubby legume) (Heithaus *et al.*, 1974). In this system the larger-bodied *Phyllostomus* empties a flower of nectar in one visit, whereas the smaller *Glossophaga* can return to a flower at least a dozen times. *Phyllostomus* forages in groups, perhaps to reduce the chance that an individual will visit empty flowers. *Glossophaga* forages along individual trap lines. In this system coevolution between nectar production by the plant and the foraging behavior of bats might be difficult because the feeding behavior of the two bat species imposes different selective pressures on the plant.

Predator avoidance behavior by bats may influence seed dispersal and pollination, which also complicates coevolutionary responses by plants (Howe, 1979). Of particular importance to plants is the generally observed behavior of bats leaving fruiting or flowering trees to consume fruit and pollen. Movement from food sources to feeding roosts occurs commonly among phyllostomids and less often among pteropodids. The selective basis for this behavior may be related to the risks of predation near predictable and attractive flowers and fruits (Chapter 1).

3.2. Flexibility and Diffuse Coevolution

3.2.1. Plants versus Other Sources of Food

There are at least two levels of flexibility with respect to bats incorporating fruits or nectar into their diets. First, bats may eat fruits, nectar, pollen, insects, vertebrates, or a combination of these. At this level of choice there are large differences among food items in nutritional content, abundance, and ease of location and capture. Very different requirements for obtaining and handling food items in these different classes complicate potential trade-offs in selection for profitable foraging patterns. Breadth of feeding on different species within these general classes defines the second level of flexibility. For example, do bats show fidelity to a particular plant species, once fruits, nectar, or pollen are sought as food?

Old World pteropodids are essentially restricted to eating fruits, nectar, or leaves, but the phyllostomids are rarely confined to plant diets (Gardner, 1977). Pteropodids may switch their diets to include leaves when fruits or nectar are scarce, and only occasionally do insects show up in the diet. Phyllostomid flexibility is underscored by the fact that only three species apparently eat only fruits; no species are restricted entirely to pollen and nectar, one species eats only insects, and one species eats only vertebrates (Table I).

Much of the flexibility in consuming different major food groups is a response to seasonal changes in their availability. *Leptonycteris sanborni* can switch to fruit if flowers are rare (Howell, 1980). *Micronycteris hirsuta, M. megalotis, M. brachyotis,* and *Trachops cirrhosus* only eat fruits when the fruits are extremely abundant (Wilson, 1971; Bonaccorso, 1979). Otherwise, these species are foliage-gleaning insectivores (*Micronycteris* spp.) or generalist gleaners (*Trachops*). These incentives have different implications for potentially coevolving plants. In the first case, switching permits flexibility in the reproductive timing of plants. Seasonal flowering or fruiting may occur without causing the preferred dispersal agents to go extinct during reproductive dormancy. The attraction of bats by very abundant fruiting can influence the evolution of mass-flowering or mass-fruiting phenologies (McKey, 1975; Howe and Estabrook, 1977; Howe, 1979).

TABLE I
Food Classes Used by Phyllostomids, Based on Records of Stomach or Fecal Contents and Pollen Collected from Pelage[a,b]

	Number of species using			
Additional components	Fruit	Flowers	Insects	Vertebrates
None	3[c]	0[d]	1[e]	1[f]
Fruit		5	14	1[g]
Insects		6[h]		
Vertebrates			1[i]	
Flowers and insects	7			
Flowers, insects, and vertebrates	2[j]			

[a]Species are included only if there were at least three independent reports or at least ten individuals observed.
[b]Data are from summaries given in Gardner, 1977, Bonaccorso, 1979, and Howell, 1980.
[c]*Vampyrodes caraccioli, Vampyressa pusilla,* and *Centurio senex* (Stenodermatinae).
[d]*Leptonycteris sanborni* and *Choeronycteris mexicana* diets are usually (99%) nectar and pollen, but fruit may be used during times of flower scarcity.
[e]*Macrophyllum macrophyllum.*
[f]*Vampyrum spectrum.*
[g]All six species are of the Glossophaginae.
[h]*Chrotopterus auritus.*
[i]*Trachops cirrhosus.*
[j]*Phyllostomus hastatus* and *Artibeus lituratus.*

The assignment of species to "pigeonholes" (Table I) is intended to reflect observed variation in feeding and not ecological dependence. A bat that eats fruit and nectar does not necessarily require both, nor would fruits and nectar necessarily be interchangeable. Although *Choeronycteris mexicana* and *Leptonycteris sanborni* may eat fruits under some conditions, flowers are clearly required because in no place do they survive and reproduce on a diet comprised exclusively of fruit. Similarly, *Artibeus jamaicensis* eats nectar, but fruits (especially from the family Moraceae) are required for its survival. Observations that emphasize dependence lead to the usual classification of species as "flower bats" or "fruit bats." Such classifications are not very rigorous usually, because so little is known about the diets of most bat species (Gardner, 1977). "Dependence" also may have different connotations. For example, *Leptonycteris* is dependent on flowers most of the time, but also may depend on fruits during brief times of flower shortage. The megachiropteran bat *Epomophorus wahlbergi* may depend on leaves for essential nutrients but otherwise eat fruit as its staple diet (Wickler and Seibt, 1976).

Classifying species as fruit bats or flower bats is convenient but hides important variation in the evolutionary commitment of a species to a particular food class (Table I). For example, members of the Glossophaginae are usually classified as flower bats, but species vary in their specialization for flower feeding. Reduction in the number of teeth and increasing palatal length is correlated with increasing specialization to feed on flowers. *Glossophaga soricina, Anoura geoffroyi, Leptonycteris sanborni*, and *Choeronycteris mexicana* vary in morphological and diet specialization (Table II). Geographical flexibility in diets is seen for *G. soricina*, which is another diffusing factor in coevolution. Pollen is important to *G. soricina* in sites at 800-m elevation in Mexico, but no pollen is eaten in the lowlands (Alvarez and Gonzales Q., 1970).

3.2.2. Flexibility in Plant Feeding

Bats are fickle feeders, from a plant's perspective. In general, bats use a variety of the fruits or flowers that are available to them, including many species of plants that show no signs of being specifically coevolved with bats. For example, bats eat nectar and pollen from flowers that are pollinated by other animals (Alvarez and Gonzales Q., 1970; Heithaus *et al.*, 1975; Gardner, 1977). Pollen from a bee-pollinated *Lonchocarpus* sp. was commonly found on glossophagines in Mexico, as was pollen from several Compositae, *Cordia* sp., and *Roupala*, none of which are bat pollinated (Alvarez and Gonzales Q., 1970). *Artibeus jamaicensis*, in Veracruz, Mexico, eats fruits that are also dispersed by birds (Vasquez-Yanes *et al.*, 1975). Fruits from a "typical" bird-dispersed plant, *Acacia cornigera*, comprise a major portion of the diet of *Carollia per-*

TABLE II
Variation in Morphology and Diet for Four Species of Glossophagine Bats in Mexico[a]

Character	*G. soricina*	*A. geoffroyi*	*L. sanborni*	*C. mexicana*
Palatal length (mm)	10.8	13.9	14.4	17.9
Number of teeth	34	32	30	30
Percentage of diet consisting of				
Insects	46	90	0.5	0.5
Fruits	29	0	0	0
Nectar	—	6	75	75
Pollen	25	4	25	25

[a]Adapted from Howell, 1974b.

spicillata during the dry season in northwestern Costa Rica (E. Stashko, personal communication).

Phyllostomid bats use plants from at least 146 genera, but this represents a composit of reports over a large range of habitats and times (Gardner, 1977). Two patterns emerge with respect to feeding on plants in local areas: sequential specialization and individual generalization. "Sequential specialists" are bats that use only one plant at a time but change food sources after several days or weeks (Colwell, 1974). An "individual generalist" will visit several species of plants for nectar, pollen, or fruit in one night.

The individual-generalist pattern appears to be common among bats for which fruits and flowers are major parts of the diet. Unfortunately, detailed studies are relatively uncommon (Gardner, 1977). Some individual pteropodids visit different species of flowers in one night (Gould, 1978; Start and Marshall, 1976). Several Neotropical nectarivores visit plants along a regularly traversed route (a "trap line"), and these routes often include several species of plants (Baker, 1973; Heithaus *et al.*, 1974). Examination of pollen on the hairs of bats has shown incorporation of several kinds of flowers in a single night's foraging by phyllostomids in northwestern Costa Rica. More than 40% of the individuals from each of four species (*Artibeus jamaicensis*, *Glossophaga soricina*, *Phyllostomus discolor*, and *Sturnira lilium*) carried more than one type of pollen (Heithaus *et al.*, 1975). Those species that showed the strongest fidelity to a single plant species (*A. lituratus*, *A. phaeotis* and *Carollia perspicillata*) were the bats least likely to visit flowers (Fig. 2).

One explanation for this phenomenon is that bats are more likely to visit only a single kind of flowering plant in an evening if they are supplementing a normally non-nectarivorous diet. Foraging theory would predict that these occasional flower visitors would be attracted by the few plants providing especially rich rewards. Occasional flower visitors will act as sequential specialists with

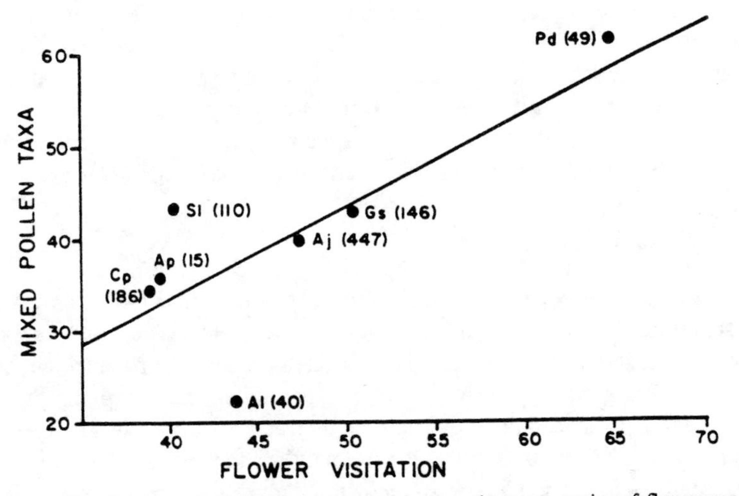

FIGURE 2. The tendency for individual bats to visit more than one species of flower per night (mixed pollen taxa) as a function of the population's use of flowers (flower visitation). Flower visitation is the proportion of bats captured that carried pollen. The intervals on the graph are the proportions transformed by the arcsine method. Sample sizes are indicated next to the data points for *Artibeus jamaicensis* (Aj), *A. lituratus* (Al), *A. phaeotis* (Ap), *Carollia perspicillata* (Cp), *Glossophaga soricina* (Gs), *Phyllostomus discolor* (Pd), and *Sturnira lilium* (Sl). The correlation for this regression is significant, $r = 0.80$ ($df = 5$, $P < 0.05$). (Data from Heithaus *et al.*, 1975.)

respect to flower feeding because the mass-flowering plants that attract them tend to offset their peak flowering times. Consequently, they may be less wasteful of pollen, from the plant's perspective, than true flower bats, which may visit many kinds of flowers in one night.

Fruit-eating bats also show both patterns of flexibility, but differences are less likely to affect the dispersal services received by the plants. Bats that normally eat insects will take fruits from a few mass-fruiting trees, so these bats tend to be narrow in their choice of fruit. In contrast, *Artibeus jamaicensis* depends on fruit and eats many species (Gardner, 1977). Most of its food comes from fruits in the Moraceae, including figs (*Ficus* spp.), *Cecropia* spp., and *Chlorophora* spp. (Heithaus *et al.*, 1975; Morrison, 1978a; Bonaccorso, 1979), but even this more-specialized bat will switch to short-term, mass-fruiting species, as *Brosimum* (Alvarez and Gonzales Q., 1970). More flexibility is seen among other phyllostomids and pteropodids. Individual *Carollia perspicillata* may switch to different fruits during a night. Early in feeding they take fruits that are preferred, but later they tend to switch to less-preferred species (E. Stashko, personal communication). Feces from bats may contain seeds from two plant species, but there are few data that allow generalizations about the incidence of bats taking different fruits in a single night. True generalists are difficult to identify because seeds are not retained as long as pollen, and individuals are difficult to follow while foraging. Switching food plants by frugivorous bats might affect the number of seeds dispersed, but changing to other fruits does not adversely affect the quality of seed dispersal as it does pollination. In fact, there may be some benefit to seeds that are deposited near a plant of a different species if specialist seed predators are less abundant there (Janzen, 1971).

3.2.3. Limits to Specialization

Nutritional requirements and susceptibility to environmental variability limit the specificity of coevolution between bats and plants. The most important limit to specialization is the general lack of plant species that produce abundant fruits or flowers each night throughout the year. Even the few plants that are aseasonal tend to have pulses of reproductive activity (Frankie *et al.*, 1974; Gentry, 1974). The few bats that are extremely specialized, such as the flower visitors *Leptonycteris* and *Choeronycteris*, must make long migrations to ensure an adequate food supply (Humphrey and Bonaccorso, 1979).

Increased specialization for flower feeding could make bats more susceptible to environmental variability. Flower feeders may switch to fruit, or to insects and fruit, but they do not switch to a diet including only insects (Humphrey and Bonaccorso, 1979). This suggests that plant-adapted digestive systems may not be able to function effectively on an insectivorous diet (Humphrey and Bonaccorso, 1979). Locating and capturing insects might also be difficult for specialized flower feeders. The glossophagine bats that are most specialized for flower feeding are less able to avoid obstacles compared to glossophagine feeding generalists (Howell, 1974b).

Nutritional deficiencies may also limit the coevolution of specialized bats to specific plants. Nectar in bat-pollinated flowers is nutritionally deficient by being low in amino acids, proteins, and lipids compared to other floral nectars (Baker and Baker, 1975; Baker, 1978). The evolution of pollen-eating behavior was critical to the development of strong flower dependence, as seen in *Leptonycteris sanborni* (Howell, 1974a); most other nectar-feeding bats must use insects or fruit to complete their diets. Other nutritional factors might limit specialization on fruits (Birney *et al.*, 1976), but little is known about the requirements of bats or the nutritional content of their foods.

Unpredictable co-occurence of interacting species is another reason that species-specific coevolution may be unusual (Howe, 1980). Associations of tropical plants can be very flexible, so bats with large geographical distributions encounter different species of plants. The problems of unpredictable co-occurence may be accentuated by shifts in local associations through time. Such shifts are especially likely if the composition of tropical communities is not in equilibrium, as claimed by Connell (1978), Hubbell (1979), Howe and Vande

Kerckhove (1979), and Huston (1979). Nonequilibrial species assemblages would show relatively high variation in space and time, reducing the chances for species to evolve very specific, coevolved characteristics.

3.3. The Search for Order: Pollination and Dispersal Syndromes

The earliest studies of interactions between bats and plants emphasized describing and documenting pollination or dispersal and were organized by searching for characteristics that were congruent between bats and plants. Paired characteristics that were repeated in many systems defined the syndromes of bat pollination and dispersal (Faegri and van der Pijl, 1966; Vogel, 1969; van der Pijl, 1972). The description of a few, integrated syndromes had two major benefits: they provided a general orientation, even though they were oversimplified (van der Pijl, 1972), and they were used as predictive tools. Plants that were likely to be dispersed or pollinated by bats could be identified by examining morphology; these predictions could then be tested by observations and experiments.

But defining simplified syndromes has had some disadvantages. Rigid and convenient classification obscures potentially important variation. Syndromes tend to be defined by the taxon of the visitor, as in bee-, butterfly-, hummingbird-, or bat-pollination systems, yet plants are often visited by more than one taxon. Saguaro is most effectively pollinated by bats, but bees and white-winged doves also pollinate its flowers, with losses of only 16% and 28% of potential seed set compared to flowers pollinated by *Leptonycteris sanborni* (Alcorn *et al.*, 1961; McGregor *et al.*, 1962). Potential overlap between bats and other taxa as pollinators has been indicated by others (e.g., Sazima and Sazima, 1978; Sussman and Raven, 1978). Dispersal systems are even less exclusive. In particular, many plants produce fruits that are taken by bats, birds, monkeys, and nearly anything else that can get into the canopy (McKey, 1975; Howe and Estabrook, 1977). Even fruits that are alleged to be specialized bat fruits, such a *Piper*, are sometimes taken by birds or monkeys (personal observation).

This variation is important because it is the basis for evolutionary flexibility. Additionally, species that are effective pollinators or dispersal agents for other systems can be sustained during periods of food scarcity (Baker *et al.*, 1971). Most importantly, this variation reduces the dependence of a plant on a particular animal species that might vary in density or be unpredictable in its food choice. At the same time, receiving services from many species with very different traits can limit the ability of plants to respond through coevolution to the special traits of one or more species.

The other disadvantage to describing syndromes that were first proposed is that the number of known exceptions is becoming increasingly large so the benefits derived from generalizations are being obscured. Baker (1973) cautioned against the uncritical acceptance of "old dogmas," and Start and Marshall (1976) found that former assumptions about bat behavior were too broad. A revision of bat-pollination and dispersal syndromes should reconcile the extensive variation now known and the need to define predictive generalities.

3.3.1. The Bat-Pollination Syndrome

The classic view of the plant characteristics most often associated with bat pollination is reviewed by Jaeger (1954), Baker and Harris (1959), Baker (1963, 1973), Vogel (1958, 1968–1969), and Faegri and van der Pijl (1966) and is partially summarized in Table IIIA. Bats are most likely to pollinate flowers that are open for one night only. The visual features of these flowers reflect characteristics of fruit and flower bats, which see well but are color-blind. The odor may be important for matching a bat's olfactory sense, if there is convergence between the specific odor of flowers and fruits and a bat's pheromones (Baker, 1963). Two general shapes of flowers appear to be appropriate for bat pollination. The stamens, pistil, and nectar may be fully exposed as in the "shaving brush" or penicillate flower (Figs. 3A and B; Vogel, 1958; Baker, 1973). Flowers of the campanulate type have strong petals that form a bell shape or cup that may be more restrictive. To obtain nectar from a restrictive, campanulate flower, a bat must insert its head or body into the flower (Fig. 3C). Small, campanulate flowers, however, may be presented in easily accessible clusters, as in *Ceiba pentandra* (Fig. 3D).

Flowers are presented where they are likely to be encountered by bats. Usually, bat-pollinated plants are trees or vines but not herbs. To increase access to flowers trees may be leafless during flowering and flowers are held free of obstructions (Fig. 3). At least six morphological mechanisms have evolved to increase access to flowers (Vogel, 1969), examples of which are long stalks to suspend flowers (as in *Kigelia*), presentation of flowers at the tips of branches or on trunks and branches under the canopy, and accentuated layering of branches.

Modifications of the classic view include dropping some generalizations that appear to be too sweeping and the incorporation of congruence between plant phenology and the foraging characteristics of bats. An early generalization was that bat-pollinated flowers had strong support systems to accomodate large, clinging bats (Faegri and van der Pijl, 1966). But members of the Neotropical Glossophaginae and the Paleotropical Macroglossinae are excellent hoverers (Fig. 3B; Yalden and Morris, 1975), so many flowers would not require strong stems. Often, Neotropical species that do fit the early generalization have pantropical distributions and are pollinated by large, clinging pteropodids in the Old World (Start and Marshall, 1976). Another generalization was that large quantities of pollen were alleged to be a response to protein requirements of bats, but it is unlikely that this is the only or even most important factor. More direct and stronger selective factors for high modes of pollen production are selection for optimal pollen/ovule ratios (Cruden, 1975) and compensation for inefficiency caused by pollen eating and by having large pollinators relative to stigma size (Heithaus *et al.*, 1974). Faegri and van der Pijl (1966) assumed that pollen would be the only source of protein for flower bats, but this assumption holds for relatively few bats (Sections 3.2.1 and 3.2.2).

Phenological "syndromes" (Table IIIB) can be added to the general characteristics of plants that are pollinated primarily by bats (Start and Marshall, 1976). The trade-off between allocating reproductive effort to long flowering periods versus large numbers of flowers per night is thought to result in several possible phenological patterns for flowering (Heinrich and Raven, 1972; Gentry, 1974). Using Gentry's terminology, one extreme is the "big-bang" production of masses of flowers for a few days (e.g., *Ceiba pentrandra*), and at the other extreme is the "steady-state" production of a few flowers for all or most of the growing season (e.g., *Crescentia cujete* in central Panama).

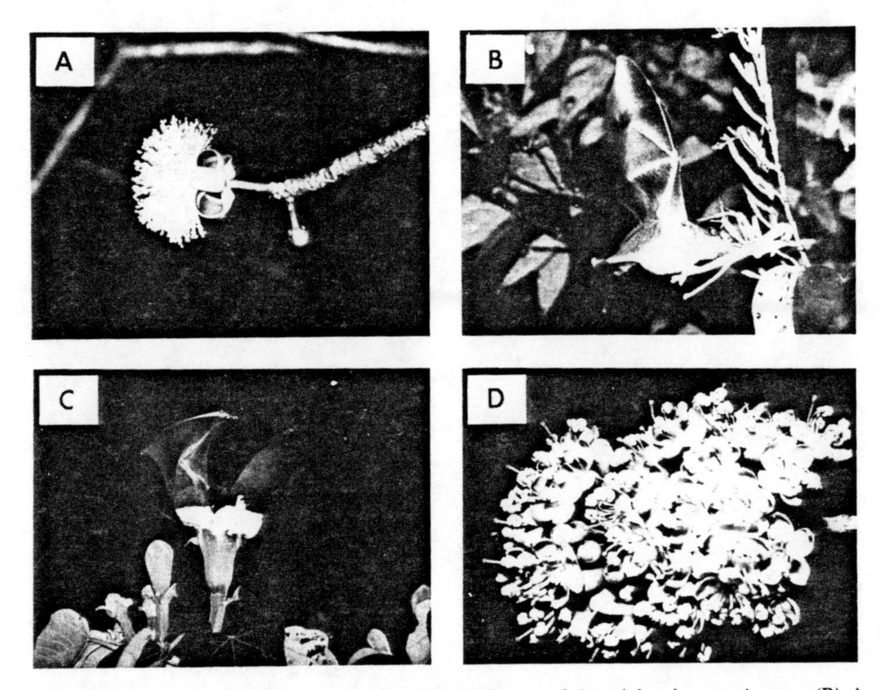

FIGURE 3. (A) The penicillate, "shaving brush" flower of *Pseudobombax septinatum*. (B) A brief hover visit by *Glossophaga soricina* to *Bauhinia ungulata*. Pollen is specifically deposited on the abdomen of the bat. (C) A phyllostomid (probably *Phyllostomus discolor*) landing on the flower of *Ochroma lagopus*. (D) A cluster of campanulate flowers (*Ceiba pentandra*), pollinated by bats now but probably pollinated by nonflying mammals before the evolution of flower-visting bats.

TABLE III
Characteristics of Plants That Are Correlated with Pollination by Bats

Character of plant	Consequences to plant
A. The "classical" syndrome[a]	
1. Flowers open at night	1. Excludes most diurnal visitors
2. Flowers usually can be pollinated for only one night	2. Reduces effects of diurnal visitors that collect residual nectar
3. Flowers sometimes whitish	3. Attracts nocturnal, visually orienting feeders
4. Colors drab, when present	4. Reduces attraction to diurnal feeders
5. Often a strong odor when flowers are open	5. Promotes attraction of nocturnal feeders from longer distances than possible with visual cues
6. Stale odor, like fermentation	6. Distinguishes flowers from those with inaccessible nectar? Mimics bat's odors or pheromones?
7. Large quantities of pollen	7. Decreases impact of pollen loss to feeding by bats and dilution of pollen over large surfaces
8. Flowers positioned away from foliage	8. Increases detectability and accessibility to volant feeders

B. Phenological syndromes	
B1. The "Steady-state" syndrome	
1. Flowers produced for many nights	1. Visitors learn locations and visit repeatedly
2. Few flowers produced each night	2. Visitors move between plants frequently; may have low flower constancy
3. More restrictive morphology or timing of nectar flow	3. Reduces impact of low flower constancy among pollinators
B2. The "Big-bang" syndrome	
1. Many flowers produced each night	1. Attracts many visitors, including species that usually are not flower visitors
2. Flowers produced for few nights	2. Visitors forage opportunistically
3. Low probability of flowering simultaneously with other "big-bang" plants[b]	3. Low probability of hybridizing pollination, even with opportunistic foragers as pollinators
4. Nonrestrictive flower shapes	4. Attracts many visitors, excluding few potential pollinators

[a]Modified from Faegri and van der Pijl, 1966.
[b]This may be a coevolved trait, but it could occur with randomly timed flowering periods that are short.

Between these extremes is the "cornucopia" pattern (moderate numbers of flowers for 2–4 weeks (e.g., *Hymenaea courbaril* and *Ochroma* sp. in northwestern Costa Rica) and the "multiple-bang" pattern of short pulses of flowering scattered through the year (e.g., *Crescentia alata* in northwestern Costa Rica). The phenological pattern for a plant may depend on coevolution with its pollinators (Janzen, 1967, 1971b; Heinrich and Raven, 1972).

The steady-state pattern promotes trap-line foraging by bats (Janzen, 1971b; Baker, 1973). Small numbers of flowers do not attract opportunistic flower feeders, so most pollination is by bats that rely on flowers for a major portion of their diet. As shown earlier, these visitors have low flower constancy, increasing the problems of incorrect pollination. Campanulate flower structure and special timing for nectar production can restrict the diversity of visitors or the chances of transferring pollen between different species of plants. *Oroxylum indicum,* an extremely specialized plant (for pollination by *Eonycterus spelaea*) shows both responses. Its flowers reduce visitation by other bats, and it starts producing nectar later than nearby bat-pollinated plants (Gould, 1978).

Large numbers of flowers tend to attract bats that are foraging opportunistically. Big-bang and cornucopia phenologies attract bats that may forage in large groups (Baker, 1973; Ayensu, 1974) but which may move between plants less frequently. These phenologies also attract bats that usually do not eat nectar but which may provide consistent pollen transfer among conspecifics. Floral morphologies should not be restrictive; flowers should be penicillate, or very large if they are campanulate (Figs. 3A and C). Cornucopia flowering, by having relatively long flowering periods, should attract bats that are trap liners as well as opportunistic foragers. Multiple-bang plants, by being unpredictable, would probably follow the patterns seen in single-burst, big-bang species.

These phenological "syndromes" (Table IIIB) should be viewed strictly as new working hypotheses, not new dogma. An added caution is that predictions based on phenology apply to specific foraging patterns, not to specific bats. Several bats are known to change foraging patterns to match resource distributions. *Eonycteris spelaea* can behave as a trap liner or opportunistic forager (Gould, 1978). *Phyllostomus discolor* also varies its behavior (Sazima and Sazima, 1975). *Leptonycteris sanborni* forages in groups in North American deserts (Howell, 1979, 1980) but forages alone when feeding at *Ceiba acuminata* in forests (Baker *et al.*, 1971). These patterns illustrate the diffuse nature of coevolution between bats and plants and the importance of foraging behavior to determining the effects of bats on plants.

3.3.2. The Bat-Dispersal Syndrome

Fruits that are specialized for dispersal by bats share many visual, olfactory, and positional characteristics with bat-pollinated flowers (Table IIIA; van der Pijl, 1972). In the Old World bat fruits often are large compared to other dispersal systems; this is correlated with the large body size of many pteropodids. Smaller bat-dispersed fruits in the Neotropics are associated with small frugivorous bats (Heithaus *et al.*, 1975). Bat-dispersed fruits tend to be more accessible to other consumers than are most bat-pollinated flowers. The ripening process is not a discrete event, as is the opening of a flower, so fruits cannot restrict visits by being "available" just during the night. Many animals, such as monkeys and birds, do not even bother to wait for fruits to ripen (McKey, 1975; Howe and Estabrook, 1977; personal observation), and this adds to the sharing of fruits between bats and other animals.

The phenological patterns of fruiting parallel those of flowering, with steady-state and big-bang patterns being common. Intermediate forms are also observed (Fig. 1). For example, *Piper amalago* and *Solanum hayesii* have at least two fruiting periods, each 2–4 weeks in duration.

Foraging responses of bats to the extreme phenologies are similar for fruits and flowers. Predator-avoidance behavior by bats may affect seed dispersal, too. In the Neotropics many phyllostomids move seeds away from the canopy, regardless of the plant's phenological pattern, by transporting fruits to feeding roosts to reduce risks of predation at the food source (Chapter 1). "Lunar phobia" is another possible predator-avoidance behavior that affects seed dispersal (Morrison, 1978b; Heithaus and Fleming, 1978; Humphrey and Bonaccorso, 1979). These behaviors to avoid predators may reduce the intensity of selection that bats exert on plant fruiting phenologies (Howe, 1979).

4. ECOLOGICAL CONSEQUENCES OF BAT–PLANT INTERACTIONS

Bats and plants usually benefit from their interactions, but this mutualism includes many kinds of benefits, including nutrition, reduction of environmental variability, fertilization of ovules, promotion of gene exchange, escaping specialized seed predators, and enchanced colonizing potential. This diversity of possible benefits shows that the effects of mutualism may be realized within a short time or over a long time period. A major question is whether and how their interactions influence the population dynamics of mutualists. Usually, ecologists have studied competition, predation, or the effects of physical limitations in attempts to understand patterns of population growth and species coexistence (Risch and Boucher, 1976). This is understandable, because intuition suggests that populations should be limited by factors having negative effects, not by mutualism, which is a system of positive feedback. More formal models of coevolution also show that selection for mutualism does not require that interactions effect population dynamics (e.g., Gilbert, 1975, 1977).

Recently several investigators have shown that mutualism may have important, short-term effects on population regulation (May, 1976; Handel, 1978;

Vandermeer and Boucher, 1978; Heithaus *et al.*, 1980). Whether these potential effects are realized depends on the benefits derived from mutualism. The large variation in the interactions between bats and plants makes generalizations difficult. Variation is seen in the kind of benefits received by bats and plants and in the way that reproduction is influenced.

4.1. Variation in Effects

4.1.1. Effects on Bats

For bats, nutrition is an obvious benefit of mutualism with plants, but a less obvious and more variable benefit is the potential reduction of fluctuations in food supplies. Fluctuation in food supplies has been an important selective force for bats, as seen in responses in basal metabolic rates (McNab, 1980) and the timing of reproduction by bats to avoid lactation during times of food scarcity (Heithaus *et al.*, 1975; Wilson, 1979; Bonaccorso, 1979). Although no studies have demonstrated that variability in food supply affects the size of bat populations, these responses show that such an effect is possible.

A "succession of flowering or fruiting periods" has become a cliché in describing the constancy of food supply for tropical flower feeders and fruit feeders. The term describes the overlapping and sequential periods of fruit or flower production by different species of plants (Snow, 1966). The pattern provides a continuous supply of food to specialized consumers, even though no single plant is necessarily active all year. Sequential flowering and fruiting periods are required to sustain bats such as the Pteropodidae that are restricted to feeding on plants (Faegri and van der Pijl, 1966).

The incidence of sequential flowering may be overestimated, especially in the Neotropics. While Paleotropical pteropodids are dependent on plants, the phyllostomids of the Neotropics are more flexible. Especially in the widespread seasonal Neotropics, sequential flowering may not be found. On Barro Colorado Island (BCI), Panama Canal Zone, floral foods for bats are absent from mid-March through early December (Bonaccorso, 1979). When the four bat-pollinated flowers on BCI are active, their flowering is simultaneous rather than sequential (Bonaccorso, 1979). Similarly, only *Crescentia* produces flowers from late June through early September in the northwestern Costa Rican lowlands, and its flowers are produced in discontinuous pulses (Heithaus *et al.*, 1975). Bats in the New World respond to flower scarcity by migrating (Heithaus *et al.*, 1975; Humphrey and Bonaccorso, 1979) and by switching to other food items.

In contrast to flowers, fruits from some species are available at any time, even in seasonal Neotropical areas. On BCI two species have abundant fruit even in the leanest times, but variation in fruit availability is still observed, because 19 species of plants fruit during the peak fruiting season (Bonaccorso, 1979). Similar patterns were observed in northwestern Costa Rica (Heithaus *et al.*, 1975). The effect of this variation on bats is a function of their competitive abilities and dependence on fruit.

4.1.2. Effects on Plants

4.1.2a. Pollination. The benefit that plants receive when they are pollinated by bats depends on the quantity and source of pollen deposited. Large amounts of pollen increase the proportion of ovules that develop. A second possible benefit of depositing large amounts of pollen is that competition is encouraged among the growing pollen tubes; competition among pollen grains may increase the quality of mating for plants (Mulcahy, 1979). The quantity of pollen that a bat carries is influenced by its size, its tendency to groom pollen from its pelage, the number of flowers visited, and the amount of pollen produced by a flower. All of these factors vary, depending on the species of bat and the plant. Field studies are needed to test the hypotheses that this variation affects the reproductive potential of the plant and its embryo vigor.

Pollen must be carried from a conspecific, and usually from a different individual. Most noncoastal, tropical trees and shrubs are either dioecious or self-incompatible (Bawa, 1974; Bawa and Opler, 1975; Zapata and Arrozo, 1978), so pollen must be carried away from the source plant. Some bats may carry pollen several kilometers from source plants (Heithaus *et al.*, 1975). The high diversity of tropical plants had promoted the view that tropical plants are usually spaced far apart (e.g., Ashton, 1969; Janzen, 1970), but tropical plants tend to be clumped together or randomly spaced rather than hyperdispersed (Hubbell, 1979). Consequently, long-distance pollination is not always required for fertilization, but it is needed when plant densities are low and individuals randomly spaced. Another effect, presumably a beneficial one, of long-distance pollination is the increase in the effective breeding size of plant populations. Unfortunately, detailed genecological studies of bat-pollinated plants are lacking. Such studies would be especially important, as tropical habitats are becoming more subdivided by human development (Myers, 1980).

Movement of pollen between individuals is promoted when plants produce small amounts of nectar or a few flowers per individual so that bats must visit more than one individual to satisfy foraging requirements (Baker, 1973; Gould, 1978). The requirement of promoting outcrossing must be balanced by the need to attract pollinators (often in competition with simultaneously flowering plants). This conflicting requirement calls for high predictability or relatively large amounts of nectar. Often, maximal fruit set and outbreeding occur when nectar levels are intermediate (Carpenter, 1976). For example, *Agave palmeri* produces nectar in quantities that promote movement among individual plants but which also attract bats that choose rich sites in which to forage (Howell, 1979). As indicated by different phenological syndromes (Table IIIB), not all plants have reached the same compromise.

Bats can have detrimental effects when they feed at flowers. Large-bodied pteropodids can damage flowers (Faegri and van der Pijl, 1966), although this does not usually affect the ovaries (Baker, 1973). More importantly, bats do not always carry pollen to new plants of the same species. As described above, bats are notoriously unfaithful to particular plant species during foraging (Faegri and van der Pijl, 1966; Heithaus *et al.*, 1975). In addition, some bats eat pollen before it can be transferred to a new flower. *Leptonycteris sanborni* may visit flowers for 20 min, then remove accumulated pollen from their pelage (Howell, 1979). Also, pollen may be "lost" for pollination if it is scattered over parts of a bat's body that will probably not touch the stigma. To reduce this loss some flowers place pollen on small specific areas of visitors (Fig. 3B). Losses of pollen sometimes promote unusual mechanisms for providing high ratios of pollen/ovule numbers. One example is seen in *Bauhinia pauletia*, a Neotropical legume. Many flowers drop their pistils just after opening, creating male as well as hermaphroditic flowers and a surplus of pollen (Heithaus *et al.*, 1974).

4.1.2b. Seed Dispersal. The most likely benefits of seed dispersal are promoting gene flow, making seeds less obvious to seed predators, reducing parent–offspring competition, and moving seeds to favorable sites for colonization by a seedling. The importance of gene exchange, also promoted by bat pollination, among tropical plants is poorly understood. That gene exchange is favored by selection is indicated by the large frequency of species that cannot self-pollinate (Bawa, 1974; Bawa and Opler, 1975). Janzen (1970) suggests that constantly changing interactions with seed predators and competitors in tropical systems should promote the maintenance of genetic variability.

Dispersal, including dispersal by bats, can reduce the impact of seed predators, which might otherwise kill large fractions of tropical seed crops (Janzen, 1971a, 1971b, 1975, 1976; Moore, 1978; Vandermeer, 1975). Large seeds, often associated with bat dispersal, are rich food sources for insects, birds, and mammals. Small seeds that are dispersed after passing through the digestive tracts of bats may also be damaged by seed predators. Lygaeid bugs often destroy small fig seeds that are not carried beyond the canopy of the fig tree (Slater, 1972). The small seeds of *Cecropia peltata*, *Chlorophora tinctoria*, and *Piper amalago* were removed from experimental sites by ants (*Pheidole* sp.) and small rodents in northwestern Costa Rica; seeds were destroyed by captive rodents, and ants were presumed to be predators (Perry and Fleming, 1980). The effect of ants on these seeds is still unclear because ants could be secondary seed dispersal agents (Bullock, 1974; Horvitz, 1980). Whatever the specific effect of *Pheidole*, however, many seeds are likely to be lost to generalist predators (rodents, lygaeids, and some ants) and to more specific seed predators.

Avoiding seed predators should be important, however; merely moving seeds away from the parent tree does not necessarily reduce the chances that they will be eaten by seed predators. Janzen (1970) originally proposed that specialist seed predators would focus their search for seeds near parent trees. But the hypothesis that the seeds nearest to parent trees are subject to high predation rates has not been supported (e.g., Hubbell, 1980). Generalist seed eaters are even less likely than specialists to restrict their searching area to sites near parent trees; thus moving seeds away from the parent canopy will not necessarily reduce the predation effects. The experiments by Perry and Fleming (1980) showed little reduction in losses with distance, as would be expected when predators are generalist feeders. When bats deposit seed in large concentrations far from parent trees, as under feeding roosts, the seeds can still be located by predators (Janzen *et al.*, 1976). Whether bats provide the benefit of reducing seed predation depends on the type of predator, whether seeds are clumped after dispersal, and the distance that seeds are moved. Avoiding parent–offspring competition is just as likely to be a benefit of moving away from the parent as avoiding seed predators.

Aiding the colonization of new sites is another potential benefit of bat dispersal. In tropical forests, colonizing and utilizing small light gaps is the major way that new plants get started (Bazzaz and Pickett, 1980; Brokaw, 1980). Long-distance seed dispersal by bats may spread seeds over a relatively large area, increasing the chances that they will be associated with light gaps, which are irregularly generated (Brokaw, 1980). In support of this, Hubbell (1979) found larger mean distances between plants dispersed by birds and bats than between plants dispersed by other means. Many Neotropical bats eat seeds from pioneer (shade-intolerant) and persistent (shade-tolerant) species (classified by Brokaw, 1980) in the same night, often creating mixtures of seed types in one location (Fleming and Heithaus, 1981). This may also enhance the movement of late successional tree species to sites appropriate for colonization.

A major conclusion from considering the benefits received by plants is that these benefits are slow in feeding back to the bats. Better pollination or seed dispersal will increase the amount of nectar or fruit only after one or more plant generations. If the major benefit to plants were maintaining genetic variability, time lags between bats providing pollination or dispersal and the generation of new food supplies would be especially long. At best, the time lag is several bat generations long.

The benefits to a plant that are provided by any particular dispersal agent depend on the number of seeds taken from the parent tree ("quantity" effects) and the "quality" of treatment (McKey, 1975; Howe and Estabrook, 1977). The quality of treatment depends on whether an animal damages seeds, the distance seeds are moved, the environment in which they land. For example, a pteropodid that sits in one tree all night may drop hundreds of seeds below the parent plant; it would provide high-quantity but low-quality dispersal. The trade-offs between the quantity and quality of dispersal are a focal point for studying coevolution between plants and their dispersers (McKey, 1975; Howe and Estabrook, 1977; Howe, 1977, 1979, 1980; Howe and De Steven, 1979). For bird-dispersed plants there appears to be a negative correlation between the number of seeds produced and the average quality of dispersal received.

Bats can disperse large numbers of seeds. A single *Artibeus jamaicensis* on BCI eats approximately 2300 figs per year; with an average density of six individuals per hectare, this single species of bat disperses at least 7% of the annual fig crop (Morrison, 1978a). Individual *Carollia perspicillata* in northwestern Costa Rica eat approximately 35 fruits per night, which would involve the dispersal of 350–2500 seeds per night for plants such as *Ficus ovata* or *Muntingia calabura* (Lockwood *et al.*, 1982). Pteropodids in the Old World tropics consume even larger numbers of fruits (van der Pijl, 1972).

Dispersal quality varies with the species involved. For example, 16 species of mammals eat fruits of *Dipteryx panamesis* (Papilionaceae) on BCI (Bonaccorso *et al.*, 1980). Only three species regularly carry fruits away from the canopy of the parent tree. These include *Artibeus lituratus*, the agouti (*Dasyprocta punctata*), and a squirrel (*Sciurus granatensis*). Three other bats, *Carollia castanea*, *C. perspicillata*, and *Artibeus jamaicensis*, eat the fruits at the tree but are too small to carry the fruits. The agouti and squirrel eat many of the seeds, as well as fruits, which leaves *Artibeus lituratus* as the single high-quality dispersal agent among the 16 fruit eaters. *Carollia perspicillata* is a poor disperser of *Dipteryx* and of *Anacardium*, whose seeds it eats (Bonaccorso, 1979), but the same bat is a principle disperser for *Piper* spp. and *Solanum hayesii* (Heithaus and Fleming, 1978). Thus a given bat may vary in the quality of dispersal it provides to different plants.

In many bird-dispersal systems the best quality of dispersal is provided by species that are specialized to eat fruits: species that occasionally eat fruits tend to be poor quality dispersers (McKey, 1975; Howe and Estabrook, 1977). This trade-off does not appear clearly in bat-dispersal systems. Quality of dispersal varies more within species at different times, as with *Carollia perspicillata*, than it varies between generalist and specialist feeders. It seems that any frugivorous bat can provide high-quality dispersal. Consequently, opportunistic bats that temporarily switch to mass-fruiting trees do not necessarily provide low-quality dispersal. A lesson here is that coevolutionary patterns between plants and avian dispersers will not always serve as models for dispersal systems that include bats.

4.1.3. Optimal Foraging and Bats, Effects on Plants

Considerable attention has been given to the question of whether pollinators or seed dispersers forage optimally (Chapter 8). Optimal foraging is exceedingly difficult to demonstrate, although *Leptonycteris sanborni* appears to do so in Arizona (Howell, 1979; Howell and Hartl, 1980), and *Artibeus jamaicensis* may forage optimally on BCI (Morrison, 1978a, 1978c). "Optimality" must be defined in terms of specific constraints, such as maximizing energy return over a given period (such as within one night or through a reproductive period), minimizing the time spent foraging, or minimizing the risk of predation. There are even optimal compromises for combinations of contraints (Sih, 1980). Foraging bats can respond to any of these contraints. For example, *Leptonycteris* forages optimally in the context of maximizing energy return per night (Howell, 1979, 1980) and *A. jamaicensis* may optimize energy over a period of days or weeks (Morrison, 1978b, 1978c, 1979). *Carollia perspicillata* does not optimize with respect to energy return but minimizes its exposure to predators (Heithaus and Fleming, 1978; Lockwood *et al.*, 1982). These variations may complicate coevolutionary responses by a plant population if the relevant constraints for a bat include factors insensitive to the resources provided by the plant.

Optimal foraging by bats may not provide optimal benefits to plants, a fact often overlooked by those claiming that optimal foraging theory is the key to studying mutualism (e.g., Pyke *et al.*, 1977). Predictability is the relevant characteristic of selection for the presentation of flowers or fruits. For example, the movement of seeds away from parent plants is greatly enhanced when bats carry fruits to feeding roosts. This behavior is energetically costly, as it can account for over 6% of the daily energy budget of *Carollia perspicillata* (Heithaus, unpublished data). Maximizing energy return or minimizing foraging time would call for eliminating this behavior, which would be detrimental to the plant. Similarly, aggressive encounters among pteropodids enhances seed and pollen movement (Gould, 1978) but reduces the foraging efficiency of bats. Rather than looking for optimal foraging *per se* studies of mutualism between bats and plants should focus on all factors that make bats' behavior predictable.

4.2. Demographic Effects?

A major unanswered question is whether the mutualism between bats and plants influences the equilibrium population size or recruitment rates for either group. It is likely that food supply influences these population characteristics for bats, but mechanisms for population regulation among bats are mostly speculative. Even with a general relationship between food and population regulation, one cannot assume that any specific plant will have a significant effect. The flexible feeding patterns of most bats would reduce a single plant's effect on the population size of any given bat.

Bats might have large effects on seed production, but few systems have been studied with this question in mind. It should be obvious that if plants require cross-pollination, they will suffer when pollinators are absent. The point in question is whether normal variation in bat services affects reproductive success, especially considering the other factors that might have demographic effects. The study of one bat-pollinated shrub, *Bauhinia ungulata* (Heithaus *et al.*, 1982), shows that mutualism can have demographic effects. In seven isolated stands seed production by *B. ungulata* was reduced by five factors: herbivores eating flowers, incomplete pollination, spontaneous abortion, herbivores eating developing seed pods, and seed predation. Of these possible interactions, variable pollination by bats and seed predation accounted for most of the spatial variation in seed production. Generalizations about the effect of mutualism on plant demography will require studies of many other systems, with consideration of effects other than bats.

4.3. Community Effects

Plant-visiting bats influence community diversity by contributing to the survival of plants (direct effects), and through plants bats indirectly influence the diversity of herbivores and their predators. The influence of bats is not always diversifying, because mutualistic dependence on bats involves risks to the plants. A mutualist that loses its ability to persist without the interaction (an obligate mutualist) is especially vulnerable to variation in pollination or dispersal services. If a plant is specialized as well as obligate in its mutualism, it may follow its partner to extinction (Temple, 1977). Specialized and obligately mutualistic bats would be exposed to similar risks. The flexibility of bat feeding strategies reduces the risks of mutualism to bats. Plants also show flexibility in pollination and dispersal systems; often pollination or dispersal can be accomplished by interacting with several species of bats, or with birds and nonflying mammals. The relatively long generation times of trees and shrubs also buffers obligate pollination and dispersal systems against unusual population crashes by bats. An added consequence of the long generation times of plants is that pollination or dispersal failure in one year will not disrupt food supplies for bats; individual plants are likely to continue reproductive activity whether or not offspring are successful in a particular year.

A potential problem with increasing disturbance to tropical ecosystems is the progressive loss of flexibility in mutualistic systems. Destruction of natural habitats with the elimination of many roosting sites may reduce the number of species available to interact in mutualistic systems. Because flexibility is a stabilizing influence in obligate mutualism, each loss of a potential mutualist increases the chance that another species will follow.

The potential impact of population reduction, or extinction, extends beyond the interacting bats and plants. Plants support assemblages of herbivores, and food webs are based on these. The fates of species in different food webs are linked by the mutualistic services of pollinators and seed dispersers (Futuyma, 1973; Gilbert, 1980). Sometimes the persistence of many mutualists depends upon a "keystone mutualist" that supports species at critical times (Gilbert, 1980). The bat-pollinated plant *Ceiba acuminata* is an example of a keystone mutualist, because at times its flowers provide the only nectar available for many animals that are mutualists for other plants (Baker *et al.*, 1971). We do not know whether keystone mutualists are found frequently, but they would make communities especially vulnerable to the disturbances of mutualistic interactions.

An important question is whether interactions between bats and plants affect the dynamics of relatively undisturbed communities. Whether this mutualism affects community stability depends on whether the interactions influence variation in reproductive rates or carrying capacities. Data on this critical question are few, but preliminary results suggest that these influences should be considered. Bats are likely to effect the demography of *Agave* (Howell, 1979) and *Bauhinia* (Heithaus *et al.*, 1982). In addition, bats influence the colonizing ability of plants, the escape from parent–offspring competition, and the rate of seed predation; each of which could influence reproductive rates.

Recently mathematical models have been developed to explore the consequences of mutualistic interactions that influence the reproductive rates and carrying capacities of populations (May, 1976; Vandermeer and Boucher, 1978; Heithaus *et al.*, 1980). In these models mutualists can vary in the amount of benefit received and in their dependence on mutualism for survival. One caution is that these models assume that benefits of mutualism are expressed when plants and animals interact. In contrast, we know there is considerable variation in nature and there are delays of one generation or more. Introducing time lags to some models, surprisingly, fails to alter their predictions (MacDonald, 1978). Theoretical models are worth considering, because they include realistic variation and are not overly sensitive to the unrealistic assumption that time lags are absent.

Mutualism, according to these models, may result in three patterns: (1) one or both mutualists will go extinct, which is most likely to happen when a mutualist is obligate. (2) Both mutualists will increase in number until limits are set by some factor like predators, competitors, or catastrophe, which is external to the mutualists' systems. This is the positive feedback that led May (1973) to conclude that mutualism is a destabilizing force in communities. This result is called persistence by Vandermeer and Boucher (1978), but this term obscures the fact that such growth tendencies may alter competitive or predator–prey balances. (3) The balance between benefits from mutualism and self-regulating factors may set stable equilibria for mutualist populations.

Stable populations of mutualists are promoted by two factors. One factor is a significant effect of predators on one mutualist. Predators can increase the range of benefits that allow stable coexistence, even when predators alone are not a limiting factor (Heithaus *et al.*, 1980). The juxtaposition of seed-predator and bat-pollinator effects in the *Bauhinia ungulata* system may promote system stability (Heithaus *et al.*, 1982). Future studies of mutualism would benefit from tests for the effects of predators or herbivores. The second factor that promotes stable mutualism is an increase in mutualist density (density-dependent effects). This would occur in pollination and dispersal systems when the proportion of pollinated flowers or dispersed seeds fails to increase consistently with increasing numbers of mutualists. This situation was also observed in the *B. ungulata* bat system (Heithaus *et al.*, 1982).

5. DOES COEVOLUTION "MATTER"?

Either coevolution or one-sided selection can lead to diversification, the complicated congruence of traits, or mutualism between bats and plants. The alternate hypotheses of coevolution and simple selection are difficult to distinguish, yet others have called for rigorous attempts to differentiate them (e.g., Baker and Hurd, 1968; Baker, 1973; Janzen, 1980). Three reasons to pursue this approach include (1) simple accuracy, (2) alternate hypotheses suggesting different possible levels of selection for mutualism, and (3) models of coevolution and simple selection giving different predictions about evolutionary changes in stability, optimal fitness, and density.

It is very easy to fall into the trap of attributing all mutualism to coevolution, but if all mutualistically congruent systems were called coevolution, the term would lose meaning (Janzen, 1980). Especially for bats and plants, specific matches between species are unlikely to be caused by species-specific coevolution; the demonstrable aspects of coevolution are diffuse responses and counter-responses involving groups of species. I will draw upon my own work to illustrate the trap. The beneficial effects of *Glossophaga soricina* or *Phyllostomus discolor* foraging at the flowers of *Bauhinia pauletia* do not show specific coevolution, as implied by Heithaus *et al.* (1974). One can now appreciate the fact that the foraging characteristics of bat pollinators did not evolve through a specific relationship with *B. pauletia;* only the behavioral response was specific to the distribution of the flowers of this plant species. This behavioral response is based on genetic potential that evolved with many plants, and over a much wider range than is occupied by *B. pauletia*. A similar problem is seen in Howell's

(1979) analysis of specific "coevolution" between *Leptonycteris sanborni* and *Agave*. In both cases plants appear to respond to selective pressures that are defined in part by the bats. A genetic counter-response to these specific plants is unclear.

Discussions of plant reproductive phenology typify the different levels of selection that are invoked to explain mutualism. In tropical areas bats that depend on plants for food are sustained because some fruits or flowers are always available (Allen, 1939). While many plant species are seasonal, a lack of synchrony among them produces a pattern often called "sequential" flowering or fruiting, one kind of flower or fruit being replaced by another. Is this the result of coevolution, as often claimed (e.g., Humphrey and Bonaccorso, 1979)? Speculations abound, although no one has yet demonstrated that the "pattern" for bat plants is anything other than random, or if it is nonrandom, that the pattern is related to a "need" to sustain bats. The alternate hypothesis of random flowering or fruiting times is rarely discussed (but see Heithaus *et al.*, 1974; Poole and Rathcke, 1979). For statistical reasons the "random" hypothesis is especially difficult to test when plants have different lengths of reproductive activity or multiple periods of fruiting, which are common patterns for bat-visited plants. One approach is to calculate the ratio between the mean number of plants active per month and the variance for this over a year. Mean/variance ratios near unity suggest the random co-occurence of active plants in a month, while the clumping or even spacing of activity are implied by ratios above and below unity, respectively. Analyses of the fruiting of bat-visited plants in Costa Rica (Heithaus *et al.*, 1975) and Panama (Bonaccorso, 1979) give ratios of 0.91 and 1.50, respectively, suggesting random or overlapping patterns rather than evenly spaced, sequential production.

Assuming for the moment that phenological patterns have evolved in response to variation in the behavior of bats, selection at nearly every level has been proposed. Intraspecific competition for dispersers may lead to the evolution of phenological patterns such as steady-state or big-bang fruiting (Howe and Estabrook, 1977). McKey (1975) proposed that competition between species could produce these patterns. Neither of these models necessarily involves coevolution between mutualists. Baker (1973) notes that community efficiency is increased by sequential phenologies, and Wilson (1980) provides formal models for mutualists coevolving as weak altruists, both approaches suggesting that selection acts on species assemblages, in other words, community-level selection. We do not know whether mutualism exerts selective pressures at the intraspecific, interspecific, or community levels, or all of these.

Mathematical models of coevolution clearly emphasize the importance of distinguishing the ways that organisms interact. Although these models are early in their development, they agree in several conclusions (Levin and Udovic, 1977; Roughgarden, 1979; Slatkin and Maynard-Smith, 1979). One conclusion is that coevolution at the population level does not necessarily lead to increased community stability. In fact, coevolution can greatly destabilize a community if interactions affect gene frequencies and population density for both species. Also, when gene frequencies and population density vary in response to interactions, natural selection does not necessarily maximize fitness; and even if fitness is maximized (as it will under certain conditions), this does not imply that population sizes will also be maximized. Models of response by single species to independent factors lead to simpler and different predictions. Unfortunately, these models of coevolution are not necessarily appropriate for diffuse systems, nor is further development likely to make predictions simpler. The clear message now is that coevolution should be considered separately from a one-sided evolutionary response and that we need good tests of the hypothesis that mutualism influences variation in population density.

6. SUMMARY

Diffuse coevolution has promoted diversification among bats and plants, but species-specific systems are rare. The initial adaptive radiation of bats was probably promoted by flowers and fruits that were preadapted for chiropteran feeding. Such preadaptation would have occurred among plants pollinated or dispersed by nonflying mammals. Because preadaptation characterizes several modern bat–plant systems, one should avoid assuming that mutualism between bats and plants is coevolved; coevolution occurs only with response and counterresponse. Generalist feeding strategies provided the bridge from insectivory to plant feeding (Gillette, 1975). With the evolution of more specialized fruit- and nectar-feeding bats, plant species produced fruits and flowers that were modified for interacting especially with bats.

Bats that eat fruits, nectar, or pollen show much variation in specialization for plant diets. Variation is seen in the morphology of teeth, palate, and pelage, in the physiology of digestive systems and sonar systems, and in behavior. "Unspecialized" bats are occasional visitors to plants, but the benefits they confer may be as important and dependable as the benefits provided by bats that are specialized and dependent on plants. Specialized should not be equated with "best." Even the most specialized bats have flexible feeding patterns, as seen in frequent changes in the plants that are visited. A consequence of feeding flexibility is that bats rarely evolve parallel to one or even a few plant species.

The ecological consequences of interactions between bats and plants are also variable, and often poorly documented. Bats tend to reproduce when food is most abundant, but no studies have shown clearly that population sizes are influenced by food supply. Some plants may be keystone mutualists; a small proportion of the plants used by bats may have a relatively large effect on populations of bats by providing food during times of low food diversity. There is conjecture, but no proof, that flowering or fruiting may be timed through coevolution to coincide with bats' requirements. The benefits conferred on plants depend on the foraging patterns of the bats, which are quite variable. Likely benefits to plants include promoting gene flow, reducing parent–offspring competition, avoiding seed predators, and scattering seeds to await the creation of light gaps in the forest canopy. Only a few species of plants are obligately dependent on one or two species of bat. As a result, one cannot predict *a priori* the effects of varying populations of bats on the ecology of most plant populations.

An unanswered question is whether diffuse coevolution between bats and plants has resulted in a mutualistic system that will persist. Species coexistence and system stability are likely to be affected by (1) whether the mutualists are obligate or facultative, (2) whether the mutualists may be served by several alternate species, (3) the amount of benefit received, (4) whether these amounts are density dependent, (5) the presence of a third species acting as a predator or competitor, and (6) whether the mutualism affects population growth, carrying capacity, or neither. Mutualism will affect community dynamics only if a species' population growth or carrying capacity is influenced by this interaction. Much work remains to be done to show that this is true or whether other conditions necessary for persistence are observed in natural ecosystems.

ACKNOWLEDGEMENTS. Thanks go to Thomas Kunz, Nicholas Brokaw, and Patricia Heithaus for their comments on this manuscript. Thanks also to Peter Raven, Andrew Beattie, and David Culver for their encouragement and advice. Portions of this work were financially supported by NSF DEB75-23450 and Northwestern University.

7. REFERENCES

Alcorn, S. M., S. E. McGregor, and G. Olin. 1961. Pollination of Saguaro cactus by doves, nectar-feeding bats, and honey-bees. Science, **133**:1594–1595.

Allen, G. M. 1939. Bats. Harvard University Press, Cambridge, 368 pp.

Alvarez, T., and L. Gonzalez Q. 1970. Analisis polinico del contenido gastrico de Murcielagos Glossophaginae de Mexico. An. Esc. Nac. Cienc. Biol., Mexico City, **18**:1–77.

Ashton, P. S. 1969. Speciation among tropical forest trees: Some deductions in light of recent evidence. Biol. J. Linn. Soc., **1**:155–96.

Ayensu, E. S. 1974. Plant and bat interactions in West Africa. Ann. Mo. Bot. Gard., **61**:702–727.

Baker, H. G. 1963. Evolutionary mechanisms in pollination biology. Science, **139**:877–883.

Baker, H. G. 1973. Evolutionary relationships between flowering plants and animals in American and African forests. Pp. 145–159, *in* Tropical forest ecosystems in Africa and South America: A comparative review. (B. J. Meggers, E. S. Ayensu, and W. D. Duckworth, eds.). Smithsonian Institution Press, Washington, 350 pp.

Baker, H. G. 1978. Chemical aspects of the pollination biology of woody plants in the tropics. Pp. 57–82, *in* Tropical trees as living systems. (P. B. Tomlinson and M. H. Zimmermann, eds.). Cambridge University Press, Cambridge, 675 pp.

Baker, H. G., and I. Baker. 1975. Studies of nectar-constitution and pollinator–plant coevolution. Pp. 100–140, *in* Coevolution of animals and plants. (L. E. Gilbert and P. H. Raven, eds.). University of Texas Press, Austin. 246 pp.

Baker, H. G., and B. J. Harris. 1959. Bat-pollination of the silk cotton tree, *Ceiba pentandra* (L.) Gaertn. (Sensu Lato), in Ghana. J. West Afr. Sci. Assoc., **5**:1–9.

Baker, H. G., and P. D. Hurd. 1968. Intrafloral ecology. Ann. Rev. Entomol., **13**:385–414.

Baker, H. G., R. W. Cruden, and I. Baker. 1971. Minor parasitism in pollination biology and its community function. The case of *Ceiba acuminata*. Bioscience, **21**:1127–1129.

Bartholomew, G. A., P. Leitner, and J. E. Nelson. 1964. Body temperature, oxygen consumption, and heart rate in three species of Australian flying foxes. Physiol. Zool., **37**:179–198.

Bawa, K. S. 1974. Breeding systems of tree species of a lowland tropical community. Evolution, **28**:85–92.

Bawa, K. S., and P. A. Opler. 1975. Dioecism in tropical forest trees. Evolution, **29**:167–179.

Bazzaz, F. A., and S. T. A. Pickett. 1980. Physiological ecology of tropical succession: A comparative review. Ann. Rev. Ecol. Syst., **11**:287–310.

Benedict, F. A. 1957. Hair structure as a generic character in bats. Univ. Calif. Publ. Zool., **59**:285–548.

Birney, E. C., R. Jenness, and K. M. Ayaz. 1976. Inability of bats to synthesise L-ascorbic acid. Nature, **260**:626–628.

Bonaccorso, F. J. 1979. Foraging and reproductive ecology in a Panamanian bat community. Bull. Fla. State Mus. Biol. Sci., **24**:359–408.

Bonaccorso, F. J., W. E. Glanz, and C. M. Sandford. 1980. Feeding assemblages of mammals at fruiting *Dipteryx panamensis* (Papilionaceae) trees in Panama: Seed predation, dispersal, and parasitism. Rev. Biol. Trop., **28**:61–72.

Bradbury, J. W., and S. L. Vehrencamp. 1976a. Social organization and foraging in emballonurid bats. I. Field studies. Behav. Ecol. Sociobiol., **1**:337–381.

Bradbury, J.W., and S.L. Vehrencamp. 1976b. Social organization and foraging in emballonurid bats. II. A model for the determination of group size. Behav. Ecol. Sociobiol. **1**:383–404.

Brokaw, N. V. 1980. Gap-phase regeneration in a tropical forest. Unpublished Ph.D. dissertation, University of Chicago, Chicago, 175 pp.

Bullock, S. H. 1974. Seed dispersal of *Dendromecon* by the seed predator *Pogonomyrmex*. Madrono, **22**:378–379.

Carpenter, F. L. 1976. Plant-pollinator interactions in Hawaii. Ecology, **57**:1125–1144.

Colwell, R. K. 1974. Competition and coexistence in a simple tropical community. Am. Nat., **107**:737–760.

Connell, J. H. 1978. Diversity in tropical rain forests and coral reefs. Science, **199**:1302–1310.

Cruden, R. W. 1975. Fecundity as a function of nectar production and pollen–ovule ratios. Pp. 171–178, *in* Tropical trees: Variation, breeding and conservation. Linnean Society Symposium Number 2. (J. Burley and B. T. Styles, eds.). Academic Press, New York, 243 pp.

Daniel, M. J. 1979. The New Zealand short-tailed bat, *Mystacina tuberculata:* A review of present knowledge. N. Z. J. Zool., **6**:357–370.

Darwin, Charles. 1872. The origin of species. 6th ed. Crowell-Collier, New York, 512 pp.

Ehrlich, P. R., and P. H. Raven. 1964. Butterflies and plants: A study in coevolution. Evolution, **18**:586–608.

Faegri, K., and L. van der Pijl. 1966. The principles of pollination ecology. Pergamon, Elmsford, 248 pp.

Findley, J. S., E. H. Studier, and D. E. Wilson. 1972. Morphologic properties of bat wings. J. Mammal., **53**:429–444.

Fleming, T. H., and E. R. Heithaus. 1981. Frugivorous bats, seed shadows and the structure of tropical forests. Reprod. Bot., Suppl. Biotropica, **13**:45–53.

Forman, G. L., C. J. Phillips, and C. S. Rouk. 1979. Alimentary tract. Pp. 205–227, *in* Biology of bats in the New World family Phyllostomatidae. Part III. (R. J. Baker, J. K. Jones, Jr., and D. C. Carter, eds.). Spec. Publ. Mus. Texas Tech Univ., Lubbock, **16**:1–441.

Frankie, G. W., H. G. Baker, and P. A. Opler. 1974. Comparative phenological studies of trees in tropical wet and dry forests in the lowlands of Costa Rica. J. Ecol., **62**:881–919.

Futuyma, D. J. 1973. Community structure and stability in constant environments. Am. Nat., **107**:443–446.

Gardner, A. L. 1977. Feeding habits. Pp. 293–350, *in* Biology of bats in the New World family Phyllostomatidae. Part II. (R. J. Baker, J. K. Jones, Jr., and D. C. Carter, eds.). Spec. Publ. Mus. Texas Tech Univ., Lubbock, **13**:1–364.

Gentry, A. H. 1974. Flowering phenology and diversity in tropical Bignoniaceae. Biotropica, **6**:64–68.

Gilbert, L. E. 1971. Butterfly–plant coevolution: Has *Passiflora adenopoda* won the selectional race with heliconiine butterflies? Science, **172**:585–586.

Gilbert, L. E. 1975. Ecological consequences of a coevolved mutualism between butterflies and plants. Pp. 210–240, *in* Coevolution of animals and plants. (L. E. Gilbert and P. H. Raven, eds.). University of Texas Press, Austin, 246 pp.

Gilbert, L. E. 1977. The role of insect–plant coevolution in the organization of ecosystems. Colloq. Int. CNRS, **265**:399–413.

Gilbert, L. E. 1980. Food web organization and the conservation of Neotropical diversity. Pp. 11–33, *in* Conservation biology an evolutionary–ecological perspective. (M. E. Soulé and B. A. Wilcox, eds.). Sinauer Assoc., Sunderland, 394 pp.

Gillette, D. D. 1975. Evolution of feeding strategies in bats. Tebiwa, **18**:39–48.

Gould, E. 1978. Foraging behavior of Malaysian nectar-feeding bats. Biotropica, **10**:184–193.

Gould, E. 1977. Foraging behavior of *Pteropus vampyrus* on the flowers of *Durio zibethinus*. Malay. Nat. J., **30**:53–57.

Grant, V., and K. A. Grant. 1965. Flower pollination in the phlox family. Columbia University Press, New York, 180 pp.

Handel, S. N. 1978. The competitive relationship of three woodland sedges and its bearing on the evolution of ant-dispersal of *Carex pedunculata*. Evolution, **32**:151–163.

Harris, B. J., and H. G. Baker. 1958. Pollination in *Kigelia africana* Benth. J. West Afr. Sci. Assoc., **4**:25–30.

Harris, B. J., and H. G. Baker. 1959. Pollination of flowers by bats in Ghana. Niger. Field, **24**:151–159.

Heinrich, B., and P. H. Raven. 1972. Energetics and pollination. Science, **176**:597–602.

Heithaus, E. R., and T. H. Fleming. 1978. Foraging movements of a frugivorous bat, *Carollia perspicillata* (Phyllostomatidae). Ecol. Monogr., **48**:127–143.

Heithaus, E. R., P. A. Opler, and H. G. Baker. 1974. Bat activity and pollination of *Bauhinia pauletia:* plant–pollinator coevolution. Ecology, **55**:412–419.

Heithaus, E. R., and T. H. Fleming, and P. A. Opler. 1975. Foraging patterns and resource utilization in seven species of bats in a seasonal tropical forest. Ecology, **56**:841–854.

Heithaus, E. R., D. C. Culver, and A. J. Beattie. 1980. Models of some ant–plant mutualisms. Am. Nat., **116**:347–361.

Heithaus, E. R., E. Stashko, and P. Anderson. 1982. The cumulative effects of plant–animal interactions on seed production by *Bauhinia ungulata*, a Neotropical legume. Ecology, (in press).

Horvitz, C. C. 1980. Seed dispersal and seedling demography of *Calathea microcephala* and *Calathea ovandensis*. Unpublished Ph.D. dissertation, Northwestern University, Evanston, 162 pp.

Howe, H. F. 1977. Bird activity and seed dispersal of a tropical wet forest tree. Ecology, **58:**539–550.

Howe, H. F. 1979. Fear and frugivory. Am. Nat., **114:**925–931.

Howe, H. F. 1980. Monkey dispersal and waste of a Neotropical fruit. Ecology, **61:**944–959.

Howe, H. F., and D. De Steven. 1979. Fruit production, migrant bird visitation, and seed dispersal of *Guarea glabra* in Panama. Oecologia, **59:**1–12.

Howe, H. F., and G. F. Estabrook. 1977. On intra-specific competition for avian dispersers in tropical trees. Am. Nat., **111:**817–832.

Howe, H. F., and G. A. Vande Kerckhove. 1979. Fecundity and seed dispersal of a tropical tree. Ecology, **60:**180–189.

Howell, D. J. 1974a. Bats and pollen: physiological aspects of the syndrome of chiropterophily. Comp. Biochem. Physiol., **48:**263–276.

Howell, D. J. 1974b. Feeding and acoustic behavior in glossophagine bats. J. Mammal., **55:**293–308.

Howell, D. J. 1979. Flock foraging in nectar-eating bats: Advantages to the bats and to the host plants. Am. Nat., **115:**23–49.

Howell, D. J. 1980. Adaptive variation in diets of desert bats has implications for evolution of feeding strategies. J. Mammal., **61:**730–733.

Howell, D. J., and D. Burch. 1974. Food habits of some Costa Rican bats. Rev. Biol. Trop., **21:**281–294.

Howell, D. J., and D. L. Hartl. 1980. Optimal foraging in glossophagine bats: When to give up. Am. Nat., **114:**23–49.

Howell, D. J., and N. Hodgkin. 1976. Feeding adaptations in the hairs and tongue of nectar-feeding bats. J. Morphol., **148:**329–336.

Hubbell, S. P. 1979. Tree dispersion, abundance, and diversity in a tropical deciduous forest. Science, **203:**1299–1309.

Hubbell, S. P. 1980. Seed predation and the coexistence of tree species in tropical forests. Oikos, **35:**214–219.

Humphrey, S. R., and F. J. Bonaccorso. 1979. Population and community ecology. Pp. 409–441, *in* Biology of bats of the New World Family Phyllostomatidae. Part III. (R. J. Baker, J. K. Jones, Jr., and D. C. Carter, eds.). Spec. Publ. Mus. Texas Tech Univ., Lubbock, **16:**1–441.

Huston, M. 1979. A general hypothesis of species diversity. Am. Nat., **113:**81–101.

Jaeger, P. 1954. Les aspects actuels du problème de la cheiroérogamie. Bull. Inst. Fondam. Afr. Noire. Ser. A, **16:**796–821.

Janzen, D. H. 1967. Synchronization of sexual reproduction of trees within the dry season in Central America. Evolution, **21:**620–637.

Janzen, D. H. 1970. Herbivores and the number of tree species in tropical forests. Am. Nat., **111:**657–675.

Janzen, D. H. 1971a. Seed predation by animals. Ann. Rev. Ecol. Syst., **2:**465–492.

Janzen, D. H. 1971b. Euglossine bees as long-distance pollinators in tropical plants. Science, **171:**203–205.

Janzen, D. H. 1975. Intra- and inter-habitat variation in *Guazuma ulmifolia* (Sterculiaceae) seed predation by *Amblycerus cistelinus* (Bruchidae) in Costa Rica. Ecology, **56:**1009–1013.

Janzen, D. H. 1976. Two patterns of pre-dispersal seed predation by insects on Central American deciduous forest trees. Pp. 179–188. *in* Tropical trees: Variation, breeding and conservation. (J. Burley and B. T. Styles, eds.). Linnean Society Symposium No. 2., Academic Press, London, 243 pp.

Janzen, D. H. 1980. When is it coevolution? Evolution, **34:**611–612.

Janzen, D. H., G. A. Miller, J. Hackforth-Jones, C. M. Pond, K. Hooper, and D. P. Janos. 1976. Two Costa Rican bat-generated seed shadows of *Andira inermis* (Swartz) HBK, Leguminosae. Ecology, **57:**1068–1076.

Jepsen, G. L. 1970. Bat origins and evolution. Pp. 1–64, *in* Biology of bats. Vol. 1. (W. A. Wimsatt, ed.). Academic Press, New York, 406 pp.

Jones, J. K., Jr., and H. H. Genoways. 1970. Chiropteran systematics. Pp. 3–21, *in* About bats. (B. H. Slaughter and D. W. Walton, eds.). Southern Methodist University Press, Dallas, 339 pp.

Klite, P. D. 1965. Intestinal bacterial flora and transit time of three Neotropical bat species. J. Bacteriol., **90**:375–379.

Levin, S. A., and J. D. Udovic. 1977. A mathematical model of coevolving populations. Am. Nat., **111**:657–675.

Lockwood, R., E. R. Heithaus, and T. H. Fleming. 1982. Fruit choices among free-flying captive *Carollia perspicillata*. Unpublished manuscript.

MacDonald, N. 1978. Time lags in biological models. Lecture notes in biomathematics. Vol. 27. Springer-Verlag, New York.

McGregor, S. E., S. M. Alcorn. and G. Olin. 1962. Pollination and pollinating agents of the saguaro. Ecology, **43**:259–267.

McKey, D. 1975. The ecology of coevolved seed dispersal systems. Pp. 159–191, *in* Coevolution of animals and plants. (L. E. Gilbert and P. H. Raven, eds.). University Texas Press, Austin, 246 pp.

McNab, B. K. 1969. The economics of temperature regulation in Neotropical bats. Comp. Biochem. Physiol., **31**:227–268.

McNab, B. K. 1980. Food habits, energetics, and the population biology of mammals. Am. Nat., **116**:106–124.

May, R. M. 1973. Stability and complexity in model ecosystems. Princeton University Press, Princeton, 235 pp.

May, R. M. 1976. Models for two interacting populations. Pp. 49–70, *in* Theoretical ecology. (R. M. May, ed.). W. B. Saunders, Philadelphia, 317 pp.

Mode, C. J. 1958. A mathematical model for the coevolution of obligate parasites and their hosts. Evolution, **12**:158–165.

Moore, L. R. 1978. Seed predation in the legume *Crotalaria*. I. Intensity and variability of seed predation in native and introduced populations of *C. pallida* Ait. Oecologia, **34**:185–202.

Morrison, D. W. 1978a. Foraging ecology and energetics of the frugivorous bat *Artibeus jamaicensis*. Ecology, **59**:716–723.

Morrison, D. W. 1978b. Lunar phobia in a neotropical fruit bat, *Artibeus jamaicensis* (Chiroptera: Phyllostomidae). Anim. Behav., **26**:852–855.

Morrison, D. W. 1978c. On the optimal searching strategy for refuging predators. Am. Nat., **112**:925–934.

Morrison, D. W. 1979. Apparent male defense of tree hollows in the fruit bat, *Artibeus jamaicensis*. J. Mammal., **60**:11–15.

Mulcahy, D. L. 1979. The rise of the angiosperms: A genecological factor. Science, **206**:20–23.

Myers, N. 1980. Conversion of tropical moist forests. National Academy of Science, Washington, D.C., 205 pp.

Novick, A. 1977. Acoustic orientation. Pp. 73–287, *in* Biology of bats. Vol. 3. (W. A. Wimsatt, ed.). Academic Press, New York, 651 pp.

Perry, A. E., and T. H. Fleming. 1980. Ant and rodent predation on small animal-dispersed seeds in a dry tropical forest. Brenesia, **17**:11–22.

Pirlot, P. 1977. Wing design and the origin of bats. Pp. 375–410, *in* Major patterns in vertebrate evolution. (M. K. Hecht, P. C. Goody, and B. M. Hecht, eds.). Plenum Press, New York, 908 pp.

Poole, R. W., and B. J. Rathcke. 1979. Regularity, randomness, and aggregation in flowering phenologies. Science, **203**:470–471.

Pyke, G. H., H. R. Pulliam, and E. L. Charnov. 1977. Optimal foraging: A selective review of theory and tests. Q. Rev. Biol., **52**:137–154.

Raven, P. H. 1977. A suggestion concerning the Cretaceous rise to dominance of the angiosperms. Evolution, **31**:451–452.

Regal, P. J. 1977. Ecology and evolution of flowering plant dominance. Science, **196**:622–629.

Risch, S., and D. Boucher. 1976. What ecologists look for. Bull. Ecol. Soc. Am., **57**:8–9.

Roughgarden, J. 1979. Theory of population genetics and evolutionary ecology: An introduction. MacMillan Publ. Co., New York, 634 pp.

Sazima, M., and I. Sazima. 1975. Quiropterofilia en *Lafoensia pacari* St. Hil. (Lythraceae), na serra do cipó, minas gerais. Rev. Cienc. Cult., **27**:405–416.

Sazima, I., and M. Sazima. 1977. Solitary and group foraging: Two flower-visiting patterns of the lesser spear-nosed bat *Phyllostomus discolor*. Biotropica, **9**:213–215.

Sazima, M., and I. Sazima. 1978. Bat pollination of the passion flower, *Passiflora mucronata*, in southeastern Brazil. Biotropica, **10**:100–109.

Sih, A. 1980. Optimal behavior: Can foragers balance two conflicting demands? Science, **210**:1041–1043.

Simmons, J. A., M. B. Fenton, and M. J. O'Farrell. 1979. Echolocation and pursuit of prey by bats. Science, **203**:16–21.

Slater, J. A. 1972. Lygaeid bugs (Hemiptera: Lygaeidae) as seed predators of figs. Biotropica, **4**:145–151.

Slatkin, M. and J. Maynard-Smith. 1979. Models of coevolution. Q. Rev. Biol. **54**:233–263.

Smith, J. D. 1976. Chiropteran evolution. Pp. 49–70, *in* Bats of the New World family Phyllostomatidae. Part I. (R. J. Baker, J. K. Jones, Jr., and D. C. Carter, eds.). Spec. Publ. Mus. Texas Tech Univ., Lubbock, **10**:1–218.

Smith, J. D. 1977. Comments on flight and the evolution of bats. Pp. 427–437, *in* Major patterns in vertebrate evolution. (M. K. Hecht, P. C. Goody, and B. M. Hecht, eds.). Plenum Press, New York, 908 pp.

Smith, J. D., and A. Starrett. 1979. Morphometric analysis of chiropteran wings. Pp. 229–316, *in* Biology of bats of the New World family Phyllostomatidae. Part III. (R. J. Baker, J. K. Jones, Jr., and D. C. Carter, eds.). Spec. Publ. Mus. Texas Tech Univ., Lubbock, **16**:1–441.

Snow, D. W. 1965. A possible selective factor in the evolution of fruiting seasons in tropical forest. Oikos, **15**:274–281.

Start, A. N., and A. G. Marshall. 1976. Nectarivorous bats as pollinators of trees in West Malaysia. Pp. 141–140, *in* Tropical trees: Variation, breeding, and conservation. (J. Burley and B. T. Styles, eds.). Linnean Society Symposium Series No. 2., Academic Press, New York, 243 pp.

Sussman, R. W., and P. H. Raven. 1978. Pollination by lemurs and marsupials: An archaic coevolutionary system. Science, **200**:731–736.

Temple, S. A. 1977. Plant–animal mutualism: Coevolution with Dodo leads to near extinction of plant. Science, **197**:885–886.

Van Valen, L. 1979. The evolution of bats. Evol. Theory, **4**:103–121.

Vandermeer, J. H. 1975. A graphical model of insect seed predation. Am. Nat., **109**:147–160.

Vandermeer, J. H., and D. H. Boucher. 1978. Varieties of mutualistic interaction in population models. J. Theor. Biol., **74**:549–558.

van der Pijl, L. 1956. Remarks on pollination by bats in the genera *Freycinetia*, *Duabanga* and *Haplophragma*, and on chiropterophily in general. Acta Bot. Neerl., **5**:135–144.

van der Pijl, L. 1972. Principles of dispersal in higher plants. Springer-Verlag, New York, 161 pp.

Vazquez-Yanes, C., A. Orozco, G. Francois, and L. Trejo. 1975. Observations on seed dispersal by bats in a tropical humid region in Veracruz, Mexico. Biotropica, **7**:73–76.

Vogel, S. 1958. Fledermausblumen in Sud Amerika. Oesterr. Bot. Z., **104**:491–530.

Vogel, S. 1968. Chiropterophilie in der neotropischen Flora. Neue Mitt. I. Flora Abt. B, **157**:562–602.

Vogel, S. 1969. Chiropterophilie in der neotrophischen Flora. Neue Mitt. II, III. Flora Abt. B, **158**:185–222, 289–323.

Voss, R., M. Turner, R. Inouye, M. Fisher, and R. Cort. 1980. Floral biology of *Markea neurantha* Hemsley (Solanaceae), a bat pollinated epiphyte. Am. Midl. Nat., **103**:262–268.

Wickler, W., and U. Seibt. 1976. Field studies of the African fruit bat, *Epomophorus wahlbergi* (Sunderall), with special reference to male calling. Z. Tierpsychol., **40**:345–376.

Wilson, D. E. 1971. Food habits of *Micronycteris hirsuta* (Chiroptera: Phyllostomatidae). Mammalia, **35**:107–110.

Wilson, D. E. 1979. Reproductive patterns. Pp. 317–378, *in* Biology of bats in the New World family Phyllostomatidae. Part III. (R. J. Baker, J. K. Jones, Jr., and D. C. Carter, eds.). Spec. Publ. Mus. Texas Tech Univ., Lubbock, **16**:1–441.

Wilson, D. S. 1980. The natural selection of populations and communities. Benjamin/Cummings, Menlo Park, 186 pp.

Yalden, D. W., and P. A. Morris. 1975. The lives of bats. Quadrangle/The New York Times Book Co., New York, 247 pp.

Zapata, T. R., and M. T. K. Arroyo. 1978. Plant reproductive ecology of a secondary deciduous tropical forest in Venezuela. Biotropica, **10**:221–230.

Chapter 10

Ecology of Insects Ectoparasitic on Bats

Adrian G. Marshall

Department of Zoology
University of Aberdeen
Aberdeen AB9 2TN, Scotland

I. INTRODUCTION

About 6000 species of insects belonging to seven orders are known to be external parasites as adults of warm-blooded vertebrates. Of these, 687 described species of four orders are known to parasitize bats, with members of all the bat families except for three small ones (Craseonycteridae, Mystacinidae, and Myzopodidae) acting as hosts (Tables I and II). The six insect families involved (Fig. 1) are all exclusively associated with bats except for the Cimicidae: of the 89 species of cimicids known, 61 (68%) are parasitic upon bats (including the two species of bedbugs, *Cimex hemipterus* (F.) and *C. lectularius* L., commonly associated with man), and the remainder upon birds. Members of five of the six families have piercing–sucking mouthparts and feed on blood, and there is no doubt that they are truly parasitic; however, arixeniids have chewing mouthparts and feed on solid material such as skin detritus and host feces and may in fact be commensals rather than parasites. Three unusual families of Diptera, each containing a single species, have been reported from bat roosts: the Chiropteromyzidae in Finland, Mormotomyiidae in Kenya, and Mystacinobiidae in New Zealand (Holloway, 1976; Maa, personal communication). But in this review I assume these to be commensals and do not consider them further. For a more complete discussion of ectoparasite taxonomy see Marshall (1981a).

Bat ectoparasites spend their entire lives either on the body or in the roosts of their hosts. Because bats have a largely tropical distribution, most ectoparasites live in a comparatively warm, moist, and equable environment. However, a few species are associated with bats that hibernate, and these must

TABLE I
Insects Ectoparasitic on Bats

Division	Order (vernacular name)	Suborder	Family	Number of genera	Number of species
Hemimetabola	Dermaptera (earwigs)	Arixeniina	Arixeniidae	2	5
	Hemiptera (bugs)	Heteroptera	Cimicidae	13	61
			Polyctenidae	5	32
Holometabola	Diptera (flies)	Cyclorrhapha	Nycteribiidae	12	256
			Streblidae	31	224
	Siphonaptera (fleas)	—	Ischnopsyllidae	17	107[a]

[a]Includes both species and subspecies.

TABLE II
The Geographical Distribution and Hosts of the Families of Insects Ectoparasitic on Bats

Insect family	Geographical distribution	Host families
Arixeniidae	Oriental	Molossidae
Cimicidae	Cosmopolitan	Emballonuridae, Molossidae, Noctilionidae, Pteropodidae, Rhinolophidae, Vespertilionidae
Polyctenidae	Cosmopolitan (not Palearctic)	Emballonuridae, Megadermatidae, Molossidae, Nycteridae, Rhinolophidae
Nycteribiidae	Cosmopolitan	Emballonuridae, Phyllostomidae, Pteropodidae, Rhinolophidae, Thyropteridae, Vespertilionidae
Streblidae	Cosmopolitan (not Australian)	Desmodontidae, Emballonuridae, Furipteridae, Megadermatidae, Molossidae, Mormoopidae, Natalidae, Noctilionidae, Nycteridae, Phyllostomidae, Pteropodidae, Rhinolophidae, Rhinopomatidae, Vespertilionidae
Ischnopsyllidae	Cosmopolitan	Desmodontidae, Emballonuridae, Megadermatidae, Molossidae, Noctilionidae, Pteropodidae, Rhinolophidae, Rhinopomatidae, Vespertilionidae

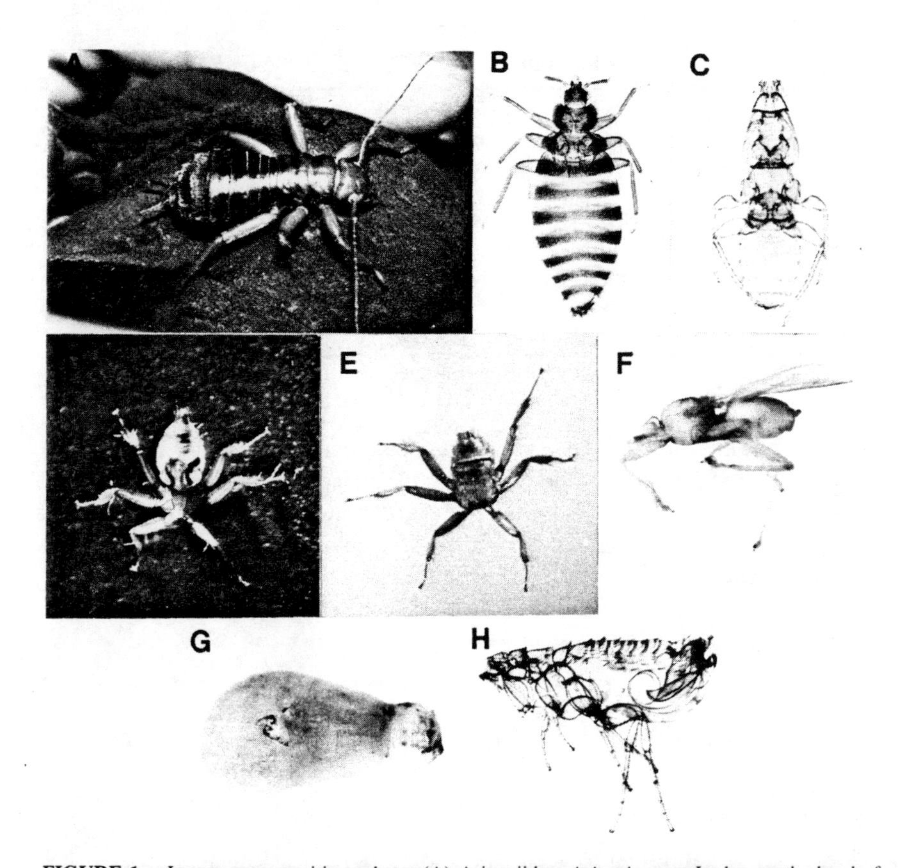

FIGURE 1. Insects ectoparasitic on bats. (A) Arixeniidae: *Arixenia esau* Jordan on the head of *Cheiromeles torquatus* Horsfield. (B) Cimicidae: *Cimex lectularius* L. (C) Polyctenidae: *Eoctenes spasmae* (Waterhouse). (D,E) Nycteribiidae: *Basilia hispida* Theodor, pregnant female (dorsal view) and female immediately after offspring deposition (ventral view). (F) Streblidae: *Brachytarsina amboinensis* Rondani. (G) Streblidae: neosomic female of *Ascodipteron phyllorhinae* Adensamer (anterior to left). (H) Ischnopsyllidae: *Myodopsylla insignis* (Roths.)

survive considerable climatic variations. For lives associated with the bodies of active, furred mammals, ectoparasites possess a number of conspicuous adaptations. Although they range in size from the tiny louselike streblid *Mastoptera minuta* (Lima), less than 1.0 mm long, to the arixeniid *Arixenia esau* Jordan, which can reach a length of 27 mm, most ectoparasites are small and flattened (Table III) so that they can readily move through dense pelage, press themselves close to the host's body, or hide in narrow crevices within the roost. Their integuments are generally well sclerotized and streamlined by setae or combs that

TABLE III
Some Adaptations among Insects Ectoparasitic on Bats[a]

Insect family	Body flattening	Wings	Functional eyes	Combs
Arixeniidae	None	−	r	−
Cimicidae	Dorsoventral	−	+	−
Polyctenidae	Dorsoventral	−	−	+
Nycteribiidae	Dorsoventral	−	r, −	+, −
Streblidae	Some lateral	+, r, −	r, −	+, −
Ischnopsyllidae	Lateral	−	−	+

[a]Note: + = present, r = reduced, − = absent.

aid in locomotion, reduce abrasive damage, and facilitate escape if caught while the host is grooming. (For a discussion of the function of combs see Marshall, 1980b). Backward-pointing setae and powerful claws allow them to maintain a firm grip on the body or home of the host. And finally, the behavioral responses of ectoparasites to a wide variety of stimuli lessen the chance of dislodgement and ensure that they pass much of their lives at sites where the risks of mortality are minimal.

2. LIFE CYCLES

2.1. Patterns

In the six families of bat ectoparasites a variety of life-cycle patterns can be recognized:

Oviparous families

Cimicidae—five nymphal instars and adult bugs mostly in the roost.

Ischnopsyllidae—three larval instars and pupae in the roost, adult fleas mostly on the host.

Viviparous families

Arixeniidae—four nymphal instars and adult earwigs sometimes on the host, sometimes in the roost.

Polyctenidae—Three nymphal instars and adult bugs always on the host.

Nycteribiidae and Streblidae—three larval instars within the adult female, pupae in the roost, adult flies mostly on the host.

Among the oviparous families, cimicids leave their hosts to deposit eggs, gluing them in small cracks or crevices in the host's roost (Usinger, 1966), whereas ischnopsyllids deposit eggs wherever they are, the eggs falling haphazardly to the ground beneath the roost, where they hatch and the larvae develop.

Arixeniids and polyctenids both exhibit "pseudoplacental viviparity" (Hagan, 1951), wherein the first-instar nymph is nourished within the female via a pseudoplacenta. Arixeniids deposit first-instar nymphs largely in the roost, whereas polyctenids deposit them only on the body of their host. Both nycteribiids and streblids undergo "adrenotrophic viviparity," the complete larval life occurring within the maternal uterus, being nourished there by secretions from special "milk glands." Females of both families deposit third-instar larvae immediately prior to pupation, nearly always leaving the host to attach the offspring to the roost substrate. However, Old World streblids of the genus *Ascodipteron*, in which the female is permanently embedded in the skin of the host, drop their offspring freely in the roost, and this behavior may also occur in certain New World streblids that do not embed (Muir, 1912; R. L. Wenzel, personal communication). Both nycteribiids and streblids deposit offspring at sites away from the immediate vicinity of the host, the puparia thus being protected from host-induced mortality through soiling or damage. Furthermore, offspring deposition in some nycteribiids is stimulated by rising temperatures, so that it occurs only when the hosts are in the daytime roost (Leong and Marshall, 1968; Marshall, 1970). This may also occur in certain streblids (Overal, 1980; Ross, 1961).

In hemimetabolous species the last nymphal instar molts to the adult stage in the habitat in which the nymphal life has occurred, and there are no reported requirements for any specific stimuli. Thereafter females rapidly reach maturity: if males are present, cimicids can deposit their first egg within about 5 days; and polyctenids may reach sexual maturity and mate even before molting to the adult (Hagan, 1951; Usinger, 1966). However, in holometabolous species, in which immature stages and adults have completely different life-styles, adults sometimes remain quiescent within their puparia until stimulated to emerge by some external trigger, perhaps provided by a nearby host. These triggers can be a rise in temperature (ischnopsyllids), a rise in the concentration of CO_2 (nycteribiids), or mechanical shock or vibration (nycteribiids and ischnopsyllids) (Hurka, 1963a, 1963b; Marshall, 1970; Ryberg, 1947).

Reproductive rates vary widely both within and among species, influenced by such factors as temperature and feeding. For example, a single female cimicid *Cimex lectularius* may produce up to 5 eggs per day, with a maximum of 541 in a lifetime (Davis, 1964; Usinger, 1966), whereas a female nycteribiid *Basilia hispida* Theodor produces only a single offspring every 9 days, with a maximum of 16 in a lifetime (Marshall, 1970). These rates appear typical, but other viviparous species presumably reproduce more slowly, although little data are available (Balcells Rocamora, 1956; Funakoshi, 1977; Hurka, 1964; Joyeux, 1913; Leong and Marshall, 1968; Maa, 1965; Overal, 1980; Rodhain and Bequaert, 1916; Ryberg, 1947; Schulz, 1938; Theodor and Moscona, 1954). No data on reproductive rates are available for ischnopsyllids, but knowledge of the biology of other fleas indicates that they should be able to lay many eggs in a day and hundreds in a lifetime.

2.2. Food and Feeding

The only food ingested by cimicids, polyctenids, nycteribiids, and streblids is the host's blood, which is imbibed directly by all stages, except the immature instars of viviparous species which are nourished within the female. Blood is also the sole food of adult ischnopsyllids, but larval fleas live in the host's guano and feed there on a variety of organic matter, sometimes including other fleas (e.g., *Myodopsylla* species; Hase, 1931b). The larvae of certain rodent fleas can survive only if they ingest host blood derived from the dried fecal pellets of adult fleas, but whether this is true of bat fleas is not known. Arixeniids possess biting mouthparts: *Arixenia* spp. are believed to feed on skin debris and the skin and body exudates of their host, whereas it is possible that members of the second genus, *Xeniaria*, may feed largely apart from their hosts, perhaps consuming other roost-dwelling insects (Marshall, 1977a; Nakata and Maa, 1974; Popham, 1962).

As would be expected, species that live permanently on their hosts feed more frequently than those which spend much of their lives in the roost. For example, among the Hemiptera polyctenids feed every few hours, whereas cimicids feed every few days and can survive many months without food (Marshall, 1982; Overal and Wingate, 1976; Ryckman, 1956; Usinger, 1966). Nycteribiids and streblids feed as soon as they find a host, and thereafter at intervals of a few hours throughout their lives (except on hibernating hosts; see Section 2.3); among streblids, those confined to their hosts feed more frequently than those that often leave their hosts (Marshall, 1970; Overal, 1980; Wenzel *et al.*, 1966). Adult ischnopsyllids presumably feed soon after first reaching a host and thereafter at intervals of a few hours on active hosts, but no precise information is available.

2.3. Influence of Climate and Host Hibernation

Temperature profoundly affects many aspects of the lives of ectoparasites. For example, in *Cimex lectularius* temperature may affect copulation, oviposition, hatching of the egg, duration and survival of the egg and immature instars, molting, adult longevity, fecundity, activity, and the feeding rate (Johnson, 1942; Omori, 1941; Usinger, 1966). Humidity generally has little effect on the timing of life cycles, but it may affect survival (Section 2.4.5).

Cimicids, nycteribiids, streblids, and ischnopsyllids are known to occur in association with hibernating bats, but no studies have been made on cimicids under such circumstances. In Europe nycteribiids apparently cease to reproduce on hibernating hosts in northern regions, whereas further south reproduction continues at a reduced rate. In northern Europe *Nycteribia kolenatii* Theodor and Moscona, *Penicillidia monoceros* Speiser, and *Basilia nattereri* (Kolenati) cease to deposit offspring but remain active and continue to feed mostly below the hosts' wings (Ryberg, 1947; Schulz, 1938). But further south in Czechoslovakia, Hurka (1964) noted that nycteribiids such as *Nycteribia latreillii* (Leach) continue to produce offspring slowly throughout the winter. In Japan *Penicillidia jenynsii* (Westwood) ceases to reproduce on hibernating hosts (Funakoshi, 1977).

The Streblidae is a largely tropical family, and only two North American *Trichobius* species are known to occur on hibernating hosts (Kunz, 1976; Reisen *et al.*, 1976; Ross, 1960, 1961; Zeve, 1959). *Trichobius major* Coquillett overwinters on hibernating *Myotis velifer* (Vespertilionidae), where it is found exclusively on the external pinnae, but in summer it is found on membranes and in the pelage of its host by day and in the roost by night. *Trichobius corynorhini* Cockerell overwinters on hibernating *Plecotus townsendii* (Vespertilionidae), living permanently on the wings and external pinnae in winter and on the wings in summer. In both species of streblid, feeding is greatly reduced during winter (November–March) and copulation and offspring deposition cease.

K. Hurka (1963a, 1963b, 1970), L. Hurka (1965) and Hurka and Hurka (1964) examined the ischnopsyllids of hibernating bats in central Europe and noted that different genera bred at different times. Six species of *Ischnopsyllus* bred only on active hosts; for example, *I. intermedius* (Rothschild) and *I. simplex* Rothschild ceased to breed in midwinter on *Myotis* species, which hibernate deeply, whereas *I. hexactenus* (Kolenati) continued to deposit eggs slowly throughout the winter on *Plecotus auritus*, which is more active. *Rhinolophopsylla unipectinata* (Taschenberg) appeared also to be a "summer flea," breeding only on active hosts, as is *Myodopsylla collinsi* Kohls in North America (Reisen *et al.*, 1976). In contrast, four species of *Nycteridopsylla* were "winter fleas," undergoing a number of generations on hibernating hosts but being absent at other times (see Labunets and Bogdanov, 1959).

2.4. Causes of Mortality

2.4.1. The Host

The grooming activities of the host are the major cause of mortality for permanent ectoparasites such as polyctenids and are a serious threat to all insects while on the host's body. The effectiveness of grooming may alter with age, health, and temperature. If grooming activities are thwarted through ill health or deformity, parasite populations can rapidly rise. Such high populations on sick individuals are seldom, if ever, the direct cause of ill health but, rather, a symptom of it. Overal and Wingate (1976) noted that sick bats were more heavily infested with the cimicid *Stricticimex antennatus* Ferris and Usinger than were healthy ones; on healthy hosts they estimated a blood loss of 84 mg per bat per day due to the feeding of 220 cimicids per bat and suggested that while this would not harm healthy hosts, a greater daily loss could hasten the death of unhealthy ones.

2.4.2. Starvation

An ectoparasite may be exposed to starvation by failing to locate the correct host, by being removed from the host (e.g., by grooming or accidentally falling off), or by being separated from the host by short-term or seasonal host movements. As would be expected, permanent parasites are less resistant to starvation than temporary ones. It is apparent that arixeniids can withstand separation from their hosts for at least several weeks (Marshall, 1977a), and this is also true of cimicids: the maximum recorded longevity of *Cimex lectularius* without food is 562–572 days, although this varies with climate, mating, and feeding (Johnson, 1942). For *Stricticimex antennatus*, Overal and Wingate (1976) showed that at 21 ± 1°C and 75% RH the mean survival in days of unfed, recently molted, first- to fifth-instar nymphs and adult males and females were, respectively, 11.2, 15.5, 18.6, 24.3, 30.1, 38.6, and 47.4 days. In contrast, polyctenids have poor powers of survival without food: *Eoctenes spasmae* (Waterhouse) can survive starvation only for less than 30 hr as adults and less than 7 hr as nymphs (Marshall, 1982).

Newly emerged nycteribiids and streblids can live for about 1–3 days without food, but flies that have already fed survive starvation only for less than one day. In general, mature males resist starvation better than mature females, and low temperatures are more favorable than high ones (Funakoshi, 1977; Hase, 1931a; Leong and Marshall, 1968; Marshall, 1970; Overal, 1980; Ryberg, 1947; Schulz, 1938; Theodor and Moscona, 1954). Like the flies, fleas tend to resist starvation for longer if newly emerged and unfed than if they have already fed, and low temperatures and high humidities are favorable. However, few data are available specifically on ischnopsyllids, although Hurka and Doskočil (1961) showed that at 20°C and 55% RH unfed *Ischnopsyllus intermedius* could live for up to 4 days without food.

2.4.3. Competition

An individual bat may be associated with a community of ectoparasitic insects (a faunule) of some complexity. For example, in west Malaysia, out of 86 bat species, 28 (33%) are known to be commonly infested with more than one species of ectoparasitic insect, five of these bats possessing five insect species each, and one having seven insect species, these faunules being composed of members of up to four families (Marshall, unpublished).

The factors that influence the diversity of the faunule associated with one bat species compared to another include the geographical distribution, behavior, and size of the bat, and the form of the roost. Thus a bat species with a broad geographical range, which occurs in dense populations, or one that is large in size will tend to be associated with a greater diversity of ectoparasites than a closely related species which is more narrowly confined, well dispersed, or smaller.

At the level of the individual host within a confined geographical area, the question arises as to what limits the size and diversity of the ectoparasitic community. When an individual harbors more than a single ectoparasite, the possibility of interaction between these insects exists. Intraspecific competition may sometimes occur in crowded populations, and this may result in mortality. For example, repeated matings in crowded laboratory colonies of certain cimicids may cause death (Usinger, 1966), and disturbance caused by crowding can cause abortion and death in some female nycteribiids (Ryberg, 1947). But in the wild, crowding is unlikely to occur except on unhealthy or injured bats. Although some studies of bat ectoparasites have indicated an apparent lack of interspecific competition (Marshall, 1977a; Overal, 1980; Overal and Wingate, 1976), it is notable that among New World, but not Old World, molossid bats, species with polyctenids have no nycteribiids or streblids, and nycteribiids and streblids do not occur on the same host (Wenzel and Tipton, 1966; R. L. Wenzel, personal communication). A good example of competitive exclusion among streblids has been given by Wenzel and Tipton (1966), who examined members of the genus *Strebla* on the bat *Phyllostomus hastatus* (Phyllostomidae) occurring in Central and South America. In conclusion, our knowledge of competitive interactions between ectoparasites remains so slight that little can definitely be said about their effects upon mortality and distribution.

Although the nature of competition between species and the factors that regulate community composition are of major interest to ecologists, there has been virtually no discussion of this with regard to ectoparasites. And yet in certain ways ectoparasite communities provide a fine system for the study of competition, for they emulate isolated island communities and are of a size suitable for experimental manipulation. Of notable value would be a study of a host species that has different faunules in different parts of its range (the vespertilionid genus *Miniopterus* would be of particular interest), with emphasis on the theories of island biogeography (MacArthur and Wilson, 1967; Price, 1980).

2.4.4. Parasites, Pathogens, and Predators

As Table IV indicates, ectoparasites may be attacked by a variety of parasitic, phoretic, or predatory organisms; however, there is no indication that these

TABLE IV
Parasitic, Phoretic, and Predatory Organisms Recorded for Ectoparasitic Insects[a]

Insect family	Microorganisms	Fungi	Protozoa	Cestoda	Nematoda	Phoretic Acarina	Parasitic Hymenoptera	Predatory arthropods	Predatory birds
Arixeniidae						√			
Cimicidae	√	√	√					√	
Polyctenidae									
Nycteribiidae	√	√	√			√	√		
Streblidae	√	√						√	
Siphonaptera	√	√	√	√	√	√	√	√	√

[a]Records for all Cimicidae and Siphonaptera included, not just for bat-infesting species.

cause significant mortality in natural populations. Among microorganisms, rickettsia and bacteria have been recorded from cimicids (Jenkins, 1964), and rickettsia from nycteribiids and streblids (Aschner, 1931). Among fungi, members of the order Laboulbeniales have been reported from nycteribiids and streblids (Benjamin, 1971; Blackwell, 1980a, 1980b), and members of the Fungi Imperfecti from cimicids (Cockbain and Hastie, 1961; Usinger, 1966). Flagellate and sporozoan protozoa are known from cimicids, flagellates from nycteribiids, and possibly sporozoans from both nycteribiids and streblids (Anciaux de Faveaux, 1965; Bearup and Lawrence, 1947; Jenkins, 1964; Tamsitt and Valdivieso, 1970; Wallace, 1966). Among phoretic Acarina, members of the Laelaptidae have been reported from arixeniids and nycteribiids, and members of the Dermanyssidae from nycteribiids (Domrow, 1961; Maa, 1971a; Marshall, 1970, 1976; Nakata and Maa, 1974). Various predatory insects of the orders Dictyoptera, Hemiptera, Hymenoptera, and Lepidoptera, and predatory Chelicerata (mites, pseudoscorpions, spiders, and solpugids) are known to attack cimicids (Overal and Wingate, 1976; Usinger, 1966; Wattal and Kalra, 1960), and spiders can capture streblids (Ross, 1961; Marshall, unpublished). And parasitic hymenopterans (Chalcidoidea) have been recorded from the puparia of nycteribiids (Urich *et al.*, 1922). Although a variety of organisms have been recorded in association with fleas (Table IV), there are no records specifically for ischnopsyllids.

2.4.5. Climate

Unsuitable temperatures and humidities can be major causes of mortality among ectoparasitic insects, their influence being particularly profound on those stages whose conditions are not ameliorated by close contact with the host. However, because many bat ectoparasites are tropical in distribution and, even in temperate regions, pass their lives partly within a sheltered bat roost and partly on the body of their warm-blooded host, adverse climatic conditions probably rarely cause significant mortality in natural populations.

2.5. Number of Generations per Year

Seasonal changes in temperature usually cause seasonal fluctuations in reproductive rates, which are reflected in seasonal population cycles. Such population fluctuations are known in all families other than Arixeniidae and Polyctenidae, which are exclusively tropical and presumably breed at much the same rate throughout the year. No data are available for either of these families, but members of the Hemimeridae, which is closely related to Arixeniidae but ectoparasitic upon bats in Africa, take about 2 months from egg deposition to adult emergence, allowing about six generations per year (Ashford, 1970). At constant temperatures the length of the life cycles of Cimicidae vary greatly: for example, at 28°C *Bucimex chilensis* Usinger completes its development in 123 days, and *Cimex lectularius* does so in 30 days, the number of generations per year thus varying from about 3 to 12 (Johnson, 1942; Usinger, 1966). Even in precisely controlled environments variations within a species may be great: for example, at 21 ± 1°C and 75% RH *Stricticimex antennatus* completed its life cycle in anything from 52 to 96 days (mean, 80 days) (Overal and Wingate, 1976).

The Nycteribiidae and Streblidae generally complete their life cycles in about a month, allowing 12 generations per year (Balcells Rocamora, 1956; Funakoshi, 1977; Hurka, 1964; Leong and Marshall, 1968; Maa, 1965; Marshall, 1970; Muir, 1912; Overal, 1980; Rodhain and Bequaert, 1916; Ryberg, 1947; Schulz, 1938; Theodor and Moscona, 1954). However, in temperate regions there are fewer generations owing to the reduction or cessation of reproduction on hibernating bats (Section 2.3). Data on ischnopsyllids are sparse, but European bat fleas, whether "winter" or "summer" species (Section 2.3), may undergo four or five generations during their breeding season (Hurka, 1963a, 1963b, 1970).

3. HOST ASSOCIATIONS

3.1. Introduction

The role of the host is of utmost importance in the life of any parasitic organism, and as a result parasites all show some degree of host limitation. From the parasite's point of view host specificity is generally associated with the degree to which that parasite is confined to its host's body; for example, polyctenids are more specific than cimicids. From the host's point of view differences in behavior are important; for example, among bats of the genus *Rhinolophus* (Rhinolophidae) in Malaysia, colonially roosting species such as *R. robinsoni* and *R. stheno* can be infested with two species of nycteribiids and three species of streblids, whereas solitary roosters such as *R. luctus* have no insect parasites (Marshall, 1980a, unpublished). Furthermore, on a particular host a species of parasite may be less abundant or even absent in differing geographical areas or habitats, on different subspecies or races of that host, on different sexes or ages, or even upon different individuals that are close neighbors and of identical race, sex, and age.

Clearly, the elucidation of the exact nature of host associations is not easy. Furthermore, past discussions have often been based on inaccurate data culled from museum collections, or have been bedeviled by uncertainties over host or parasite taxonomy. Ideally it should be possible to identify true hosts (those which in nature provide the parasite with conditions suitable for indefinitely continued reproduction) from secondary or accidental hosts and to classify parasites into monoxenous (only one host species), oligoxenous (two or more host species of the same genus), pleioxenous (two or more host genera in the same subfamily), or polyxenous species (many hosts).

3.2. Patterns

The five species of Arixeniidae are all monoxenous, being associated only with the hairless molossid *Cheiromeles torquatus*, although also found accidentally on other fully furred molossids of the genus *Tadarida*, which share the roosts of *C. torquatus* (Nakata and Maa, 1974).

The Cimicidae of bats are mainly associated with the Molossidae and Vespertilionidae and are not particularly host specific. For example, *Stricticimex antennatus* lives in association with a diversity of cavernicolous bats in southern Africa, feeding on hosts of at least five species of four genera and three families (Nycteridae, Rhinolophidae, and Vespertilionidae), and in the laboratory this species will also reproduce when fed upon man, rabbits, or rats (Overal and Wingate, 1976). A number of bat cimicids can feed on humans, and the two so-called human bedbugs, *Cimex hemipterus* and *C. lectularius*, are regularly found in bat roosts (e.g., Roer, 1969; Usinger, 1966).

Polyctenidae spend their entire lives on the bodies of their hosts, which are always colonial species, usually those roosting in shallow caves or tree holes (Maa, 1964). They transfer from one host to another only by contact, and, because bats seldom come into physical contact with other species, it is not surprising that most species are monoxenous (Maa, 1964; Marshall, 1982;

Ueshima, 1972). Indeed, reports of species with a wider host range, for example certain *Hesperoctenes* species in the New World (Ueshima, 1972), probably arise from our current poor understanding of polyctenid taxonomy.

Increasing knowledge of the Nycteribiidae and Streblidae is showing, despite earlier reports to the contrary (e.g., Jobling, 1951; Theodor, 1957), that they are nearly always confined to a single species or genus of host. Maa (1971a), in a discussion of the host associations of Australian bat flies, noted that of the 27 species and subspecies of nycteribiids and six of streblids, seven and one, respectively, were monoxenous, 16 and five oligoxenous, and two and 0 pleioxenous, the hosts of two nycteribiids remaining unknown. Marshall (1976, 1980a) observed a similar pattern in the New Hebrides and in Malaysia. The majority of Streblidae (68% of 224 species) are from the New World, where the complexity of the taxonomy of bats, particularly the Phyllostomidae, has made true host relationships difficult to unravel. Of the 94 New World species discussed by Wenzel *et al.* (1966), 70% were apparently monoxenous, 14% oligoxenous, 15% pleioxenous, and 1% unassignable. In general, non-monoxenous bat flies tend to have clear host preferences, although this may vary with locality (Hurka, 1964; Maa, 1965, 1971a, 1971b; Overal, 1980; Ross, 1961; Stefanelli, 1942a, 1942b; Wenzel *et al.*, 1966).

Fleas are basically parasites of mammals, only about 6% of the 2000 known species and subspecies parasitizing birds. The hosts of fleas tend to be associated with distinctive homes, for these are generally necessary for the survival of the immature stages. Among mammals, rodents are by far the most common hosts, harboring 74% of all known forms, whereas fewer than 5% are known from bats. Indeed, bat fleas are seldom common: for example, in Malaysia 86 species of bats in 33 genera and seven families have been reported, but fleas are associated with only two species of pteropodids and three molossids (Marshall, 1980a); and in Panama, with 95 species of bats in 40 genera and seven families, fleas are associated with only one vespertilionid and three molossids (Tipton and Mendez, 1966). Ischnopsyllids typically have a variety of host relationships, some species being monoxenous and others polyxenous. Hurka (1963a, 1963b), examining the ischnopsyllids of Europe, noted how host preference sometimes varied from one geographical area to another, as is also true for other bat ectoparasites; and he believed that these differences could be accounted for by differences in the roosting conditions of the bats.

Two genera of lagomorph fleas, *Cediopsylla* and *Spilopsyllus*, contain species that are highly dependent on their host's hormonal cycle, these species therefore being narrowly host specific (Rothschild and Ford, 1973). Ischnopsyllids of the genus *Ischnopsyllus* are known to crowd onto female hosts before they depart from mixed-sex hibernacula for female-only maternity roosts (Hurka, 1963a). It is interesting to speculate whether this might also be associated with host hormonal levels.

It has frequently been suggested that knowledge of the distribution of certain ectoparasites, particularly lice, might be useful in elucidating the phylogeny of their hosts. Although it has been suggested that ectoparasites can throw little light on the systematics and phylogeny of bats (e.g., Patterson, 1957), both Maa (1968) and Wenzel *et al.* (1966) believed that increasing knowledge of the biology and taxonomy of both nycteribiids and streblids might provide useful evidence. Machado-Allison (1967) and Phillips *et al.* (1969) have presented some supporting data from the Acarina, which are generally less mobile than ectoparasitic insects.

3.3. Reasons

When considering reasons why a species of insect occurs on one host species but not on another closely related one, a number of questions become apparent. Is it possible for the parasite to reach the potential host? Is the climate associated with that host suitable for the insect? And, if the answers to the above questions are affirmative, are there morphological, behavioral, or physiological barriers to successful establishment on that host?

Physical isolation has often been assumed to be of great importance (e.g., Theodor, 1957), and it is certainly true that the chance of many ectoparasites moving from one host species to another are slim. But sometimes no physical isolation exists, or isolation is disrupted experimentally, and yet establishment still fails to occur. For example, experiments with nycteribiids indicate that when flies are transferred from a true host to a bat belonging to another suborder, they succumb to starvation, whereas flies introduced onto bats of the same suborder succumb to host predation (Marshall, 1971; Schulz, 1938; Stefanelli, 1942a). Overal (1980) believed that enhanced predation also prevented the establishment of the flightless streblid *Megistopoda aranea* (Coquillett) on unusual hosts of the same suborder.

As climate can limit the distribution of bats (see Chapter 4), so it can also limit the distribution of their ectoparasites (see Sections 2.3 and 2.4.5). For example, in a discussion of the altitudinal concordance between hosts and parasites, Wenzel and Tipton (1966) concluded that it was generally close in host-limited forms such as lice and bat flies, but less so in forms such as fleas, which spend much time away from their hosts. Streblids are largely tropical in distribution, being confined within the +10°C winter isotherms (Jobling, 1951). Of the 66 species of streblids recorded in Panama by Wenzel and Tipton (1966), 74% are completely or virtually restricted to the tropical zone, 14% to the subtropical, 6% to the lower montane, none to the upper montane, and 6% have a broad range from sea level to 1700 m.

Finally, ectoparasites may find even bats of a species closely related to their true hosts unsuitable for a variety of morphological, behavioral, and physiological reasons. Thus on unusual hosts their claws may be unsuitable for attachment, their body form, setae, and combs inappropriate for movement through the hair, and their mouthparts unsuited for feeding. Furthermore, the parasites may be less capable of avoiding host predation, and the composition of the host's blood may be nutritionally inadequate.

4. HOST LOCATION AND DISPERSAL

4.1. Locomotion

Bat ectoparasites move by walking, jumping, and flying. Nearly all species have unusually well-developed and flexible legs, which allow them both to walk rapidly and to cling successfully even to flying hosts; here well-developed tarsal claws play a major role. *Arixenia esau* can walk at up to 6 cm/sec on a roost substrate (Marshall, 1977a), and *Cimex lectularius* at over 2 cm/sec (Usinger, 1966). Polyctenidae have highly modified forelegs with weakly developed claws, which are held forwards and perhaps part the host's hair to aid locomotion and/or feeding. The two hind pairs of legs are large and backward pointing. These insects move rapidly through hair with a swimming motion, but off their hosts they are incapable of locomotion (Marshall, 1982). Both nycteribiids and streblids are generally capable of rapid movement forwards, backwards, or sideways on the host's body or roost. In the nycteribiid *Basilia hispida* males are more active than females, and away from the host old flies are more active than newly emerged ones, perhaps accounting for the latter's greater resistance to starvation (Section 2.4.2) (Marshall, 1970). Ischnopsyllids move mainly by walking and, compared with other fleas, are generally poor jumpers but good walkers. For example, *Ischnopsyllus octactenus* (Kolenati) can jump a maximum horizontal distance of 66 mm, compared with 330 mm in the human flea *Pulex irritans* L. (Rothschild, 1973; Rothschild *et al.*, 1975). As with polyctenids, most propulsion comes from the two hind pairs of legs, the forelegs being modified for parting the host's hair.

Although all bat ectoparasites belong to the insect subclass Pterygota, which are fundamentally winged, only certain streblids still retain functional wings (Table III); nonetheless, halteres (modified hind wings used for balance and allowing orientation in total darkness, typical of the Diptera) are found in nycteribiids as well as streblids. There are five species of streblids in one genus that are entirely wingless, 27 species in 10 genera have reduced wings and are flightless, and in the 18 species of *Ascodipteron* the females are caducous, shedding their wings upon reaching a suitable host. The remaining species, about 78% of the total, have functional wings and can fly. Flight can be quite powerful and rapid but is seldom sustained.

4.2. Initial Location and Transference between Hosts

Current knowledge of the responses of ectoparasites to a variety of external stimuli is fragmentary. Although eyes are present in four families (Table III), they are reduced and presumably play little part in host location. In both nycteribiids and streblids the eyes are variable. In streblids there is no correlation between eye reduction and the presence or absence of flight; some species with normal wings have no eyes, others with reduced wings have six to eight facets (the largest in the family), and the wingless *Paradyschiria* species has eyes of a single facet. In nycteribiids eye reduction is more pronounced, the maximum number of facets being four in those species of *Cyclopodia* that parasitize non-cave-dwelling megachiropterans.

Stimuli used by ectoparasites for host location include the warmth emanating from a host, the odor of a host, CO_2 respired by a host, and vibrations or air movements caused by a host. Arixeniids respond to odor (Marshall, 1977a), and cimicids to odor, CO_2, and warmth (Usinger, 1966). Both nycteribiids and streblids respond to odor and CO_2, and streblids to warmth (Hase, 1931a; Marshall, 1970; Overal, 1980; Schulz, 1938; Wenzel *et al.*, 1966). In general, fleas respond to all the stimuli listed above, but we know little specifically about ischnopsyllids, although *Ischnopsyllus* species are known to respond to air movements (Hurka, 1963a, 1963b).

Arixeniids live in close association with the molossid bat *Cheiromeles torquatus*, which roosts in cracks or caves, in hollow trees, or in the roofs of buildings, forming dense colonies that may contain many thousands of bats. Young arixeniids are born alive within the colony, so neither they nor older individuals should experience much difficulty in host location or transference. Transport to other roosts is due to host movements (Section 6.2). All stages of *Arixenia esau* are rapid walkers and respond to host odor (Marshall, 1977a, 1977b).

Polyctenids have never been found other than on the bodies of their hosts, and they appear incapable of locomotion elsewhere; thus they face no problems of initial host location, and transference between hosts must be by body contact between hosts. Consequently, these insects are confined to socially roosting bats, because contact between solitary roosting bats is presumably too infrequent (e.g., only at copulation) to permit survival (Marshall, 1982).

Cimicids are born and spend most of their lives in the roosts of their hosts, and so initial host contact should present few difficulties. These parasites are particularly associated with colonial bats and are carried from one colony to another by their hosts. Cimicids have seldom been recorded from flying hosts, but *Aphrania barys* Jordan and Rothschild, *Cimex adjunctus* (Barber), and *C. pilosellus* (Horvath) have been seen attached to bats captured in flight (Bradshaw and Ross, 1961; Usinger, 1966; Ubelaker and Kunz, 1971). *Cimex lectularius* is stimulated to seek a host by hunger; temperature affects the rate of digestion and therefore the frequency of host seeking, provides a stimulus for feeding, and influences the rate of movement. Host seeking occurs largely at night, initially by erratic searching, but in the final approach to the host both heat and odor (including CO_2) play a part (Marx, 1955; Mellanby, 1939). This also appears to be true for *Stricticimex antennatus* (Overal and Wingate, 1976). Like other members of the family, these cimicids possess a metathoracic scent apparatus—the odor of other bugs and that of their "nests" can be detected at distances greater than 75 cm, and favors aggregation (Levinson and Bar Ilan, 1971; Marx, 1955; Usinger, 1966).

Host location by nycteribiids is probably accomplished by random movements, and, when in proximity or in contact, by olfaction (Maa and Marshall, 1982; Marshall, 1970). Transference from one host to another generally occurs by body contact between hosts, and nycteribiids disperse to new roosts on the bodies of their hosts. Winged streblids can sense their hosts by olfaction over distances of at least a few meters (Wenzel *et al.*, 1966). The flightless species *Megistopoda aranea* parasitizes *Artibeus jamaicensis* (Phyllostomidae), which at roost hang close together, thus facilitating transference between individuals. When removed from the vicinity of its host, this fly waves its legs in the air, presumably to detect a host by olfaction; in the laboratory it can discriminate between different bat species, as can winged *Trichobius* species (Overal, 1980; W. L. Overal, personal communication).

Few observations have been made on host seeking by ischnopsyllids. Adult *Ischnopsyllus elongatus* (Curtis) are negatively geotactic and strongly positively anemotactic, waving the forelegs to receive stimuli (Hurka, 1963a, 1963b). A. M. Hutson (personal communication) made an interesting observation on *I. simplex* in a cave roost containing about 40 *Myotis nattereri* (Vespertilionidae). During the breeding season about five young bats fell 6 m to the floor each day, and these were picked up by adults each evening. On the floor the young bats each passively collected about five newly emerged fleas, and thus over the 3-week maternity period about 500 fleas would be carried up to the roost in this fashion. Because there was no indication that *I. simplex* attempted to climb up the walls, this may be the normal way in which these fleas reach their hosts.

Phoresy, a transitory association between two species of animals that involves transportation but not feeding, is known for one ischnopsyllid. *Lagaropsylla turba* Smit has been recorded from the arixeniid *Arixenia esau*, both insects parasitizing the molossid *Cheiromeles torquatus* (Medway, 1958). This association is not accidental, for 69% of 201 *A. esau* examined by Marshall (1977a) carried fleas, with up to 41 fleas per arixeniid. The association probably allows the much smaller flea to contact a host more readily, for *A. esau*, although flightless, is an active walker. This may be of value either to newly emerged fleas in the guano beneath the roost or to mature adult fleas that have spent the night in the empty roost and are seeking breakfast.

5. BEHAVIOR ON OR NEAR THE HOST

5.1. Introduction

Ectoparasites show a distinct preference for certain sites on their hosts; this apparently arises largely from an attempt to reduce the likelihood of predation by the host or dislodgement due to other host activities such as flight. Thus, where grooming occurs on resting bats, insects may seek those areas the host finds most difficult to reach, and, if grooming activity is related to total parasite burden, selective pressures may result in closely related parasites selecting different sites. When a host gets ready to fly, parasites may either leave it or move to a site where dislodgement is least likely. Repeated bites by blood-sucking ectoparasites are known to result in the development of acquired resistance in certain hosts, and this resistance is sometimes site specific. Although this has never been observed in bats, if it occurs it might explain aspects of site specificity. Other factors that may be important in site selection are the structure and length of the hair, and the skin structure which may affect ease of feeding. Ectoparasites may also favor certain sites on the host roost substrate, perhaps clustering in cracks where they are safe from dislodgement and where humidity is high.

5.2. Patterns

In roosts of *Cheiromeles torquatus* the arixeniid *Arixenia esau* walks rapidly, responding negatively to light and gravity and tending to cluster in crevices or below stones, perhaps due to an aggregating pheromone. When on its host, this large insect walks actively over the hairless body from anus to face, favoring no particular site and apparently not disturbing the host. *Xeniaria jacobsoni* (Burr) has frequently been collected from the same roosts as *A. esau*, but little is known of its behavior (Marshall, 1977a).

Cimicids live mostly in the roost and tend to inhabit cracks or crevices, preferring rough dry surfaces and being negatively phototactic and geotactic, and positively thigmotactic. They particularly favor old harborages and places that smell of other bugs (Johnson, 1942; Overal and Wingate, 1976). When feeding they can visit any part of a host's body, although furred areas are generally avoided. If disturbed by touch or air movements, *Stricticimex antennatus* may fall from the host bat, thus avoiding predation or being carried from the roost when the host leaves (Overal and Wingate, 1976).

Polyctenids appear to wander widely through the pelage of their hosts. *Eoctenes spasmae,* whose host *Megaderma spasma* (Megadermatidae) has long silky hair, moves actively throughout the pelage and never attaches for prolonged periods, but it favors the long hair in the hollow between the shoulder blades when the host is at rest, an area difficult for the host to groom (Marshall, 1982). Ischnopsyllids are also active walkers and probably do not favor any particular site, but data are lacking.

Nycteribiids can be roughly divided into two groups, large species (such as *Cyclopodia* and *Penicillidia* species) which live on the surface of the pelage, and small species (such as *Nycteribia* species) which live within the pelage. Unlike streblids, no nycteribiids favor membranous areas. On roosting hosts large species congregate in areas protected from the host's grooming, and where the hair is comparatively upright, not sleek: between the shoulder blades, below the chin, and in the axilla. On flying hosts they group around the tail, usually above, on the back, where perhaps the skin is not continually creased through flight. They can move rapidly over the surface of the hair, forwards, backwards, or sideways, generally facing forwards and rapidly burrowing into the hair if disturbed. Small species move rapidly within the pelage with a swimming motion reminiscent of polyctenids or ischnopsyllids, and there is no indication of site preference (Leong and Marshall, 1968; Marshall, 1970; Rodhain and Bequaert, 1916).

Many species of streblids occupy specific sites on their host's body, and these sites may change with the season and with the roosting habits of their hosts. For example, among the *Trichobius* species of North America, some prefer the pelage and others the membranous areas of their hosts, these sites perhaps changing seasonally (Section 2.3) (Bradshaw and Ross, 1961; Kunz, 1976; Ross, 1961; Wenzel *et al.*, 1966). Streblids are generally very mobile on their hosts, running with equal facility in all directions as do nycteribiids. Whereas flightless species, such as the hair-dwelling *Megistopoda aranea,* seldom leave their hosts (Overal, 1980), some of the winged and membrane-dwelling species feed less frequently and leave their hosts more often. Thus the less specialized winged *Brachytarsina amboinensis* (Rondani), *Joblingia schmidti* Dybas and Wenzel, and *Trichobius yunkeri* Wenzel, Tipton, and Kiewlicz may be found in large swarms in the roost itself (Wenzel *et al.*, 1966; Marshall, unpublished).

The streblid genus *Ascodipteron,* which contains 18 species in the Old World tropics, exhibits "neosomy" (Audy *et al.*, 1972). Upon emergence from the puparium both sexes resemble normal winged streblids (except for the greatly enlarged proboscis of the female), and the male remains in this condition. However, the female, upon reaching a suitable host, sheds legs, wings, and usually halteres and burrows subdermally, the abdomen then swelling to invaginate the head and thorax. The site chosen for embedding is species specific: for example, at the base of the ears, on the wings, forearms, wing–body margin, or tail membrane, or in the urogenital region (Jobling, 1939; Marshall, 1980a; Muir, 1912; Stiller, 1976; Theodor and Moscona, 1954). To date, no detailed biological studies have been made on this interesting genus.

5.3. Ectoparasites and Host Health

Ectoparasitic insects may injure bats by causing irritation, with consequent loss of condition, by physical damage to the skin or hair, by blood loss, or by the injection of toxins or transmission of pathogenic organisms. Ectoparasite populations are generally low, no physical damage has ever been reported, and our knowledge of bat pathogens is negligible. However, the drain that irritation from ectoparasites may impose upon the daily energy budgets of bats (Burnett and August, 1981) needs to be examined. Presumably the more time and energy that individuals allocate to grooming means less time and energy available for other activities such as growth, reproduction, and fat storage. Rarely, high ectoparasite populations are encountered, but these are generally the result, and not the cause, of ill health and incapacity to groom, although such high populations may then hasten death (Section 2.4.1). In fact, one of the more remarkable aspects of examining living bats is the apparent equanimity with which they view their ectoparasites, some of which are equivalent to an object the size of a dinner plate scuttling about on a man!

6. POPULATION DYNAMICS

6.1. Introduction

Populations of ectoparasitic insects can be divided into those insects on the host's body (the corporeal or body population) and those elsewhere, usually in the home (the extracorporeal population). Thus population studies of polyctenids will be concerned solely with body populations, whereas those of ischnopsyllids must consider both the body population of adults and the extracorporeal population of immature stages and adults in the roost.

Body populations are defined by two parameters, the incidence rate (the percentage of hosts infested) and the infestation rate (the mean number of parasites per host examined). These two rates, together with the number of hosts examined, form a basic description of a body population upon any set of hosts (age, sex, habitat, etc.). Such populations generally show a contagious distribution, being neither regularly nor randomly dispersed, but with the majority of parasites on relatively few hosts (Southwood, 1978). The detailed analysis of the population dynamics of any insect with overlapping generations is difficult, the data being laborious to collect and difficult to interpret, and no such analysis has yet been attempted for any bat ectoparasite.

6.2. Patterns and Limits

Data on populations of arixeniids are sparse. In roosts of *Cheiromeles torquatus*, *Arixenia esau* can number many hundreds, even thousands of individuals, as roosts may contain tens of thousands of bats (Marshall, 1977a). Up to 12 *A. esau* have been removed from a single host taken at a roost (Nakata and Maa, 1974), although this may be the result of disturbance. On free-flying bats *A. esau* is rare: Nakata and Maa (1974) recorded four out of 31 *C. torquatus* netted in forest infested with a total of seven individuals, and Medway and Yong (1969) found four out of 18 *Tadarida johorensis* (a molossid that can share roosts with *C. torquatus*) netted in forest infested with five individuals.

The size of populations of cimicids is so variable that little can be said of general relevance. Rarely are they found on the bodies of their hosts, but populations in roosts may be very high. Overal and Wingate (1976) estimated that the total population of *Stricticimex antennatus* in a mine shaft, with an average population of 250 bats (150–900, depending upon the season), was 5500, or 22 cimicids per bat. Mitchell (1970) made some estimates of the numbers of both cimicids and adult ischnopsyllids in the guano below a cave roost of the molossid *Tadarida brasiliensis* in Texas, making his estimates in July, when insect populations would be expected to be high. Using capture–recapture methods for bats and adult cimicids, and quadrat sampling for all stages of cimicids and adult fleas, he estimated bat numbers at about 5000, the number of *Primicimex cavernis* Barber at about 7900 (with adults accounting for about 17%), and the number of the flea *Sternopsylla distincta* (Rothschild) at about 1,113,100. The relationship of these numbers in the guano to the total populations in the cave was unknown, but it is tempting to suggest that a high percentage of ischnopsyllids would be in the guano, whereas many cimicids would be on the walls or roof of the cave itself.

Only one population estimate has been made for polyctenids: for *Eoctenes spasmae* on the bat *Megaderma spasma* (Megadermatidae) in Malaysia (Marshall, 1982; Table V). Of the 27 hosts examined, 85% were infested, with no significant difference between males (10 out of 11 infested) and females (13 out of 16). The average number of polyctenids per bat was 13.7; the median was 9; females and nymphs were about as common, and both outnumbered males. There was no indication of seasonal variations in the population from this equatorial region.

In populations of bats infested with nycteribiids the majority of hosts are usually uninfested, and numbers average less than one fly per host and rarely exceed ten (Funakoshi, 1977; Hase, 1931a; Hurka, 1964; Maa and Marshall,

TABLE V
Frequency Distribution and Population Structure of the Polyctenid *Eoctenes spasmae* on *Megaderma spasma* (Megadermatidae) in Malaysia[a]

Number of hosts examined		27
Number infested		23
Percentage infested		85
Total number of polyctenids		370
Mean number of polyctenids per host		13.7
(maximum)		(43)
Male polyctenids: percentage (mean per host)		25 (3.4)
Female polyctenids: percentage (mean per host)		38 (5.2)
Nymphal polyctenids: percentage (mean per host)		37 (5.0)
Percentage of infested hosts with number of polyctenids indicated	1–10	48
	11–20	26
	21–30	9
	31–40	13
	41–50	4

[a]Data from Marshall, 1982.

1982; Marshall, 1971, unpublished; Table VI). In general, numbers per host appear to be correlated with the relative size of the parasite, the highest infestations being recorded for the smallest species. For example, *Basilia* and *Penicillidia* species (Table VI) are considerably larger than *Nycteribia* species: Hurka (1964) recorded that *N. kolenatii* infested 90% of 217 *Myotis daubentoni* (Vespertilionidae) examined in northern Europe, with an average of 4.7 flies per bat, ranging from 4.0 in winter to 6.5 in summer, with a maximum of 24 flies from a single host.

On hibernating bats nycteribiids may cease to deposit offspring if the temperatures are very low, or they may continue to do so at a greatly reduced rate (Section 2.3). Thus it is not surprising that in temperate regions infestations of bats are lowest in winter (Funakoshi, 1977; Hurka, 1964). Elsewhere, population fluctuations may be due to seasonal changes in host biology, rather than parasite biology. In equatorial Malaysia *Basilia hispida* shows two periods of declining infestation rates on the vespertilionid bats *Tylonycteris* species (Marshall, 1971). The first occurs in April–May, associated with the annual birth season of the bats when the numbers of potential hosts greatly increase, and the second in October–November during the period of copulation when these bats (particularly males) are more gregarious. At other times of the year males tend to roost alone, thus more readily being infested with newly emerged flies (the presence of a host bat triggers adult fly emergence; see Section 2.1), and also in this way avoiding mutual grooming by a conspecific bat; therefore males are more heavily infested than females (Table VII). Indeed, 60 out of 80 (75%) of all infestations of more than one fly occurred on male hosts.

In nycteribiids an unusually high proportion of infestations of more than a single parasite consists of one male and one female together (Hurka, 1964; Maa and Marshall, 1982; Marshall, 1971). This situation occurs because flies copulate at the time of offspring deposition, a time when much host transference must occur. With streblids, which are generally more mobile and transfer more readily, this combination is not unusually common. Even with the flightless *Megistopoda aranea*, Overal (1980) recorded that of 91 infestations of more than

TABLE VI
Frequency Distribution of Certain Nycteribiidae and Streblidae on the Bodies of Their Hosts[a]

		Nycteribiidae			Streblidae	
Parasite		*Basilia hispida*	*Penicillidia buxtoni*	*Penicillidia jenynsii*	*Megistopoda aranea*	*Trichobius corynorhini*
Host		*Tylonycteris pachypus*	*Miniopterus australis*	*Miniopterus schreibersii*	*Artibeus jamaicensis*	*Plecotus townsendii*
Number of hosts examined		707	268	3555	412	308
Number infested		246	125	902	224	237
Percentage infested		35	47	25	54	77
Total number of parasites		370	213	1261	347	809
Mean number of parasites per host (maximum)		0.5 (8)	0.8 (9)	0.4 (5)	0.8 (6)	2.6 (12)
Percentage of infested hosts with number of parasites indicated	1	67	57	72	59	22
	2	24	29	20	30	20
	3	5	7	5	8	21
	4	2	6	2	1	13
	5	<1	—	<1	—	6
	6	<1	<1	—	1	7
	7	<1	—	—	—	5
	8	<1	—	—	—	4
	9	—	<1	—	—	1
	10	—	—	—	—	<1
	11	—	—	—	—	<1
	12	—	—	—	—	<1

[a] *Basilia hispida* in Malaysia (from Marshall, 1971); *Penicillidia buxtoni* Scott in the New Hebrides (from Maa and Marshall, 1982); *Penicillidia jenynsii* in Japan (from Funakoshi, 1977); *Megistopoda aranea* in Panama (from Overal, 1980); *Trichobius corynorhini* on hibernating hosts in the United States (from Kunz, 1976, personal communication).

TABLE VII
Variations in the Abundance of Certain Nycteribiidae and Streblidae on Different Sexes of Their Hosts[a]

	Nycteribiidae				Streblidae	
Parasite	*Basilia hispida*		*Penicillidia buxtoni*		*Trichobius corynorhini*	
Host	*Tylonycteris pachypus*		*Miniopterus australis*		*Plecotus townsendii*	
	♂	♀	♂	♀	♂	♀
Number of hosts examined	296	411	145	123	92	216
Percentage infested	45	27	41	54	58	85
Mean number of parasites per host	0.8	0.3	0.7	0.9	1.3	3.2

[a]Same sources as Table VI.

one fly per bat, 68 (75%) consisted of two flies, but only 27 (30%) of two flies of the opposite sex. Equivalent figures for the nycteribiid *Basilia hispida* are 80, 58 (73%), and 39 (49%) (Marshall, 1971).

The Nycteribiidae as a family have a broader distribution than the Streblidae and are better able to maintain themselves on hosts that roost in small groups in relatively restricted sites. Streblids generally prefer hosts that roost in large, well-established colonies, and they usually avoid those in exposed situations. Generally, they seem to occur in greater numbers on their hosts than do nycteribiids: in tropical regions two to five streblids per host is not uncommon (Maa and Marshall, 1982; Marshall, 1976, unpublished; Wenzel *et al.*, 1966). Flightless species have possibly somewhat smaller populations than fully winged species, although data are sparse (e.g., *Megistopoda aranea* in Table VI). Among *Ascodipteron* species nothing is known about populations of winged males, but the neosomic females (Section 5.2) usually number only one per infested host, with a maximum of five recorded (Maa, 1965; Muir, 1912; Stiller, 1976; Theodor and Moscona, 1954).

Most streblids are tropical, and only two species are known to overwinter on hibernating hosts (Section 2.3). Kunz (1976), in his study of *Trichobius corynorhini* on *Plecotus townsendii*, noted that in winter female hosts were significantly more heavily infested than males (Table VII), and males that roosted in clusters were more heavily infested than solitary males, but both solitary and clustered females were equally heavily infested. Kunz suggested that two factors may account for higher populations on females: males are more active and flies tend to leave active hosts, and females may produce a host factor that enhances host-finding success. The favoring of females is beneficial to the fly, because in summer males tend to roost singly, whereas females form maternity colonies, and there is some evidence that a host density of not less than 50 bats is necessary to maintain a population of *T. corynorhini* (Beck, 1969).

Details about population size of ischnopsyllids are sparse. A single host species may harbor a number of flea species, all of which are present throughout the year, or else different species may be abundant at different seasons, thereby possibly avoiding competition. For example, in Europe *Ischnopsyllus* species are "summer fleas," breeding only on active hosts and therefore virtually nonbreeding in winter; and *Nycteridopsylla* species are "winter fleas," confined to the host's hibernacula (Section 2.3). Observed differences in the infestation levels of two *Plecotus* species with *N. pentactena* (Kolenati) (Table VIII) are due to differences in their hibernacula; and the decline in numbers of *I. hexactenus* between spring/summer and winter on *P. austriacus* at a time when the numbers of *N. pentactena* increase may be due to competition. Certainly, differences in roosting sites may affect ischnopsyllids: for example, Hurka (1963b) found that *Rhinolophopsylla unipectinata* had an average infestation rate of more than one flea per host in colonies of *Rhinolophus* (Rhinolophidae) in buildings, but only a tenth of that in colonies in caves.

TABLE VIII
Variations in the Abundance of Two Species of Ischnopsyllidae in the Spring/ Summer Roosts and Winter Roosts of Their Hosts in Czechoslovakia[a]

Host species	*Plecotus auritus*		*Plecotus austriacus*	
Host roosts	Spring/summer	Winter	Spring/summer	Winter
Number of hosts examined	66	126	78	243
Ischnopsyllus hexactenus				
Total number of fleas	66	144	92	99
Mean number of fleas per host	1.0	1.1	1.2	0.4
Nycteridopsylla pentactena				
Total number of fleas	0	15	0	298
Mean number of fleas per host	0	0.1	0	1.2
Percentage of fleas: *I. hexactenus*	100	91	100	25
N. pentactena	0	9	0	75

[a]Data from Hurka and Hurka, 1964.

6.3. Age Structure

Detailed studies on the age structure of a population of bat ectoparasites have never been made, but some information is available about the percentage of nymphs and of adults in populations of certain of the Hemimetabola. Of 459 individuals of the arixeniid *Arixenia esau* removed from roosts by Marshall (1977a), 34% were first- to third-instar nymphs, and 66% fourth-instar nymphs and adults. Overal and Wingate (1976) collected 2733 cimicids, *Stricticimex antennatus*, of which 83% were first- to fifth-instar nymphs (14, 11, 14, 20, and 24%, respectively) and 17% adults; adults also accounted for about 17% of a population of *Primicimex cavernis* examined by Mitchell (1970). Equivalent values for 1646 *Cimex lectularius* removed from a human dwelling were 72% nymphs (21, 17, 15, 10, and 9%) and 28% adults (Mellanby, 1939). Among polyctenids, Ueshima (1972) noted that of 148 individuals of *Hesperoctenes fumarius* (Westwood), 34% were nymphs and 66% adults. Equivalent values for *Eoctenes spasmae* are 24% nymphs and 76% adults for 102 individuals in museum collections (Maa, 1964), and 37 and 63% for 370 individuals removed directly from 27 hosts (Marshall, 1982). Thus in an active population of the viviparous arixeniids and polyctenids perhaps two-thirds are adults, whereas in the oviparous cimicids only one-quarter are adults. The adults of certain ectoparasitic members of the Diptera and the Siphonaptera have been aged by examination of their reproductive organs (e.g., Klein, 1966; Saunders, 1964), but this has not yet been done with bat ectoparasites.

6.4. Sex Ratio

In a survey of published data Marshall (1981b) recorded the sex ratio of 379 collections, each consisting of more than 100 individuals, of 250 different species of ectoparasites belonging to seven orders of insect. In general, ectoparasites emerge as adults in equal numbers of each sex, but thereafter the ratio is usually imbalanced, females predominating. Out of 359 collections from the host's body or home, the sex ratio of 30% showed no significant departure from unity, 63% contained significantly more females, and only 7% significantly more males. The data for bat ectoparasites are sparse (Table IX), but there is no reason to believe that they differ fundamentally from the general pattern. However, compared with other groups, a surprising number of streblid collections showed a significant preponderance of males. All 19 of these collections were made by Wenzel (1976, personal communication), and he suggested that this might be an artifact of collecting methods; each collection came from hosts' bodies, whereas females might be more abundant in the roost. The usual preponderance of females occurs, presumably, because males are generally shorter lived.

TABLE IX
Sex Ratio of Insects Ectoparasitic on Bats: The Numbers of Collections of Each Family Recorded and Their Relative Balance[a]

	Number of species recorded	Collections recorded			
		Sexes equal	Significantly more males	Significantly more females	Total
At emergence					
Cimicidae	1	1			1
Nycteribiidae	1	1			1
Ischnopsyllidae	1	1			1
From body or roost					
Arixeniidae	1			1	1
Cimicidae	1		1		1
Polyctenidae	1			1	1
Nycteribiidae	13	12		4	16
Streblidae	46	27	19	1	47
Ischnopsyllidae	9	1		8	9

[a]Data from Marshall, 1981b.

6.5. Changes in Abundance with Space and Time

As noted previously (Section 3), all ectoparasites are to some extent host limited, and even those species that possess a number of true hosts may reach greater numbers in association with one host species than with another. Furthermore, within a given host population parasite abundance may differ on hosts of different age, sex, reproductive condition, and health, it may differ with differences in host behavior and abundance, with differences in habitat and climate, and thus with season and time. The most critical factor seems to be the size of the host roosting population. Thus it is clear that if the ecology of an ectoparasite is to be thoroughly understood, the roosting ecology and behavior of its host must be examined in detail (see Chapter 1).

7. CONCLUSIONS

Obviously, current knowledge of the ecology of bat ectoparasites is fragmentary. A handful of species have been studied in some detail, but whole families, such as the ischnopsyllids, and many fascinating groups, such as streblids of the genus *Ascodipteron*, have virtually never been studied alive. Although it is apparent that host and climate affect the population size of ectoparasites, the reasons why numbers are often so low remain unknown. Furthermore, reasons for variations in numbers from one true host species to another, from one individual to another, and from one region to another are poorly understood.

Bat ectoparasites apparently cause little, if any, harm to their hosts, although differences in infestations between host individuals in a population may possibly affect their relative fitness. None are of significance to the welfare of man. Furthermore, as long as their hosts exist, there is little danger of insect extinction: I know of only one proposal for an "ectoparasite sanctuary" (Marshall, 1977b) and it is in fact not biologically valid, for it involved the conservation of only those insects that had become irreversibly separated from their hosts! But these remarkable insects are of great interest in their own right, and bat ecologists could well find that a little time spent studying them (see Appendix) would add a new and fascinating dimension to the study of the bats themselves.

ACKNOWLEDGMENTS. I would like to express my gratitude to all those who have helped me study bat ectoparasites and their hosts over the past years; particularly to the Earl of Cranbrook, who introduced me to the marvelous world of tropical bats, and to T. C. Maa for his unfailing interest in the ectoparasites I collected. I am also most grateful to A. M. Hutson, W. L. Overal, and R. L. Wenzel for allowing me to use unpublished material in this chapter, and to T. H. Kunz for his careful criticism of the manuscript.

APPENDIX: METHODS FOR A SIMPLE STUDY OF BAT ECTOPARASITES

Bat ecologists can, with little extra effort, collect data of interest on the ectoparasites of the bats they are studying. The following is a simple guide.

1. At all stages great care must be taken to avoid contamination of one host by ectoparasites from another.
2. Each host should be carefully searched and all negative hosts should be recorded. Insects from positive hosts should be placed in vials containing 80% ethyl alcohol, using a separate vial for each bat. Insects may also be collected from the roost.
3. Each vial should have clearly associated with it (on a label placed *inside*) as much of the following information as possible: (a) the host species, sex, age, reproductive condition, state of health, and collection number; (b) the site where found (part of body or of roost); (c) the locality and date of collection; and (d) the name of the collector and the ectoparasite collection number.

4. The vials from each study should be clearly associated with data on precautions taken against contamination and the number of hosts examined (including negative ones), with species, sex, age, reproductive condition, and state of health. Later, identification and sexing of the insects will provide the remaining data for a discussion of ectoparasite populations.

For further discussions on ecological techniques for the study of ectoparasites and on methods of maintaining laboratory populations, see Marshall (1981a).

9. REFERENCES

Anciaux de Faveaux, M. F. 1965. Les parasites des Chiroptères. Rôle épidémiologique chez les animaux et l'homme au Katanga. Ann. Parasitol. Hum. Comp., **40**:21–38.

Aschner, M. 1931. Die Bakterienflora der Pupiparen (Diptera). Z. Morphol. Ökol. Tiere, **20**:368–442.

Ashford, R. W. 1970. Observations on the biology of *Hemimerus talpoides* (Insecta : Dermaptera). J. Zool., **162**:413–418.

Audy, J. R., F. J. Radovsky, and P. H. Vercammen-Grandjean. 1972. Neosomy: radical intrastadial metamorphosis associated with arthropod symbioses. J. Med. Entomol., **9**:487–497.

Balcells Rocamora, E. 1956. Datos para el estudio de la fauna Pupipara de los quirópteros en España. Speleon, Oviedo, **6**:287–312.

Bearup, A. J., and J. J. Lawrence. 1947. A search for the vector of *Plasmodium pteropi* Breinl. Proc. Linn. Soc. New South Wales, **71**:197–200.

Beck, A. J. 1969. New records and notes on streblid flies in California (Diptera: Pupipara). Wasmann J. Biol., **27**:115–119.

Benjamin, R. K. 1971. Introduction and supplement to Roland Thaxter's contribution towards a monograph of the Laboulbeniaceae. Bibl. Mycol., **30**:1–155.

Blackwell, M. 1980a. Incidence, host specificity, distribution and morphological variation in *Arthrorhynchus nycteribiae* and *A. eucampsipodae* (Laboulbeniomycetes). Mycologia, **72**:143–158.

Blackwell, M. 1980b. Developmental morphology and taxonomic characters of *Arthrorhynchus nycteribiae* and *A. eucampsipodae* (Laboulbeniomycetes). Mycologia, **72**:159–168.

Bradshaw, G. V. R., and A. Ross. 1961. Ectoparasites of Arizona bats. J. Ariz. Acad. Sci., **1**:109–112.

Burnett, C. D., and P. V. August. 1981. Time and energy budgets of dayroosting in a maternity colony of *Myotis lucifugus*. J. Mammal., **62**:758–766.

Cockbain, A. J., and A. C. Hastie. 1961. Susceptibility of the bedbug, *Cimex lectularius* Linnaeus, to *Aspergillus flavus* Link. J. Insect Pathol., **3**:95–97.

Davis, M. T. 1964. Studies of the reproductive physiology of Cimicidae (Hemiptera). I. Fecundation and egg maturation. J. Insect Physiol., **10**:947–963.

Domrow, R. 1961. New and little known Laelaptidae, Trombiculidae and Listrophoridae (Acarina) from Australia. Proc. Linn. Soc. New South Wales, **86**:60–95.

Funakoshi, K. 1977. Ecological studies on the bat fly, *Penicillidia jenynsii* (Diptera : Nycteribiidae), infesting the Japanese long-fingered bat, with special reference to their adaptations to the life cycle of their hosts (in Japanese). Jap. J. Ecol., **27**:125–140.

Hagan, H. R. 1951. Embryology of the viviparous insects. Ronald Press, New York, 472 pp.

Hase, A. 1931a. Uber die Lebensgewohnheiten einer Fledermausfliege in Venezuela, *Basilia bellardii* Rondani (fam. Nycteribiidae—Diptera Pupipara). Z. Parasitenkd., **3**:220–257.

Hase, A. 1931b. Uber die Eier und über die Larven des Fledermausflohes *Myodopsylla*. Z. Parasitenkd., **3**:258–263.

Holloway, B. A. 1976. A new bat fly family from New Zealand (Diptera : Mystacinobiidae). N. Z. J. Zool., **3**:279–301.

Hurka, K. 1963a. Bat fleas (Aphaniptera, Ischnopsyllidae) of Czechoslovakia. Contribution to the distribution, morphology, bionomy, ecology and systematics. Part I. Subgenus *Ischnopsyllus* Westw. Acta Fauna Entomol. Mus. Nat. Pragae, **9**:57–120.

Hurka, K. 1963b. Bat fleas (Aphaniptera, Ischnopsyllidae) of Czechoslovakia. II. Subgenus *Hexactenopsylla* Oud., genus *Rhinolophopsylla* Oud., subgenus *Nycteridopsylla* Oud., subgenus *Dinycteropsylla* Ioff. Acta Univ. Carol. Biol., **1963**:1–73.

Hurka, K. 1964. Distribution, bionomy and ecology of the European bat flies with special regard to the Czechoslovak fauna (Dip., Nycteribiidae). Acta Univ. Carol. Biol., **1964**:167–234.

Hurka, K. 1970. Systematic, faunal and bionomic notes on the European and Asiatic flea species of the family Ischnopsyllidae (Aphaniptera). Acta Univ. Carol. Biol., **1969**:11–26.

Hurka, K., and J. Doskočil. 1961. Influence of relative atmospheric humidity on the survival of bat-fleas (Aphaniptera, Ischnopsyllidae). Čas. čs. Spol. Ent., **58**:111–116.

Hurka, K., and L. Hurka. 1964. Zum flohbefall der Beiden Europaischen *Plecotus* - arten: *auritus* L. und *austriacus* Fisch in der Tschechoslowakei (Aphaniptera: Ischnopsyllidae). Acta Soc. Zool. Bohem., **28**:155–163.

Hurka, L. 1965. Experimentaluntersuchungen zum befall der Wirte mit dem floh *Nycteridopsylla pentactena* (Kolenati) (Aphaniptera: Ischnopsyllidae). Acta Soc. Zool. Bohem., **29**:239–243.

Jenkins, D. W. 1964. Pathogens, parasites and predators of medically important arthropods. Annotated list and bibliography. Bull. W.H.O. **30**, Suppl., 150 pp.

Jobling, B. 1939. On the African Streblidae (Diptera, Acalypterae) including the morphology of the genus *Ascodipteron* Adens., and a description of a new species. Parasitology, **31**:147–165.

Jobling, B. 1951. A record of the Streblidae from the Philippines and other Pacific islands, including the morphology of the abdomen, host–parasite relationship and geographical distribution, and with descriptions of five new species (Diptera). Trans. R. Entomol. Soc. London, **102**:211–246.

Johnson, C. G. 1942. The ecology of the bed-bug, *Cimex lectularius* L., in Britain. J. Hyg., **41**:345–361.

Joyeux, C. 1913. Biologie de *Cimex boueti*. Arch Parasitol., **1913**:140–146.

Klein, J. M. 1966. Données écologiques et biologiques sur *Synopsyllus fonquerniei* Wagner et Roubaud, 1932 (Siphonaptera), puce du rat péridomestique, dans la région de Tananarive. Cah. ORSTOM Ser. Entomol. Med. Paris, **4**:3–29.

Kunz, T. H. 1976. Observations on the winter ecology of the bat fly *Trichobius corynorhini* Cockerell (Diptera: Streblidae). J. Med. Entomol., **12**:631–636.

Labunets, N. F., and O. P. Bogdanov. 1959. Winter species of fleas from bats in Uzbekistan (in Russian). Zool. Zh., **38**:221–228.

Leong, M. C., and A. G. Marshall. 1968. The breeding biology of the bat-fly *Eucampsipoda sundaicum* Theodor, 1955 (Diptera: Nycteribiidae). Malay. Nat. J., **21**:171–180.

Levinson, H. Z., and A. R. Bar Ilan. 1971. Assembling and alerting scents produced by the bedbug *Cimex lectularius* L. Experientia, **27**:102–103.

Maa, T. C. 1964. A review of the Old World Polyctenidae (Hemiptera: Cimicoidea). Pac. Insects, **6**:494–516.

Maa, T. C. 1965. Ascodipterinae of Africa (Diptera: Streblidae). J. Med. Entomol., **1**:311–326.

Maa, T. C. 1968. On Diptera Pupipara from Africa, part 1. J. Med. Entomol., **5**:238–251.

Maa, T. C. 1971a. Revision of the Australian batflies (Diptera: Streblidae and Nycteribiidae). Pac. Insects Monogr., **28**:1–118.

Maa, T. C. 1971b. Review of the Streblidae parasitic on megachiropteran bats. Pac. Insects Monogr., **28**:213–243.

Maa, T. C., and A. G. Marshall. 1982. Diptera Pupipara of the New Hebrides: Taxonomy, zoogeography, host associations and ecology. Q. J. Taiwan Mus., **34.**

MacArthur, R. H., and E. O. Wilson. 1967. The theory of island biogeography. Princeton University Press, Princeton, 203 pp.

Machado-Allison, C. E. 1967. The systematic position of the bats *Desmodus* and *Chilonycteris*, based on host–parasite relationships (Mammalia; Chiroptera). Proc. Biol. Soc. Wash., **80**:223–226.

Marshall, A. G. 1970. The life cycle of *Basilia hispida* Theodor 1967 (Diptera: Nycteribiidae) in Malaysia. Parasitology, **61**:1–18.

Marshall, A. G. 1971. The ecology of *Basilia hispida* (Diptera: Nycteribiidae) in Malaysia. J. Anim. Ecol., **40**:141–154.

Marshall, A. G. 1976. Host specificity amongst arthropods ectoparasitic upon mammals and birds in the New Hebrides. Ecol. Entomol., **1**:189–199.

Marshall, A. G. 1977a. Interrelationships between *Arixenia esau* (Dermaptera) and molossid bats and their ectoparasites in Malaysia. Ecol. Entomol., **2**:285–291.

Marshall, A. G. 1977b. The earwigs (Insecta: Dermaptera) of Niah Caves, Sarawak. Sarawak Mus. J., **25**:205–209.

Marshall, A. G. 1980a. The comparative ecology of insects ectoparasitic upon bats in West Malaysia. Pp. 135–142, *in* Proceedings of the Fifth International Bat Research Conference. (D. E. Wilson and A. L. Gardner, eds.). Texas Tech Press, Lubbock, 434 pp.

Marshall, A. G. 1980b. The function of combs in ectoparasitic insects. Pp. 79–87, *in* Fleas—Proceedings of the International Conference on Fleas, June 1977. (R. Traub, ed.). Balkema, Rotterdam, 420 pp.

Marshall, A. G. 1981a. The ecology of ectoparasitic insects. Academic Press, London, 459 pp.

Marshall, A. G. 1981b. The sex ratio in ectoparasitic insects. Ecol. Entomol., **6**:155–174.

Marshall, A. G. 1982. The ecology of *Eoctenes spasmae* (Hemiptera: Polyctenidae) in Malaysia. Biotropica, **14**:50–55.

Marx, R. 1955. Uber die Wirtsfindung und die Bedeutung des artespezifischen Duftstoffes bei *Cimex lectularius* Linné. Z. Parasitenkd., **17**:41–72.

Medway, Lord. 1958. On the habit of *Arixenia esau* Jordan (Dermaptera). Proc. R. Entomol. Soc. London A, **33**:191–195.

Medway, Lord, and G. C. Yong. 1969. Parasitic earwigs (*Arixenia esau*) on the wrinkle-lipped bat. Malay. Nat. J., **23**:33.

Mellanby, K. 1939. The physiology and activity of the bed-bug (*Cimex lectularius* L.) in a natural infestation. Parasitology, **31**:200–211.

Mitchell, R. W. 1970. Total number and density estimates of some species of cavernicoles inhabiting Fern Cave, Texas. Ann. Speleol., **25**:73–90.

Muir, F. 1912. Two new species of *Ascodipteron*. Bull. Mus. Comp. Zool. Harvard, **54**:351–366.

Nakata, S., and T. C. Maa. 1974. A review of the parasitic earwigs (Dermaptera: Arixeniina; Hemimerina). Pac. Insects, **16**:307–374.

Omori, N. 1941. Comparative studies on the ecology and physiology of common and tropical bed bugs, with special references to the reactions to temperature and moisture. J. Med. Assoc. Formosa, **40**:555–729.

Overal, W. L. 1980. Host-relations of the batfly *Megistopoda aranea* (Diptera: Streblidae) in Panama. Univ. Kan. Sci. Bull., **52**:1–20.

Overal, W. L. and L. R. Wingate. 1976. The biology of the batbug *Stricticimex antennatus* (Hemiptera: Cimicidae) in South Africa. Ann. Natal Mus., **22**:821–828.

Patterson, B. 1957. Mammalian phylogeny. Pp. 15–49, *in* First Symposium on host specificity among parasites of vertebrates. (J. G. Baer, ed.). Institut de Zoologie, Université de Neuchâtel, Neuchâtel, 324 pp.

Phillips, C. J., J. K. Jones, Jr., and F. J. Radovsky. 1969. Macronyssid mites (*Radfordiella*) in the oral mucosa of long-nosed bats (*Leptonycteris*): Notes on occurrence and associated pathology. Science, **165**:1368–1369.

Popham, E. J. 1962. The anatomy related to the feeding habits of *Arixenia* and *Hemimerus* (Dermaptera). Proc. Zool. Soc. London, **139**:429–450.

Price, P. W. 1980. Evolutionary biology of parasites. Princeton University Press, Princeton, 237 pp.

Reisen, W. K., M. L. Kennedy, and N. T. Reisen. 1976. Winter ecology of ectoparasites collected from hibernating *Myotis velifer* (Allen) in southwestern Oklahoma (Chiroptera: Vespertilionidae). J. Parasitol., **62**:628–635.

Rodhain, J., and J. Bequaert. 1916. Observations sur la biologie de *Cyclopodia greeffi* Karsch (Dipt.), nyctéribiide parasite d'une chauve-souris Congolaise. Bull. Soc. Zool. Fr., **40**:248–262.

Roer, H. 1969. Uber vorkommen und lebensweise von *Cimex lectularius* und *Cimex pipistrelli* (Heteroptera, Cimicidae) in Fledermaus-quartieren. Bonn. Zool. Beitr., **20**:355–359.

Ross, A. 1960. Notes on *Trichobius corynorhini* on hibernating bats (Diptera: Streblidae). Wasmann J. Biol., **18**:271–272.

Ross, A. 1961. Biological studies on bat ectoparasites of the genus *Trichobius* (Diptera: Streblidae) in North America, north of Mexico. Wasmann J. Biol., **19**:229–246.

Rothschild, M. 1973. The flying leap of the flea. Sci. Am., **229**:92–100.

Rothschild, M., and B. Ford. 1973. Factors influencing the breeding of the rabbit flea (*Spilopsyllus cuniculi*): A spring-time accelerator and a kairomone in nestling rabbit urine (with notes on *Cediopsylla simplex,* another "hormone bound" species). J. Zool., **170**:87–137.

Rothschild, M., J. Schlein, K. Parker, C. Neville, and S. Sternberg. 1975. The jumping mechanism of *Xenopsylla cheopis*. III. Execution of the jump and activity. Phil. Trans. R. Soc. London Ser. B, **271**:499–515.

Ryberg, O. 1947. Studies on bats and bat parasites. Bokforlaget Svensk Natur, Stockholm, 330 pp.

Ryckman, R. E. 1956. Parasitic and some nonparasitic arthropods from bat caves in Texas and Mexico. Am. Midl. Nat., **56**:186–190.

Saunders, D. S. 1964. Age changes in the ovaries of the sheep kid, *Melophagus ovinus* (L.) (Diptera: Hippoboscidae). Proc. R. Entomol. Soc. London A, **39**:68–72.

Schulz, H. 1938. Uber fortpflanzung und vorkommen von Fledermausfliegen (Fam. Nycteribiidae—Diptera Pupipara). Z. Parasitenkd., **10**:297–328.

Southwood, T. R. E. 1978. Ecological methods, with particular reference to the study of insect populatons. 2nd ed. Chapman and Hall, London, 524 pp.

Stefanelli, A. 1942a. Affinita sistematiche dei chirotteri e parassitismo dei Nycteribiidae, Diptera Pupipara. II. I parassiti. Riv. Parassitol., **6**:61–86.

Stefanelli, A. 1942b. La specificita ectoparassitaria studiata sui Ditteri Pupipari parassitidei chirotteri, con speciale riferimento alla *Nycteribosca africana* (Walk.). Med. Biol. Roma, **1**:55–77.

Stiller, D. 1976. Report on a new bat fly of the genus *Ascodipteron* Adensamer, 1896 (Diptera: Streblidae) parasitic on *Hipposideros diadema* (Geoffroy) in Malaysia. J. Med. Entomol., **13**:374–375.

Tamsitt, J. R., and D. Valdivieso. 1970. Los murcielagos y la salud publica. Estudio con especial referencia a Puerto Rico. Boln. Ofic. Sanitaria Panam., **69**:122–140.

Theodor, O. 1957. Parasitic adaptation and host–parasite specificity in the pupiparous Diptera. Pp.

50–63, *in* First symposium on host specificity among parasites of vertebrates. (J. G. Baer, ed.). Institut de Zoologie, Université de Neuchâtel, Neuchâtel, 324 pp.

Theodor, O., and A. Moscona. 1954. On bat parasites in Palestine. 1. Nycteribiidae, Streblidae, Hemiptera, Siphonaptera. Parasitology, **44**:157–245.

Tipton, V. J., and E. Mendez. 1966. The fleas (Siphonaptera) of Panama. Pp. 289–385, *in* Ectoparasites of Panama. (R. L. Wenzel and V. J. Tipton, eds.). Field Museum of Natural History, Chicago, 824 pp.

Ubelaker, J. E., and T. H. Kunz. 1971. Parasites of the evening bat, *Nycticeus humeralis*. Tex. J. Sci., **22**:425–427.

Ueshima, N. 1972. New World Polyctenidae (Hemiptera), with special reference to Venezuelan species. Brigham Young Univ. Sci. Bull. Biol. Ser., **17**:13–21.

Urich, F. W., H. Scott, and J. Waterston. 1922. Note on the dipterous bat parasite *Cyclopodia greeffi* Karsch, and on a new species of hymenopterous (chalcid) parasite bred from it. Proc. Zool. Soc. London, **1922**:471–477.

Usinger, R. L. 1966. Monograph of Cimicidae (Hemiptera—Heteroptera). Entomological Society of America, Thomas Say Foundation Vol. 7, New York, 585 pp.

Wallace, F. G. 1966. The trypanosomatid parasites of insects and arachnids. Exp. Parasitol., **18**:124–193.

Wattal, B. L., and N. L. Kalra. 1960. *Pyralis pictalis* Curt. (Pyralidae: Lepidoptera) larvae as predators of the eggs of bed bug, *Cimex hemipterus* Fab. (Cimicidae: Hemiptera). Indian J. Malariol., **14**:77–79.

Wenzel, R. L. 1976. The streblid batflies of Venezuela (Diptera: Streblidae). Brigham Young Univ. Sci. Bull. Biol. Ser., **20**:1–177.

Wenzel, R. L., and V. J. Tipton. 1966. Some relationships between mammal hosts and their ectoparasites. Pp. 677–723, *in* Ectoparasites of Panama. (R. L. Wenzel and V. J. Tipton, eds.). Field Museum of Natural History, Chicago, 824 pp.

Wenzel, R. L., V. J. Tipton, and A. Kiewlicz. 1966. The streblid batflies of Panama (Diptera Calypterae: Streblidae). Pp. 405–675, *in* Ectoparasites of Panama. (R. L. Wenzel and V. J. Tipton, eds.). Field Museum of Natural History, Chicago, 824 pp.

Zeve, V. H. 1959. Notes on the biology and distribution of *Trichobius* in northwest Oklahoma (Diptera, Streblidae). Proc. Okla. Acad. Sci., **39**:44–49.

Author Index

*Italics indicate Reference section.

Species Index

Subject Index